Great Transformations in
Vertebrate Evolution

Great Transformations in
VERTEBRATE EVOLUTION

Edited by

KENNETH P. DIAL, NEIL SHUBIN, AND ELIZABETH L. BRAINERD

THE UNIVERSITY OF CHICAGO PRESS | CHICAGO AND LONDON

Kenneth P. Dial is professor of biology at the University of Montana and founding director of the university's Flight Laboratory and Field Station at Fort Missoula.
Neil Shubin is senior advisor to the president and the Robert R. Bensley Distinguished Service Professor of Anatomy at the University of Chicago.
Elizabeth L. Brainerd is professor of medical science and director of the XROMM Technology Development Project at Brown University.

The University of Chicago Press, Chicago 60637
The University of Chicago Press, Ltd., London
© 2015 by The University of Chicago
All rights reserved. Published 2015.
Printed in the United States of America

24 23 22 21 20 19 18 17 16 15   1 2 3 4 5

ISBN-13: 978-0-226-26811-8   (cloth)
ISBN-13: 978-0-226-26825-5   (paper)
ISBN-13: 978-0-226-26839-2   (e-book)
DOI: 10.7208/chicago/9780226268392.001.0001

Illustration credit: Robert Petty created the cover, frontispiece, and part illustrations.

Library of Congress Cataloging-in-Publication Data

Great transformations in vertebrate evolution / edited by Kenneth P. Dial, Neil Shubin, and Elizabeth L. Brainerd.
      pages cm
   Includes bibliographical references and index.
      ISBN 978-0-226-26811-8 (cloth : alkaline paper) — ISBN 978-0-226-26825-5 (paperback : alkaline paper) — ISBN 978-0-226-26839-2 (ebook)   1. Vertebrates—Evolution.   I. Dial, Kenneth Paul, editor. II. Shubin, Neil, editor. III. Brainerd, Elizabeth L., editor.
      QL607.5.G74 2015
      596—dc23

2014047532

We dedicate

Great Transformations in Vertebrate Evolution

to Farish A. Jenkins Jr.

1940–2012

FARISH A. JENKINS JR.

Master of the art of discovery through the language of bones

Professor Jenkins's influence on the field of vertebrate evolutionary biology is evident throughout this volume. Never satisfied with just one slant on thorny problems, he deftly wove anatomy, physiology, biomechanics, and paleontology into a lexicon for the language of bones. His joy in scientific discovery, and his equal joy in conveying that passion through teaching, inspired and informed all he touched.

Contents

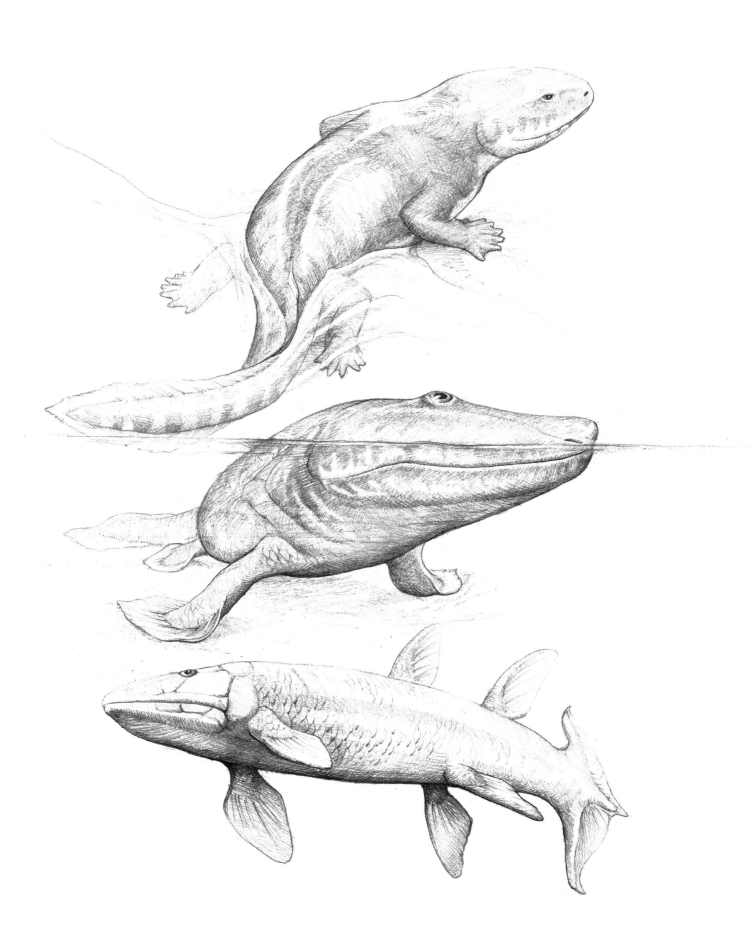

Introduction

Kenneth P. Dial, Elizabeth L. Brainerd, Neil Shubin, and Farish A. Jenkins Jr.

At a large plenary session in the 1960s, one of the great luminaries in the field of vertebrate paleontology pronounced that all of the major transitions in the history of vertebrate life are now well understood. While this declaration may have seemed like a capstone moment for those who had spent long careers in vertebrate paleontology, you can imagine the despair it caused in the hearts of students in the audience. What was there left to do? Were they to build their fledgling scientific careers on carefully stacked minutiae? The good news for students in the 21st century is that his statement couldn't have been further from the reality we are experiencing today. Students entering the field of vertebrate evolution are now armed with a wonderful combination of big intellectual problems and powerful new technologies that extend our reach every day.

In an age when genomes of species can be sequenced in a matter of hours, bones and joints can be nondestructively and dynamically visualized in three dimensions, and new transitional fossils are pulled from far-flung places in the world, we are routinely confronted with vast amounts and diverse kinds of new data. Our challenge as scientists, young and old, is not acquiring new data—that step is getting more inexpensive and easier with each passing day. Our real challenge lies in defining and identifying *the important questions to ask*. Questions drive discovery and are the means to guide us through the darkness of the unknown. Indeed, a key feature of scientific breakthroughs is not deriving answers, but rather posing meaningful, precise, and testable questions leading toward an integrative and synthetic scientific platform.

How did birds evolve flight, fish transform to walk on land, and whales recolonize the seas from terrestrial ancestors? The major transformations in the history of vertebrate life give rise to some of the classic questions of evolutionary biology. Progress in this venture has, and will continue to, come from understanding how complex adaptations emerge and diversify in a landscape of ever-changing environments. Over 150 years ago, the Darwinian notion of descent with modification provided the means to develop testable hypotheses about evolutionary origins. In the ensuing century and a half, the study of great events in vertebrate evolution has itself been influenced by new discoveries, technologies, and analytic methods. A scientist today can feel genuinely intimidated to keep abreast of fields and approaches that are diverse and ever-expanding. The study of evolutionary transformations is now itself being transformed; this is the spirit that motivates this volume and the contributions herein.

Embedded within any complex anatomical structure are fundamental questions. What is old? What is new? What has changed during history and why? The twin problems of homology and novelty are the yin and the yang of the study of morphological evolution. While phylogenetic analysis is the backbone for disentangling patterns of evolutionary descent, a deeper understanding of "why" features evolved entails integration of multiple lines of inquiry. Large shifts in evolution come about through genetic perturbation of developmental systems that ultimately yield a change in the function of species in its ecosystem. Students today have at their fingertips a tool kit to analyze transformations that extends from fossil discovery to phylogenetic analysis, developmental genetics, life history ecology, functional morphology, and skeletal kinematics, to name a few.

At first glance, the goal of making any progress in understanding evolutionary transformations seems daunting: the endpoints often seem entirely different. Life in water is vastly different from life on land, and virtually every anatomical structure is affected by adaptation to these different environments. The list of features needing to change for animals to successfully colonize and diversify on land is intimidatingly long: gill-based respiration in water simply couldn't work on land, nor could the modes of excretion, locomotion, feeding, and reproduction work in a terrestrial environment without significant modification. Start linking fields as different

as natural history, developmental biology, paleontology, and functional anatomy, and intermediates can be found *everywhere.*

Look to living relatives of tetrapods to find fish with lungs and homologues of upper arm and cranial bones. What do we see? Fish evolved lungs to adapt to diverse aquatic environments long before the transition to land. One of the key transitions in the origin of vertebrate terrestriality was not the origin of lungs, but the ultimate reduction and loss of gills. The same is true with appendages. Extant lungfish and basal ray-finned fish have fins that contain homologies of upper arm bones and aspects of the genetic apparatus that tetrapods use to form digits. Fish adapting to diverse aquatic ecosystems developed a range of morphologies and developmental mechanisms that were later to become useful during the transition to land. The paleontological approach reveals a range of even more dramatic intermediates. When paleontologists target rocks of the right age, type, and exposure quality to find fossils, they can discover taxa with intermediate morphologies that are absent in extant taxa. Integration of ecology with the study of morphological transformation allows morphological change to be understood in the context of its biological role in the environments in which transformations occurred. By recognizing the abundance of transitional microhabitats within most ecological communities, one begins to appreciate the endless opportunities for incremental and stepwise adaptive evolution. When we know how to look, we find intermediates around us today and see ways to uncover even more in the rocks.

Linking studies of the fossil world with the living one opens another realm of powerful new scientific questions. Knowledge of behavior and natural history changes how one interprets fossil bones. When neontologists encounter goats that are arboreal or fish fins used in aerial locomotion, they see a range of possibility that would be absent if only the skeleton were preserved. Indeed, seeing the kinematics of bone motions in living animals reveals functions that would have been unknowable solely by relying on analyses of joint surfaces. Linking to genetic data also reveals challenges: analyses of molecular phylogenies in certain clades reveal massive independent evolution of anatomical traits. How do these discoveries apply to phylogenies inferred from anatomical landmarks of fossil taxa? While these findings provide cautionary notes to analysis of fossils, the paleontological record reveals intermediates and ancient environments that would have been otherwise invisible to neontologists. For example, if we did not have fossils, it would be likely that neontologists would have viewed birds and mammals as sister groups. As endothermic vertebrates with four-chambered hearts, there are numerous characters that birds and mammals share that are absent in lizards, turtles, crocodiles, and snakes. The addition of nonmammalian synapsids and non-avian archosaurs reveals a landscape of evolutionary changes that may otherwise have been unseen.

Because of this need for interdisciplinary thinking, we assembled 22 chapters, each including diverse perspectives, to explore great transformations in the history of vertebrates. Chapters are organized into two broad themes: an opening section on "specific exemplars" and a concluding one on general "perspectives and approaches" to transformation. This dichotomy among chapters is not absolute because a number of chapters on exemplars extract general principles and vice versa. The organizational scheme was chosen because it provides a temporal and phylogenetic framework for the succession of chapters on specific transformations while allowing the volume to extract general themes for discussion and analysis in the closing chapters.

Each of the authors takes a classic issue from vertebrate evolutionary biology and explores it by linking it to different fields: developmental biology and genetics (Abzhanov; Burke; K. Smith; M. Smith and Johanson; Stringham and Shapiro), ecology (Dial et al.), physiology (Brainerd; Claessens; Owerkowicz et al.), biomechanics and functional morphology (Biewener; Brainerd; Claessens; Dial et al.; Fleagle and Lieberman; Gatesy and Baier; Lauder), comparative analysis and paleontology (Claessens; Crompton et al.; Gingerich; Hopson; Luo; M. Smith and Johanson; Shubin et al.; Sullivan), animation (Brainerd; Gatesy and Baier), comparative and phylogenetic analysis of extant taxa (Lauder; K. Smith; M. Wake; D. Wake et al.), faunal analysis (Padian and Sues), and field paleontology (Gingerich; Shubin et al.).

Typical caricatures of great transformations, such as the march from chimp to human or a fish crawling out on land, reveal a persistent bias in what major shifts in evolution are and how they are interpreted. A historically dominant approach to great transformations has been the tendency to envision that anatomical, functional, and ecological changes go in lockstep from inefficient, primitive systems to more efficient, advanced ones. One of the emergent themes of virtually every chapter of this volume is that the temporal pace, morphological and functional patterns, and ecological landscape of great transformations have rich textures of independent evolution, reversal, and evolutionary and ecological experimentation.

For example, the transformation of sprawling limbs of early tetrapods to the upright postures of mammals and archosaurs, and the idea that upright postures are somehow superior, has perennial appeal. New data reveal a richness and complexity to these transformations. The broad outlines of the scenarios remain intact, but closer scrutiny, new fossil finds, biomechanical analysis of extant animals, and new analysis of fossilized trackways suggest that progress from sprawling to upright was certainly not efficient nor linear (see chapters by Sullivan and Hopson in this volume), although one might reasonably conclude that upright postures are superior, if defined as allowing mammals and dinosaurs to diversify into the realm of very large body size (see Biewener, this volume). Kathleen Smith, using an entirely different data set, exposes the same theme. Marsupial development has been interpreted as a primitive and inefficient version of the placental pattern, arising because of constraints placed upon marsupial biology. When placental diversity is analyzed in a phylogenetic and ecological framework, primitive or intermediate-looking characters of marsupial placentation become complex adaptations in their own right.

Complex patterns of gain and loss are revealed when we explore the genetic and developmental basis of transformations, and use molecular phylogenies to interpret the origins of integrated anatomical systems. Quantitative trait locus (QTL) mapping, transgenic analysis, genome sequencing, and experimental embryology have the power to reveal the genetic basis of characters (see Abzhanov; Burke; and Stringham and Shapiro in this volume). Armed with this information, the rates and patterns of evolution of diverse features, such as bird beaks and fish fins, reveal complex patterns of gain and loss, including parallel evolution. Using new molecular phylogenies of plethodontid salamanders, D. Wake and colleagues show that the multiple origins of complex tongue projection systems are prime examples of rampant homoplasy. Parallel evolution of morphological traits may be the rule rather than the exception, with thought-provoking implications for our interpretation of the fossil record. The biological basis of homoplasy is a starting point for M. Wake who shows that understanding the biological basis of reproductive characters reveals the mechanisms by which features evolve in parallel.

A persistent fallacy in interpreting the great transformations is the idea that novel features evolve to anticipate the invasion of new environments. For example, the notion that fish evolved lungs and limbs in order to breathe and walk on land is one such fallacy. Early hypotheses for the evolution of tetrapod limbs and lungs featured ponds drying up and the need to migrate over land to find more water. To the contrary, we now know that aerial gas exchange via lungs and many features of tetrapod limbs evolved and persisted for tens of millions of years in aquatic environments before eventually being co-opted for fully terrestrial lifestyles (see Brainerd; Shubin et al., this volume). In the novelties that underlie many great transformations, changes in the functions of structures become as important as the origins of the structure itself.

These discoveries provide the grist for new kinds of questions. What were the selective pressures and the genetic and ecological factors involved with the transformation? It is here, in these questions, that interdisciplinary approaches become paramount. For example, with a rich fossil record and robust phylogenetic framework, we can be confident that bird wings evolved from the forelimbs of bipedal, non-avian dinosaurs, but it is hard to envision how natural selection acting on intermediate conditions could have bridged the gap. What can be the selection pressure for aerodynamic wings if the limbs cannot yet be used for flight? What was the function of structures present in species that have been long extinct? Breakthroughs come from adding concepts and data from other lines of inquiry. By adding an understanding of the life histories, functional anatomy, and ecology of extant archosaurs to the fossil record, testable hypotheses about the origin of flight can be generated (see Dial et al., this volume).

Among all of the data, questions, and ideas in this volume, the scientific spirit reigns paramount. The greatest pleasures of a life in science ultimately derive from the joy of discovery. When discovery hits you, you're looking at something and you don't see it, and then all of a sudden, you do see it. Those moments of discovery, large and small, are times of highest elation in the life of a scientist. But getting to the point of seeing what was previously hidden requires forcing ourselves out of our comfort zone. We cannot be rigidly tied to specific hypotheses, tools, or methods. Techniques improve, knowledge grows, and some of today's seemingly intractable problems will yield to the pressure of new data, technologies, and approaches. Nature proves again and again to be more beautiful and complex than we had imagined, but we can only access this beauty and complexity by staying humble, curious, and hungry to find ever new and important questions.

We thank all of the chapter authors and peer reviewers for their outstanding contributions to this volume. We also thank Christie Henry and Logan Ryan Smith at the University of Chicago Press for shepherding this project forward with such care and efficiency, Robert Petty for creating the cover, frontispiece, and section illustrations, Jared Leeds for the photo on the dedication page, Erika Tavares for proofreading the whole volume, the Helen R. Whiteley Center for two residencies that allowed ELB to focus on this project, and finally we thank our spouses, Karen Dial, Tim Hiebert, Michele Seidl, and Eleanor Jenkins, for all of their support throughout our careers.

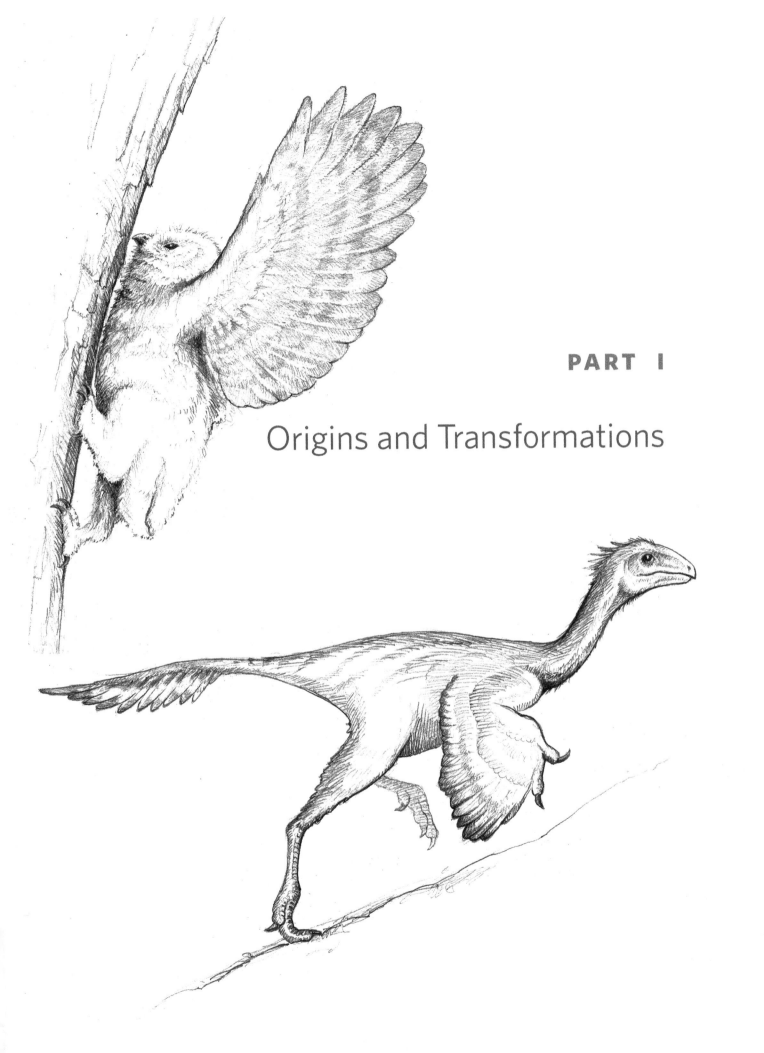

Origins and Transformations

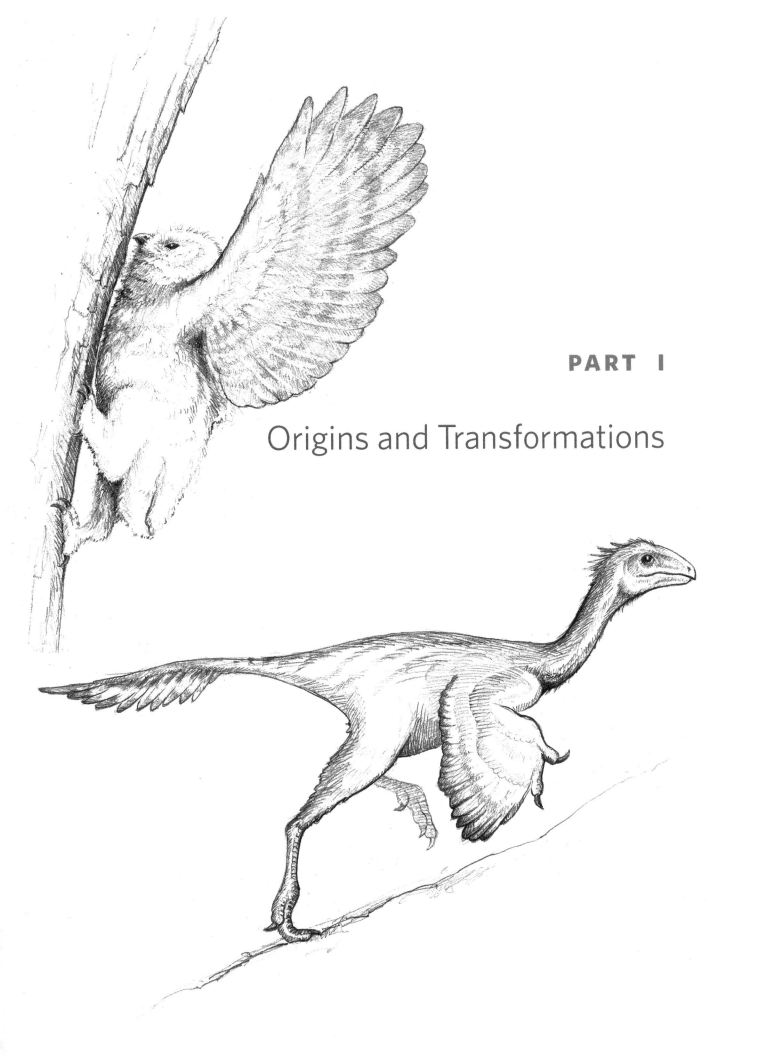

PART I

Origins and Transformations

1

Origin of the Vertebrate Dentition:
Teeth Transform Jaws into a Biting Force

Moya Meredith Smith* and Zerina Johanson[†]

Introduction

How did the great transformation from "jawless suckers" to the "predatory monsters" of the oceans occur, involving the evolution of organized dentitions and jaws? It might be expected that jaws and teeth evolved together, but both are considered as independent modules with separate developmental pathways and separate evolutionary origins.

The emphasis of this chapter is on the organization of dentitions (regulation of separate teeth into ordered sets), how they evolved, and how they are patterned in development. The origins of the vertebrate dentition are currently controversial, contrasting tooth origins in evolution either from internal denticles lining the gill arches (seen to be anatomically separate from bony gill rakers), or from external body denticles (scales). Essentially these theories consider how teeth became organized into functional dentitions, and discussion of these controversies will be the main focus of this chapter. An essential premise that we will develop is that structures like teeth and jaws evolve via co-option of existing developmental modules of other structures (e.g., gill arch denticles vs. external denticles), by tinkering with the developmental genes that are also co-opted with the module (e.g., Raff

* Craniofacial Development and Stem Cell Biology, King's College London and Earth Sciences, Natural History Museum, London
† Earth Sciences, Natural History Museum, London

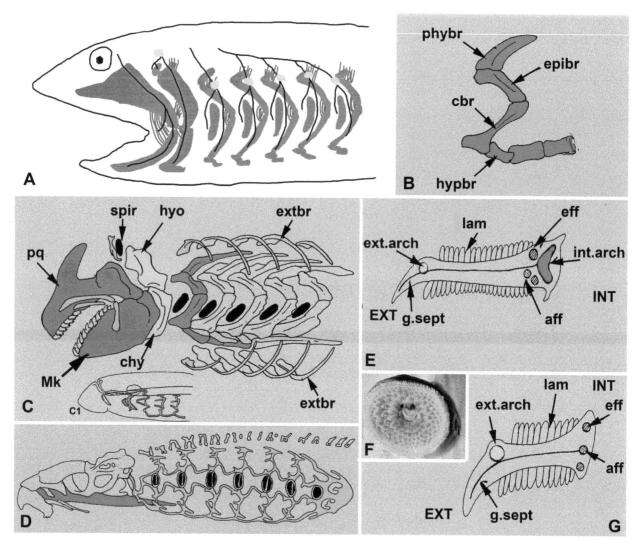

FIG. 1.1 (A) Serially repeating pharyngeal structures of a generalized vertebrate. First two arches, mandibular and hyoid, form jaw supports, five pharyngeal arches posteriorly are structurally similar cartilages (red); gill clefts (gray); nerves (black); epibranchial placodes (green); joint ligaments (blue). (B, C) *Squalus acanthas* (Chondrichthyes), (B) pharyngeal arch with jointed elements. (C) Mandibular arch, red; hyoid arch, green; pharyngeal arches more posteriorly, pink; extrabranchial cartilages, blue. (Inset C1) Developing lamprey skeleton, with putative equivalents of mandibular arch, red; hyoid arch, green; external branchial basket, blue (from de Beer 1971). (D) Adult lamprey with medial piston cartilage from mandibular arch, red; medial support cartilage from hyoid arch, green; extrabranchial cartilages, blue (from de Beer 1971). (E, G) Schematic drawings, individual pharyngeal arches with gills, cartilage, and vascular supply; (E) generalized gnathostome with internal pharyngeal arch (pink), external pharyngeal supports (blue), and (G), generalized jawless vertebrate with external arches only. (F) Lamprey oral sucker disk with keratinous toothlets (orange). (Acknowledgments to Mike Coates for B–E, G, and to Anthony Graham for A.)

1996; but see Kuratani 2012). First, however, we discuss the evolution of the vertebrate jaws, a topic of considerable research interest in recent years.

Evolution of Vertebrate Jaws

The vertebrate jaw and gill arches can be considered serially homologous structures (Gegenbaur 1878; Gillis et al. 2013); classically, jaw evolution was believed to

involve a modification of the first two arches into the jaw, with all structures in the pharyngeal walls arranged as an iterative series (fig. 1.1A). The skeletal structures of the most anterior mandibular arch include a dorsal palatoquadrate articulating to a ventral Meckelian arch, with the functional jaw articulation including the second hyoid arch as support, also with dorsal and ventral elements (fig. 1.1A, C). This is a generalized arrangement found, for example, in sharks (fig. 1.1C, *Squalus*) where behind the jaws five dorsal and ventral paired pharyngeal cartilages (fig. 1.1B, C) are preserved. Support for the oral dentition is provided by the first two arches, while more posterior pharyngeal arches (three to five, or six) are cartilaginous gill supports, but may also support teeth (arch five) in fish with well-ordered pharyngeal teeth functioning in food processing (zebrafish, cichlids). These more posterior arches are internal to the gills and also jointed, with a rostral bend to the dorsal part, then reflexed (fig. 1.1B, C, E). These contrast with the branchial basket of extant jawless vertebrates such as the lamprey (with a mouth as a sucker with rings of keratinous toothlets, supported by a circular cartilage; fig. 1.1D, F), where the branchial basket cartilages surround the gill openings and are external to the gills, neither jointed nor reflected in a rostral to caudal direction (fig. 1.1D, G). However, the modern shark visceral skeleton also has an iterative set of extrabranchial cartilages (fig. 1.1C, E, blue), equivalent topographically and with a similar passive respiratory function (recoil) to the lamprey branchial basket (Mallatt 1996; fig. 1.1D, blue, acting as gill supports, but inner branchial arch cartilages are absent; cf. fig. 1.1E, G). Therefore, the iterative cartilages of the lamprey branchial basket cannot have given rise to the inner cartilages of the jaws and gill supports, because both coexist in the modern shark (extrabranchial cartilages homologous to the branchial basket; Mallatt 1996), establishing the condition for Patterson's congruence test for nonhomology (Patterson 1988).

Although the iterative nature of the jaws, pharyngeal arches, and associated structures is not in doubt, and these can be compared as serial homologues in each of the major clades of jawed vertebrates, the way in which jaws evolved is still a question being studied by several research groups. Whether any structures in the lamprey gave rise to mandibular and hyoid skeletons of the jaws in chondrichthyans and osteichthyans is the subject of active debate and new experimental data (cf. fig. 1.1C, D; see Kuratani 2012 for a summary). It is interesting that in de Beer's famous treatise on development of the vertebrate skull in the lamprey he showed paired medial structures that were thought to represent the mandibular and hyoid precursors of the cartilaginous jaws (fig. 1.1C1; from de Beer 1971: pl. 9-6; anlagen for mandibular [red] and hyoid [green] cartilages); in the adult they become the piston cartilages

and support (fig. 1.1D, same coding as C1), but there are no caudal medial cartilages in series with them as gill supports (pink cartilage as in fig.1.1C, E).

In jawed vertebrates, development of the jaw cartilages and hyoid precedes that of the pharyngeal arches. Also, earlier work in the chick embryo (Veitch et al. 1999) has shown that in the serial organization of the pharyngeal arches (see figs. 1.1D, 1.2A for general organization of the jawless vertebrate pharyngeal region) neural crest-dependent and -independent patterning mechanisms operate. The pharyngeal arches are regionalized and have a sense of identity, antero-posterior position in the pharynx, different from that of pharyngeal segmentation (such as pouches, fig. 1.2D, E), which predates in evolutionary terms the role of neural crest in this process that is largely contributing to the cartilaginous supports. This suggested that the serial, medial pharyngeal cartilaginous skeleton is a secondary evolutionary acquisition. Perhaps it can be proposed that the inner series of pharyngeal cartilages (medial) are a neomorphic evolutionary step (neural crest-dependent) modeled on the mandibular but in particular the hyoid arch, as an iteration of this structure, using the hyoid set of genes and developmental processes to make gill supports (fig. 1.1B, C, E). In some ways the ideas of Kuratani (2012) reflect this, as he suggests evolution of a new developmental program based on novel gene regulatory networks, assumed in the context of a stepwise evolution of jaw patterning. In the classic model (fig. 1.1A), putative jaw supports developed from the unmodified mandibular and hyoid arches, as anterior supports of the series in the pharynx that were proposed to have already existed in jawless vertebrates (see fig. 1.1C1, D); however, medial, iterative pharyngeal skeletal arches were absent. Consequently, these could not have given rise in evolution to the medial skeletal supports of the more anterior jaws and hyoid (cf. fig. 1.1E, G), hence their origin in evolution was proposed as from the existing, more rostral, medial arches (mandibular and hyoid, fig 1.1C1, D).

The evolution of developmental processes that gave rise to articulating jaws are very controversial and in need of developmental, experimental data, some of which is outlined here. Linking these iterative structures with molecular information about a proposed patterning mechanism, nested Hox genes provide anterior-posterior positional information to the gill arches, and determine their identity, but are absent

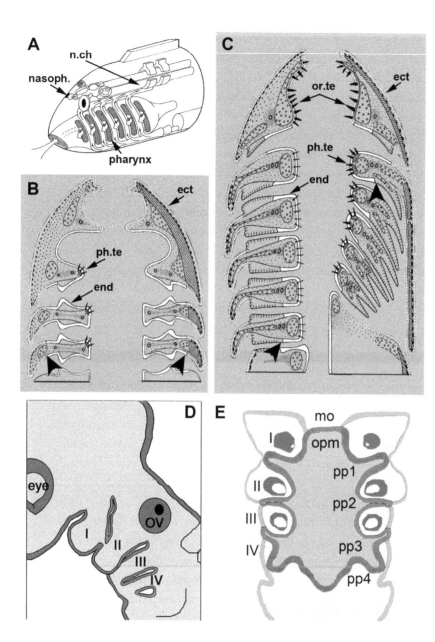

FIG. 1.2 (A) Generalized agnathan body plan showing water flow through the gill slits of the pharynx, only five branchial arches shown (six openings; number more extensive in both extant and fossil forms). (B, C) Schematic horizontal sections, body plans of (B) jawless and (C) jawed vertebrates (Jollie 1968, figs. 3, 4; Smith and Coates 1998, fig. 9), showing classic ideas regarding the extent of the endoderm (end, white) and ectoderm (ect, green), junction between these deep into adult gill slits (arrowhead). (B, C) Taxa with a solid dermal armor (macromeric, hatched) represented on the right, taxa with separate scales (denticles; micromeric) and no dermal bone represented on left. Pharyngeal teeth (ph. te) in the micromeric agnathan with open gill slits represent the thelodont condition (left); lack of teeth in the macromeric agnathan represents the osteostracan condition (right). (C) Open gill slits (chondrichthyans) represented on left, bony operculum covering gill slits represented on right (osteichthyans). Teeth (or. te) are present in the oral cavity, but also in the pharyngeal cavity of the osteichthyan (ph. te). (D, E) Organization of developing pharyngeal pouches in chick, red: (D) lateral view, (E) horizontal section (endoderm red, ectoderm green). Before the openings break through from pharyngeal cavity, endoderm lines each pouch completely to contact the barely indented ectoderm, and the shallow mouth ectoderm mo/opm (E). Additional abbreviations: mo, mouth; nasoph; nasopharyngeal opening; n.ch, notochord; opm, oropharyngeal membrane; pp1–4, pharyngeal pouches; I–IV, gill arches (D, E by permission, Anthony Graham).

from the first arch (jaw) (Cohn 2002), and, intriguingly, also from the first pharyngeal arch in lampreys (Takio et al. 2007). Homeobox genes such as the Dlx family have been implicated in the dorsal-ventral patterning of gill arches, while Bapx1 and Gdf5/6/7 are involved in joint development (Depew et al. 2002, 2005; Cerny et al. 2010; Kuraku et al. 2010; Gillis et al. 2013; Takechi et al. 2013). Nested Dlx genes are present in jawed vertebrates (*dlx1, 2* expressed dorsally in all arches, *dlx1–6* ventrally, *dlx1, 2, 5, 6* expressing in an intermediate position; Gillis et al. 2013), and have also been identified in lampreys (Cerny et al. 2010), however, whether these dlx genes are orthologous to those in jawed vertebrates remains uncertain (Gillis et al. 2013; Takechi et al. 2013).

Bapx1 was also absent from the lamprey arches (Cerny et al. 2010).

Researchers studying jaw evolution have focused on comparisons between living jawless vertebrates such as the lamprey and living jawed vertebrates. However, there are several groups of fossil jawed vertebrates that are more closely related to extant jawed vertebrates, including the Galeaspida and Osteostraci. New nondestructive methods, such as synchrotron radiation X-ray tomographic microscopy, are now available to study the internal anatomy of these fossils in substantial detail, providing crucial data on the question of jaw evolution. One current theory, the "heterotopy theory of jaw evolution," developed by Shigeru Kuratani and his colleagues suggests that the jaw evolved via a posterior (heterotopic) shift of the genes involved in jaw development and the change from monorhyny (single dorsal nasohypophyseal opening) to the diplorhyny (two nasal openings and a separate hypophyseal opening) characterizing jawed vertebrates. This latter change, seen in the Galeaspida among jawless vertebrates, allowed the migration of tissues that would form the anterior parts of the braincase and jaws, previously blocked by the nasohypophyseal opening in jawless vertebrates such as the lamprey and Osteostraci (Gai et al. 2011; Gai and Zhu 2012).

Also, a novel theory for a heterochronic mechanism of jaw evolution, using fossil data together with developmental data, proposed that inductive mechanisms for making dermal bone were recruited from the region below the eye. These ideas were different because they emphasized the phylogenetic origin of the bones of the jaws (those that surround the cartilage supports), together with a developmental mechanism that may have been recruited to make these bones (Long et al. 2010). Long et al. (2010) proposed that because sclerotic bones (surrounding the eyes) appear in fossil species before bones of the jaw and are present in both the jawless group Osteostraci and in extant jawed forms, they are potential candidates to develop bone in the lower jaw. This is justified by developmental data in the chick showing that scleral mesenchyme can be invoked to produce bone by mandibular epithelium. Long et al. (2010) propose that this shared developmental process is co-opted from scleral bone development in jawless osteostracans to jawed vertebrates, to make bone in the jaw apparatus.

Classic and Novel Theories of Tooth Evolution

Canonical (Reif 1982) and more recent hypotheses (Smith and Coates 1998) of tooth origins have come to be known as "outside to inside" (i.e., dermal scales/denticles, located outside the mouth, in the skin, evolving into teeth along the jaws inside the mouth) as opposed to "inside to outside"; our preferred phrase is "teeth from the inside" (i.e., gill arch [pharyngeal] denticles evolving into teeth along the jaws in the mouth; fig. 1.2B, C). However, these gill arch denticles never extend outside to give rise to skin denticles (restricted instead to the internal arches), but "outside to inside" and "inside

to outside" have been a convenient, if inaccurate, shorthand for "teeth only from the inside." These hypotheses have provoked much debate and a search for new and more accurate data (molecular, palaeontological) to test both. As we emphasize, W.-E. Reif (1982), a pioneer in the study of the evolutionary origins of teeth and dentitions, recognized that both the outer dermal system of skin denticles, or scales, and the inner one of oral teeth represent separate developmental and evolutionary systems. Later, Smith and Hall (1993) proposed a developmental model for the evolution of the dermal skeleton independent of teeth, although strongly linked with neural crest in both processes, as a vertebrate skeletogenic innovation (Hall and Gillis 2013).

Current Hypotheses

In the canonical theory of origins of teeth making a dentition (Hertwig 1874; "outside to inside," Smith and Coates 1998), skin denticles, or placoid scales, were utilized for dentitions once jaws evolved, when the ability to make odontodes (see box 1.1) was transferred from the external skin of the body to the edge of the jaws and into the mouth (Ørvig 1967; Peyer 1968; Kemp 1999). This theory was largely based on the observation that the skeleton of the earliest, jawless vertebrates was only external, comprised of extensive shields of dermal bone (macromeric), but also scales (figs. 1.2B, 1.5A). Tubercles (odontodes) covered the entire surface of the bone, made of dentine with a tissue structure comparable to that of dentine in the human tooth (fig. 1.5B). Other external skeletons were composed of dentine as separate denticles (placoid scales) but not joined by bone, as in the Chondrichthyes (figs. 1.2B, C, 1.5E–G) and Thelodonti (figs. 1.2B, 1.5C, D; micromeric). A challenge to this classic theory occurred when similar denticles were recognized, for the first time in a whole body fossil, and located inside the oropharyngeal cavity in the fossil jawless vertebrate group Thelodonti (figs. 1.5C, 1.6A–E; Van der Bruggen and Janvier 1993). As well, there was a realization that the pharyngeal covering of denticles was considerably more extensive in extant jawed vertebrates, in both chondrichthyan and osteichthyan fish than conventionally thought, comparable to the external scale covering. Moreover, in some taxa these were organized into distinct rows on the gill arches (Nelson 1970).

BOX 1.1 DEFINITIONS

Odontode: the unit of development for both teeth and scales, involving interaction of an epithelium with mesenchymal tissue (potentially neural crest-derived). Odontodes are comprised of combinations of enamel, dentine, and bone (fig. 1.3A, B1–3) and include denticles, tubercles, and teeth.

Ectoderm: external embryonic layer (fig. 1.2B–C, E).

Endoderm: internal layer lining the gut and oropharynx (fig. 1.2B–C, E).

Neural crest: derived from ectoderm as the dorsal neural tube invaginates and marginal cells form freely migrating mesenchymal cells.

Oropharyngeal cavity: oral entrance leading to the pharynx (gill arch cavity; fig. 1.2B–E).

Patterned: a term referring to development as regulated by genes and resulting in phenotypic spatial organization related to function.

Macromeric versus micromeric skeleton: external dermal skeleton composed of larger, fused bony units versus smaller, unfused units (fig. 1.2B, C).

Dental lamina: double sheet of epithelial cells, typically invaginated from the oral epithelium as characterized in sharks. Always associated with spatially ordered teeth and linked with their initiation. Present in all dentitions, where successive teeth are made (figs. 1.3A, 1.4A, B).

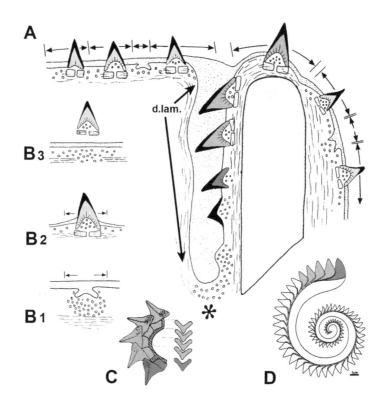

FIG. 1.3 (A) Tooth development in sharks, from a dental lamina (d. lam.; older teeth, mauve; newer teeth, pink), compared to superficial skin and oral denticles (orange). Denticle cover to skin is added to by growth only with expansion of the epithelium, and superficial development of new denticle germs in the space created (see fig. 1.4C, D). By comparison, new tooth (protogerm, asterisk) develops deep in the oral cavity (see fig. 1.4A, B; and fig. 2, Smith and Coates 1998). (B1–3) Denticle development is from interaction between neural crest-derived ectomesenchyme (cell cluster below expanding epithelium, and bud) and inductive epithelium (Reif 1982). (C) Extant shark tooth whorl with separate bases: (D) fossil tooth whorl of edestid shark *Helicoprion* (newest tooth, pink). (From Reif 1982, and fig. 2, Smith and Coates 1998.)

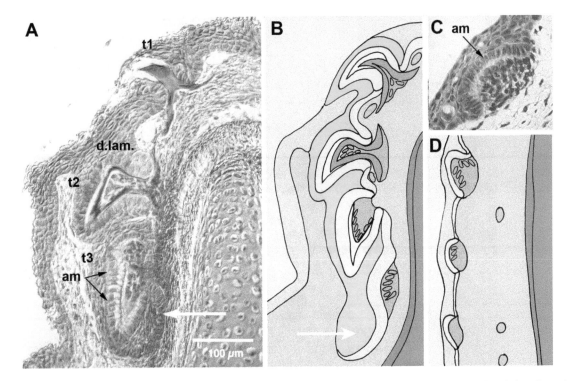

FIG. 1.4 *Scyliorhinus canicula* (catshark) tooth and denticle development. (A, B) Dental lamina for tooth development deep below the oral epithelium is compared to (C, D) superficial development of skin denticles in sharks. (A, B) New teeth (t1, t2, t3) developing from a dental lamina (d. lam.; sea blue/pale blue; protogerm, white arrow); (A) teeth have not yet emerged into mouth, with enameloid cap (white space in A; yellow in B), dentine (A, red stain; B, pink) formed within each pulp cavity, with differentiated cell layer of ameloblasts contributing to enameloid cap (am; green in B) (A–D, from figs. 2 and 4C, Smith et al. 2009).

Like sharks, thelodonts had a dermal skeleton composed of separate denticles (see size of these relative to internal denticles in fig. 1.5C), but more importantly, the internal oropharyngeal denticles were organized into small whorls (fig. 1.6A, B, D, E, cf. to a shark tooth family, figs. 1.3A, C, D, 1.4A, B, and 1.6 F), but assumed to be located on the (unpreserved) gill supports (fig. 1.2B). This new data suggested an alternative to the "outside to inside" evolutionary scenario, one in which these pharyngeal denticle sets (and the developmental controls regulating their patterning into whorls and spatial, temporal organization, situated on the gill arches) were co-opted to make tooth sets. These were first in the pharynx (deep inside the pharynx at the oesophageal-pharyngeal junction), but denticle set structure and their developmental controls were then co-opted to the jaw margin. This internal skeleton of thelodonts (figs. 1.2B, 1.6A–E) was unknown at the time the original "outside to inside" hypothesis had been developed, so the novel "inside to outside" theory was proposed (Smith and Coates 1998). Thelodont scales and denticles are made of tubu-

lar dentine and assumed to form in the same way as all dentine-based structures (fig. 1.5 B–D).

Current Debate

Both the "outside to inside" and "inside to outside" theories have been vigorously contested (as discussed in Blais et al. 2011; Rücklin et al. 2012; and Murdock et al. 2013). Amendments to these theories have included the recently proposed "modified outside to inside" (Huysseune et al. 2009, 2010) and "both inside and outside" theories (Fraser et al. 2010; Ohazama et al. 2010). With respect to the latter, an odontode (in the form of a denticle or tooth) could develop wherever the relevant genes, competent epithelium (ectoderm, endoderm, or a combination; Soukup et al. 2008, as demonstrated in the salamander), and neural crest-derived mesenchyme were combined. An important point made by the "both inside and outside" theory is that this combination could occur either internally or externally (Fraser et al. 2010). In other words, odontodes and teeth are

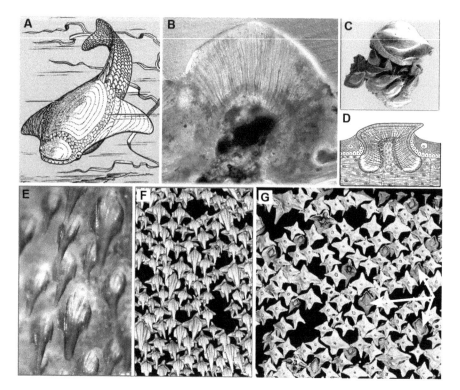

FIG. 1.5 (A) Reconstruction of the jawless vertebrate *Psammolepis* (Heterostraci, Devonian). (B) Dentine in tubercle (odontode) covering the external surface of the bony shield of *Psammolepis* (Middle Devonian Gauja Formation, Estonia; Johanson et al. 2013). (C) Dermal scale and pharyngeal denticles of the jawless vertebrate *Loganellia* (Thelodonti, Silurian), showing size difference between dorsal scale and internal denticles (Smith and Coates 2001). (D) Drawing of putative development of thelodont denticle (as large one in C), with lines in dentine formed from successive growth deposited from the pulp. (E–G) Scales of catshark *Scyliorhinus canicula*, as photomacrograph (E), and (F, G) computed tomography (mCT) of same lateral skin area as E, (F) external and (G) internal view, showing irregular pattern of new scale addition and lack of structural pattern, and different stages of growth revealed by open (arrows), or closed bases (see thelodont fig. 1.6D) (F, G from Fraser and Smith 2011, fig. 1C, D).

homologous units at the morphogenetic level, and a building block of all dentitions, a long accepted fact (Schaeffer 1975). From this standpoint, we again stress that the relevant question involves the origins of developmental patterning of teeth and their spatial organization in the jawed vertebrate dentition, rather than just the origin of odontodes themselves. For example, Fraser and Smith (2011) (Fraser 2011) demonstrated that spatial initiation of new denticles (scales) in the external skin of the catshark is very different from that of teeth on the jaws (fig. 1.5E–G). This implies that the skin denticles lacked any comparable developmental patterning mechanisms that could have been co-opted to form the sets of teeth in the dentition. This topic is also discussed with respect to a molecular model for genetic pattern information in dentitions versus skin-derived structures (see below; also Smith and Johanson 2012).

The Role of the Epithelium: Ectodermal, Endodermal Origins

One difference between the "outside to inside" and "inside to outside" theories has to do with the type of epithelium proposed to be involved in the developmental patterning of the dentition. All odontodes develop from the interaction between a layer of epithelium that overlies, and interacts with, mesenchyme (neural crest-derived). Ectodermal epithelium is generally external, while endodermal epithelium is located internally (see box 1.1, fig. 1.2B–E). Boundaries between these epithelia occur early in embryology within the pharyngeal (gill) slits and inside the mouth; here the two epithelia form the oropharyngeal membrane (fig. 1.2). This membrane, and a comparable membrane associated with each of the gill slits, gets thinner and fragments during development and eventually disappears.

Currently, the extent of migration of ectoderm inward, or endoderm outward (see lack of green ectoderm, or white endoderm; fig. 1.2B, C) is not yet fully known (Richardson et al. 2011; fig. 1.2D, E). But previously, the pharyngeal pouches of both jawless (fig. 1.2A, B) and jawed (fig. 1.2C) vertebrates were interpreted as being lined by endoderm internally, with ectoderm also extending deep into the pouches (Jollie 1968). By comparison, new in vitro labeling studies using transgenic Sox17-GFP zebrafish show instead that the endoderm pouches outward to meet the ectoderm at the outer margin of the gill slit (fig 1.2E; Richardson et al. 2011; Graham and Richardson 2012) so that gills may be entirely endodermal in origin; likewise, the denticles forming there. It seems that this is a conserved pattern within jawed vertebrates, as also shown in the chick (fig. 1.2D, E). These observations are a key part of our discussion of the ectodermal or endodermal origin of pharyngeal denticles and teeth and the "outside to inside" and "inside to outside" theories and recent modifications of these theories. This new research is ongoing (Shone and Graham 2014) as it was found that pharyngeal morphogenesis in the shark involves growth of endoderm outward to impinge on the superficial ectoderm where the opening occurs. This is of interest for establishing boundaries that initiate regulatory genes for denticle and tooth initiation, and it could be an ancestral character of gnathostomes (Fraser et al. 2010; Fraser and Smith 2011; Shone and Graham 2014).

"Outside to Inside" Hypothesis and Modifications

In the "outside to inside" hypothesis, only external ectoderm originally has the competence to produce odontodes; this ectoderm is transferred internally, via migration into the mouth. Huyssuene et al. (2010) proposed a modification of the "outside to inside" hypothesis such that competent ectoderm migrated into the mouth, and also internally via the gill slits, to account for odontodes/denticles in the pharyngeal region (Huysseune et al. 2009, 2010). Contact between the endoderm and ectoderm was required, with tooth competence being transferred to the endoderm (Huysseune et al. 2010).

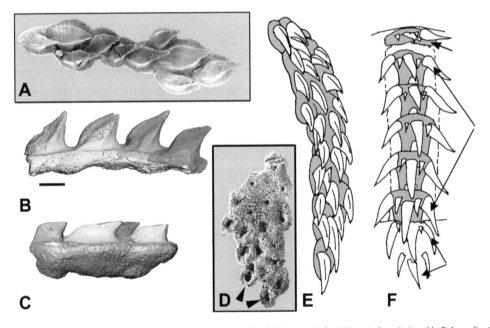

FIG. 1.6 *Loganella scotica* (Thelodonti). (A) Scanning electron micrograph of pharyngeal denticle set, dorsal view (A, D from fig. 14.1A, E, Smith and Coates 2001; E, F from fig. 2f, g, Smith 2003). (B, C) MNHN GBP384 (Museum national d'Histoire Naturelle, Paris), synchrotron radiation X-ray tomography (SRXTM), volume color rendered. (B) Pharyngeal denticle whorl, lateral view of sequential teeth and their bases. (C) Oral denticles with base, lateral view (individual denticles colored same as B, in order of timing of addition). In the oral denticles, newer ones are added to margin of first (original) denticles (Rücklin et al. 2011). (D) visceral view of base of similar set of pharyngeal denticles to (A) with timed addition, as seen from open pulps of newest (black arrowheads) compared with closed in oldest (top) (see open base of skin denticles, fig. 1.5G). (E, F) comparison between thelodont pharyngeal denticle whorl (E) and (F) shark tooth family (Smith 2003).

In this revised theory, internal pharyngeal denticles occur only in taxa with gill slits opening externally, providing a point of entrance for the competent ectoderm; this would include the denticle whorls of thelodonts (fig. 1.6A, B, D–E). As noted above however, in zebrafish the endoderm extends to the external margin of the gill slit, excluding internal migration of the ectoderm (Richardson et al. 2011; Graham and Richardson 2012). Also problematic is that some of the fossil taxa with pharyngeal denticles are not sufficiently known to determine if the clefts in the pharyngeal wall are open or not (see Smith and Coates 2001).

"Inside to Outside" Hypothesis and Modifications

By comparison, the "inside to outside" hypothesis stressed the importance of endoderm in developmental, genetic patterning of the internal dentition (as compared to external ectoderm, which as noted above lacks developmental patterning capacity, as demonstrated in the catshark; Smith and Coates 1998, 2001; Fraser and Smith 2011). The importance of boundaries between ectoderm and endoderm was also stressed in this hypothesis. These ideas were taken up by Ohazama et al. (2010) to also propose a modified "both inside and outside" theory. They suggested that heterodont dentitions in mammals such as the mouse (different teeth along the jaw) evolved from two different gene regulation mechanisms in pharyngeal/foregut endoderm and oral ectoderm. These mechanisms related to the molariform teeth and incisors, where the molecular characteristics of endoderm were coincident with the molar tooth-forming region, but not that of the incisors. They suggested that the two putatively different mechanisms active in fish (from ectoderm and endoderm) "provided the developmental and genetic diversity on which evolution has acted to produce heterodont dentitions in mammals" (Ohazama et al. 2010, 382). However and importantly, a new study using Sox17-2A-iCre/Rosa26 reporter mice showed that mouse teeth were derived in early development from epithelium of the oral ectoderm (Rothova et al. 2012). Heterodont dentitions also exist in many fish species (fig. 1.6A), suggesting that these hypotheses and observations could be tested, particularly once the molecular fingerprint of teeth versus denticles/scales is known in more detail.

An important paper by Soukup et al. (2008; also Chibon 1966) recently showed that in the Mexican salamander *Ambystoma* teeth developed with contributions from ectoderm, endoderm, or a combination of both. This led Fraser et al. (2010) to also propose another "both inside and outside" theory. Here, the type of epithelium involved was not relevant to odontode origins, all epithelia having odontogenic capacity, as long as competent neural crest-derived mesenchyme was also present. They also proposed the participation of an odontode gene regulatory network, shared among all odontodes. This network evolved from a combination of a gene regulatory network associated with the epithelium and a mesenchymal network. Fraser et al. (2010) also noted that teeth themselves shared a particular, specific dental gene regulatory network, including genes that were absent from external scale/denticle development, based on their work on cichlids. This can be related to the work of Ohazama et al. (2010), where all mouse teeth are derived from oral ectoderm, with differences between molars and incisors based on differences in expression of endodermal genes or ectodermal genes, respectively (but see earlier comment on ectodermal/endodermal boundaries, [Graham and Shone 2014]).

Dental Lamina

Role in Tooth Development in Extant and Fossil Forms

Along with Tor Ørvig (1967), Wolf-Ernst Reif was one of the first researchers to seriously debate the origin of teeth and dentitions. As noted above, he considered teeth and scales to represent evolutionarily and developmentally distinct units, with teeth forming within a structure called the dental lamina (figs. 1.3A, 1.4A, B). Working primarily with shark dentitions, Reif (1982) described the dental lamina as an epithelial tissue invaginated from the oral epithelium, with this region alone producing teeth for the dentition. Teeth differed from skin denticles (figs. 1.3A–C, 1.4A–D) because they were formed, and in a patterned order, before they became functional for use, in contrast to skin denticles that only develop (without a spatiotemporal pattern) when a space arises (Fraser and Smith 2011). Importantly, teeth developed at a fixed location (fig. 1.3A*), within the deep dental lamina along the jaw, with a pattern set up and maintained throughout life. This dental lamina is a facility for making teeth to order deep to the skin and independent of skin denticles, thus providing successor teeth for continuous replacement of the whole dentition. In other words, all necessary tissues (epithelium, mesenchyme) and genes (morphogenesis, temporal and spatial organization) were housed in, restricted to, and associated with, the dental lamina. The presence of a dental lamina has also been recognized in fossil sharks such as *Doliodus* (Miller et al. 2003; Maisey et al. 2013), with organized tooth families present along the jaw, but interestingly, also on the surface of the anterior braincase, with a separate cartilage from those of the upper jaws. As Smith emphasised (2003, 394; fig. 1), "As classically proposed, all phenotypic patterns of the dentition in the gnathostome crown group derive from a single developmental model with a persistent dental lamina (Reif 1982)."

Dental Lamina in Non-Chondrichthyan Fish

The dental lamina came to be used in phylogenetic analyses as a character to define relationships of jawed vertebrates (Goujet 2001; Goujet and Young 2004) and was inferred to exist in fossils if groups of teeth, known as tooth whorls (e.g., fig. 1.3D), were associated with the jaws. Using this definition, involving spatially and temporally timed tooth formation, Reif (1982) was able to identify a dental lamina in the extinct jawed vertebrate group Acanthodii, but absent in the Placodermi, a fossil group subsequently said to lack "true teeth" (both groups resolved as paraphyletic in recent analyses; Brazeau 2009; Davis et al. 2012; Zhu et al. 2013).

However, while the concept of an invaginated, tooth-producing, and tooth-organizing dental lamina may be applied to sharks, recent research indicates that the dental lamina develops in a different way in osteichthyan fish (fig. 1.7A, C), being superficial above the bone (extramedullary) in some taxa, or absent altogether (Smith et al. 2009; Huysseune et al. 2009). In others (e.g., the bream [Teleostei, fig. 1.7A]), separate dental laminae extend deep within the bone itself (intramedullary), connecting the newly developing tooth germs to the functional teeth in the oral cavity. As noted, the putative absence of a dental lamina in placoderms was suggested to separate the group from more derived jawed vertebrates. Alternatively, tooth development and addition in placoderms may be more comparable to teleost teeth (Smith et al. 2009; Rücklin et al. 2011, 2012), with a more restricted dental lamina associated with each functional tooth, providing a developmental model different from that of chondrichthyans. Therefore, absence of tooth development in a dental lamina should no longer be used as a character separating "Placodermi" from other jawed vertebrates (Rücklin et al. 2012). It should also be noted that Reif's scenarios were based on a belief that sharks (chondrichthyans) represented the most primitive jawed vertebrates, such that the dental lamina was also primitive. Phylogenetically, chondrichthyans and osteichthyans are sister taxa in the "crown group" Gnathostomata (Cunningham et al. 2012), and indeed, the deep dental lamina may be a derived character (synapomorphy) of the chondrichthyans, rather than primitive.

Odontogenic Genes and Timing of Interactions

Reif (1982, 288) was one of the first to clarify and define the concepts of an odontode by proposing the "Odontode Regulation Theory": "Each germ of a tooth or denticle is the result of an inductive stimulus. All germs are equivalent to each other; they [scale/denticle or tooth] simply differ with respect to their differentiation program." Within this theory, teeth and scales had separate evolutionary and developmental origins, with teeth developing within the dental lamina, as noted above. All subsequent theories discussed are essentially modifications of Reif's proposals (Smith and Coates 1998, 2001; Smith 2003; Fraser et al. 2009, 2010; Huysseune et al. 2009, 2010; recently reviewed by Smith and Johanson 2012).

One important advance on Reif's work has been a better understanding of the genes involved in development of the dentition of nonmammalian vertebrates. For example, reciprocal inductive gene expression for a single tooth position and its subsequent morphogenesis has been shown for the trout and cichlids (fig. 1.8D–F; Fraser et al. 2004; Fraser, Berkovitz, et al. 2006; Fraser, Graham, and Smith 2006; Fraser et al. 2009) and in reptiles (Buchatová et al. 2008; Handrigan and Richman 2010; Richman and Handrigan 2011). Gene expression in the external dermal skeleton is now also better understood, although by comparison to teeth, has not yet been as well established (see below; also Smith and Johanson 2012).

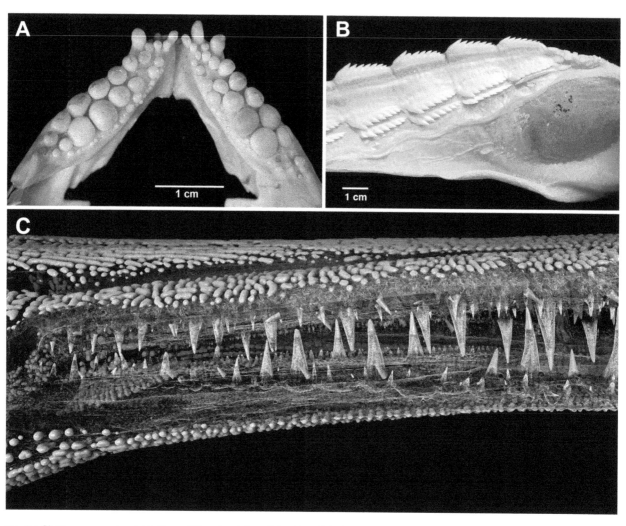

FIG. 1.7 Photo macrographs of dried jaws, (A) occlusal view of dentition of the bream *Abramis* (Actinopterygii), new teeth develop in bone below functional teeth. (B) lingual view of lower jaw right dentition of the six-gill shark (proximal part), new teeth of 2–3 generations are within the lingual epithelium, in depressions (bullae) in the cartilage (*Hexanchus griseus*; Chondrichthyes; Cleveland Museum of Natural History). (C) dentition of the gar *Lepisosteus* (Actinopterygii), computed tomography (μCT) with Drishti (http://anusf.anu.edu.au/Vizlab/drishti/index .shtml: colored volume density rendering red, hypermineralized acrodin) tips of teeth as acrodin and new teeth at an angle to the vertical functional teeth; dermal tubercles with hypermineralized cover on the dermal bones.

Co-option of an Odontode Regulatory Network

In the canonical theory, it was understood that skin denticles, when co-opted to make teeth, would involve co-opting the preexisting skin morphogenetic mechanisms regulated by a specific set of odontogenic genes. These genes would, therefore, be homologous, an archaic set shared by all odontode modules (Fraser et al. 2010). This would include fossil sharks such as Doliodus as well as other taxa now considered to be part of the lineage of stem-group chondrichthyans including the "acanthodians" Ptomacanthus, all possessing organized tooth families (e.g., Brazeau 2009; Maisey et al.

2013; Zhu et al. 2013). Those genes for pattern of spatial organization in the dentition might be different from the "odontogenic set," but as noted, currently less is known about genes associated with scale distribution (skin odontodes, or denticles). Those that have been described include *shh* (zebrafish scales, Sire and Akimenko 2004; Harris et al. 2008; *Scyliorhinus canicula* tail scales, Johanson et al. 2008), *bmp2b* (zebrafish scales, Harris et al. 2008), *eph4a* (*S. canicula*, Freitas and Cohn 2004), *eda/edar* (zebrafish scales, Harris et al. 2008; medaka scales, Kondo et al. 2001), members of the dlx family (*S. canicula* tail scales, Debiais-Thibaud et al. 2011), and *runx2* and *runx3* (*S. canicula*, Hecht et al. 2008). Harris et al.

(2008) suggested that expression of *edar, shh*, and *bmp2b* was coordinated to organize epithelial signaling centers associated with scale development, and are thus responsible for patterning the external dermal skeleton. Fraser et al. (2010) established a more extensive list of genes involved in the scales of cichlids, including *ß-cat, bmp2/4, eda/edar, fgf10, fgf3, notch2, runx2, shh,* and *wnt10a*. Fraser et al. (2010) noted that several genes involved in tooth development as part of a dental gene regulatory network were not expressed in the scales, supporting Reif's contention that teeth and scales were evolutionarily and developmentally distinct. Nevertheless, based on current knowledge, there does appear to be conservation in the eda network (*edar, shh, bmp2b*) between scales and teeth, with Fraser et al. (2009) noting that *edar* and *eda* were responsible for patterning both oral and pharyngeal teeth in the cichlid. Other genes suggested to be involved in patterning of teeth include

Shh, Wnt, and *Sostdc1* in the mouse (Cho et al. 2011). Experiments in progress (by the Fraser lab, University of Sheffield) will identify additional genes involved in developmental patterning of dentitions (in terms of initial positioning along the jaw, but also replacement) versus those for external scales. In particular, these should be across a broader phylogenetic range (e.g., more work on establishing the presence of the odontode and dental gene regulatory networks in chondrichthyans).

Odontogenic Signaling and Relative Timing

Early odontogenic signaling shifts from the epithelium to the neural crest-derived mesenchyme; this primed mesenchyme can then go on to induce non-odontogenic epithelia (Mina and Kollar 1987; Lumsden 1988) to form teeth. Also, the specific molecules, those that function in this specification of tooth organ fate, are well

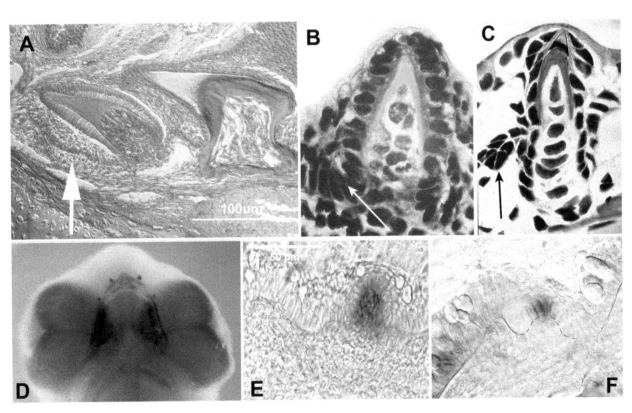

FIG. 1.8 (A) Histological section through developing dentition of the catshark *Scyliorhinus canicula*, white arrow indicates youngest developing tooth germ (organic enameloid matrix pink), older tooth germ to the right (space where mineralized enameloid matrix was). (B) Developing tooth (odontode) in the extant lungfish *Neoceratodus forsteri* with site of new tooth bud (arrow) (Smith et al. 2009, fig. 1b). (C) Developing tooth (odontode) in the axolotl *Ambystoma mexicanum* (dental lamina, arrow; Smith 1967). (D) *shh* expression showing expression loci of initial tooth sites in the premaxilla and maxilla of the trout *Onychorhynchus mykiss*, one on each dentate bone (from fig. 3C, D, Fraser et al. 2004). (E) Section through jaw with *bmp* expression in mesenchyme cells of a new tooth bud of *O. mykiss*. (F) Initial expression of *shh* in the dental oral epithelium of similar stage tooth bud, *O. mykiss* (superimposed red lines mark basal lamina between the two layers) (from fig. 2e, Fraser et al. 2004).

known and can be ectopically positioned to test odontogenic potential, not normally realized, in non-tooth-making oral epithelia, such as the chick (Chen et al. 2000). These authors (Chen et al. 2000) concluded that while there may be latent early odontogenic signaling pathways in chick epithelium, there has been an evolutionary loss of *Bmp4* expression in the epithelium that could account for arrest of tooth development and the loss of teeth in birds. Another set of experiments has replaced the neural crest tissue of birds with that of the mouse to show "that avian oral epithelium is able to induce a non-avian oral developmental program in mouse neural crest-derived mesenchymal cells" (Mitsiadis et al. 2003, 6541).

We see here that relative timing of gene expression is essential for both layers (epithelium and ectomesenchyme) to induce tooth formation, hence the experiments on chick embryos are revealing, given that birds have been missing teeth for at least 60 million years. A paper comparing tooth formation in *talpid2* chicken mutant with that of tooth formation in alligators suggested that changes in the relative position of the aboral-oral boundary (putative ectoderm/endoderm junction) relative to the timing of competent odontogenic mesenchyme led to loss of teeth in avians, despite the tooth developmental program being retained in the epithelium (Harris et al. 2006). They believed that the developmental processes in embryonic birds were homologous with those in the formation of rudimentary teeth in alligators.

Role of Neural Crest in Tooth Induction

Dentine-based units, wherever they occur in the vertebrate skeleton are assumed to be derived from the co-operative activity between an inductive epithelium and neural crest-derived mesenchyme (ectomesenchyme). The contribution of neural crest to teeth has been well established in humans and mouse (e.g., Chai et al. 2000; Janebodin et al. 2011), and also in the lungfish *Neoceratodus forsteri* (Kundrát et al. 2009). Recent research indicating that both ectoderm and endoderm contribute to *Ambystoma* teeth led Soukup et al. (2008; Fraser et al. 2010) to downplay the importance of the type of epithelium involved in tooth development (with mouse teeth deriving from ectoderm alone; Rothova et al. 2012). Instead, it was neural crest mesenchyme

that was the important tissue, and that at least the "developmental machinery" associated with tooth development evolved within the oropharyngeal cavity "driven by a neural crest signal" (Soukup et al. 2008, 798).

Future research testing the "outside to inside" and "inside to outside" hypotheses of tooth origins should focus on establishing neural crest contributions to the external skeleton, particularly in taxa with dentine in the external skeleton (fig. 1.7C, example in *Lepisosteus* of acrodin/dentine covered dermal skeleton), such as the primitive ray-finned fish *Polypterus*; current research has only tested neural crest contribution to bony skeletons lacking dentine, as in the teleosts (Lee et al. 2013) and internal skeleton (gill arch denticles). For example, the contribution of neural crest to the gill arches themselves is well established (see Graham and Richardson 2012 for review), but if the ectoderm brings competence to the epithelium does the neural crest contribute to denticles developing there, where endoderm is the covering epithelium?

Controversies and Criticism

Phylogenetic Criticisms

As noted above, the "inside to outside" hypothesis of dentition evolution was proposed to include new fossil data demanding an alternative to the original "outside to inside" hypothesis. This incorporated new evidence from the novel occurrence of denticles in the pharyngeal region of fossil jawless vertebrates (thelodonts) as homologous (proposed by Van den Bruggen and Janvier 1983) with those of sharks. Since then, several modified theories have been proposed, as described above, along with other criticisms of the "inside to outside" theory. For example, Rücklin et al. (2011) focused on the phylogenetic relationships of the Thelodonti among the vertebrates. They studied the oro-nasal and pharyngeal denticles in thelodont *Loganellia,* supporting previous suggestions that the pharyngeal whorls alone were spatially ordered, with a specific time order of addition to the tooth whorl, absent in oro-nasal denticles (fig. 1.6C vs. fig. 1.6A, B, D, E). However, given currently accepted phylogenies, this character was convergent between *Loganellia* and the gnathostomes. Agnathan taxa more closely related to the gnathostomes such as Osteostraci, and Galeaspida lack pharyngeal denticles. Also, Rücklin et al. (2011) noted that *Loganellia* is a derived thelodont, while other phylogenetically basal thelodonts lack organized denticle whorls (Donoghue et al. 2000; Wilson and Märss 2009; Rücklin et al. 2011). However, a range of denticles were described for a second thelodont, *Phlebolepis elegans*, including those suggested to have been associated with the pharyngeal cavity (Märss and Wilson 2008). These include fused denticles forming an ordered series (Märss and Wilson 2008, figs. 3T, 4T–U), very similar to those described for *Loganellia*, including differentially closed denticle bases, indicative of the relative timing of denticle addition (Märss and Wilson 2008, fig. 3T; Van der Bruggen and Janvier 1993). Rücklin et al. noted that these scales in *Phlebolepis* lined the external afferent branchial ducts (Rücklin et al. 2011, 529), but this was only one of several alternatives suggested by Märss and Wilson, including the lining of incurrent branchial ducts, or positioned more internally within the pharynx. However, *Loganellia* and *Phlebolepis* are derived taxa in recent phylogenies (Wilson and Märss 2009), and the phylogenetic gap between thelodonts and gnathostomes (Donoghue et al. 2000) is still a barrier to the homology of these denticles and organized dentitions. Organization of conodont assemblages in the oro-pharyngeal cavity (e.g., Donoghue and Purnell 1999) has also been presented as evidence for the "inside to outside" theory (Smith and Coates 1998). However, these conodonts belong to the Euconodonta, a derived group; more basal conodonts are simple and lack a comparable organization (Murdock et al. 2013). Therefore, they also are not now considered homologous with teeth in gnathostomes.

Fossils Relevant to the "Inside to Outside" Theory—Postbranchial Lamina

Other spatially ordered elements associated with the posterior pharyngeal region have been observed in association with the postbranchial lamina (part of the anterior-facing pectoral girdle), forming the back wall of the gill arch cavity. These ordered elements have been noted for the "Placodermi" (Johanson and Smith 2003, 2005), as well as sarcopterygian (lobe-finned) fishes, including *Psarolepis* (Zhu et al. 1999) and *Guiyu* (Zhu et al. 2009), and actinopterygian (ray-finned) fishes such as *Lepisosteus* and *Amia* and several fossil taxa (Liem and Woods 1973; Patterson 1977).

Certain placoderms (e.g., Arthrodira) possess teeth, but more primitive placoderms do not (Antiarchi; Smith and Johanson 2003; Rücklin et al. 2012). However, nearly all placoderms were shown to have ordered elements in the posterior pharyngeal cavity, as part of the post-branchial lamina of the anterior trunkshield wall (Johanson and Smith 2003, 2005). This corresponded well with the "inside to outside" hypothesis, that structurally ordered dentitions originated in the oro-pharyngeal cavity, and that these denticles were distinct from external denticles in that their assumed developmental pattern involved the internal endoderm tissue, rather than ectoderm.

However, recent papers have questioned the presence of patterned elements in the placoderm oro-pharyngeal cavity as relevant to the "inside to outside" hypothesis. For example, Huysseune et al. (2009), in their "modified outside to inside" hypothesis, questioned whether placoderm pharyngeal denticles developed as independent structures, when they are in an area of the pharyngeal cavity that can be reached by internally migrating ectoderm (with putative odontogenic capacity). However, as described by Johanson and Smith (2003, 2005), one placoderm, *Weejasperaspis*, lacks ordered denticles at the rear of the pharyngeal cavity, but instead shows disorganized denticles that are continuous with those on the external surface of the trunkshield bone. Johanson and Smith (2003) suggested that this indicated that the boundary between ectoderm and endoderm had migrated inward in *Weejasperaspis*, and that this ectoderm lacked the capacity for ordering denticles on the bone of the pharyngeal cavity, present in other placoderms, presumably under the influence of endoderm. This lack of ordering capacity in the ectoderm corresponds to previous observations for shark body scales (Johanson et al. 2007, 2008; Fraser and Smith 2011). Another recent paper, criticizing the patterning of these placoderm denticles, is provided by Rücklin et al. (2012). Relevant denticles from the arthrodire *Compagopiscis* were scanned using synchrotron X-ray tomography to show that these were composed of spongy bone and represented simple outgrowths of layers of bone, rather than individually added denticles. These observations need to be tested with respect to other placoderms, where denticles appear to be added as individual units, with associated bone of attachment (Johanson and Smith 2003, fig. 9). An example would be the Ptyctodontida, where individual, well-spaced denticles are added to the trunkshield margin (Johanson and Smith 2005, fig. 14A–C).

Fossils Relevant to the "Outside to Inside" Theory

Recent papers have also provided evidence from the fossil record to support the "outside to inside" theory. Blais et al. (2011) described acanthodian taxa from the Devonian of Canada showing transitional scale morphologies between external head scales and tooth whorls along the jaw margin. These "cheek scales" included cusps with pulp cavities fused to a bony base. The morphology of these cusps was tooth-like near the jaw margin, but more like the head scales further from the margin. These cheek scales were more similar to true teeth than the thelodont denticles used by Smith and Coates (1998, 2001) to develop the "inside to outside" theory, and also showed that external scales could transition into oral teeth, rather than being distinct evolutionary units (as suggested by Reif 1982 and the "inside to outside" theory). These acanthodian taxa supported the transfer of tooth-forming competence from the external scales to the oral cavity and the classical "outside to inside" theory. However, Blais et al. (2011, 1197) also state "this developmental machinery may not have been limited to the mouth but may have been expressed in a field that included the external margins of the mouth, and acted on the scales in that area." This seems a more likely explanation, particularly given that two of the three taxa described possess tooth whorls in the oral cavity (Blais et al. 2011, figs. 4–6); developmental patterning was present in the mouth in acanthodians, coincidentally with a similar pattern in the outer cheek. In other words, the relevant "machinery" is internal, both developmentally and evolutionarily, not derived externally. These acanthodian taxa may be comparable to the teleosts where teeth can be produced outside the mouth, and are described as extra-oral (Sire et al. 1998).

A recently published paper (Qu et al. 2013) examined the scales of the early bony fish *Andreolepis hedei* (Osteichthyes) using synchrotron X-ray tomography, and found that the odontodes comprising the external scales were added in a regular pattern across the scale base. However, while Rücklin et al. (2012, figs. 2, 3) demonstrated successional tooth addition with strong overlap and association from a primordial tooth in arthrodiran placoderms (*Compagopiscis*), in *Andreolepis*, linearly developing odontodes (2nd to 5th) are separated from the first-developing odontode and show minimal overlap. Subsequent odontodes develop on top of these, and somewhat randomly. There is no indication that this second series of odontodes is replacing the first series in a manner comparable to teeth in a dentition (e.g., resorption and/or loss of the older teeth), and in this way is more similar to odontode-bearing dermal bones generally (e.g., Zhu et al. 2010).

Future Developmental and Phylogenetic Research

The origin and evolution of the vertebrate dentition continues to be a subject of considerable interest, reflected by the large number of papers published in the last few years, addressing both palaeontological and genetic/developmental aspects of the question.

Pharyngeal Development

For example, the "modified outside to inside" hypothesis of Huysseune et al. (2009, 2010) suggests that ectoderm transferred tooth-producing competence into the oral cavity but also into the pharyngeal region via the gill slits. New research by Graham and Richardson (2012) and Shone and Graham (2014) suggests that the pharyngeal pouches lined with endoderm meet the ectoderm at the outer margin of the pouch during development, potentially excluding the ectoderm from a contribution to the epithelial lining of the gills (pharyngeal outpocketings; fig. 1.2E). Future research should further determine the relative positions of ectoderm and endoderm within the pharyngeal arches, especially where the two types may touch at the origin of the gill slits and the site of boundaries that may be shown to be signaling centers.

Conserved Order of Tooth Pattern

Bilateral symmetrical addition of teeth sequentially along the jaws is normal for development of any vertebrate dentition, but there is a stereotypic developmental pattern of tooth addition order along the dentary in bony fishes (Osteichthyes). New research is needed to determine if this basic pattern order is conserved, using other spatially expressed genes, and other dentate bones of the jaws. Remarkably, not only is this present in trout (Fraser et al. 2004), but also the pufferfish (Fraser et al. 2012) and in the hatchling Australian lungfish (Smith et al. 2009), the first an evolutionary novelty, the second a living link between sarcopterygians and tetrapods. The early pattern of differentially timed tooth loci is derived from a single initiator tooth in jaw position two of each dentary field (with sequential teeth added in both adjacent tooth positions but in minimally staggered times; Smith et al. 2009). The sequence was identified for the dentary bone in the rainbow trout from the expression pattern of the gene *omshh* up-regulated as focal loci for tooth initiation (fig. 1.8D–F; Fraser et al. 2004). This sequence of development of teeth may be a real difference between these and the external scale pattern order, a topic to focus on in the future.

Scale Development

Additionally, research should be directed toward establishing the similarities, or differences, between scales and teeth with respect to genes involved (e.g., the ancient gene network of Fraser et al. 2009), and the contribution of neural crest to the dermal mesenchyme. Teeth develop from ectomesenchyme, but its contribution to external scales has not been established. The contribution of neural crest to other postcranial structures is currently controversial (Smith et al. 1994; Matsuoka et al. 2005; Freitas et al. 2006; Gilbert et al. 2007; Kague et al. 2012; Lee et al. 2013). It would be expected that neural crest also contributes to scale development, but this needs to be tested further, and particularly in taxa that have not lost dentine from their postcranial skeletons.

Fossils and Phylogeny

Apart from development, there should be research into articulated, fossil jawless vertebrates, particularly with respect to their internal anatomy such as the gill arches, and related presence or absence of denticles. Research is especially needed into those taxa phylogenetically separating thelodonts such as *Loganellia* from those with jaws. Taxa of interest would be phylogenetically basal thelodonts, as well as the Galeaspida and Osteostraci. Reanalysis of the current phylogenies that separate thelodonts from jawed vertebrates should also be a focus.

* * *

Acknowledgments

We would like to thank Elizabeth Brainerd, Neil Shubin, and Ken Dial for the invitation to participate in this volume honoring Farish Jenkins, and for their editorial suggestions. We are also grateful to Kate Trinjastic and an anonymous reviewer for their comments.

References

Blais, S. A., L. A. Mackenzie, and M. V. H. Wilson. 2011. Tooth-like scales in early Devonian eugnathostomes and the "outside-in" hypothesis for the origins of teeth in vertebrates. J. Vert. Paleo. 31:1189–1199.

Brazeau, M. D. 2009. The braincase and jaws of a Devonian "acanthodian" and modern gnathostome origins. Nature 457:305–308.

Buchtová, M., G. R. Handrigan, A. S. Tucker, S. Lozanoff, L. Town, K. Fu, V. M. Diewert, C. Wicking, and J. M. Richman. 2008. Initiation and patterning of the snake dentition are dependent on Sonic hedgehog signaling. Dev. Biol. 319:132–145.

Cerny, R., M. Cattell, T. Sauka-Spengler, M. Bronner-Fraser, F. Yu, and D. M. Medeiros. 2010. Evidence for the prepattern/cooption model of vertebrate jaw evolution. Proc. Natl. Acad. Sci. USA 107:17262–17267.

Chai, Y., X. Jiang, Y. Ito, P. Bringas Jr., J. Han, D. H. Rowitch, P. Soriano, A. P. McMahon, and H. M. Sucov. 2000. Fate of the mammalian cranial neural crest during tooth and mandibular morphogenesis. Development 127:1671–1679.

Chen, Y., Y. Zhang, T. X. Jiang, A. J. Barlow, T. R. St Amand, Y. Hu, S. Heaney, P. Francis-West, C. M. Chuong, and R. Maas. 2000. Conservation of early odontogenic signaling pathways in Aves. Proc. Natl. Acad. Sci. USA. 97:10044–10449.

Chibon, P. 1966. Analyse expérimentale de la régionalisation et des capacités morphogénétiques de la crête neurale chez l'amphibien urodeles *Pleurodeles waltlii* Michah. Mém. Soc. Zool. Fr. 36:1–107.

Cho, S. W., S. Kwak, T. E. Woolley, M. J. Lee, E. J. Kim, R. E. Baker, H. J. Kim, J. S. Shin, C. Tickle, P. K. Maini, and H. S. Jung. 2011. Interactions between Shh, Sostdc1 and Wnt signaling and a new feedback loop for spatial patterning of the teeth. Development 138:1807–1816.

Cloutier, R. 2010. The fossil record of fish ontogenies: insights into developmental patterns and processes. Sem. Cell Dev. Biol. 21:400–413.

Cobourne, M. T., and T. Mitsiadis. 2006. Neural crest cells and patterning of the mammalian dentition. J. Exp. Zool. B Mol. Dev. Evol. 306:251–260.

Cohn, M. J. 2002. Lamprey Hox genes and the origin of jaws. Nature 416:386–387.

Cunningham, J. A., M. Rucklin, H. Blom, H. Botella, and P. C. J. Donoghue. 2012. Testing models of dental development in the earliest bony vertebrates, *Andreolepis* and *Lophosteus*. Biol. Lett. 8:833–837.

Davis, S. P., J. A. Finarelli, and M. I. Coates. 2012. *Acanthodes* and shark-like conditions in the last common ancestor of modern gnathostomes. Nature 4867:247–250.

Davit-Beal, T., H. Chisaka, S. Delgado, and J.-Y. Sire. 2007. Amphibian teeth: current knowledge, unanswered questions, and some directions for future research. Biol. Rev. Camb. Philos. Soc. 8:49–81.

de Beer, G. 1971. Development of the Vertebrate Skull. Oxford: Clarendon Press (1937, 1st edition).

Debiais-Thibaud, M., S. Oulion, F. Bourrat, P. Laurenti, D. Casane, and V. Borday-Birraux, V. 2011. The homology of odontodes in gnathostomes: insights from Dlx gene expression in the dogfish, *Scyliorhinus canicula*. BMC Evol. Biol. 11:307.

Depew, M. J., T. Lufkin, and J. L. R. Rubenstein. 2002. Specification of jaw subdivisions by Dlx genes. Science, 298:381–385.

Depew, M. J., C. A. Simpson, M. Morasso, and J. L. R. Rubenstein. 2005. Reassessing the Dlx code: the genetic regulation of branchial arch skeletal pattern and development. J. Anat. 207:501–561.

Donoghue, P. C. J. 2002. Evolution of development of the vertebrate dermal and oral skeletons: unraveling concepts, regulatory theories, and homologies. Paleobiology 28:474–507.

Donoghue, P. C. J., P. L. Forey, and R. J. Aldridge. 2000. Conodont affinity and chordate phylogeny. Biol. Rev. 75:191–251.

Donoghue, P. C. J., and M. A. Purnell 1999. Mammal-like occlusion in conodonts. Paleobiology 25:58–74.

Eames, B. F., N. Allen, J. Young, A. Kaplan, J. A. Helms, and R. A. Schneider. 2007. Skeletogenesis in the swell shark *Cephaloscyllium ventriosum*. J. Anat. 210:542–554.

Fraser, G. J., B. K. Berkovitz, A. Graham, and M. M. Smith. 2006. Gene deployment for tooth replacement in the rainbow trout (*Oncorhynchus mykiss*): a developmental model for evolution of the osteichthyan dentition. Evol. Dev. 8:446–457.

Fraser, G. J., R. Britz, A. Hall, Z. Johanson, and M. M. Smith. 2012. Replacing the first-generation dentition in pufferfish with a unique beak. Proc. Natl. Acad. Sci. USA 109:8179–8184.

Fraser, G. J., R. Cerny, V. Soukup, M. Bronner-Fraser, and J. T. Streelman. 2010. The odontode explosion: the origin of tooth-like structures in vertebrates. Bioessays 32:808–817.

Fraser, G. J., A. Graham, and M. M. Smith. 2004. Conserved deployment of genes during odontogenesis across osteichthyans. Proc. R. Soc. Lond. B Biol. Sci. 271:2311–2317.

Fraser, G. J., A. Graham, and M. M. Smith. 2006. Developmental and evolutionary origins of the vertebrate dentition: molecular controls for spatio-temporal organization of

tooth sites in osteichthyans. J. Exp. Zool. B Mol. Dev. Evol. 306:183–203.

Fraser, G. J., C. D. Hulsey, R. F. Bloomquist, K. Uyesugi, N. R. Manley, and J. T. Streelman. 2009. An ancient gene network is co-opted for teeth on old and new jaws. PLoS Biol. 7(2):e31.

Fraser, G. J., and M. M. Smith. 2011. Evolution of developmental pattern for vertebrate dentitions: an oro-pharyngeal specific mechanism. J. Exp. Zool. B Mol. Dev. Evol. 316B:99–112.

Freitas, R., and M. J. Cohn. 2004. Analysis of EphA4 in the lesser spotted catshark identifies a primitive gnathostome expression pattern and reveals co-option during evolution of shark-specific morphology. Dev. Genes Evol. 214:466–472.

Freitas, R., G. Zhang, and M. J. Cohn. 2006. Evidence that mechanisms of fin development evolved in the midline of early vertebrates. Nature 442:1033–1037.

Gai, Z., P. C. J. Donoghue, M. Zhu, P. Janvier, and M. Stampanoni. 2011. Fossil jawless fish from China foreshadows early jawed vertebrate anatomy. Nature 476:324–327.

Gai, Z., and M. Zhu. 2012. The origin of the vertebrate jaw: intersection between developmental biology-based model and fossil evidence. Chinese Sci. Bull. 57:3819–3828.

Gegenbaur, C. 1878. Elements of Comparative Anatomy. MacMillan and Co., London.

Gilbert, S. F., G. Bender, E. Betters, M. Yin, and J. A. Cebra-Thomas. 2007. The contribution of neural crest cells to the nuchal bone and plastron of the turtle shell. Integr. Comp. Biol. 47:401–408.

Gillis, J. A., M. S. Modrell, and C. V. Baker. 2013. Developmental evidence for serial homology of the vertebrate jaw and gill arch skeleton. Nat. Comms. 4:1436. Doi: 10.1038 /ncomms2429.

Goujet, D. F. 2001. Placoderms and basal gnathostome apomorphies. Pp. 209–222 in P. E. Ahlberg, ed., Major Events in Early Vertebrate Evolution: Palaeontology, Phylogeny and Development. London: Systematics Association.

Goujet, D., and G. C. Young. 2004. Placoderm anatomy and phylogeny: new insights. Pp. 109–126 in G. Arratia, M. V. H. Wilson, and R. Cloutier, eds., Recent advances in the origin and early radiation of vertebrates. Stuttgart: Verlag Dr. Pfeil.

Graham, A., and J. Richardson. 2012. Developmental and evolutionary origins of the pharyngeal apparatus. EvoDevo 3:24.

Grande, L., and W. E. Bemis. 1998. A comprehensive phylogenetic study of amiid fishes (Amiidae) based on comparative skeletal anatomy: an empirical search for interconnected patterns of natural history. J. Vert. Paleontol. 18(suppl.):1–681.

Grevellec, A., and A. S. Tucker. 2010. The pharyngeal pouches and clefts: development, evolution, structure and derivatives. Sem. Cell Dev. Biol. 21:325–332.

Hall, B. K., and J. A. Gillis. 2013. Incremental evolution of the neural crest, neural crest cells and neural crest-derived skeletal tissues. J. Anat. 222:19–31.

Handrigan, G. R., and J. M. Richman. 2010. A network of Wnt, hedgehog and BMP signalling pathways regulates tooth replacement in snakes. Dev. Biol. 348:130–141.

Harris, M. P., S. M. Hasso, M. W. Ferguson, and J. F. Fallon. 2006. The development of archosaurian first-generation teeth in a chicken mutant. Curr. Biol. 16:371–377.

Harris, M. P., N. Rohner, H. Schwarz, S. Perathoner, P. Konstantinidis, P., and C. Nusslein-Volhard. 2008. Zebrafish *eda* and *edar* mutants reveal conserved and ancestral roles of ectodysplasin signaling in vertebrates. PLoS Genet. 4(10):e1000206.

Hecht, J., S. Stricker, U. Wiecha, A. Stiege, G. Panopoulou, L. Podsiadlowski, A. J. Poustka, C. Dieterich, S. Ehrich, J. Suvorova, S. Mundlos, and V. Seitz. 2008. Evolution of a core gene network for skeletogenesis in chordates. PLoS Genet. (43):e1000025.

Hertwig, O. 1874. UeberBau und Entwickelung der Placoidschuppen und der Zähne der Selachier. Jenaische Zeitschrift für Naturwissenschaft 8:331–404.

Huysseune, A., J. Y. Sire, and P. E. Witten. 2009. Evolutionary and developmental origins of the vertebrate dentition. J. Anat. 214:465–476.

Huysseune, A., J. Y. Sire, and P. E. Witten. 2010. A revised hypothesis on the evolutionary origin of the vertebrate dentition. J. Appl. Ichth. 26:152–155.

Huysseune, A., and I. Thesleff. 2004. Continuous tooth replacement: the possible involvement of epithelial stem cells. Bioessays 26:665–671.

Huysseune, A., and P. E. Witten. 2006. Developmental mechanisms underlying tooth patterning in continuously replacing osteichthyan dentitions. J. Exp. Zool. B Mol. Dev. Evol 306:204–215.

James, C. T., A. Ohazama, A. S. Tucker, and P. T. Sharpe. Tooth development is independent of a Hox patterning programme. Dev. Dyn. 225:332–335.

Janebodin, K., O. V. Horst, N. Ieronimakis, G. Balasundaram, K. Reesukumal, B. Pratumvinit, and M. Reyes. 2011. Isolation and characterization of neural crest-derived stem cells from dental pulp of neonatal mice. PLoS One 6:e27526.

Janvier, P. 1985. Les Cephalaspides du Spitsbergen. Editions de CNRS, Paris.

Janvier, P. 1996. Early Vertebrates. Oxford: Oxford University Press.

Johanson, Z., and M. M. Smith 2003. Placoderm fishes, pharyngeal denticles, and the vertebrate dentition. J. Morph. 257:289–307.

Johanson, Z., and M. M. Smith 2005. Origin and evolution of gnathostome dentitions: a question of teeth and pharyngeal denticles in placoderms. Biol. Rev. 80:303–345.

Johanson, Z., M. M. Smith, and J. M. P. Joss. 2007. Early scale development in *Heterodontus* (Heterodontiformes; Chondrichthyes): a novel chondrichthyan scale pattern. Acta Zool. 83:249–256.

Johanson, Z., M. M. Smith, A. Kearsley, P. Pilecki, E. Mark-Kurik, and C. Howard. 2013. Origins of bone repair in the armour of fossil fish: response to a deep wound by cells depositing dentine instead of dermal bone. Biol. Lett. 9(5):20130144.

Johanson, Z., M. Tanaka, N. Chaplin, and M. M. Smith. 2008. Early Palaeozoic dentine and patterned scales in the embryonic catshark tail. Biol. Lett. 4:87–90.

Jollie, M. 1968. Some implications of the acceptance of a delamination principle. Pp. 89–107 in T. Ørvig, ed., Current Problems of Lower Vertebrate Phylogeny. Nobel symposium 4, Stockholm.

Kague, E., M. Gallagher, S. Burke, M. Parsons, T. Franz-Odendaal, and S. Fisher. 2012. Skeletogenic fate of zebrafish cranial and trunk neural crest. PLoS One 7(11):e47394.

Kemp, N. E. 1999. Integumentary system and teeth. Pp. 43–68 in W. C. Hamlett, ed., Shark, Skates, and Rays: The Biology of Elasmobranch Fishes. Baltimore: Johns Hopkins University Press.

Kondo, S., Y. Kuwahara, M. Kondo, K. Naruse, H. Mitani, Y. Wakamatsu, K. Ozato, S. Asakawa, N. Shimizu, N., and A. Shima. 2001. The medaka rs-3 locus required for scale development encodes ectodysplasin-A receptor. Curr. Biol. 111:1202–1206.

Kundrát, M., J. M. Joss, and M. M. Smith. 2008. Fate mapping in embryos of *Neoceratodus forsteri* reveals cranial neural crest participation in tooth development is conserved from lungfish to tetrapods. Evol. Dev. 10:531–536.

Kuraku, S., Y. Takio, F. Sugahara, M. Takechi, and S. Kuratani. 2010. Evolution of oropharyngeal patterning mechanisms involving Dlx and endothelins in vertebrates. Dev. Biol. 341:315–323.

Kuratani, S. 2012. Evolution of the vertebrate jaw from developmental perspectives. Evol. Dev. 14:76–92.

Lee, R. T. H., J. P. Thiery, and T. J Carney. 2013 Dermal fin rays and scales derive from mesoderm, not neural crest. Curr. Biol. 23:R336–337.

Liem, K. F., and L. P. Woods 1973. A probable homologue of the clavicle in the holostean fish *Amia calva*. J. Zool. Soc. Lond. 170:521–531.

Long, J. A., B. K. Hall, K. J. McNamara, and M. M. Smith. 2010. The phylogenetic origin of jaws in vertebrates: developmental plasticity and heterochrony. Kirtlandia 57:46–52.

Lumsden, A. G. S. 1988. Spatial-organization of the epithelium and the role of neural crest cells in the initiation of the mammalian tooth germ. Development 103:155–169.

Maisey, J. G., S. Turner, G. J. Naylor, and R. F. Miller. 2013. Dental patterning in the earliest sharks: implications for tooth evolution. J. Morphology 275:586–596.

Mallatt, J. 1996. Ventilation and the origin of jawed vertebrates: a new mouth. Zool. J. Linn. Soc. 117: 329–404.

Märss, T., and M. V. H. Wilson 2008. Buccopharyngo-branchial denticles of *Phlebolepis elegans* Pander (Thelodonti, Agnatha). J. Vert. Paleo. 28:601–612.

Matsuoka T., P. E. Ahlberg, N. Kessaris, P. Iannarelli, U. Dennehy, W. D. Richardson, A. P. McMahon, and G. Koentges. 2005. Neural crest origins of the neck and shoulder. Nature 436(7049):347–355.

Miller, R. F., R. Cloutier, and S. Turner. 2003. The oldest articulated chondrichthyan from the Early Devonian period. Nature 425:501–504.

Miletich, I., and P. T. Sharpe. 2004. Neural crest contribution to mammalian tooth formation. Birth Def. Res. C Embryo Today 72:200–212.

Mina, M., and E. J. Kollar. 1987. The induction of odonto-genesis in non-dental mesenchyme combined with early murine mandibular arch epithelium. Arch. Oral Biol. 32:123–127.

Mitsiadis, T. A., Y. Cheraud, P. Sharpe, and J. Fontaine-Perus. 2003. Development of teeth in chick embryos after mouse neural crest transplantations. Proc. Nat. Acad. Sci. USA 100:6541–6545.

Murdock, D. J. E., X-P. Dong, J. E. Repetski, F. Marone, M. Stamponi, and P. G. Donoghue. 2013. The origin of conodonts and of vertebrate mineralized skeletons. Nature 502:546–549.

Nelson, G. J. 1970. Pharyngeal denticles (placoid scales) of sharks, with notes on the dermal skeleton of vertebrates. Am. Mus. Novit. 2415:1–26.

Ohazama, A., K. E. Haworth, M. S. Ota, R. H. Khonsari, and P. T. Sharpe. 2010. Ectoderm, endoderm, and the evolution of heterodont dentitions. Genesis 48:382–389.

Ørvig, T. 1967. Phylogeny of tooth tissues: evolution of some calcified tissues in early vertebrates. Pp. 45–110 in A. E. W. Miles, ed., Structural and Chemical Organization of Teeth. New York: Academic Press.

Ørvig, T. 1977. A survey of odontodes ("dermal teeth") from developmental, structural, functional, and phyletic points of view. Pp. 53–75 in S. M. Andrews, R. S. Miles, and D. a. Walker, eds., Problems in Vertebrate Evolution. Linnean Society Symposium no. 4. London: Academic Press.

Patterson, C. 1977. Cartilage bones, dermal bones and membrane bones, or the exoskeleton versus the endoskeleton. Pp. 77–121 in S. M. Andrews, R. S. Miles, and A. D. Walker, eds., Problems in Vertebrate Evolution. London: Academic Press.

Patterson, C. 1988. Homology in classical and molecular biology. Mol. Biol. Evol. 5:603–625.

Peyer, B. 1968. Comparative Odontology. Chicago: University of Chicago Press.

Qu, Q., S. Sanchez, H. Blom, P. Tafforeau, and P. E. Ahlberg. 2013. Scales and tooth whorls of ancient fishes challenge distinction between external and oral "teeth." PLoS One 8(8): e71890.

Raff, R. A. 1996. The Shape of Life: Genes, Development, and the Evolution of Animal Form. Chicago: University of Chicago Press.

Reif, W.-E. 1982. Evolution of dermal skeleton and dentition in vertebrates: the odontode-regulation theory. Evol. Biol. 15:287–368.

Reif, W.-E. 1984. Pattern regulation in shark dentitions. Pp. 603–621 in G. M. Malacinski and S. V. Bryant, eds., Pattern Formation: A Primer in Developmental Biology. New York: Macmillan Publishing Company.

Richman, J. M., and G. R. Handrigan 2011. Reptilian tooth development. Genesis 49:247–260.

Richardson, J., T. Shono, M. Okabe, and A. Graham. 2011. The presence of an embryonic opercular flap in amniotes. Proc. Biol. Sci. 279:224–229.

Rothova, M., H. Thompson, H. Lickert, and A. S. Tucker. 2012. Lineage tracing of the endoderm during oral development. Dev. Dyn. 241:1183–1191.

Rücklin, M., P. C. Donoghue, Z. Johanson, K. Trinajstic, F. Marone, and M. Stampanoni. 2012. Development of teeth and jaws in the earliest jawed vertebrates. Nature 491:748–751.

Rücklin, M., S. Giles, P. Janvier, and P. C. J. Donoghue. 2011. Teeth before jaws? Comparative analysis of the structure and development of the external and internal scales in the extinct jawless vertebrate Loganellia scotica. Evol. Dev. 13:523–532.

Takio, Y., S. Kuraku, Y. Murakami, M. Pasqualetti, F. M. Rijli, Y. Narita, S. Kuratani, and R. Kusakabe. 2007. Hox gene expression patterns in Lethenteron japonicum embryos—insights into the evolution of the vertebrate Hox code. Dev. Biol. 308:606–620.

Schaeffer, B. 1975. Comments on the origin and basic radiation of the gnathostome fishes with particular reference to the feeding mechanism. Colloq. Int. CNRS 218:101–109.

Shone, V., and A. Graham 2014. Endodermal/ectodermal interfaces during pharyngeal segmentation in vertebrates. J. Anat. 225:479–491. doi:10.1111/joa.12234.

Sire, J. Y., and M. A. Akimenko 2004. Scale development in fish: a review, with description of sonic hedgehog (shh) expression in the zebrafish Danio rerio. Int. J. Dev. Biol. 48:233–247.

Sire, J. Y., and I. Arnulf 1990. The development of squamation in four teleostean fishes with a survey of the literature Japan. J. Ichthy. 37:133–143.

Sire, J. Y., S. Marin, and F. Allizard. 1998. Comparison of teeth and dermal denticles (odontodes) in the teleost Denticeps clupeoides (Clupeomorpha). J. Morph. 237:237–255.

Smith, M. M. 1967. Studies on the structure and development of Urodele teeth. PhD thesis, University of London (fig. 5.7, unpublished data).

Smith, M. M. 1988. The dentition of Palaeozoic lungfishes: a consideration of the significance of teeth, denticles and tooth plates for dipnoan phylogeny. Mem. Mus. Nat. d'Hist. Natur. Paris C 53:177–193.

Smith, M. M. 2003. Vertebrate dentitions at the origin of jaws: when and how pattern evolved. Evol. Dev. 5:394–413.

Smith, M. M., and M. I. Coates. 1998. Evolutionary origins of the vertebrate dentition: phylogenetic patterns and developmental evolution. Eur. J. Oral Sci. 106:482–500.

Smith, M. M., and M. I. Coates. 2001. The evolution of vertebrate dentions: phylogenetic pattern and developmental models. Pp. 223–240 in P. E. Ahlberg, ed., Major Events in Vertebrate Evolution. London: Taylor & Francis.

Smith, M. M., G. J. Fraser, and T. A. Mitsiadis. 2009. Dental lamina as source of odontogenic stem cells: evolutionary origins and developmental control of tooth generation in gnathostomes. J. Exp. Zool. B Mol. Dev. Evol. 312B:260–280.

Smith, M. M., and B. K. Hall. 1990. Development and evolutionary origins of vertebrate skeletogenic and odontogenic tissues. Biol. Rev. 65:277–373.

Smith, M. M., and B. K. Hall. 1993. A developmental model for evolution of the vertebrate exoskeleton and teeth: the role of cranial and trunk neural crest. Evol. Biol. 27:387–447.

Smith, M., A. Hickman, D. Amanze, A. Lumsden, and P. Thorogood. 1994. Trunk neural crest origin of caudal fin mesenchyme in the zebrafish Brachydanio rerio. Proc. R. Soc. Lond. B 256:137–145.

Smith, M. M., and Z. Johanson. 2003. Separate evolutionary origins of teeth from evidence in fossil jawed vertebrates. Science 299:1235–1236.

Smith, M. M., and Z. Johanson. 2012. A molecular guide to regulation of morphological pattern in the vertebrate dentition and the evolution of dental development. In R. J. Asher and J. Müller, eds., From Clone to Bone: The Synergy of Morphological and Molecular Tools in Palaeobiology. Cambridge: Cambridge University Press.

Smith, M. M., Z. Johanson, C. Underwood, and T. Diekwisch. 2013. Pattern formation in development of chondrichthyan dentitions: a review of an evolutionary model. Hist. Biol. 25:127–142.

Smith, M. M., M. Okabe, and J. Joss. 2009. Spatial and temporal pattern for the dentition in the Australian lungfish revealed with sonic hedgehog expression profile. Proc. Roy. Soc. B Biol. Sci. 276:623-631.

Soukup, V., H. H. Epperlein, I. Horacek, and R. Cerny. 2008. Dual epithelial origin of vertebrate oral teeth. Nature 455:795-798.

Stock, D. W. 2001. The genetic basis of modularity in the development and evolution of the vertebrate dentition. Phil. Trans. Roy. Soc. London B Biol. Sci. 356:1633-1653.

Takechi, M., Adachi, N., Hirai, T., Kuratani, S., and Kuraku, S. 2013. The Dlx genes as clues to vertebrate genomics and craniofacial evolution. Semin. Cell Dev. Biol. 24:110-118.

Van der Bruggen, W., and P. Janvier. 1993. Denticles in thelodonts. Nature 364:107.

Veitch, E., J. Begbie, T. F. Schilling, M. M. Smith, and A. Graham. 1999. Pharyngeal arch patterning in the absence of neural crest. Curr. Biol. 9:1481-1484.

Wilson, M. V. H., and T. Marss. 2009. Thelodont phylogeny revisited, with inclusion of key scale-based taxa. Estonian J. Earth Sci. 58:297-310.

Zhu, M., W. Wang, and X. Yu. 2010 *Meemannia eos*, a basal sarcopterygian fish from the Lower Devonian of China—expanded description and significance. Pp. 119-214 in D. K. Elliott, J. G. Maisey, X. Yu, and D. Miao, eds., Morphology, Phylogeny and Paleobiogeography of Fossil Fishes. Munich: Verlag Dr. Friedrich Pfeil.

Zhu, M., X. Yu, P. E. Ahlberg, B. Choo, J. Lu, T. Qiao, Q. Qu, W. Zhao, L. Jia, H. Blom, and Y. Zhu. 2013. A Silurian placoderm with osteichthyan-like marginal jawbones. Nature 502:188-193.

Zhu, M., X. Yu, and P. Janvier. 1999. A primitive fossil fish sheds light on the origin of bony fishes. Nature 397:607-610.

Zhu, M., W. Zhao, L. Jia, J. Lu, T. Qiao, and Q. Qu. 2009. The oldest articulated osteichthyan reveals mosaic gnathostome characters. Nature 458:69-474.

2

Flexible Fins and Fin Rays as Key Transformations in Ray-Finned Fishes

George V. Lauder*

The aquatic world has been home to a remarkable diversity of fishes for at least 600 million years. Although there is certainly enormous variation in the strategies used by fishes to feed and capture prey, there is at least an equal diversity of locomotor modes and associated morphologies that fishes use to navigate the three-dimensional aquatic realm. Locomotion is essential for reaching patchy prey resources, for reproduction and displaying to potential mates, for migration, for escape from predators, and for maneuvering through complex habitats.

These many functional demands on fish locomotor systems have given rise to considerable variation in the shapes of fishes and their fins, as well as to a variety of movement patterns that fishes use to swim and maneuver (Lauder 2006; Webb 1975). At least some of these functional demands may conflict with one another, so that fish body shapes may be a compromise between, for example, fin and shape patterns that enhance acceleration and those that increase steady swimming performance (Webb 1975). Fish fins are highly variable in shape, and movement of fins varies considerably among different swimming behaviors within a species. Also, evolutionary trends in the functional design of fish fins have been documented since the beginnings of scientific comparative anatomy and paleontology (Agassiz 1833–1843; Owen 1854), and a number of historical patterns of transformation in

* Department of Organismic and Evolutionary Biology, Harvard University

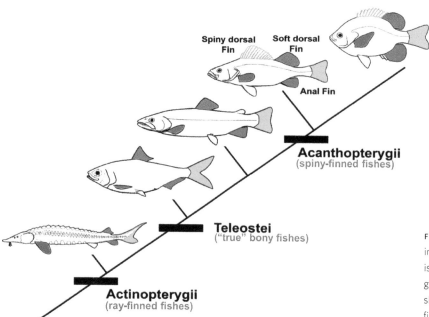

FIG. 2.1 Major patterns of fin position and shape in the evolution of ray-finned fishes. This diagram is a highly simplified phylogeny of selected major groups of ray-finned fishes to show changes in shape and position of the tail fin (green), pectoral fin (purple), pelvic fin (yellow), dorsal fin (red), and anal fin (blue).

fish locomotor design have been thoroughly discussed in the literature (Affleck 1950; Benton 1997; Breder 1926; Goodrich 1930; Lauder 1989; Lauder 2000; Romer and Parsons 1986; Rosen 1982).

This chapter will focus on a major transformation that characterizes the evolution of ray-finned fishes (Actinopterygii), the largest and most diverse clade of the general (non-monophyletic) group known as "fishes": functional design of fins. Ironically, the anatomy and function of fins is one of the least studied features of ray-finned fishes, despite being one of their most obvious characteristics and the one that gave the group its name! In particular, ray-finned fish fins have four key design features that are critical to locomotion in complex underwater habitats, and these four traits form an underlying theme to this chapter: (1) fins are flexible, (2) most fins in ray-finned fishes are collapsible, (3) fins are supported by flexible jointed fin rays with a unique bilaminar structure that allows active control of fin surface conformation, and (4) ray-finned fishes possess multiple sets of fins that allow them to take advantage of hydrodynamic interactions among fins, and to execute complex swimming behaviors requiring multifin control such as swimming backward.

Figure 2.1 summarizes schematically the overall pattern of ray-finned fish fin shape and position, although within any individual clade different species may exhibit a variety of fin locations and shapes. The caudal fin in basal ray-finned fish clades such as sturgeon (as well as in many outgroup clades such as elasmobranchiomorphs) is heterocercal in shape, with an elongate upper lobe containing the vertebral column and a shorter lower lobe containing fin rays and some musculature but not vertebral elements. Most derived clades possess a homocercal tail in which the external shape of the upper and lower lobes is symmetrical around a horizontal axis, although the internal skeletal anatomy of the fin is somewhat asymmetrical in nature reflecting the ancestral heterocercal condition.

A general trend in pectoral fin placement is evident with basal clades typically possessing ventrally located fins with a predominant horizontal orientation (fig. 2.1). The pectoral fins of basal clades, while certainly mobile and used to adjust body position, display a considerably smaller range of motion than pectoral fins in more derived clades that are more wing-like and located on the lateral body surface.

The pelvic fins of ray-finned fishes (fig. 2.1) differ primarily in their location along the body and in their attachment to the axial skeleton and pectoral girdle. For the most part, the pelvic fins of basal clades are located posteriorly on the body in the tail region and have skeletal supports that are embedded in the body wall, not attached to the axial skeleton. In derived clades, the pelvic fins are typically located more anteriorly, and may be attached to the pectoral girdle.

Dorsal and anal fins display a great variety of positions and shapes in ray-finned fishes, and broad evolutionary trends can be seen in fin position (fig. 2.1) as well as in changes in internal skeletal supports of these fins. In both basal and derived clades, dorsal and anal fin positions can mirror each other, with these fins having similar shapes and in similar longitudinal locations on the body. In other clades, the positions of dorsal and anal fins can be offset with the dorsal fin most commonly located more anterior to the anal fin. Dorsal and anal fins also can possess extremely elongate fin rays in some species of fishes, and show considerable variation in length and shape (fig. 2.1).

One general feature of fins in ray-finned fishes is so obvious that it often goes unstated: most fish fins are collapsible, or at least can undergo substantial active changes in surface area. This allows fish to adjust the area of fins that interacts with the water, and to fold fins against the body to minimize drag forces or to alter the balance of forces and torques on the body.

Another feature of ray-finned fish evolution that is relevant to understanding locomotor patterns is the swim bladder. Swim bladders (homologous to lungs) serve as organs that counter the weight of the skeleton and scales and allow for near-neutral buoyancy. Gas-filled inclusions in the body are present in the earliest fishes, and are retained as either lungs or swim bladders in the vast majority of ray-finned fishes (Brainerd, this volume; Liem 1989; Liem et al. 2001). By reducing the need to counter gravitational effects by producing lift forces, the swim bladder may have played a key role in permitting evolutionary diversification in locomotor modes and fin and body shapes in ray-finned fishes. Swim bladders have also become specialized in a number of clades and show an array of interesting functional designs that include oil-filled bladders and bladders encapsulated in bone, and have been lost in many benthic fish clades and deep-sea species where air bladders may be a liability. Studies of the correlated evolution of swim bladder function and fin and body shapes as fishes diversified into the three-dimensional aquatic habitat is an exciting area for future research.

Fin Rays of Ray-Finned Fishes: A Key Transformational Character

One would think that given a clade of over 30,000 species named for a key trait, the ray-fin, that this feature would have been studied in some detail, and the structural and mechanical properties of fin rays would be understood. But this is not the case. In fact, it is remarkable how little we know about both the structure and function of actinopterygian ray-fins.

Figure 2.2 illustrates a number of the features of ray-finned fish fin rays to provide an overview of the structural design of fin rays in this clade. The fin rays of sharks and living lungfish have a different design with a single rod-like element for each fin ray (Goodrich 1904; Goodrich 1906), and many of the features present in actinopterygian ray-fins are missing or greatly reduced in outgroup clades. For example, fin flexibility and collapsibility is greatly increased in most extant ray-finned fish clades as compared to most outgroups, although the contrast is greatest when teleost fish fins are compared to those of other extant outgroup taxa.

Perhaps most significant, however, is the bilaminar design of ray-finned fish fin rays, which contrasts with the organization of outgroup taxa. The fin rays of actinopterygians have a bilaminar structure with two halves (hemitrichia) comprising a single ray or lepidotrich (Alben et al. 2007; Geerlink and Videler 1987; Lauder, Madden, et al. 2011; Taft and Taft 2012). Fin rays may be fused into spines to support the anterior regions of dorsal and anal fins in some clades via a variety of developmental patterns, but the fin rays themselves are flexible structures that may branch toward their distal end (fig. 2.2). An unsegmented basal region of ray-finned fish fin rays is typical, while the middle and distal regions are segmented, which may increase the flexibility of rays (Alben et al. 2007; Taft and Taft 2012). The bilaminar and segmented design of actinopterygian fin rays has a particular functional significance: differential activity of muscles attaching at the base of these fin rays produces curvature of the fin ray surface (fig. 2.2F), allowing the fin to actively resist fluid loading. This feature can be of considerable importance during locomotion where fish may extend their fins into oncoming flow and stiffen the fin during a maneuver (Lauder and Madden 2006; Lauder and Madden 2007;

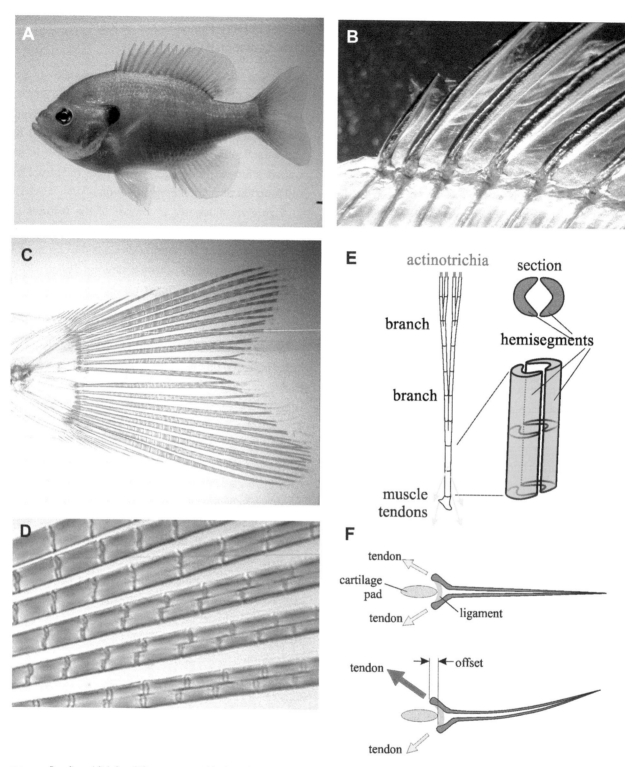

FIG. 2.2 Ray-finned fish fins (A) are supported by bony fin rays. In some clades rays are fused into spines supporting the anterior regions of the dorsal and anal fins (B). Fin rays (C) are composed of flexible jointed bony segments (D, E), each of which is itself divided into two hemisegments (E). Muscles at the base of each fin ray can produce a sliding motion (E, F) that allows fish to actively bend each fin ray and control the surface conformation of the fin. See Alben et al. (2007), Lauder et al. (2011b), and Lauder (2007).

Flammang et al. 2013). Active fin surface control is absent in outgroup taxa such as sharks and more basal groups like lamprey, placoderms, and acanthodians. These clades lack the bilaminar fin ray structure that is a prerequisite for active surface control.

The ability to actively control the conformation of a propulsive surface also contrasts with the functional design of propulsors such as bird feathers and insect wings that can be positioned as a whole in space, but cannot have their surface conformation altered actively by the animal.

Students wishing to make a simple model that illustrates the bilaminar design of ray-finned fish fin rays can use a commercial ziplock bag. Using scissors, cut off the bag just below the ziplock closure at the top to remove it, and then cut the remaining ziplock closure section in half transversely. Separate the two halves from one piece of the ziplock and you will see that the two halves are attached at the side. Zipping them together and then holding the open base between your thumb and forefinger will allow you to slide one half relative to the other half. When you do this, you will see the ziplock surface curve in space. With a little practice, considerable curvature of the ziplock closure can be achieved by sliding the two halves relative to each other. The ability to actively control the fin surface with muscles at the base of the fin has the additional advantage of not requiring the occurrence of muscles out along the fin length to achieve surface conformational changes. This in turn enables the fin to be thinner and lighter, and is a key innovation in the functional design of ray-finned fishes.

Recent functional studies of ray-finned fish fin rays have used techniques allowing control of individual hemitrichs as well as classical three-point bending and computational modeling to quantify mechanical properties of lepidotrichs (Alben et al. 2007; Flammang et al. 2013; Lauder, Madden, et al. 2011; Taft 2011; Taft and Taft 2012). At least for the relatively few species studied so far, actinopterygian fin rays vary in mechanical properties along the length of the ray with proximal regions being stiffer than distal (probably due to the unsegmented proximal region of the ray), and whole fin rays have an elastic modulus roughly equivalent to that of collagen. Actinopterygian fin rays are effective displacement transducers, with a small displacement at the base (0.1 mm) generating a large tip displacement of approximately 4.0 mm.

We still understand very little about the functional diversity of fin rays within the actinopterygian clade. There is clearly substantial variation among species and among fins within any individual (Taft 2011), and yet study of the extent of this variation and its implications for locomotor function and habitat use have barely been considered.

Pectoral Fin Function

Pectoral fins in fishes show considerable diversity in both structure and function, with changes in fin area and shape, location on the body, and attachment angle all playing an important role in governing the effect of fin use on body position. Figure 2.3 schematically illustrates some of the changes in pectoral fin shape and position that occur during locomotion, and this flexibility is important for controlling the direction of forces generated by the fin during swimming and maneuvering (Drucker and Lauder 2003). Basal clades tend to have pectoral fin positions that are relatively ventral and with shallow attachment angles to the body (fig. 2.1). In many derived clades, especially those in which pectoral fins are used to generate thrust during swimming, the fin is larger in area, and the base of the fin is attached higher on the body at a greater angle.

The underlying anatomical basis of changes in attachment angle of pectoral fins has not been subject to detailed study, but certainly changes in the angles and connections between the radial bones that support pectoral fin rays and also between the radials and the scapula (fig. 2.3A) could allow for considerable variation among species in pectoral fin motion. Basal ray-finned clades have more numerous radial bones than derived clades and different patterns of skeletal connection between the radials and the pectoral girdle; this suggests that there may be differences in ability to reorient the pectoral fin among these groups of ray-finned fishes, although this issue has not been investigated. Most fish species can reorient the whole pectoral fin via changes in the angle of the base (fig. 2.3C), and reorientation of the fin base during different behaviors has the effect of altering the direction of pectoral fin forces.

In relatively basal actinopterygian clades such as salmoniform fishes with a relatively horizontal pectoral fin base, the pectoral fin generates considerable torque around the center of mass during braking. For example,

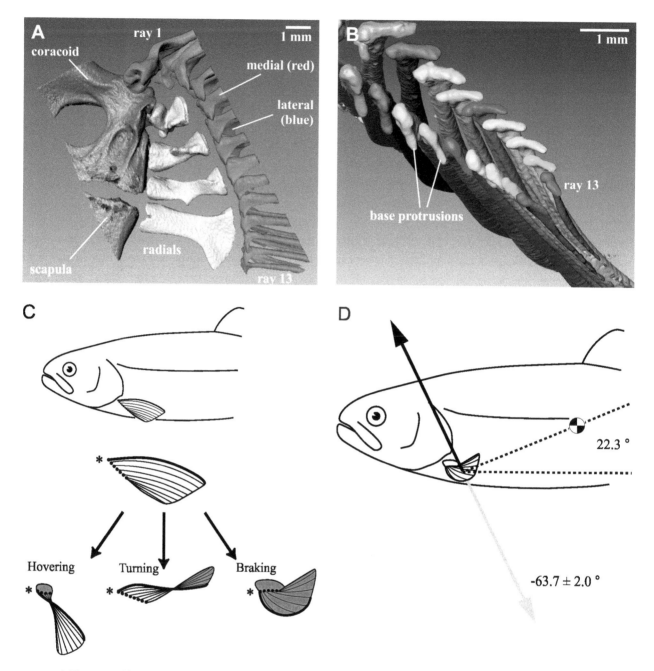

FIG. 2.3 A, The pectoral fin rays of ray-finned fishes articulate with rod-like radials that are attached to the scapula and coracoid of the pectoral girdle. B, Fin rays have small basal processes that serve as the locations of muscle attachment. C, The pectoral fin base can be actively reoriented and changes position during different locomotor behaviors. The asterisk marks the location of the base of the first fin ray for reference, and the dorsal surface is colored red. D, The pectoral fin of basal teleost fishes like trout generates forces during braking that generate torques around the center of mass: the gray arrow shows the direction of the reaction force that is almost orthogonal to the direction of the center of mass. See Drucker and Lauder (2003) and Lauder et al. (2011b).

in trout (fig. 2.3D), braking behavior to stop forward motion involves curling up the distal pectoral fin margin to generate force directed forward. But due to constraints on motion and shape of the pectoral fin, the direction of net force is anterodorsal, and hence the reaction force to this is directed posteroventrally. This results in con-

siderable torque around the center of mass (fig. 2.3D), and as a result trout move ventrally when braking and rotate counterclockwise around their center of mass (Drucker and Lauder 2003).

In other more basal clades such as sturgeon, experimental study of pectoral fin function has shown that the ventrally located pectoral fins are used for maneuvering and to induce pitch moments to change body position, and have relatively little effect on flow during steady horizontal swimming (Wilga and Lauder 1999). Elevation and depression of the posterior fin margin generates a vortex wake that induces pitching moments that initiate swimming motions up or down in the water column.

In derived ray-finned fishes, there are a number of clades that use their pectoral fins as primary thrusters to generate locomotor forces and to maneuver and swim through the water (Drucker and Lauder 2000; Walker and Westneat 2002; Westneat 1996). Fishes such as bluegill sunfish, *Lepomis macrochirus*, are an excellent example of this type of pectoral fin use, and experimental work on the hydrodynamic function of the pectoral fins (Drucker and Lauder 1999) has shown that during slow swimming, each pectoral fin generates ring-like vortex structures in the wake that represent momentum added to the fluid and that propel the fish forward. The structure of the vortex wake changes with speed, and pectoral fin motion can generate more complex linked-ring configurations reflecting additional fluid momentum as swimming speed increases (Drucker and Lauder 2000). The structure of the vortex wake produced by fishes swimming with their pectoral fins has been the subject of a considerable number of both experimental and computational studies as biologists seek to use fish fins as a model for generating new types of underwater propulsors that might replace propeller systems for future small underwater vehicles (Bozkurttas et al. 2009; Dong et al. 2010; Drucker and Lauder 1999; Lauder and Madden 2007; Lauder et al. 2006; Mittal et al. 2006; Ramamurti et al. 2002).

Derived ray-finned fish clades have a pectoral fin base more vertically oriented and placed higher on the side of the body than in basal clades (fig. 2.1), and this has an important effect on the forces and torques generated during braking as compared to basal groups with horizontally oriented fins. For example, when bluegill sunfish execute a braking maneuver to stop swimming, they extend the pectoral fins from the body to increase drag and stop forward motion (Higham et al. 2005). The forces generated during this behavior are oriented anteroventrally and at such an angle that the reaction force passes through the center of mass (Drucker and Lauder 2002). This means that in bluegill sunfish the pectoral fins can be used in braking without generating

pitch torques and thus altering body position, and stand in contrast to the function of more horizontally oriented pectoral fins where the body pitches as the fins are used in braking.

Caudal Fin Function

A very large literature exists on the anatomy and function of fish body (myotomal) musculature during swimming, but only recently has the important role of intrinsic tail musculature in fish locomotion been fully recognized. This is a key feature of the functional design of ray-finned fish fins: the fins possess intrinsic musculature distinct from the body muscles, which allows control over fin ray motion and position in space, and allows most ray-finned fishes to collapse their fins and reduce the surface area exposed to incident flow. Given the flexible surface of fins and the thin collagenous membrane that extends between adjacent fin rays, intrinsic fin musculature allows fish great control over fin posture and position.

Within the tail itself, and distinct from body musculature, are a series of intrinsic muscles that act to control tail conformation (fig. 2.4A–C). In derived ray-finned clades such as the Perciformes, there is an extensive complement of intrinsic tail musculature that controls adduction and abduction of individual fin rays (via the interradialis and supra- and infracarinalis muscles), motion of fin rays to each side of the tail (the flexor ventralis and dorsalis muscles), and movement of the upper tail lobe relative to the lower (the hypochordal longitudinalis) (fig. 2.4C). In basal ray-finned clades such as gar (*Lepisosteus*), intrinsic caudal musculature is much less extensive and consists only of a broad ventral flexor muscle on each side (fig. 2.4A). In bowfin (*Amia*), there is more extensive intrinsic musculature with most muscles focused on control of the dorsal portion of the tail (see fig. 2.4B; Flammang and Lauder 2008, 2009; Lauder 1989; Lauder et al. 2003).

These intrinsic tail muscles permit fine control of tail conformation and alteration of tail function during different locomotor behaviors. For example, the tail may be held in a relatively flat shape during steady swimming (fig. 2.4D), may assume an S shape during braking (fig. 2.4E), may be extended into a broad blade during acceleration (fig. 2.4F), or may be compressed during the glide phase following rapid forward motion (fig. 2.4G).

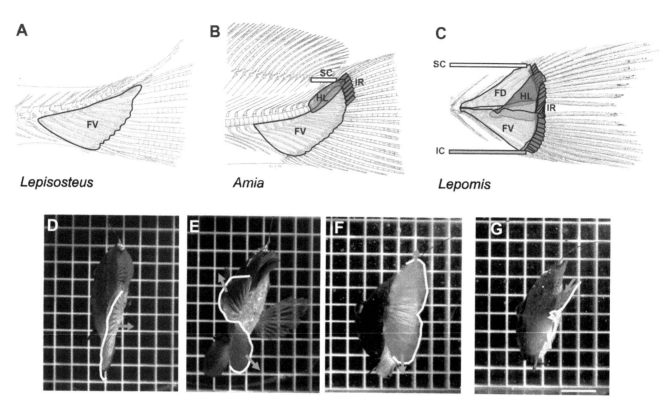

A *Lepisosteus* **B** *Amia* **C** *Lepomis*

FIG. 2.4 Ray-finned fishes have control of tail conformation via intrinsic musculature that is distinct from the myotomal body muscles. Panels A, B, and C schematically diagram the diversity of intrinsic tail muscles in three ray-finned fish species, to show the diversity of intrinsic musculature and the substantial increase in muscular control of the tail in derived taxa such as bluegill sunfish (*Lepomis*) compared to basal clades such as gar, *Lepisosteus*. Intrinsic caudal muscles shown are the flexor dorsalis (FD, green), flexor ventralis (FV, blue), hypochordal longitudinalis (HL, purple), infracarinalis (IC, gray), interradialis (IR, red), and supracarinalis (SC, yellow). Note that *Lepisosteus* lacks all intrinsic caudal musculature except for a broad FV, and *Amia* lacks the FD, IC, and all ventral IR muscles. Panels D, E, F, and G show tail conformation in a bluegill sunfish from posterior view (the posterior tail margin is outlined in yellow) during steady swimming at 1.2 L/s (D), braking (E), and kick (F) and glide (G) behaviors. Red arrows illustrate the direction of tail motion during these behaviors. See Flammang and Lauder (2008, 2009).

The hydrodynamic function of the tail in ray-finned fishes has been studied as a means of understanding how different tail shapes influence patterns of momentum in the wake as reflected in vortex structure, to estimate the forces and torques generated by fish tails, and to evaluate how the wake from the dorsal and anal fins could influence the function of the caudal fin. Studies of caudal fin function using flow visualization with the particle image velocimetry technique have revealed the typical counterrotating centers of vorticity in the wake that reflect slices through caudal fin vortex rings, and enabled reconstruction of the vortex wake of fishes with homocercal tails. Particle image velocimetry has shown that the homocercal tail generates a largely symmetrical chain of linked rings during steady swimming in which side (lateral) forces are typically nearly double that of thrust force values (Standen and Lauder 2007).

Swimming by undulatory propulsion necessarily generates substantial side forces. The vortex wake of fishes with homocercal tails generates a central momentum jet that propels fish forward (Flammang et al. 2011b; Lauder and Tytell 2006; Nauen and Lauder 2001).

In basal ray-finned clades such as sturgeon, which have heterocercal tails, analysis of the vortex wake has shown that the sturgeon tail also produces a chain of linked vortex rings, and that the angle of the momentum jet is such that the reaction force passes through the center of mass (Liao and Lauder 2000). This means that the heterocercal tail of sturgeon does not generate body torques during steady horizontal locomotion, and that no rotational moments are produced by the heterocercal tail. This differs from the function of the shark heterocercal tail where both lift forces and pitch torques are produced by asymmetrical motion of the dorsal and ventral lobes (Flammang et al. 2011a; Wilga and Lauder 2002).

Dorsal and Anal Fin Function

A key feature of ray-finned fish functional design is the presence of multiple fins that can produce interacting fluid flows. So far in this chapter we have treated fins as though they function independently, but this is clearly not always the case. In particular, the wake from the dorsal, anal, and pelvic fins passes downstream along the swimming fish and can at least potentially influence water flow over the tail and hence the vortex wake produced there. This process of wake interaction in fishes has been demonstrated both experimentally and computationally, as well as with simple robotic motels (e.g., Akhtar et al. 2007; Drucker and Lauder 2001; Flammang et al. 2011b; Lauder, Lim, et al. 2011; Standen 2008; Standen and Lauder 2007; Tytell 2006).

Figure 2.5A, B illustrates the motion of the dorsal fin relative to the tail in rainbow trout, *Oncorhynchus mykiss*, and shows the out-of-phase motion that the trailing edge of the dorsal fin exhibits relative to the leading edge of the caudal fin (Drucker and Lauder 2005). The flexible dorsal fin in trout is actively moved and generates a vortex wake that the tail moves through (fig. 2.5C). The wake can be quite dramatic, and substantially alters the fluid environment around the tail. For example, as shown in figure 2.5C, the dorsal fin produces a wake of strong alternating lateral jets, and the path taken by the tail goes through the vortex centers. One important consequence of the dorsal fin wake is that much of the time the tail experiences an incident flow direction that is nearly perpendicular to the free stream local flow that the fish is swimming through. Hydrodynamic models of fish tail function rarely consider the greatly altered flow environment experienced by the tail as a result of upstream fin action.

The production of strong lateral wake jets by the dorsal and anal fins of swimming fish has important consequences for the overall force balance during swimming. Figure 2.5D illustrates that these dorsal and anal fin fluid jets produce torques on the fish body that must be balanced for steady swimming, or can be modulated to induce maneuvers. Dorsal and anal fins produce opposite sign roll torques, and yaw moments produced by motion of the pectoral and tail fins must all be integrated to produce an overall stable force balance during steady swimming.

In the fish locomotion literature one often sees reference to the distinction between "body and caudal fin locomotion" and "median fin locomotion" as though these were two distinct locomotor modes. But in light of recent results showing that the median dorsal and anal fins are actively used even during steady swimming and that they generate significant hydrodynamic forces used to control the roll torque balance, this distinction seems artificial at best. Fish, even those swimming steadily without maneuvering, use their median fins actively, and use of median fins is integral to understanding the overall force and torque balance on fishes. The active use of multiple fins to control body posture during swimming is a hallmark of ray-finned fishes.

Median fins also play an important role in unsteady locomotor behaviors, and the ability of fishes to collapse and extend their fins during swimming behaviors is a critical feature of fin functional design. As fishes swim faster, the dorsal and anal fins are often depressed, which reduces their surface area (fig. 2.6A, B). But these fins are erected rapidly when an unsteady maneuver or a C-start escape response is initiated (fig. 2.6C, D). Median fins of fishes play an important role in unsteady locomotion, and perform numerous functions including increasing surface area near the center of mass, controlling roll and yaw torques, and adding momentum during escape responses (Chadwell et al. 2012a; Chadwell et al. 2012b; Tytell and Lauder 2008). Despite the recent increase in data on median fin function in fishes, there is much more to be learned about how these fins function during diverse locomotor behaviors, and how median fin function is integrated by the fish nervous system with input from the body and caudal fin musculature to control body position.

Fins as Sensors for Complex Locomotor Tasks

The median and paired fins characteristic of ray-finned fishes are also important when fish execute complex behaviors such as locomotion through obstacles, or move backward. The entire array of fins may move in concert to achieve complex locomotor behaviors. Although there is only very limited research on fish moving in

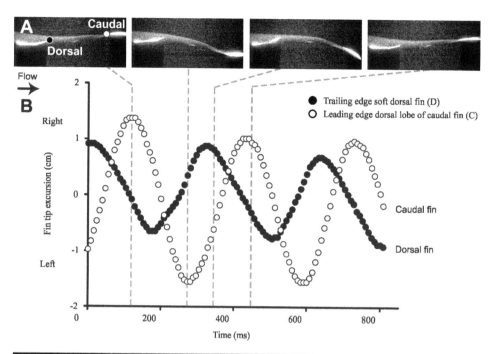

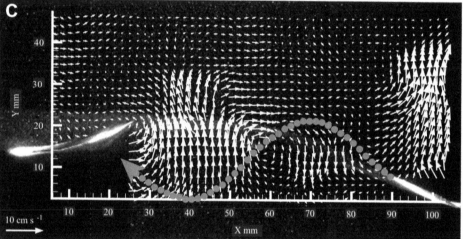

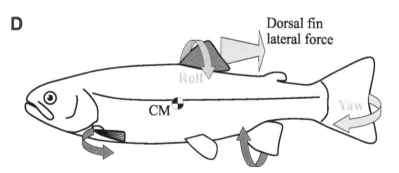

FIG. 2.5 Dorsal fin function in teleost fishes. Fin rays within the dorsal fin are under active control by muscles intrinsic to the fin, and generate an active wake flow that is encountered by the tail. A, Dorsal view video frames of the relative movement of the dorsal and caudal fin. B, Plot to show the out-of-phase motion of the dorsal and caudal fin. C, Path (red dots) taken by the tail through the wake generated by the dorsal fin (yellow arrows). Note the strong lateral jet flows generated by the dorsal fin. Data are from rainbow trout, *Oncorhynchus mykiss*. D, Both dorsal and anal fins are important for contributing to the balance of torques during locomotion, and movement of these fins (as well as the pectoral fin and tail) is controlled to balance roll, pitch, and yaw torques during steady swimming. See Drucker and Lauder (2003, 2005).

an obstacle-filled environment, such situations are very common for many fishes especially in lacustrine or riverine situations (Ellerby and Gerry 2011; Liao et al. 2003; Flammang et al. 2013). Study of fish moving through obstacles can also reveal unexpected functions for fish fins, and this is a rich area for future research.

A recent study of bluegill sunfish swimming through an array of posts (fig. 2.7) showed that fish did not avoid touching the posts as initially expected, but instead made contact with posts numerous times with their pectoral fins. Fish did not push off the posts, but instead used post contact as a means of sensing the obstacles and for

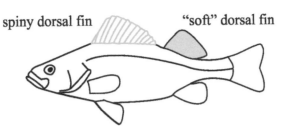

spiny dorsal fin   "soft" dorsal fin

FIG. 2.6 The dorsal fin of many teleost fishes is separated into distinct anterior spiny and posterior "soft" or flexible regions, and the dorsal fin has important hydrodynamic functions during both steady swimming and rapid maneuvers. A, B, The dorsal and anal median fins are depressed (yellow arrows) and have a smaller surface area as steady swimming speed increases. C, D, During rapid escape behaviors such as a C-start response, the dorsal and anal fins are erected (yellow curved arrow). See Tytell et al. (2008) and Chadwell et al. (2012a, 2012b).

directing navigation through the array (Flammang et al. 2013). Study of post navigation in darkness (filmed using infrared light) and under conditions in which the lateral line sense has been eliminated (using a cobalt chemical treatment) showed that fish increased the number of fin taps on nearby posts. These data suggest strongly that ray-finned fishes use their fins for sensing the environment, and not just for propulsion (Flammang et al. 2013). Furthermore, recent physiological studies of ray-finned fish fin rays have shown that sensory nerves in the rays have the capability of acting as proprioceptors and generating action potentials in afferent fin ray nerves in response to both the amplitude and velocity of ray bending (Williams et al. 2013). The most important lesson from this recent work is that ray-finned fish fin rays can act as

both propulsors and sensors, and this additional role for flexible fins is potentially of considerable importance in allowing fish to navigate the aquatic realm.

Of course, most fish cannot only swim forward: fish frequently back up and reposition themselves in the water column. When fish back up, all fins are active at the same time in patterned motion to hold position, correct for unbalanced torques, and to generate backward thrust. In order to present a significant locomotor challenge to fish, we induced bluegill sunfish to move backward through an array of obstacles (fig. 2.7E, F). Backward thrust is generated with the pectoral, dorsal, and anal fins, and each of these fins in addition to the caudal fin makes contact with the posts. This suggests that all the fins are used as both sensors and propulsors

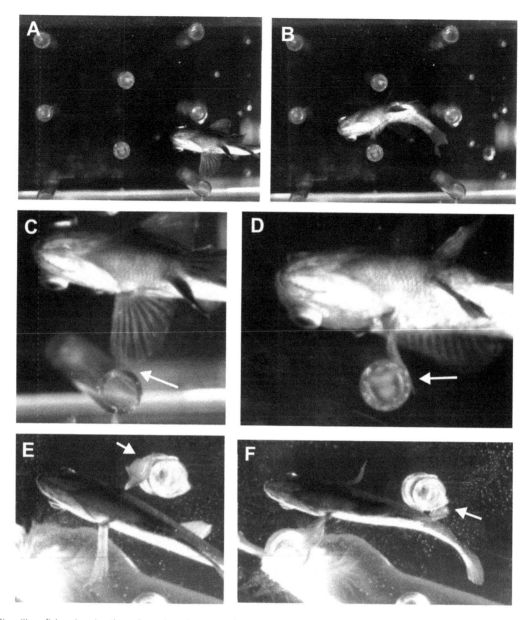

FIG. 2.7 Bluegill sunfish swimming through an obstacle course of posts in the forward direction (A) and (B), with enlarged views of the contact between the pectoral fins and posts (C) and (D). Panels (E) and (F) show bluegill swimming backward through the obstacles, with contact between the pectoral fins (E) and the dorsal and anal fins (F). These data suggest that ray-finned fishes can use their flexible fins for sensing the environment in addition to propulsion.

during the execution of complex swimming behaviors in cluttered environments.

Future Studies of Fish Fin Function

Although study of the kinematics and hydrodynamics of fish locomotion has revealed many aspects of how fishes interact with their fluid environment, there are inherent limitations to studies of live animals. Live fishes can be induced to perform only a relatively lim-

ited range of behaviors and movements, and there can be considerable variation among sequences. Many alterations of interest from a purely experimental and fluid dynamic perspective are difficult or impossible to perform in live fishes, such as changing the stiffness of fins or the body. And measuring three-dimensional forces during swimming and estimating the efficiency of swimming is extremely challenging if we are limited to working with live fish. Overall, when working on living animals, it is difficult to alter only one experimental variable, while simultaneously maintaining control of all others (Lauder et al. 2012).

One area of research in fish biology that has recently seen considerable activity, in part due to an interest in escaping many of the limitations of working on live fishes, is the use of robotic models. Robotic models allow precise manipulation of movement patterns, ease in changing stiffness and shape of the swimming object, and direct measurement of forces and torques during locomotion. In addition, robotic models of morphology that exists now only in fossil taxa allow us to explore how fossils may have functioned. As a result of these advantages, research has progressed on robotic fish pectoral fins (Phelan et al. 2010; Tangorra et al. 2010) caudal fins (Esposito et al. 2012; Lauder et al. 2012), and on whole-fish robotic models, both simple (Alben et al. 2012; Flammang et al. 2011a; Flammang et al. 2013; Lauder et al. 2012; Lauder, Lim, et al. 2011) and complex (Long 2012; Long et al. 2006; Low and Chong 2010; Wen et al. 2012).

Another important approach that has played a key role in understanding how fishes swim and the effect of different movement patterns on swimming efficiency and patterns of force production is the use of computational models. Using a computer model of a swimming fish or their fins and computing three-dimensional flows and forces allows us to investigate how both fins and fishes of different shapes generate force on the fluid environment. Asking a computational eel to swim like a mackerel, for example, allows us to isolate the effect of movement pattern on swimming efficiency (Borazjani and Sotiropoulos 2010). Computationally dissecting fin kinematics allows us to ask which specific features of fin motion are responsible for components of force generation, and hence provides a deeper understanding of how fins generate locomotor forces (Bozkurttas et al. 2009; Dong et al. 2010; Mittal et al. 2006).

With both robotic and computational approaches as important components of future research on the locomotion of ray-finned fishes as well as a wide array of unsolved biological problems, future students of ray-finned fishes have much to look forward to as new techniques and comparative approaches reveal ever more intriguing aspects of locomotor diversity and fin function in this remarkable clade.

* * *

Acknowledgments

Preparation of this chapter was supported by the National Science Foundation grant EFRI-0938043, and ONR N00014-0910352 on fin neuromechanics monitored by Dr. Thomas McKenna. I thank two reviewers and the editors for their suggestions for improving the chapter. Special thanks go to my many collaborators over the years who have played a key role in generating ideas that underlie this research and the data presented here: Alice Gibb, Lara Ferry, Silas Alben, Gary Gillis, Miriam Ashley-Ross, Jen Nauen, Tim Higham, Bruce Jayne, Eliot Drucker, Cheryl Wilga, Brad Chadwell, Em Standen, Brooke Flammang, Klea Kalionzes, Jimmy Liao, Rajat Mittal, Haibo Dong, Meliha Bozkurttas, Eric Tytell, Peter Madden, Erik Anderson, Chuck Witt, Vern Baker, Li Wen, Melina Hale, and James Tangorra. Many thanks to Chuck Pell who cleverly suggested the ziplock bag top analogy for fish fin rays.

References

Affleck, R. J. 1950. Some points in the function, development, and evolution of the tail in fishes. Proc. Zool. Soc. Lond. 120:349–368.

Agassiz, L. 1833–1843. Recherches sur les poissons fossiles. Neuchatel: Imprimerie de Petitpierre.

Akhtar, I., R. Mittal, G. V. Lauder, and E. Drucker. 2007. Hydrodynamics of a biologically inspired tandem flapping foil configuration. Theor. Comput. Fluid Dyn. 21:155–170.

Alben, S., P. G. Madden, and G. V. Lauder. 2007. The mechanics of active fin-shape control in ray-finned fishes. J. Roy. Soc. Inter. 4:243–256.

Alben, S., C. Witt, T. V. Baker, E. J. Anderson, and G. V. Lauder. 2012. Dynamics of freely swimming flexible foils. Phys. Fluids A 24:051901.

Benton, M. J. 1997. Vertebrate Paleontology, Second Edition. London: Chapman and Hall.

Borazjani, I., and F. Sotiropoulos. 2010. On the role of form and kinematics on the hydrodynamics of self-propelled body/caudal fin swimming. J. Exp. Biol. 213:89–107.

Bozkurttas, M., R. Mittal, H. Dong, G. V. Lauder, and P. G. A. Madden. 2009. Low-dimensional models and performance scaling of a highly deformable fish pectoral fin. J. Fluid Mech. 631:311–342.

Breder, C. M. 1926. The locomotion of fishes. Zoologica N. Y. 4:159–256.

Chadwell, B. A., E. M., Standen, G. V. Lauder, and M. A. Ashley-Ross. 2012a. Median fin function during the escape response of bluegill sunfish (*Lepomis macrochirus*). I: fin-ray orientation and movement. J. Exp. Biol. 215:2869–2880.

Chadwell, B. A., E. M., Standen, G. V. Lauder, and M. A. Ashley-Ross. 2012b. Median fin function during the escape response of bluegill sunfish (*Lepomis macrochirus*). II: fin-ray curvature. J. Exp. Biol. 215:2881–2890.

Dong, H., M. Bozkurttas, R. Mittal, P. Madden, and G. V. Lauder. 2010. Computational modeling and analysis of the hydrodynamics of a highly deformable fish pectoral fin. J. Fluid Mech. 645:345–373.

Drucker, E. G., and G. V. Lauder, 1999. Locomotor forces on a swimming fish: three-dimensional vortex wake dynamics quantified using digital particle image velocimetry. J. Exp. Biol. 202:2393–2412.

Drucker, E. G., and G. V. Lauder. 2000. A hydrodynamic analysis of fish swimming speed: wake structure and locomotor force in slow and fast labriform swimmers. J. Exp. Biol. 203:2379–2393.

Drucker, E. G., and G. V. Lauder. 2001. Locomotor function of the dorsal fin in teleost fishes: experimental analysis of wake forces in sunfish. J. Exp. Biol. 204:2943–2958.

Drucker, E. G., and G. V. Lauder. 2002. Experimental hydrodynamics of fish locomotion: functional insights from wake visualization. Int. Comp. Biol. 42:243–257.

Drucker, E. G., and G. V. Lauder. 2003. Function of pectoral fins in rainbow trout: behavioral repertoire and hydrodynamic forces. J. Exp. Biol. 206:813–826.

Drucker, E. G., and G. V. Lauder. 2005. Locomotor function of the dorsal fin in rainbow trout: kinematic patterns and hydrodynamic forces. J. Exp. Biol. 208:4479–4494.

Ellerby, D. J., and S. P. Gerry. 2011. Sympatric divergence and performance trade-offs of bluegill ecomorphs. Evol. Biol. 38:422–433.

Esposito, C., J. Tangorra, B. E. Flammang, and G. V. Lauder. 2012. A robotic fish caudal fin: effects of stiffness and motor program on locomotor performance. J. Exp. Biol. 215:56–67.

Flammang, B. E., S. Alben, P. G. A. Madden, and G. V. Lauder. 2013. Functional morphology of the fin rays of teleost fishes. J. Morphol. 274:1044–1059.

Flammang, B. E., and G. V. Lauder. 2008. Speed-dependent intrinsic caudal fin muscle recruitment during steady swimming in bluegill sunfish, *Lepomis macrochirus*. J. Exp. Biol. 211:587–598.

Flammang, B. E., and G. V. Lauder. 2009. Caudal fin shape modulation and control during acceleration, braking and backing maneuvers in bluegill sunfish, *Lepomis macrochirus*. J. Exp. Biol. 212:277–286.

Flammang, B. E., and G. V. Lauder. 2013. Pectoral fins aid in navigation of a complex environment by bluegill sunfish under sensory deprivation conditions. J. Exp. Biol. 216:3084–3089.

Flammang, B. E., G. V. Lauder, D. R. Troolin, and T. Strand. 2011a. Volumetric imaging of shark tail hydrodynamics reveals a three-dimensional dual-ring vortex wake structure. Proc. Royal Soc. B 278:3670–3678.

Flammang, B. E., G. V. Lauder, D. R. Troolin, and T. Strand. 2011b. Volumetric imaging of fish locomotion. Biol. Let. 7:695–698.

Geerlink, P. J., and J. J. Videler. 1987. The relation between structure and bending properties of teleost fin rays. Neth. J. Zool. 37:59–80.

Goodrich, E. S. 1904. On the dermal fin-rays of fishes, living and extinct. Quart. J. Microsc. Sci. 47:465–522.

Goodrich, E. S. 1906. Notes on the development, structure and origin of the median and paired fins of fish. Quart. J. Microscop. Sci. 50:333–376.

Goodrich, E. S. 1930. Studies on the Structure and Development of Vertebrates. London: Macmillen.

Higham, T. E., B. Malas, B. C. Jayne, and G. V. Lauder. 2005. Constraints on starting and stopping: behavior compensates for reduced pectoral fin area during braking of the bluegill sunfish (*Lepomis macrochirus*). J. Exp. Biol. 208:4735–4746.

Lauder, G. V. 1989. Caudal fin locomotion in ray-finned fishes: historical and functional analyses. Amer. Zool. 29:85–102.

Lauder, G. V. 2000. Function of the caudal fin during locomotion in fishes: kinematics, flow visualization, and evolutionary patterns. Amer. Zool. 40:101–122.

Lauder, G. V. 2006. Locomotion. Pp. 3–46 in D. H. Evans and J. B. Claiborne, eds., The Physiology of Fishes, Third Edition. Boca Raton: CRC Press.

Lauder, G. V., E. G. Drucker, J. Nauen, and C. D. Wilga. 2003. Experimental hydrodynamics and evolution: caudal fin locomotion in fishes. Pp. 117–135 in V. Bels, J.-P. Gasc, and A. Casinos, eds., Vertebrate Biomechanics and Evolution. Oxford: Bios Scientific Publishers.

Lauder, G. V., B. E. Flammang, and S. Alben. 2012. Passive robotic models of propulsion by the bodies and caudal fins of fish. Int. Comp. Biol. 52:576–587.

Lauder, G. V., J. Lim, R. Shelton, C. Witt, E. J. Anderson, and J. Tangorra. 2011. Robotic models for studying undulatory locomotion in fishes. Marine Tech. Soc. J. 45:41–55.

Lauder, G. V., and P. G. A. Madden. 2006. Learning from fish: kinematics and experimental hydrodynamics for roboticists. Internat. J. Automat. Comput. 4:325–335.

Lauder, G. V., and P. G. A. Madden. 2007. Fish locomotion: kinematics and hydrodynamics of flexible foil-like fins. Exp. Fluids 43:641–653.

Lauder, G. V., P. G. A., Madden, R., Mittal, H., Dong, and M. Bozkurttas. 2006. Locomotion with flexible propulsors I: experimental analysis of pectoral fin swimming in sunfish. Bioinsp. Biomimet. 1:S25–S34.

Lauder, G. V., P. G. A. Madden, J. Tangorra, E. Anderson, and T. V. Baker. 2011. Bioinspiration from fish for smart material design and function. Smart Mater. Struct. 20:doi:10.1088/0964-1726/20/9/094014.

Lauder, G. V., and E. D. Tytell. 2006. Hydrodynamics of undulatory propulsion. Pp. 425–468 in R. E. Shadwick and G. V. Lauder, eds., Fish Biomechanics. Vol. 23 in Fish Physiology. San Diego: Academic Press.

Liao, J., D. N. Beal, G. V. Lauder, and M. S. Triantafyllou. 2003. The Kármán gait: novel body kinematics of rainbow trout swimming in a vortex street. J. Exp. Biol. 206:1059–1073.

Liao, J., and G. V. Lauder. 2000. Function of the heterocercal tail in white sturgeon: flow visualization during steady swimming and vertical maneuvering. J. Exp. Biol. 203:3585–3594.

Liem, K. F. 1989. Respiratory gas bladders in teleosts: functional conservatism and morphological diversity. Amer. Zool. 29:333–352.

Liem, K. F., W. E. Bemis, W. F. Walker, and L. Grande. 2001. Functional Anatomy of the Vertebrates: An Evolutionary Perspective. 3rd ed. Fort Worth: Harcourt College Publishers.

Long, J. 2012. Darwin's Devices: What Evolving Robots Can Teach Us about the History of Life and the Future of Technology. New York: Basic Books.

Long, J. H., J. Schumacher, N. Livingston, and M. Kemp. 2006. Four flippers or two? tetrapodal swimming with an aquatic robot. Bioinsp. Biomimet. 1:20–29.

Low, K. H., and C. W. Chong. 2010. Parametric study of the swimming performance of a fish robot propelled by a flexible caudal fin. Bioinsp. Biomimet. 5:046002.

Mittal, R., H. Dong, M. Bozkurttas, G. V. Lauder, and P. Madden. 2006. Locomotion with flexible propulsors II: computational analysis of pectoral fin swimming in sunfish. Bioinsp. Biomimet. 1:S35–S41.

Nauen, J. C., and G. V. Lauder. 2001. Locomotion in scombrid fishes: visualization of flow around the caudal peduncle and finlets of the Chub mackerel Scomber japonicus. J. Exp. Biol. 204:2251–2263.

Owen, R. 1854. The Principal Forms of the Skeleton and the Teeth: As the Basis for a System of Natural History and Comparative Anatomy. New York: William Wood Co.

Phelan, C., J. Tangorra, G. V. Lauder, and M. Hale. 2010. A biorobotic model of the sunfish pectoral fin for investigations of fin sensorimotor control. Bioinsp. Biomimet. 5:035003.

Ramamurti, R., W. C. Sandberg, R. Lohner, J. A. Walker, and M. Westneat. 2002. Fluid dynamics of flapping aquatic flight in the bird wrasse: three-dimensional unsteady computations with fin deformation. J. Exp. Biol. 205:2997–3008.

Romer, A. S., and T. S. Parsons. 1986. The Vertebrate Body. New York: Saunders.

Rosen, D. E. 1982. Teleostean interrelationships, morphological function, and evolutionary inference. Amer. Zool. 22:261–273.

Standen, E. M. 2008. Pelvic fin locomotor function in fishes: three-dimensional kinematics in rainbow trout (Oncorhynchus mykiss). J. Exp. Biol. 211:2931–2942.

Standen, E. M., and G. V. Lauder. 2007. Hydrodynamic function of dorsal and anal fins in brook trout (Salvelinus fontinalis). J. Exp. Biol. 210:325–339.

Taft, N. K. 2011. Functional implications of variation in pectoral fin ray morphology between fishes with different patterns of pectoral fin use. J. Morphol. 272:1144–1152.

Taft, N. K., and B. N. Taft. 2012. Functional implications of morphological specializations among the pectoral fin rays of the benthic longhorn sculpin. J. Exp. Biol. 215:2703–2710.

Tangorra, J. L., G. V. Lauder, I. Hunter, R. Mittal, P. G. Madden, and M. Bozkurttas. 2010. The effect of fin ray flexural rigidity on the propulsive forces generated by a biorobotic fish pectoral fin. J. Exp. Biol. 213:4043–4054.

Tytell, E. D. 2006. Median fin function in bluegill sunfish Lepomis macrochirus: streamwise vortex structure during steady swimming. J. Exp. Biol. 209:1516–1534.

Tytell, E. D., and G. V. Lauder. 2008. Hydrodynamics of the escape response in bluegill sunfish, Lepomis macrochirus. J. Exp. Biol. 211:3359–3369.

Walker, J. A., and M. Westneat. 2002. Kinematics, dynamics, and energetics of rowing and flapping propulsion in fishes. Int. Comp. Biol. 42:1032–1043.

Webb, P. W. 1975. Hydrodynamics and energetics of fish propulsion. Bull. Fish Res. Bd. Can. 190:1–159.

Wen, L., T. M. Wang, G. H. Wu, and J. H. Liang. 2012. Hydrodynamic investigation of a self-propelled robotic fish based on a force-feedback control method. Bioinsp. Biomimet. 7:036012.

Westneat, M. W. 1996. Functional morphology of aquatic flight in fishes: kinematics, electromyography, and mechanical modeling of labriform locomotion. Amer. Zool. 36:582–598.

Wilga, C. D., and G. V. Lauder. 1999. Locomotion in sturgeon: function of the pectoral fins. J. Exp. Biol. 202:2413–2432.

Wilga, C. D., and G. V. Lauder. 2002. Function of the heterocercal tail in sharks: quantitative wake dynamics during steady horizontal swimming and vertical maneuvering. J. Exp. Biol. 205:2365–2374.

Williams, R., N. Neubarth, and M. E. Hale. 2013. The function of fin rays as proprioceptive sensors in fish. Nature Communications 4, doi:10.1038/ncomms2751.

3

Major Transformations in Vertebrate Breathing Mechanisms

Elizabeth L. Brainerd*

Vertebrate breathing mechanisms have undergone several major transformations, including (1) the transformation from breathing water with gills to breathing air with lungs, and (2) the transformation from buccal pumping, in which the lungs are ventilated entirely with bones and muscles of the head region, to aspiration breathing, in which the lungs are ventilated with bones and muscles of the trunk region.

The stories of these transformations are full of twists and turns, and almost never follow what one might expect, a priori, to be the simplest path from a water-breathing early fish to a costal-aspirating amniote. A parsimonious and satisfying scenario for the transition from water to air might be that small-bodied fishes began gradually to spend time on land, first exchanging gases with air across skin, oropharyngeal and gill surfaces, much like modern mudskippers, and then gradually evolving lungs and larger body size, and modifying ribs for costal aspiration breathing. Indeed, a fish crawling out on land is one of the most common cartoon images used to depict evolution, second only to the ubiquitous march from chimp to human. However, both paleontological and neontological evidence argue against this simple scenario, indicating instead that 1) lungs were present in early and fully aquatic bony fishes (Liem 1988); 2) early tetrapods were large-bodied and also largely aquatic (Clack 2012); 3) lungs were retained and transformed into the

* Department of Ecology and Evolutionary Biology, Brown University

swim bladder of teleost fishes (Liem 1988; Cass et al. 2013; Longo et al. 2013); and 4) some amniotes retain the ability to ventilate lungs with a buccal pump, along with costal aspiration (Owerkowicz et al. 1999; Brainerd and Owerkowicz 2006). Thus, the origin of lungs was decoupled from the advent of terrestriality, lungs transformed into the swim bladder of teleost fishes, freeing teleost fishes from any connection with air and the surface, and the transformation from buccal pumping to costal aspiration was not abrupt and remains problematic due to the current lack of definitive bony characters in the fossil record to diagnose costal aspiration breathing.

From Gills to Lungs

The evolution of lungs was necessary for vertebrates to colonize land, but was by no means coincident with the evolution of terrestriality. Comparative analysis of extant vertebrates indicates that lungs were present in the common ancestor of Osteichthyes (fig. 3.1). Some paleontological evidence suggests that lungs may have been present even earlier in a placoderm, *Bothriolepis canadensis* (Denison 1941), which would push the evolution of lungs back to the base of Gnathostomata and require loss of lungs in Chondrichthyes. Recent reanalysis of a *Bothriolepis* specimen with exceptional 3-D soft-tissue preservation confirmed the presence of paired, rounded masses of fine-grained sediment in the anterior part of the thoracic armor, ventral to the gill chamber (Janvier et al. 2007). However, this study found no additional evidence that these structures were lungs, and suggested that the sediment masses could just as likely be preservation artifacts or other, non-lung structures. Thus, at present, it is most parsimonious to conclude that lungs evolved in a fully aquatic ancestor of bony fishes, and have been retained as paired, ventral lungs in Sarcopterygii and transformed into the swim bladder in teleost fishes (fig. 3.1).

The evolution of lungs was also not coincident with the loss of gills. Bimodal gas exchange, across lungs and gills, or trimodal, across lungs, gills and skin, appears to have characterized the early evolution of actinopterygians and sarcopterygians. Both freshwater and marine environments can experience periodic or persistent oxygen depletion (Packard 1974). Modern teleost fishes in these environments have respiratory gas bladders or other accessory respiratory organs for collecting oxygen from the air, but still excrete carbon dioxide (CO_2) primarily into the water through gills. The repeated evolution of accessory air-breathing organs in teleosts (Liem 1980; Graham 1997), and the retention of gills in all of these air-breathers, suggest that selection for extracting oxygen from air can be very strong, while selection for aquatic CO_2 excretion simultaneously maintains the presence of gills.

Early tetrapods retained a branchial skeleton with grooves for large blood vessels, suggesting that they at least excreted CO_2 through internal gills, and may have facultatively extracted oxygen from air via lungs or water via gills, depending on ambient aquatic oxygen concentrations (Clack 2007; Witzmann, 2013). Only with the transition to full terrestriality did gills become superfluous, with the presumed assumption of responsibility for CO_2 excretion by the lungs and likely concomitant increase in lung ventilation rate (Perry and Sander 2004).

Extant fishes offer several degrees of specialization for extraction of oxygen from air, which together offer a likely scenario for the evolution of lungs. In oxygen-depleted water, some fishes with no accessory air-breathing organs will come to the surface and irrigate their gills with water from just below the air-water interface, which always contains some oxygen (Kramer and McClure 1982). They also sometimes suck bubbles of air into the water stream, thereby further oxygenating this water as it passes across the gills. It is easy to imagine this aquatic surface respiration (ASR) behavior leading to elaborations of the oropharyngeal region for direct oxygen extraction from the air bubbles, as is seen in fishes that extract oxygen from air in the esophagus, and in the repeated evolution of suprabranchial air-breathing organs in the opercular chamber (Graham 1997). Lungs evolved as sac-like, ventral elaborations of the pharynx, which allowed the fishes to collect a bubble of air and carry it with them, rather than being tied to the surface by the need for ASR whenever oxygen tensions dropped.

In polypterid fishes, lungfishes, and tetrapods, lungs develop as a ventrally directed diverticulum of the pharynx that branches into two lungs in all of these groups, although one of these lungs is lost in subsequent development in the Australian lungfish (*Neoceratodus fosteri*). This shared developmental pattern, along with shared histological features and developmental mechanisms of

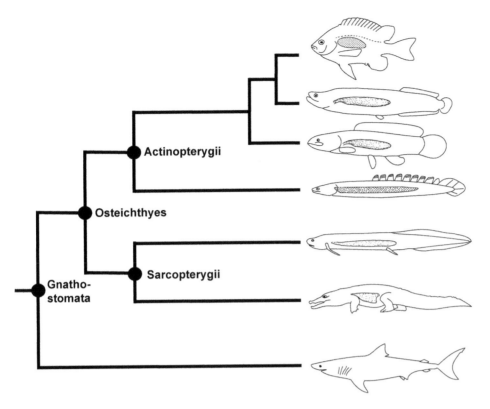

FIG. 3.1 Lungs were present in the ancestor of all bony vertebrates (Osteichthyes), and then evolved into the swim bladder of ray-finned fishes (Actinopterygii). *From bottom to top:* shark as a representative chondrichthyan with no lungs or swim bladder; alligator as representative tetrapod with paired lungs; lepidosirenid lungfish with paired lungs; *Polypterus* with paired lungs; *Amia* and *Arapaima* as representative basal actinopterygians with unpaired respiratory gas bladders; and a teleost with a nonrespiratory swim bladder.

lungs and gas bladders (Liem 1989; Cass et al. 2013) and blood supply (Farmer 1997; Longo et al. 2013), suggests that lungs evolved once in a common ancestor of Actinopterygii and Sarcopterygii, and subsequently transformed into the teleost swim bladder. A single origin is certainly most parsimonious (Liem 1988), but the possibility for homoplasy should not be underestimated (see D. Wake et al., this volume). Some features of lung histology and ventilation patterns suggest the potential for independent origins in Actinopterygii and Sarcopterygii (Brainerd 1994a), but the bulk of evidence points to a single origin of lungs at the base of Osteichthyes (fig. 3.1).

From Lungs to Swim Bladder

The transformation series from lungs in early bony fishes to a definitive swim bladder in acanthomorph teleosts is exceptionally well supported by extant diversity (fig. 3.2). Lungs allowed early bony fishes to rise to the surface and take advantage of oxygen in the air, and holding gas inside the body also added buoyancy. Muscle and bone are heavier than water, so a gas-filled organ potentially confers neutral buoyancy, thereby reducing energy expenditure for remaining in the water column. Within the ray-finned fishes, we see transformation

from lungs used for both respiration and buoyancy to a swim bladder that serves only buoyancy and is filled with gas from the circulatory system, with no connection at all to the outside. Thus, an organ that originally linked bony fishes to the surface for air breathing, eventually freed them from all contact with both the surface and the bottom.

The transformation from a lung that is filled and emptied through a pneumatic duct to the definitive physoclistous swim bladder that is filled and emptied from the circulatory system appears to have occurred in four stages (fig. 3.2): 1) reduction from ventral, paired respiratory organs (lungs) to a dorsal, unpaired respiratory gas bladder; 2) loss of respiratory function; 3) evolution of the rete mirabile for filling the swim bladder from gasses dissolved in the blood; and 4) loss of the pneumatic duct. Here I follow Cass and colleagues (2013) and Longo and colleagues (2013) in using "lung" to refer to air-filled structures that develop by ventral evagination, and "bladder" to refer to air-filled structures that develop by dorsal evagination.

This transformation series is well supported by extant diversity, but several interesting homoplasies and reversals are also evident (fig. 3.2). A single (unpaired) organ is present in the Australian lungfish (homoplasy

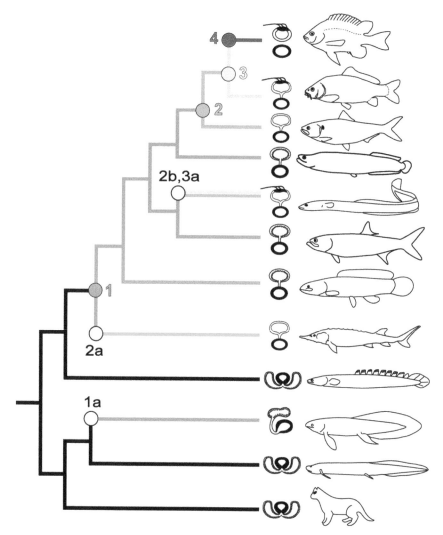

FIG. 3.2 Transformation series from lungs in early bony fishes to a definitive swim bladder in acanthomorph teleosts. Extant fish diversity provides exceptional support for four stages in this transformation: (1) reduction from paired, ventral lungs to an unpaired, dorsal respiratory gas bladder; (2) loss of respiratory function; (3) evolution of the rete mirabile for filling the swim bladder from gases dissolved in the blood; and (4) loss of the pneumatic duct (physoclistous swim bladder). See text for descriptions of homoplasies 1a, 2a, 2b, and 3a. *From bottom to top:* cat as representative tetrapod with paired lungs; lepidosirenid lungfish with paired lungs; *Neoceratodus* with single respiratory gas bladder and ventral connection to the gut tube; *Polypterus* with paired, ventral lungs; sturgeon as representative chondrostean with non-respiratory gas bladder; *Amia* and *Megalops* with unpaired respiratory gas bladders; anguilliform eel with non-respiratory bladder and a rete mirabile; *Arapaima* as a representative osteoglossomorph with a respiratory gas bladder; shad as a representative clupeid with a non-respiratory, physostomous gas bladder; carp as representative of several physostomous teleost clades with retia mirabile; a pomacentrid as a representative of many acanthomorph clades with physoclistous swim bladders. Phylogeny after Near et al. (2012).

1a). This lung develops as a ventral evagination that branches into two lungs, and then one is lost, and the other takes a dorsal position in the body cavity, similar to respiratory gas bladders. The respiratory function of the gas bladder was lost in sturgeons and paddlefishes (homoplasy 2a) and in non-elopiform elopomorphs (homoplasy 2b). Perhaps most remarkably, a rete mirabile appears to have evolved in parallel within Elopomorpha and other teleosts (homoplasy 3a). The Elopiformes

(tarpons and ladyfish) and Albuliformes (bonefish) retain the respiratory function of their gas bladders and do not have a rete mirabile; the rete mirabile appears only in Anguilliformes.

A rete mirabile is present in Ostariophysi and all other teleosts, and an anterior chamber of the swim bladder in ostariophysans also takes on a connection to the Weberian ossicles and a role in hearing. The loss of respiratory function has been reversed (O_2 uptake regained) in several isolated physostomous genera, but once the pneumatic duct is lost in Acanthomorpha, the physostomous condition and respiratory function for the gas bladder are never regained. Physostomous air-breathing fishes have evolved other accessory respiratory organs, such as suprabrachial chambers, and these are present in addition to a swim bladder.

The buoyancy effects of carrying air inside the body of an aquatic organism should not be overlooked in the consideration of lung evolution as well. We tend to focus on selection for oxygen acquisition, but the presence of gas inside the body necessarily has profound effects on buoyancy (Graham et al. 1977; Gee 1978). Some authors have even suggested that the primary selection for the evolution of lungs could have been buoyancy, particularly given the heavy body armor of Paleozoic fishes (Eaton 1960; Schaeffer 1965). But others have emphasized the difficulty of assigning relative importance to selection on respiration versus buoyancy, both of which would necessarily have been affected by the presence of air-filled lungs (Liem 1988; Burggren and Bemis 1990).

The homology of lungs and swim bladders was recognized by Richard Owen and Charles Darwin, but they assumed that lungs evolved from swim bladders. In his chapter "Difficulties on theory of descent with modification," in *On the Origin of Species* (1859), Darwin writes that any "organ of extreme perfection" would challenge his theory could it "not possibly have been formed by numerous, successive, slight modifications." He goes on to cite the lung as an organ that developed by successive modification of the swim bladder (Darwin 1859):

> all physiologists admit that the swimbladder is homologous, or "ideally similar," in position and structure with the lungs of the higher vertebrate animals: hence there seems to me to be no great difficulty in believing that natural selection has actually converted a swimbladder into a lung.

This passage reflects a mistaken assumption that fish are primitive and tetrapods are more advanced, not recognizing that teleost fishes are advanced masters of the aquatic realm, in part due to the buoyancy control afforded by a swim bladder that is disconnected from the atmosphere. Indeed, fine buoyancy control in teleosts is thought to have contributed to the success of this group and their diversification into structured habitats, such as reefs and grass beds (Steen 1970).

An issue that has received little attention in fishes is how the distribution of air within the gas bladder is controlled, particularly in elongate fishes with elongate swim bladders. If the gas bladder wall is compliant, and the head of the fish is pitched up, then hydrostatic pressure will cause gas to move into the rostral end of the gas bladder (Webb 2002). The resulting asymmetry of the gas bladder volume will unbalance the centers of buoyancy and gravity to create a net force that tends to pitch the head up more. This issue of gas distribution has received some attention in aquatic and amphibious tetrapods, with active pitch control mechanisms described for salamanders (Whipple 1906), alligators (Uriona and Farmer 2008), and manatees (Rommel and Reynolds 2000).

In fishes, this unsteady buoyancy effect could potentially be mitigated by more spherical gas bladder shape, by a stiff-walled and constantly pressurized gas bladder, or by active, muscular mechanisms to help redistribute gas. Striated muscle is present in the walls of the lungs or gas bladder in *Polypterus*, *Amia*, and *Lepisosteus*. This striated muscle comes from the striated muscle of the pharynx, from which the organs develop. The striated muscle of the lung walls contributes to exhalation in polypterid fishes, and may well help distribute air within these fishes, but a potential buoyancy control function has not been investigated. Lepidosirenid lungfishes and tetrapods have smooth muscle in their lungs, from the development of these organs as outpocketings of the foregut. Tone of this smooth muscle could well help control air distribution, but this function has not been investigated. Finally, many fishes have various accretions of striated muscle on the surface of the swim bladder that act primarily for sound production. These sonic muscles are mostly not well situated to control gas distribution, but again, the potential for these muscles to help control gas distribution and center of buoyancy have not been investigated.

From Buccal Pumping to Aspiration Breathing

How to fill a respiratory gas bladder or lung with air poses a fundamental biomechanical problem. Muscles only do work while shortening, so how can they be used to expand a lung? The answer lies in musculoskeletal levers that convert muscle shortening into expansion. In air-breathing fishes and amphibians, this musculo-skeletal lever system is in the head: contraction of the sternohyoid and/or hypaxial muscles retracts the hyoid apparatus, which is hinged to the suspensorium such that hyoid retraction is converted to hyoid depression and expansion of the mouth cavity. The mouth is then sealed and the buccal cavity compressed to pump air down into the gas bladder or lungs. In amniotes, the musculoskeletal lever system is in the trunk region of the body: hinged ribs and intercostal muscles act together to lever the ribs away from the center of the body and expand the cavity surrounding the lungs. The primitive condition was clearly to house the ventilatory lever system in the head, in the form of a buccal pump, and then gradually the responsibility for lung ventilation was transferred to the trunk for costal aspiration breathing (Brainerd and Owerkowicz 2006).

The ventilation mechanisms of extant vertebrates, combined with information from the fossil record, offer some information about the transformation from buccal pumping to aspiration breathing, but intriguing questions about rib structure and function remain. Until about 20 years ago, the standard view was that extant animals were either buccal pump or aspiration breathers, with no intermediate conditions present.

Then we discovered that salamanders use trunk musculature, specifically the *m. transversus abdominis*, to power exhalation, and retain the primitive buccal pump for inhalation (Brainerd et al. 1993; Brainerd and Dumka 1995; Brainerd 1998; Brainerd and Monroy 1998; Brainerd and Ferry-Graham 2006). This pattern suggests that aspiration breathing may have evolved in two steps: first to the use of trunk musculature for exhalation, and then to the use of trunk musculature for both inhalation and exhalation (fig. 3.3).

In further support of active, axial muscle-powered exhalation as tetrapod innovation, the presence of the *m. transversus abdominis* is a tetrapod synapomorphy. Actinopterygian fishes and lungfishes lack this muscle (Maurer 1912), and exhalation is usually a passive process, driven by hydrostatic pressure and possibly gas bladder elasticity and smooth muscle tone (McMahon 1969; Deyst and Liem 1985). In polypterid fishes, exhalation is active, but it is driven by striated muscles in the lung walls and not by axial musculature (Brainerd et al. 1989; Brainerd 1994b). The transverse abdominal muscle contributes to active exhalation during heavy breathing in all amniotes, a ubiquitous function that is somewhat surprising given the diverse body forms of mammals, turtles, squamates, crocodilians, and birds (Brainerd and Owerkowicz 2006).

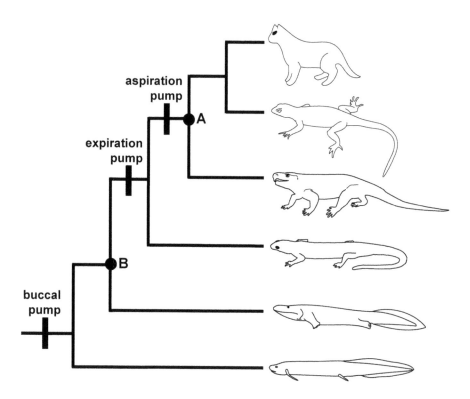

FIG. 3.3 Transformation series from buccal pump breathing to aspiration breathing. Aspiration may have evolved in two steps: first to the use of trunk musculature for exhalation (expiration pump), and then to the use of trunk musculature and ribs for both inhalation and exhalation (aspiration pump). The expiration pump likely evolved at the base of tetrapods (node B), but we only have definitive evidence from extant salamanders and amniotes. *From bottom to top:* lepidosirenid lungfish; *Acanthostega* as representative stem tetrapod; salamander as representative lissamphibian; *Limnoscelis* as representative stem amniote (after Berman et al. 2010); lizard as representative saurian; cat as representative mammal. The letter A denotes the prevailing view that costal aspiration evolved in stem amniotes, and B denotes an alternate view that costal aspiration evolved in early tetrapods and has been lost in extant lissamphibians.

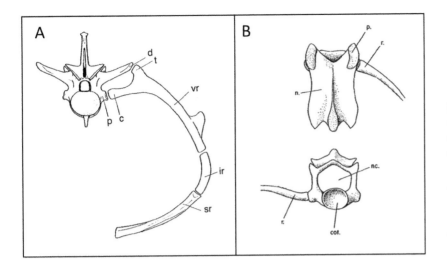

FIG. 3.4 Costovertebral joints and rib morphology in amniotes. (A) Second dorsal vertebra and rib in an alligator. (B) Unicapitate vertebrocostal articulations in a squamate (a gekkonid is figured). Abbreviations: c, capitulum; cot, cotyle; d, diapophysis; ir, intermediate rib; n, neural arch; nc, neural canal; p, parapophysis; r, rib; sr, sternal rib; t, tuberculum; vr, vertebral rib. From Hoffstetter and Gasc (1969).

Transformations leading to the recruitment of ribs and associated hypaxial intercostal muscles for inspiration are somewhat unclear. It is clear, however, that costal aspiration is the primitive aspiration mechanism for crown amniotes. Several other accessory ventilation mechanisms have evolved within amniotes, including the mammalian, testudine, and crocodilian diaphragm-like arrangements. However, all extant amniotes (except for turtles) share mobile ribs, intercostal muscles, and the ability to ventilate the lungs, at least in part, via rib motions. Some studies have focused on accessory mechanisms with little reference to this fundamental costal mechanism, but the starting assumption for extinct crown amniotes should be that they have retained the primitive costal aspiration system, and accessory mechanisms should be hypothesized in this context (Brainerd and Owerkowicz 2006).

Thus, the starting and ending points of the transformation from buccal pumping in fishes to aspiration breathing in amniotes are clear, as is the early adoption of axial muscles for exhalation (fig. 3.3), but the origin of inhalation by costal aspiration is less clear. Did the use of ribs and intercostal muscles for breathing arise during the fish to tetrapod transition, or only within stem amniotes? Available evidence comes from extant fish, lissamphibians (salamanders, frogs/toads, and caecilians), and amniotes, as well as from the fossil record. Analysis of lung ventilation in extant vertebrates places the origin of costal aspiration within stem amniotes or at the base of Amniota (fig. 3.3, node A), since lissamphibians and lungfishes do not use ribs for breathing, and amniotes do. However, it is also possible

that costal aspiration evolved in the early tetrapods, and was retained in amniotes and lost in lissamphibians (fig. 3.3, node B). This hypothesis is less parsimonious, and is not currently in favor (Janis and Keller 2001; Clack 2012), but is based in large part on evidence from the morphology of the joints between the vertebrae and ribs in early tetrapods (Gans 1970; Romer 1972). These costovertebral joints in early tetrapods are intriguing because they show the two-headed (bicapitate) structure that is also primitive for amniotes (fig. 3.4), and retained in some lissamphibians. It has been thought that this bicapitate structure is necessary for, or indicative of, rib mobility (Romer 1972), but recent work on the biomechanics of costal aspiration in squamates with unicapitate ribs suggests that this is not so (see below).

In addition to the parsimonious explanation of ventilation in extant amphibians and amniotes, the hypothesis that costal aspiration arose in stem amniotes (fig. 3.3, node A) is supported by head shape and distal rib morphology (Janis and Keller 2001). Extinct anamniotes had broader heads, relative to length, than stem amniotes, suggesting that head shape may have been constrained by buccal pumping in anamniotes, and then freed to take on a wider range of shapes in stem amniotes (Janis and Keller 2001). The distal ends of fossilized ribs may also have an interesting tale to tell. We generally only see the vertebral rib segment in the fossil record, but in most modern amniotes, additional (often cartilaginous) intermediate and sternal rib segments extend from the ossified vertebral ribs to surround the body and articulate with the sternum (fig. 3.4). Complete ribs surrounding the lungs are necessary for costal aspiration, and we are increasingly finding that the sternal ribs are particularly important for ventilation (see below). Were intermediate and sternal ribs absent in early tetrapods, or did they just not fossilize? Salamanders and frogs have only short, peg-like vertebral ribs and no intermediate or sternal ribs. The distal ends of the ribs in early tetrapods had tapering ends covered in periosteal bone, suggesting that they did not have cartilaginous extensions, whereas the ends of the ribs in some stem amniotes have a roughened appearance that may indicate a cartilaginous extension (Janis and Keller 2001). However, some bone histology on these fossilized ribs should be done to examine the distal ends more fully. Comparison with extant salamanders and frogs (no cartilaginous extensions) and squamates and crocodilians (have cartilaginous extensions) might tease out whether there is a signal in fossil record for presence of cartilaginous rib segments.

Even if we accept that inhalation by costal aspiration arose with the amniotes, and most evidence now points in that direction, we still have difficulty reconstructing the transformation series from buccal pump to aspiration breathing. A substantial barrier to understanding the early evolution of costal aspiration is poor understanding of rib and intercostal muscle function in extant amniotes. Breathing seems like one of the simplest things that we do, but in fact the morphology and motions of the ribs and intercostal muscles are enormously complex. If we are to have any hope of interpreting rib morphologies in early tetrapods and amniotes, we need a better understanding of the form-function relationships of costal aspiration in extant amniotes.

The Vexing Problem of Costal Aspiration

The respiratory actions of the ribs and intercostal muscles have been controversial for centuries (Derenne and Whitelaw 1995; De Troyer et al. 2005). The Greek physician Galen (ca. 200) described the anatomy of the intercostal musculature, and concluded that the external layer produces exhalation and the internal layer produces inhalation. Galen's works on respiration were lost until modern times (Furley and Wilkie 1984), and Renaissance scientists independently rediscovered the opposing muscle fascicle angles of the external and internal intercostal layers (fig. 3.5). Both Leonardo da Vinci (1509, cited in De Troyer et al. 2005) and Andreas

Vesalius (1543) developed theories of intercostal muscle actions that were at odds with Galen's interpretation. Da Vinci thought that the external intercostal (EI) produces inhalation and the internal intercostal (II) produces exhalation, whereas Vesalius concluded that both layers produce exhalation (Vesalius 1543). The next three centuries saw much interest in the actions of the intercostal musculature, with every possible combination of exhalation and inhalation proposed by various researchers (table 3.1).

Why is it so difficult to infer the actions of the intercostal muscles from their structure and attachments to the ribs? The answer lies in 1) the 3-D complexity of rib shape; 2) the 3-D complexity of rib motion during breathing; 3) the 3-D complexity of intercostal muscle fiber architecture; and 4) strong rostrocaudal and dorsoventral variation in all of the above. Below I review the state of knowledge in each of these four areas, and then describe a new technology, X-ray Reconstruction of Moving Morphology (XROMM), that is poised to provide musculoskeletal morphology and motion data that address all four problems simultaneously.

Rib Shape and Articulations

The primitive condition for amniotes is bicapitate ribs that form two separate joints with the vertebrae, and bicapitate ribs are retained in most amniote groups (Romer 1956; Hoffstetter and Gasc 1969). The capitulum articulates with the parapophysis on the vertebral centrum, and the tuberculum articulates with the diapophysis on the transverse process (fig. 3.4A). There is, however, substantial variation in the relative positions of these articulations. For example, in most birds the diapophysis lies almost directly dorsal to the parapophysis, with little mediolateral or craniocaudal deviation (Claessens 2009b; Schachner et al. 2009), whereas in the thoracic region of crocodilians (Claessens 2009a; Schachner et al. 2009), the diapophysis is located on the distal tip of elongated transverse processes and the parapophysis is medial and only slightly anteroventral (fig. 3.4A; mediolateral orientation becomes more exaggerated by 4th dorsal vertebra). In most mammals, the capitulae of ribs 2–10 are themselves double headed, with demi-facets that articulate across the intervertebral joints to a large facet (the parapophysis) on the same numbered vertebra, and a smaller facet on the

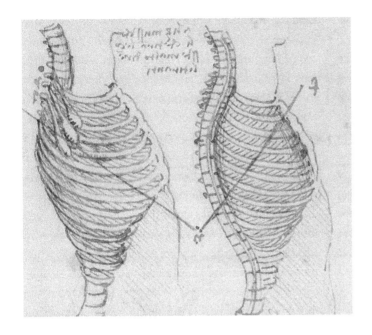

FIG. 3.5 Human ribs and intercostal musculature, figured by Leonardo da Vinci (ca. 1508). In this detail from a larger sheet, external intercostals are shown on the left, and have been removed on the right to reveal the opposing muscle fiber angles of the internal intercostals. Reproduced from verso of RCIN 91904 with permission of Royal Collection Trust / © Her Majesty Queen Elizabeth II 2014.

TABLE 3.1 Proposed intercostal (IC) muscle actions through history

Year	Author	External IC	Internal IC
ca. 200	Galen	exhalation	inhalation
1510	da Vinci	inhalation	exhalation
1568	Vesalius	exhalation	exhalation
1681	Borelli	inhalation	inhalation
1749	Hamberger	inhalation	exhalation
1786	Haller	inhalation	inhalation
1843	Beau & Massat	exhalation	exhalation
1867	Duchenne	inhalation	inhalation
current	De Troyer	inh/exh	inh/exh

Sources: Derenne and Whitelaw 1995; De Troyer et al. 2005.

next anterior vertebral centrum. The heads of ribs 1, 11, and 12 show the more typical tetrapod articulation with just one centrum. In mammalian vertebrae, the diapophysis is located lateral, dorsal, and slightly caudal to the parapophysis, with the caudal offset increasing along the thoracic series.

Squamates are unusual in having unicapitate (holocephalous) ribs that articulate with the vertebrae via single, hemispheric joints that resemble shallow ball (vertebra) and socket (rib) articulations (fig. 3.4B). The single rib facet on squamate vertebrae (synapophysis) represents the fusion of the parapophysis and

diapophysis, rather than the loss of one of them (Hoffstetter and Gasc 1969). The condition in *Sphenodon* (tuatara) is sub-bicapitate, in which the first cervical rib has completely separate capitulum and tuberculum, and the remaining ribs articulate to the vertebrae via hemi-elliptical synapophyses that are elongated in the posterodorsal-anteroventral direction (Hoffstetter and Gasc 1969). Thus, the *Sphenodon* costovertebral articulations represent an intriguing intermediate condition between the primitive bicapitate condition and derived unicapitate condition in Squamata.

In addition to variation in costovertebral articulations, amniotes also show variability in segmentation within the ribs. Crocodilians and tuataras have tripartite ribs divided into vertebral rib, intermediate rib, and sternal rib sections (fig. 3.4A). Birds and most non-serpentine squamates have bipartite ribs, but a few squamates, such as teiids, have tripartite ribs. Snakes are interesting because they have only vertebral ribs and no sternum, with the ventral scutes likely providing the ventral stiffness that would otherwise come from sternal ribs and sternum. Variation is also present in ossification of the segments. In birds, both the vertebral and sternal ribs are usually ossified, but in crocodilians and lepidosaurs, the vertebral rib segments are ossified, and the intermediate (when present) and sternal ribs are usually cartilaginous. Mammals are unusual in having no mobile intracostal joints. Long, sharply curved vertebral ribs are fused to short, robust costal cartilages that then articulate with the sternum or adjacent costal cartilages.

Finally, ribs of course vary in curvature and cross-sectional shape within each animal and across species and groups. These 3-D shapes then interact with 3-D rib motions to produce changes in the shape and volume of the thorax during breathing. This is not to say that we should consider rib shape and articulations only in the context of lung ventilation; rib and thorax morphology should also be considered in the context of locomotion, both in terms of mechanical properties of the trunk for locomotor functions (see Claessens, this volume) and in terms of respiratory-locomotor interactions (Carrier 1987).

Rib Kinematics

The measurement of rib kinematics (problem 2 above) has been particularly troublesome. Indeed, the motions of individual ribs and the relative motions between ribs have been almost as controversial as the intercostal muscle functions (Ward and Macklem 1995). Some authors describe the motion of human ribs as rotation about an axis running through the two heads of the rib (fig. 3.6A), but rotation about this single axis cannot capture the full range of motion of all 12 human ribs. There must be some sliding translation as well as rotation at the costovertebral joints, or else the ventral rib ends could not remain articulated with the sternum (Saumarez 1986). Others describe three rotations about three axes centered at the costovertebral articulations (fig. 3.6B), but they disagree about whether this axis system should be oriented to the body planes or to some other anatomical features such as the spine (Jordanoglou 1970;

Osmond 1985; Saumarez 1986; Wilson et al. 1987; Margulies et al. 1989; Kenyon et al. 1991; Ward and Macklem 1995). Roughly speaking, rotation about a dorsoventral axis is called bucket handle motion, rotation about a (more or less) mediolateral axis is called pump handle motion, and rotation about a longitudinal axis is called caliper motion (fig. 3.6B). However, authors differ in the orientations of these axes and therefore the terms pump handle, bucket handle, and caliper have somewhat inconsistent meanings.

Rib kinematics are potentially even more complex in nonmammalian amniotes with bipartite or tripartite ribs (fig. 3.4A). Little is known about the extent to which motion occurs at the joints between the vertebral, intermediate, and sternal ribs, or whether cartilaginous segments bend appreciably during rib motions associated with lung ventilation.

Intercostal Muscles

The structure of the intercostal musculature (problem 3, above) adds additional complexity to this system. The two primary layers of intercostal musculature, the external intercostal (EI) and internal intercostal (II), have opposing muscle fiber orientations across most of the rib cage of most amniotes (figs. 3.5 and 3.7A), and they vary in their angle relative to the ribs along dorsoventral and craniocaudal gradients (problem 4, above). For example, in green iguanas, the external intercostal muscle fibers become more steeply angled in a gradient from dorsal to ventral (Carrier 1989). This variation in fiber angle, combined with the fact that every intercostal muscle bundle (fascicle) attaches to two ribs and depends on the 3-D shape and 3-D relative

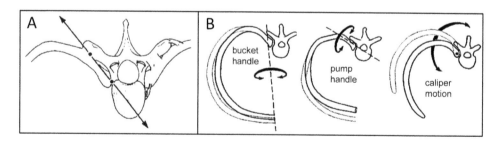

FIG. 3.6 Hypothesized motions of the ribs at the costovertebral joints. (A) Some authors describe the motion of human ribs as rotation about an axis running through the two heads of the rib, but rotation about this single axis cannot capture the full range of rib motion (from De Troyer et al. 2005). (B) Motion can be broken down into rotation about three axes, and described as a combination of bucket handle, pump handle, and caliper motion (from Osmond 1985).

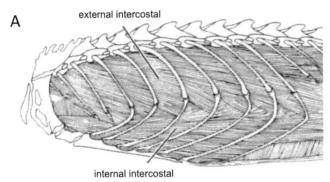

A

external intercostal

internal intercostal

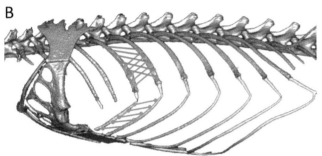

B

FIG. 3.7 Intercostal musculature and ribs of *Iguana iguana*. (A) Intercostal musculature (from Carrier 1989). (B) Rendering from a CT scan of vertebrae and vertebral ribs (gold) and sternum and sternal ribs (turquoise). Intercostal muscle fascicles from the same individual iguana have been mapped back onto the ribs with a 3-D digitizer (Microscribe). Red fascicles are external intercostal, blue fascicles are internal intercostal, and the ventrally located blue fascicles are the parasternal internal intercostal fascicles.

motion of those two ribs, means that it is not possible to determine, a priori from just the anatomy, whether a given fascicle will be lengthening or shortening during exhalation or inhalation.

This inability to determine whether a muscle should be lengthening or shortening is a strange state of affairs. Across most joints we generally can say that certain muscles shorten when the joint flexes (the flexors) and others lengthen when the joint flexes (the extensors). But intercostal muscle strain depends on the shape and motion of two adjacent ribs, combined with the angle of the muscle fascicles, making the simple actions (lengthing or shortening) of the fascicles indeterminant (Loring 1992; Wilson et al. 2001; De Troyer et al. 2005). Thus, to determine whether any given fascicle of intercostal musculature is shortening or lengthening, it is necessary to know simultaneously 3-D rib shape, 3-D rib motion, and attachment points of the fascicles to the ribs. A new technology for measuring bone morphology and motion, XROMM, combined with muscle fascicle mapping, makes it possible to measure all of these during natural breathing motions in live animals.

X-ray Reconstruction of Moving Morphology (XROMM)

X-ray Reconstruction of Moving Morphology provides information on rib shape and motion because XROMM animations combine high-resolution 3-D bone shape, from a CT or laser scan, with high-precision (± 0.1 mm) bone kinematics from biplanar X-ray videos (Brainerd, Baier, et al. 2010). Polygonal mesh bone models from an individual animal are animated with 3-D rigid-body kinematics from the same individual. The result is that XROMM animations specify precisely how every point on every bone moved through space and time in an individual animal in a specific sequence of video. XROMM is based on computer animation, but it is not a modeling or simulation technique. XROMM combines two sources of raw data, static 3-D bone scans and X-ray video, into a more powerful form of raw data—the XROMM animation (Gatesy et al. 2010).

XROMM makes it possible to measure 3-D rib kinematics as never before (fig. 3.8). We can apply a joint coordinate system to the XROMM animation and thereby separate the pump handle, caliper, and bucket handle motions. For one species, *Iguana iguana*, we find that rib motion is dominated by bucket handle motion, with a small amount of caliper motion as well (fig. 3.8).

Since XROMM measures the motion of every point on every bone, we can determine muscle fascicle strain from the relative motions of a muscle's attachment points (i.e., origin and insertion points mapped onto the mesh bone models with a 3-D digitizer). This is particularly true for the relatively simple and tendon-free fascicles of the intercostal musculature (fig. 3.7A),

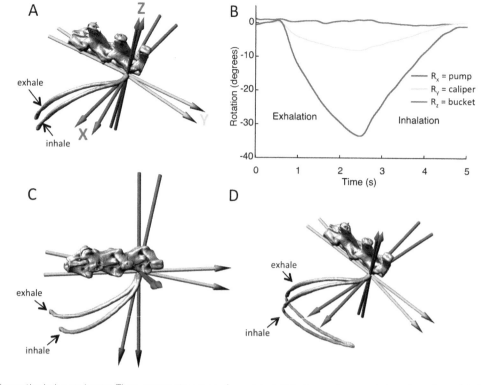

FIG. 3.8 Rib kinematics in *Iguana iguana*. Three perspective views of a vertebral rib in the maximum exhale and maximum inhale position (A, C, D). Sternal rib included in D. The colored axes show a joint coordinate system that breaks the motion down into rotations about the X (red), Y (green), and Z (blue) axes, which are equivalent to pump handle, caliper, and bucket handle motion, respectively. Maximum exhale and inhale positions are shown for the axes and the rib. Plot of rotation about the three axes (B) shows that rib kinematics during breathing in *I. iguana* are dominated by bucket handle motion (rotation about the Z axis), with a small and variable amount of caliper motion (rotation about the Y axis).

although determining fascicle shortening and lengthening for other muscles may be complicated by tendon compliance or curved fascicle pathways. Once we map the attachment points of muscle fascicles onto the rib models, we can run the XROMM animation and read out the patterns of intercostal muscle shortening and lengthening during breathing (fig. 3.7B). Finally, to infer whether the muscles are actively contributing to exhalation or inhalation, muscle activity must also be measured with electromyography (EMG). Then, for example, if a set of intercostal fascicles are electrically active and are shortening during inhalation, we can reasonably infer that they are functioning to power inhalation (De Troyer et al. 2005).

The application of XROMM to rib kinematics and intercostal muscle function has only just begun, and preliminary data are available for just two species: *Iguana iguana* and *Alligator mississippiensis* (Brainerd, Ritter, et al. 2010; Moritz 2013). However, one interesting pattern is emerging; the sternal ribs and associated intercostal muscle fascicles seem to be important drivers of inhalation. This finding fits with existing data for dogs and humans, in which the parasternal intercostals, which attach to the costal cartilages and are similar to the sternal rib intercostals, have been shown to have the best mechanical advantage for inspiration (De Troyer et al. 2005). Furthermore, these parasternal intercostals maintain a respiratory motor pattern during locomotion in dogs, when other parts of the intercostal musculature change to a locomotor pattern (Deban and Carrier 2002). Motions of the sternal ribs and sternum are also critical components of lung ventilation in birds (Claessens 2009b), and the potential contributions of gastralia (dermal belly ribs) in non-avian dinosaurs points further toward the ventral part of the trunk playing a critical role in breathing (Claessens 2004).

Thus, preliminary XROMM data combined with other existing data are pointing toward a conclusion that a lot of the inspiratory action, and maybe even the primary action, is taking place in the ventral part of the rib cage and involves the sternum, sternal ribs or costal cartilages, and associated intercostal muscles.

Putting It All Together: Fossils, Development, Phylogeny, and XROMM

The finding that a lot of the ventilatory action involves the sternum and sternal ribs is, unfortunately, bad news for our prospects for reconstructing the great transformation from buccal pump inspiration to costal inspiration. Sterna and sternal ribs are often cartilaginous and do not fossilize, and they are largely unknown from fossils of stem amniotes. However, focusing on these structures potentially will lead to the discovery of new patterns of ventilation among extant taxa, and also potentially opens up some interesting questions in the evolution of development. It has been hypothesized that the sternum may not be a homologous structure among major amniote groups, which would certainly change our perspective on the evolution of costal aspiration. Might the sternum and sternal ribs be homoplastic characters? This could potentially be tested by studying the development of the sternum in birds, crocodilians, mammals, and squamates. If developmental mechanisms are very different, that would add evidence in favor of homoplasy, but if developmental mechanisms are similar, the structures themselves could still have been acquired independently in the amniote groups by co-opting the same existing developmental mechanisms (see D. Wake et al., this volume). Finally, we should not give up entirely on the role of fossils, since the vertebral ribs do fossilize and would have been connected to the sternal ribs. Even if sternal ribs were the primary drivers of ventilation, the costovertebral joints still contain information about overall rib motion. With XROMM we can potentially use extant species to develop fundamental, form-function relationships linking vertebral rib shape, costovertebral joint structure, sternal rib structure, and rib kinematics to respiratory action. Appreciation that nonfossilized sternal ribs also contribute substantially toward breathing complicates use of vertebral rib morphology to infer breathing mechanisms in extinct animals, but it also may open up new ways of understanding the great transformations in stem amniotes leading to aspiration breathing.

Concluding Remarks

Gills, lungs, swim bladders, and breathing mechanisms are not, for the most part, preserved directly in the fossil record. Yet we are fortunate that extant diversity appears to have preserved clear transformation series from gills to lungs (fig. 3.1), lungs to swim bladders (fig. 3.2), and, to some extent, buccal pumping to aspiration breathing (fig. 3.3). Thorough analysis of the structures and biomechanics of breathing, when placed in a phylogenetic framework and combined with data from fossils and development, has yielded substantial insights into the evolution of lung ventilation mechanisms. A new method, XROMM, seems particularly promising in this regard, since XROMM provides detailed and precise information on both bone shape and motion. With XROMM we have a new tool for linking bone and joint morphology to motion, thereby to make rigorous inferences from fossilized bones about the evolution of form and function.

* * *

Acknowledgments

I am forever grateful to Professor Farish A. Jenkins Jr. for his mentorship during my doctoral work. He taught me the value of clarity and beauty in scientific illustration, and of clarity combined with stylish fun in writing. There could be no better model for an early career scientist than Farish's joyful dedication to his vocation.

Thanks to Colleen Farmer, Amy McCune, Adam Huttenlocker, and Florian Witzmann for reviewing versions of this chapter and providing excellent advice. Sections of this chapter were prepared as a Whiteley Scholar in the delicious quiet and solitude of the Helen R. Whiteley Center at the Friday Harbor Laboratories. Thanks to HRWC and FHL for providing such a valuable scholarly retreat. Development and dissemination of XROMM technology has been supported by the US National Science Foundation under three grants (DBI-0552051; IOS-0840950; DBI-1262156), and work on rib kinematics and intercostal muscle function is ongoing under grant number IOS-1120967.

References

Berman, D. S., R. R. Reisz, and D. Scott. 2010. Redescription of the skull of *Limnoscelis paludis* Williston (Diadectomorpha: Limnoscelidae) from the Pennsylvanian of Canon del Cobre, Northern New Mexico. Pp. 185–210 in Carb-Permian transition in Canon del Cobre. New Mexico Museum of Natural History and Science Bulletin 49.

Brainerd, E. L. 1994a. The evolution of lung-gill bimodal breathing and the homology of vertebrate respiratory pumps. Amer. Zool. 34:289–299.

Brainerd, E. L. 1994b. Mechanical design of polypterid fish integument for energy storage during recoil aspiration. J. Zool. Lond. 232:7–19.

Brainerd, E. L. 1998. Mechanics of lung ventilation in a larval salamander, *Ambystoma tigrinum*. J. Exp. Biol. 201:2891–2901.

Brainerd, E. L., D. B. Baier, S. M. Gatesy, T. L. Hedrick, K. A. Metzger, S. L. Gilbert, and J. Crisco. 2010. X-ray Reconstruction of Moving Morphology (XROMM): precision, accuracy and applications in comparative biomechanics research. J. Exp. Zool. 313A:262–279.

Brainerd, E. L., J. S. Ditelberg, and D. M. Bramble. 1993. Lung ventilation in salamanders and the evolution of vertebrate air-breathing mechanisms. Biol. J. Linn. Soc. 49:163–183.

Brainerd, E. L., and A. M. Dumka. 1995. Mechanics of lung ventilation in an aquatic salamander, *Amphiuma tridactylum*. Amer. Zool. 35:20A.

Brainerd, E. L., and L. A. Ferry-Graham. 2006. Biomechanics of respiratory pumps. In R. Shadwick and G. Lauder, eds., Fish Biomechanics. New York: Academic Press.

Brainerd, E. L., K. F. Liem, and C. T. Samper. 1989. Air ventilation by recoil aspiration in polypterid fishes. Science 246:1593–1595.

Brainerd, E. L., and J. A. Monroy. 1998. Mechanics of lung ventilation in a large aquatic salamander, *Siren lacertina*. J. Exp. Biol. 201:673–682.

Brainerd, E. L., and T. Owerkowicz. 2006. Functional morphology and evolution of aspiration breathing in tetrapods. Resp. Physiol. Neurobi. 154:73–88.

Brainerd, E. L., D. A. Ritter, M. M. Dawson, and A. Sullivan. 2010. XROMM analysis of rib kinematics and intercostal muscle strain during breathing in *Iguana iguana*. Integr. Comp. Biol. 50: E15.

Burggren, W. W., and W. E. Bemis. 1990. Studying physiological evolution: paradigms and pitfalls. Pp. 193–228 in M. H. Nitecki, ed., Evolutionary Innovations. Chicago: University of Chicago Press.

Carrier, D. R. 1987. Lung ventilation during walking and running in four species of lizards. Exp. Biol. 47:33–42.

Carrier, D. R. 1989. Ventilatory action of the hypaxial muscles of the lizard *Iguana iguana:* a function of slow muscle. J. Exp. Biol. 143:435–457.

Cass, A. N., M. D. Servetnick, and A. R. McCune. 2013. Expression of a lung developmental cassette in the adult and developing zebrafish swimbladder. Evol. Dev. 15:119–132.

Clack, J. A. 2007. Devonian climate change, breathing, and the origin of the tetrapod stem group. Integr. Comp. Biol. 47:510–523.

Clack, J. A. 2012. Gaining Ground: The Origin and Evolution of Tetrapods. Bloomington: Indiana University Press.

Claessens, L. P. A. M. 2004. Dinosaur gastralia; origin, morphology, and function. J. Vert. Paleontol. 24:89–106.

Claessens, L. P. A. M. 2009a. A cineradiographic study of lung ventilation in *Alligator mississippiensis*. J. Exp. Zool. A 311A:563–585.

Claessens, L. P. A. M. 2009b. The skeletal kinematics of lung ventilation in three basal bird taxa (emu, tinamou, and guinea fowl). J. Exp. Zool. A 311A:586–599.

Darwin, C. 1859. On the Origin of Species by Means of Natural Selection or the Preservation of Favoured Races in the Struggle of Life. London: John Murray.

Deban, S. M., and D. R. Carrier. 2002. Hypaxial muscle activity during running and breathing in dogs. J. Exp. Biol. 205:1953–1967.

Denison, R. H. 1941. The soft anatomy of *Bothriolepis*. J. Paleontol. 15:553–561.

Derenne, J.-P., and W. A. Whitelaw. 1995. An abbreviated history of the respiratory muscles from antiquity to the classical age. Pp. 399–411 in C. Roussos, ed., The Thorax. New York: Marcel Dekker, Inc.

De Troyer, A., P. A. Kirkwood, and T. A. Wilson. 2005. Respiratory action of the intercostal muscles. Physiol. Rev. 85:717–1756.

Deyst, K. A., and K. F. Liem. 1985. The muscular basis of aerial ventilation of the primitive lung of *Amia calva*. Resp. Physiol. 59:213–223.

Eaton, T. H., Jr. 1960. The aquatic origin of tetrapods. Trans. Kans. Acad. Sci. (1903–) 63:115–120.

Farmer, C. 1997. Did lungs and the intracardiac shunt evolve to oxygenate the heart in vertebrates? Paleobiology 23:358–372.

Furley, D. J., and J. S. Wilkie. 1984. Galen on Respiration and the Arteries: An Edition with English Translation and Commentary on *De usu respirationis. An in arteriis natura sanguis contineatur. De usu pulsuum*, and *De causis respirationis*. Princeton: Princeton University Press.

Gans, C. 1970. Strategy and sequence in the evolution of the external gas exchangers of ectothermal vertebrates. Forma et Functio 3:61–104.

Gatesy, S. M., D. B. Baier, F. A. Jenkins, and K. P. Dial. 2010. Scientific rotoscoping: a morphology-based method of 3-D motion analysis and visualization. J. Exp. Zool. 313A:244–261.

Gee, J. 1978. Respiratory and hydrostatic functions of the intestine of the catfishes *Hoplosternum thoracatum* and *Brochis splendens* (Callichthyidae). J. Exp. Biol. 74:1–16.

Graham, J. B. 1997. Air-Breathing Fishes: Evolution, Diversity, and Adaptation. New York: Academic Press.

Graham, J. B., D. L. Kramer, and E. Pineda. 1977. Respiration of the air breathing fish *Piabucina festae*. J. Comp. Physiol. B 122:295–310.

Hoffstetter, R., and J. P. Gasc. 1969. Vertebrae and ribs of modern reptiles. London: Academic Press.

Janis, C. M., and J. C. Keller. 2001. Modes of ventilation in early tetrapods: costal ventilation as a key feature of amniotes. Acta Palaeontol. Polonica 46(2):137–170.

Janvier, P., S. Desbiens, and J. A. Willett. 2007. New evidence for the controversial "lungs" of the late devonian antiarch *Bothriolepis canadensis* (Whiteaves, 1880) (Placodermi: Antiarcha). J. Vert. Paleontol. 27:709–710.

Jordanoglou, J. 1970. Vector analysis of rib movement. Respir. Physiol. 10:109–120.

Kenyon, C. M., T. J. Pedley, and T. W. Higenbottam. 1991. Adaptive modeling of the human rib cage in median sternotomy. J. Appl. Physiol. 70:2287–2302.

Kramer, D. L., and M. McClure. 1982. Aquatic surface respiration, a widespread adaptation to hypoxia in tropical freshwater fishes. Environ. Biol. Fishes 7:47–55.

Liem, K. F. 1980. Air ventilation in advanced teleosts: biomechanical and evolutionary aspects. In M. A. Ali, ed., Environmental Physiology of Fishes. New York: Plenum Press.

Liem, K. F. 1988. Form and function of lungs: the evolution of air breathing mechanisms. Amer. Zool. 28:739–759.

Liem, K. F. 1989. Respiratory gas bladders in teleosts: functional conservatism and morphological diversity. Amer. Zool. 29:333–352.

Longo, S., M. Riccio, and A. R. McCune. 2013. Homology of lungs and gas bladders: insights from arterial vasculature. J. Morphol. 274:687–703.

Loring, S. H. 1992. Action of human respiratory muscles inferred from finite-element analysis of rib cage. J. Appl. Physiol. 72:1461–1465.

Margulies, S. S., J. R. Rodarte, and E. A. Hoffman. 1989. Geometry and kinematics of dog ribs. J. Appl. Physiol. 67:707–712.

Maurer, F. 1912. Untersuchen über das Muskelsystem der Wirbeltiere: die ventrale Rumpfmuskulatur der Fische (Selachier, Ganoiden, Teleostier, Crossopterygier, Dipnoer). Jena. Zeits. Naturw. 49:1–118.

McMahon, B. R. 1969. A functional analysis of aquatic and aerial respiratory movements of an African lungfish, *Protopterus aethiopicus,* with reference to the evolution of the lung-ventilation mechanism in vertebrates. J. Exp. Biol. 51: 407–430.

Moritz, S. 2013. Rib kinematics during ventilation in *Alligator mississippiensis.* Integr. Comp. Biol. 53:E338.

Near, T. J., R. I. Eytan, A. Dornburg, K. L. Kuhn, J. A. Moore, M. P. Davis, P. C. Wainwright, M. Friedman, and W. L. Smith. 2012. Resolution of ray-finned fish phylogeny and timing of diversification. Proc. Natl. Acad. Sci. 109:13698–13703.

Osmond, D. G. 1985. Functional anatomy of the chest wall. Pp. 199–233 in C. Roussos and P. T. Macklem, eds., The Thorax. New York: Marcel Dekker, Inc..

Owerkowicz, T., C. Farmer, J. W. Hicks, and E. L. Brainerd. 1999. Contribution of gular pumping to lung ventilation in monitor lizards. Science 284:1661–1663.

Packard, G. C. 1974. The evolution of air-breathing in paleozoic gnathostome fishes. Evolution 28:320–325.

Perry, S., and M. Sander. 2004. Reconstructing the evolution of the respiratory apparatus in tetrapods. Respir. Physiol. Neurobiol. 144:125–139.

Romer, A. S. 1956. Osteology of the Reptiles. Chicago: University of Chicago Press.

Romer, A. S. 1972. Skin breathing—primary or secondary? Respir. Physiol. 14:183–192.

Rommel, S., and J. E. Reynolds. 2000. Diaphragm structure and function in the Florida manatee (*Trichechus manatus latirostris*). Anat. Rec. 259:41–51.

Saumarez, R. C. 1986. An analysis of possible movements of human upper rib cage. J. Appl. Physiol. 60:678–689.

Schachner, E. R., T. R. Lyson, and P. Dodson. 2009. Evolution of the respiratory system in nonavian theropods: evidence from rib and vertebral morphology. Anat. Rec. 292:1501–1513.

Schaeffer, B. 1965. The rhipidistian-amphibian transition. Amer. Zool. 5:267–276.

Steen, J. B. 1970. The swimbladder as a hydrostatic organ. In W. S. Hoar and D. J. Randall, eds., Fish Physiology. New York: Academic Press.

Uriona, T. J., and C. Farmer. 2008. Recruitment of the diaphragmaticus, ischiopubis and other respiratory muscles to control pitch and roll in the American alligator (*Alligator mississippiensis*). J. Exp. Biol. 211:1141–1147.

Vesalius, A. 1543. De humani corporis fabrica. Basel: Joannis Oporini.

Ward, M. E., and P. T. Macklem. 1995. Kinematics of the chest wall. Pp. 515–533 in C. Roussos, ed., The Thorax. New York: Marcel Dekker, Inc.

Webb, P. W. 2002. Control of posture, depth, and swimming trajectories of fishes. Integr. Comp. Biol. 42:94–101.

Whipple, I. L. 1906. The ypsiloid apparatus of urodeles. Biol. Bull. 10:255–297.

Wilson, T. A., A. Legrand, P. A. Gevenois, and A. De Troyer. 2001. Respiratory effects of the external and internal intercostal muscles in humans. J. Physiol. (London) 530:319–330.

Wilson, T. A., K. Rehder, S. Krayer, E. A. Hoffman, C. G. Whitney, and J. R. Rodarte. 1987. Geometry and respiratory displacement of human ribs. J. Appl. Physiol. 62:1872–1877.

Witzmann, F. 2013. Phylogenetic patterns of character evolution in the hyobranchial apparatus of early tetrapods. T. Roy. Soc. Edin. 104:145–167.

4

Origin of the Tetrapod Neck and Shoulder

Neil Shubin,* Edward B. Daeschler,[†] and Farish A. Jenkins Jr.[‡]

Introduction

The origin of terrestrial vertebrates is one of the critical transformations in the history of life. The shift from water to land involved the origin of new anatomical features, new organismal functions, and new ecosystems. In vertebrates, the shift came about through changes to a suite of anatomical and physiological features that ultimately permitted a new evolutionary radiation of creatures dwelling in air, thus enabling the origin of running, flying, arboreal, fossorial, and secondarily aquatic forms. The physical challenges placed on organisms in water and land are dramatically different: animals entering terrestrial ecosystems were faced with new physical regimes of body support, respiration, excretion, and sensory perception. Because of these demands, the skeleton is a major locus of evolutionary changes. No longer supported by the buoyancy provided by the water column, newly terrestrial, or partially terrestrial, organisms came to rely more on the musculoskeletal system of the appendages to play a greater role in body support and propulsion.

* Department of Organismal Biology and Anatomy, University of Chicago, and Field Museum of Natural History
† Department of Vertebrate Paleontology, Academy of Natural Sciences of Philadelphia, and Department of Biodiversity, Earth and Environmental Science, Drexel University
‡ Museum of Comparative Zoology, Department of Organismic and Evolutionary Biology, Harvard University (deceased)

Similarly, the differing physical constraints imposed by life in water versus those on land required modification of most other parts of the skeletal system related to sight, audition, feeding, and mobility. The wide range of novelties in limbed vertebrates (tetrapods), from digits (fingers and toes) to the middle ear bones, relate to these new functions.

The more we learn of the water to land transition, the more we come to understand that it involved interlinked morphological changes with antecedents that extend deep in vertebrate history. Importantly, the range of adaptations that arose during this transition came about as freshwater and terrestrial ecosystems themselves began to flourish. The vertebrate transition to land was only possible because of the novel environments that began to emerge in the post-Silurian world approximately 400 million years ago. The origin of land plants and the subsequent invasion of land by diverse invertebrates provided the ecological foundation for the vertebrate radiation that followed. The kinds of shallow freshwater streams and rivers in which many of key Devonian taxa dwelled are, themselves, a novelty driven by physical changes associated with vegetation (Gibling and Davies 2012). The development of soils with extensive root systems provided for stable bank margins that helped create a diversity of floodplain habitats including meandering streams, oxbow lakes, and other aquatic environments that become relatively common by the Late Devonian (Cressler et al. 2010; Gibling and Davies 2012). Against this backdrop of change in the earth's biosphere, topography, and geomorphology, vertebrates began a diverse morphological and ecological radiation starting in the Early Devonian (419 mya). Among these new groups, sarcopterygians, or lobe-finned fish, developed a number of novel traits that became associated with the vertebrate transition to land much later in time.

Our knowledge of the transition of finned sarcopterygians to limbed forms has undergone a dramatic expansion in the past two and a half decades. A main driver for this enhanced understanding came from fossil discoveries. New material of the Late Devonian limbed vertebrate, *Acanthostega gunnari,* discovered in East Greenland and described in great detail beginning in the late 1980s, revealed the degree to which the earliest tetrapods retain adaptions suggestive of an aquatic way of life (Clack 1988; Coates and Clack 1990, 1991; Coates 1996). Key, in this regard, is the presence of

internal gills, flipper-like limbs, and a broad tail. Importantly, this discovery helped to decouple the origin of limbs from the origin of terrestrial lifestyles, and to promote the idea that many tetrapod novelties were first of some use in aquatic habitats. One of the seminal, but indirect, impacts of the *Acanthostega* discovery and detailed description was its effect on the paleontological community itself: it revealed that the Devonian record was not depauperate in fossils, but instead potentially rich in diverse forms worldwide. The ensuing search for new material led to discoveries of Late Devonian tetrapods and near tetrapods in North America (e.g., Daeschler et al. 1994; Daeschler 2000), Europe and Russia (Ahlberg 1991, 1995; Ahlberg et al. 1994, 2000, 2008; Clement et al. 2004; Lebedev and Clack 1993; Lebedev and Coates 1995), and Asia (Zhu et al. 2002). In addition, the hunt for fossil vertebrates from lower in the Devonian section led to the discovery and reexamination of sarcopterygians with morphological conditions that were transitional between finned and limbed forms such as *Tiktaalik roseae* (Daeschler et al. 2006; Shubin et al. 2006) and *Panderichthys rhombolepis* (Ahlberg et al. 1996; Boisvert 2005; Boisvert et al. 2008; Brazeau and Ahlberg 2006; Vorobyeva 2000; Vorobyeva and Schultze 1991). *Tiktaalik*, *Panderichthys*, and the poorly known *Elpistostege watsoni* (Schultze and Arsenault 1985; Schultze 1996) are the finned sarcopterygians that are closest to the origin of limbed forms. These taxa can be referred to collectively as finned elpistostegalians.

The discovery of new fossils, coupled with the renewed analysis of ones discovered earlier in the 20th century, reveals the complex sequence of changes associated with the origin of limbed vertebrates. At first glance, the suite of changes required to complete the fish to tetrapod transition appear to form an almost unbridgeable gap. Like many transformational series, the endpoints of the fish to tetrapod one seem distinct: the skull, shoulders, and appendages of limbed vertebrates appear utterly novel when compared to their finned relatives. Yet, the picture becomes much clearer as fossils, particularly those from the Middle and Late Devonian, are added to the analysis. The successive groups of finned sarcopterygians related to tetrapods can show the sequence of acquisition of major functional novelties and thereby often reveal intermediate conditions in the transition. Indeed, as we learn more of Devonian finned and limbed sarcopterygians, the transition from finned

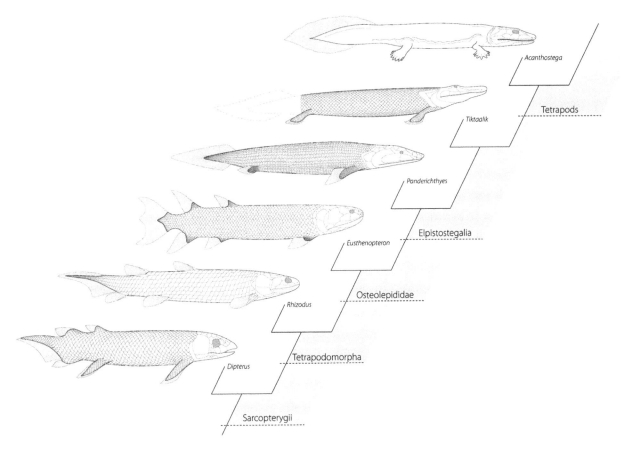

FIG. 4.1 Cladogram of key taxa associated with the origin of limbs.

to limbed forms reflects not a complete and rapid shift from water to land, but initially a shift toward a benthic and perhaps more amphibious way of life. Moreover, the bulk of this change happened in creatures we typically call "fish," that is, finned vertebrates. A complete commitment to a terrestrial lifestyle in the vertebrate realm is not witnessed until the origin of the amniotic egg in the Pennsylvanian, some 60 million years after the origin of limbed vertebrates.

Figure 4.1 is a cladogram that includes key fossil taxa associated with the origin of limbed vertebrates. There are a number of character complexes that change along the main stem of this tree, each revealing likely functional modifications. Most apparent is a clear trend toward appendage-based support and propulsion coupled with an expansion and enhanced mobility of the distal fin skeleton. Enhanced flexor surfaces for pectoral fin and limb musculature also become apparent. With the appendages playing a larger role in supporting the body, head mobility comes at a premium. The origin of a neck is thought to derive from the need to move

the head independently of the body, a condition predicated by having the body fixed against the substrate by the appendages. Changes in respiration would also be reflected in the skeleton: more efficient air breathing would require a broader skull and reduction of the branchial apparatus. Each of these modifications has ramifications across the skeleton.

Critical in the evolution of these functional complexes of the head, body, and appendages were changes to the architecture and composition of the pectoral girdle. In finned vertebrates, the pectoral girdle serves as a key link between skeletal elements of the appendages, axial skeleton, and cranial skeleton. Consequently, the girdle plays a role in diverse functions: from feeding and head mobility to respiration and locomotion. Major evolutionary shifts involving large-scale ecological and functional change entailed modifications of the pectoral girdle as well as its relations to other components of the skeleton. Accordingly, we will focus on the pectoral girdle as an important window on the fin-to-limb transformation. Our presentation is informed by high-quality fossil material from several fossil taxa along a key segment of the lineage presented in figure 4.1.

Overview of the Pectoral Girdle

The pectoral girdle of finned sarcopterygians consists of both dermal and endochondral skeletal elements. Figure 4.2 is a labeled anatomical diagram of the major bones of a finned sarcopterygian shoulder girdle. Pectoral girdle elements from each side generally meet at the dorsal and ventral midlines: dorsally beneath a bony element at the back of the skull and ventrally at an interclavicle that lies within the ventral wall of the body. In caudal view the dermal constituents of the pectoral girdle form a complete hoop composed of multiple bones that extend from the ventral elements, interclavicle and clavicle, dorsally to the cleithrum and the supracleithral and extrascapular series. This hoop-like arrangement stablilizes the two sides of the girdle against one another and with the skull and the body wall. The dorsal series forms a continuous link of dermal bones that connect the cleithrum to the skull via the interlocking bones of the supracleithral series. Limited

movement in the neck region may have been possible but, for the most part, movement of the head in space is linked to movements of the pectoral girdle and body axis, that is, when fish move their head, the body must move in three dimensional space as well. This linkage is lost in early tetrapods and is associated with the independent movement of the head from the body and the development of neck joints and musculature that stabilize and guide this action. The end result is that, in the primitive finned sarcopterygian condition, the dermal girdle is supported ventrally and dorsally as a hoop connected to the skull, whereas this hoop is broken and disconnected with the skull dorsally in tetrapods. Newly discovered material of *Tiktaalik roseae* (described below) bears directly on the evolution of this novel condition.

In most finned sarcopterygians, the cleithrum and supracleithral series are slightly overlapped along their cranial margins by large, flat bones of the opercular series that bridge the neck region between the head and pectoral girdle, adding to its rigidity. These opercular bones help to pump water across the gills. The loss of the bony opercular series in limbed sarcopterygians

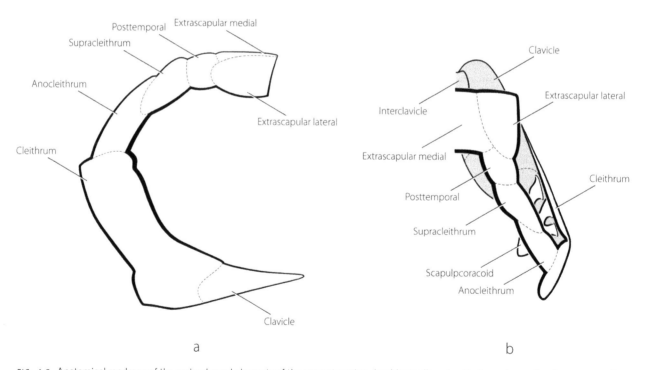

FIG. 4.2 Anatomical roadmap of the major dermal elements of the sarcopterygian shoulder girdle, using *Eusthenopteropn foordi* as an exemplar. a, Right lateral view; b, dorsal view.

(and in *Tiktaalik roseae*) suggests that the neck region was more mobile and that respiratory mechanisms were changing as well.

The endochondral component of the girdle, the scapulocoracoid, lies medial to the dermal bones and includes the glenoid—the articulation with the pectoral appendage (fig. 4.2). In most finned sarcopterygians, the scapulocoracoid is small relative to the dermal series and is attached to the medial surface of the cleithrum by three processes of bone that form a tripodal pedestal supporting the glenoid. Classically in vertebrate paleontology, the well-preserved Devonian sarcopterygian fish *Eusthenopteron foordi* has served as an exemplar of this condition (e.g., Andrews and Westoll 1970; Jarvik 1980; Clack 2012) with similar features seen in most other finned sarcopterygians.

This condition of the pectoral girdle differs, however, from that seen in Late Devonian limbed sarcopterygians such as *Acanthostega*, *Hynerpeton*, or *Ichthyostega*. In the girdle of these taxa, the dermal portion of the girdle is generally reduced, and the endochondral elements are expanded. The supracleithral series is almost entirely lost, and the scapulocoracoid becomes the bulkiest element of the girdle. Changes in the orientation of the glenoid suggest changes in posture: the glenoid of limbed sarcopterygians faces more laterally than finned ones, where it faces more caudally. The expansion of the scapulocoracoid and reorientation of the glenoid are associated with a change in the way the scapulocoracoid attaches to the dermal cleithrum. The tripodal arrangement is lost as the scapulocoracoid develops a broad, continuous surface of articulation with the cleithrum. This shift suggests a more rigid connection between the two bones, which may be associated with an appendage that transmits greater forces through the glenoid fossa.

In the comparison of finned to limbed vertebrates, finned elpistostegalians such as *Tiktaalik roseae* become vital evidence of transitional stages. *Tiktaalik roseae* dates to the early part of the Late Devonian (Frasnian stage) and is from the Fram Formation on Ellesmere Island in Nunavut, Canada. It possesses a range of intermediate conditions in the skull, girdle, and appendages (Daeschler et al. 2006; Shubin et al. 2006; Downs et al. 2008). Since the original description, well-preserved material of *Tiktaalik* has been prepared that provides important new information on which to evaluate the sequence of stages in the transition from a "fish-like" pectoral girdle to that of early tetrapods. In this paper, we describe this new material of *T. roseae* and place it in the morphological, phylogenetic, and functional context to assess the role of the pectoral girdle played in the functional transitions underlying the origin of limbed vertebrates.

Description

Major elements of the girdle of *Tiktaalik roseae*, including cleithrum, clavicle, and scapulocoracoid were described in Shubin et al. (2006). Newly discovered and prepared material includes the interclavicle and supracleithral components of the girdle, thus allowing a three-dimensional reconstruction of the complete girdle. All of the pectoral girdle material of *Tiktaalik* comes from the NV2K17 site on Ellesmere Island where it has been found in a range of states of preservation: (1) In articulation with pectoral appendages, (2) in isolation preserved as three dimensionally intact elements, and (3) as fragments of the girdle skeleton. In the latter, among the most common elements are partial cleithrum/scapulocoracoid elements preserved as surficial float. Following are descriptions based on the complete suite of pectoral girdle elements of *Tiktaalik roseae*.

Supracleithrum.—The supracleithrum is preserved in the holotype, NUFV 108, and in NUFV 110, a referred specimen with both cranium and fin (fig. 4.3). Both specimens are preserved in partial articulation, with elements in their correct positions. The supracleithrum is rectangular in shape, overlapping the anocleithrum across almost its entire length (fig. 4.3a, c). The exposed dorsal surface of the supracleithrum has a tuberculate dermal ornament similar to the ornament on the body scales in the same region. At the cranial end of the supracleithrum, there is an unornamented surface that is suggestive of a zone of overlap for other elements, either the posttemporal (not known) or scales. No *Tiktaalik* specimen shows evidence of posttemporal, extrascapular, opercular, or subopercular bones, despite specimens that are preserved in articulation in this region. This observation has led to the hypothesis that these elements were lacking in *Tiktaalik* and that, accordingly, *Tiktaalik* had a mobile neck region (Daeschler

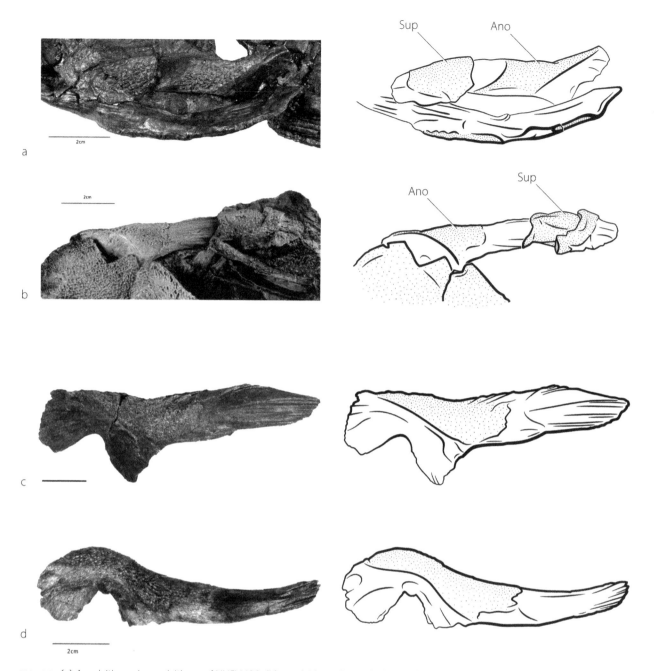

FIG. 4.3 (a) Anocleithrum/supracleithrum of NUFV 108; (b) anocleithrum/supracleithrum of NUFV 110; (c) anocleithrum of NUFV 109; (d) anocleithrum of NUFV 669.

et al. 2006). This trait is a shared derived feature with limbed sarcopterygians.

Anocleithrum.—The anocleithrum is present in articulated and partially articulated specimens (NUFV 108, 109, 110, 669). It is an elongate element overlapped ventrally by the dorsal blade of the cleithrum and dorsally by the supracleithrum (fig. 4.3). The area of overlap for the cleithrum is a smooth, C-shaped surface that is off-set at a slight angle from the rest of the element. The small, dermally exposed portion of the anocleithrum bears the same tuberculate dermal ornament as the body scales. The long area for overlap of the supra-cleithrum narrows cranially and displays ridges along its length. The smooth surfaces of overlap between these supracleithral elements suggest that there may have been mobility between these elements in *T. roseae*, allowing for some pectoral girdle shape change.

The anocleithrum and supracleithrum were oriented parallel to the rows of overlapping rhombic scales in the dorsal region of the trunk suggesting that the dorsal portion of the pectoral girdle was anchored within the integument. Such a configuration could give the girdle stability even though the dermal elements from each side of the pectoral girdle do not meet at the dorsal midline nor insert beneath an extrascapular element.

Cleithrum.—Numerous specimens preserve the cleithrum in various states of completeness (NUFV 108, 109, 110, 112, 114, 115, 669, 670) (figs. 4.4 and 4.5). The dorsal blade of the cleithrum has rounded dorsal corners and a dorsal edge suggesting that in life position

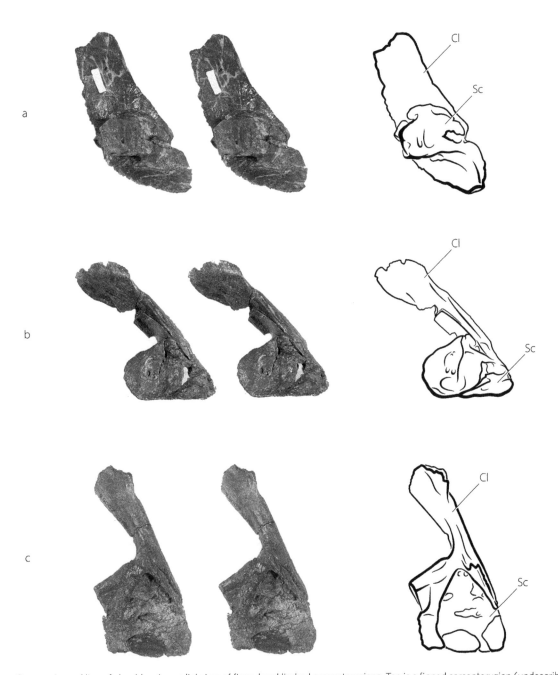

FIG. 4.4 Stereopairs and line of shoulders in medial view of finned and limbed sarcopterygians. Top is a finned sarcopterygian (undescribed tristichopterid from Nunavut), middle is *Tiktaalik roseae*, and bottom is *Hynerpeton*. Sc, scapulocoracoid; Cl, cleithrum.

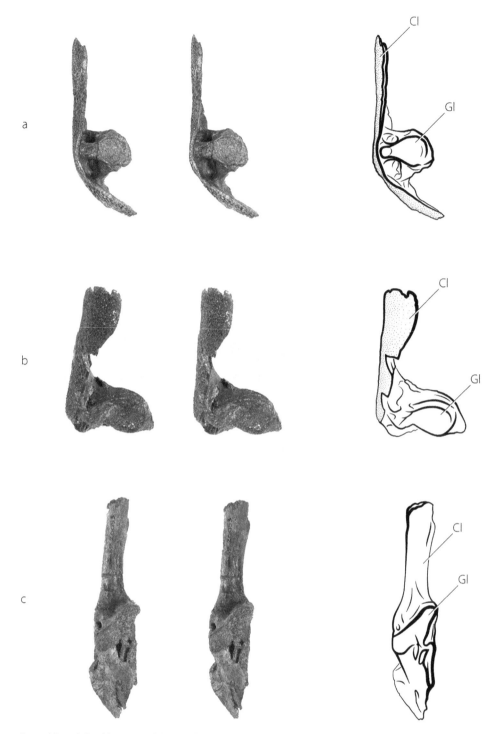

FIG. 4.5 Stereopairs and line of shoulders in caudal view of finned and limbed sarcopterygians. Top is a finned sarcopterygian (undescribed tristichopterid from Nunavut), middle is *Tiktaalik roseae*, and bottom is *Hynerpeton*. Sc, scapulocoracoid; Cl, cleithrum.

it was tilted caudally at about 45 degrees. The lateral surface is covered with tubercles that are sometimes connected to form a network of anastomosing ridges. The dorsal blade of the cleithrum tapers ventrally, and the dermal ornamentation ends at the area of overlap for the clavicle. There is no ventrally facing component of the cleithrum. The unornamented surface for overlap with the clavicle is less than 20% of the length of the entire cleithrum and is textured with a longitudinal groove or series of grooves along the tapered ventral point. The cleithrum thickens considerably in the area of articulation with the scapulocoracoid and particularly in the region that is adjacent to the glenoid. The dorsal blade of the cleithrum is relatively thin caudally and slightly thicker along the cranial margin. There is no evidence of an overlap area for a bony opercular on the cranial margin of the cleithrum.

Scapulocoracoid.—The scapulocoracoid is usually recovered tightly sutured to the cleithrum (figs. 4.4 and 4.5), but at least two isolated scapulocoracoids are known (NUFV 113 and 669—right side). The contact between the cleithrum and the scapulocoracoid is a continuous rugose surface that overlaps the entire ventral portion of the medial aspect of the cleithrum. The large scapulocoracoid includes a blade-like buttress, most likely a homologue of the supraglenoid buttress of *Eusthenopteron*, that extends dorsally along the medial surface of the cleithrum to about the midpoint in the height of the dorsal blade of the cleithrum. At the base of the supraglenoid buttress, the scapulocoracoid thickens dramatically and projects posteromedially with a massive buttress supporting the glenoid. The glenoid fossa itself is an ovoid, concave surface that faces posteroventrolaterally from a position that is on the same horizontal plane as the ventral-most region of the cleithrum (fig. 4.5). Extending cranially from the glenoid buttress is a more flattened portion of the scapulocoracoid that bends medially to form a ventral flange on the same horizontal plane as the ventral edge of the glenoid fossa and ventral-most cleithrum. A large coracoid foramen penetrates the scapulocoracoid between the glenoid buttress and this ventral flange. As discussed by Shubin et al. (2006), this coracoid foramen is oriented to direct flexor musculature from an origin on the dorsomedial surface of the scapulocoracoid to an insertion of the ventral surface of the humerus.

Clavicle.—Clavicles of *Tiktaalik roseae* are preserved in the NUFV 108 (right side), 109, and 110 (right side).

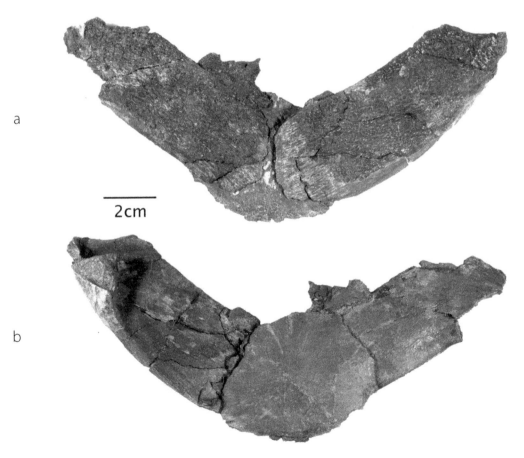

a

2cm

b

FIG. 4.6 Clavicles and interclavicle of *Tiktaalik roseae* in a. ventral and b. dorsal view.

The ventral blade of the clavicle is a thin, rectangular blade that widens slightly medially. The left and right clavicles meet at the ventral midline, but because they are very thin along the medial edge, they do not firmly abut one another. The dermal ornament on the ventral surface is composed of transverse anastomosing ridges and occasional isolated tubercles. The clavicle thickens considerably along the cranial edge as it bends to face laterally and overlap the ventral point of the cleithrum. The smooth curve of the cranial margin rolls dorsally and ends in a short but robust ascending process that conforms to the thickened cranial edge of the cleithrum (fig. 4.6).

Interclavicle.—The interclavicle of *Tiktaalik roseae* is known only in NUFV 109 (fig. 4.6). The ventral elements of the pectoral girdle of NUFV 109 are preserved in articulation with the left and right clavicle meeting at the midline along a slight crest on the ventral surface of the interclavicle (fig. 4.6). The relatively small, dermally exposed, hourglass-shaped area on the ventral surface of the interclavicle has a rugose surface texture but does not have ornamentation as seen on the dermal surface of the adjacent clavicles. The interclavicle is a large, oval-shaped element, slightly thicker in the center than at the margins. The interclavicle overlaps about 25% of the length of the ventral lamina of each clavicle. The nature of the contact surface between the clavicles and the interclavicle is not visible. The dorsal (visceral) surface of the interclavicle is smooth, although the bone surface exhibits a radiating fabric.

Discussion

With the extensive and well-preserved sample of pectoral girdle material of *Tiktaalik roseae*, we are able to reconstruct the dermal and endochondral components of the pectoral girdle in three dimensions (figs. 4.7 and 4.8). For comparative purposes, figure 4.7 also illustrates the pectoral girdle of *Eusthenopteron foordi* and *Acanthostega gunnari*. These three taxa are among the most completely known and well-preserved members of the stem tetrapod lineage and thereby illustrate the segment of the tree that holds the origins of digits, development of a neck, flattening of the head, and many other tetrapod novelties.

Several trends in the transformation of the pectoral girdle are evident: (1) loss of the connection to the skull, including complete loss of the extrascapular and opercular series and reduction of the supracleithral series; (2) enlargement of the scapulocoracoid with a correspondingly large area of attachment to the cleithrum; (3) reorientation of the glenoid from caudally facing to caudoventrolaterally and laterally facing; (4) loss of the tripodal form of the scapulocoracoid; (5) reduction of the dorsal process of the clavicle; (6) enlargement of the interclavicle and flattening of the ventral surface of the body wall. We discuss each of these six changes in more detail below.

In general, the transformation of the pectoral girdle of finned sarcopterygians to that of limbed taxa involved a reduction of dermal components and the expansion of the endochondral scapulocoracoid. In elpistostegalians, the scapulocoracoid is a relatively larger portion of the entire girdle than in more basal taxa such as *Eusthenopteron*. Given that the scapulocoracoid contains the glenoid joint and insertions for muscles that move the humerus at the shoulder, this expansion correlates to an enhanced role of the joint in appendage support and locomotion. The reduction of the dermal components of the girdle does not occur across the entire hoop: the dorsal portion of the hoop is more completely reduced, or even lost, relative to the ventral. The clavicles and interclavicle remain significant components of the girdle in limbed taxa while the cleithrum and dorsal series are reduced. Thus, ventral support for the girdle remains significant, or is enhanced, during the origin of digited appendages.

The increased size of the scapulocoracoid in *Tiktaalik* and limbed forms is associated with a new articulation between the scapulocoracoid and the cleithrum. *Eusthenopteron* retains the condition seen on many basal taxa where the scapulocoracoid is a relatively small element that is attached to the medial surface of the cleithrum by a tripod of three processes of bone. Piercing this tripod are foramina that are formed by the three bony processes. The relatively large cleithrum forms an uninterrupted surface both dorsal and ventral of the scapulocoracoid such that the scapulocoracoid is not visible in lateral view. Whereas the tripodal arrangement is generally encountered in basal forms, a number of basal taxa lack the three processes and have a scapulocoracoid that lies flush against the medial surface of the cleithrum. In

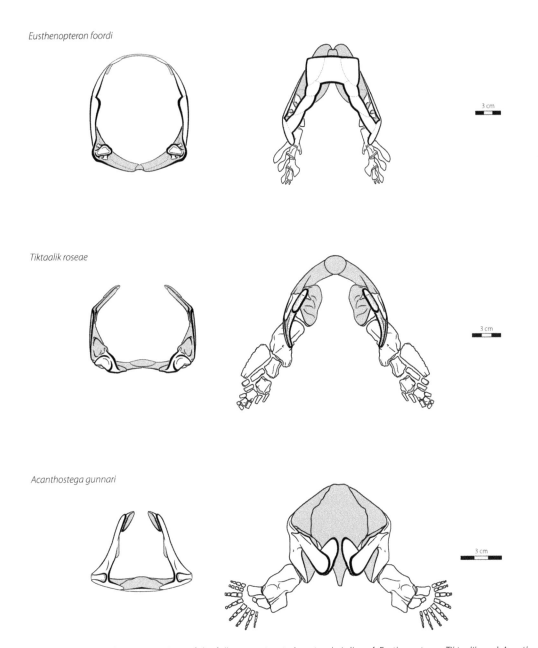

Eusthenopteron foordi

Tiktaalik roseae

Acanthostega gunnari

FIG. 4.7 Dorsal and caudal views of reconstructions of the fully reconstructed pectoral girdles of *Eusthenopteron*, *Tiktaalik*, and *Acanthostega*.

these basal taxa, the dorsal and ventral blades of the cleithrum remain expanded, and the scapulocoracoid is not exposed laterally. *Tiktaalik*, like *Acanthostega*, *Ichthyostega*, and other early tetrapods, lacks these three processes, and the tripod configuration, entirely. In these taxa, the scapulocoracoid is firmly sutured to the cleithrum across a broad, rugose surface. Moreover, the cleithrum is reduced ventrally to the point where the scapulocoracoid is exposed laterally and ventrally.

Associated with this architectural change to the cleithrum and scapulocoracoid is an alteration of the position and orientation of the glenoid articulation itself. Because the glenoid of *Tiktaalik* faces more laterally and ventrally than those of more basal taxa, the origin of a more sprawled posture is likely. The trend toward more lateral orientation of the glenoid continues in early limbed forms such as *Acanthostega*. Each of these pectoral girdles, however, preserves a degree

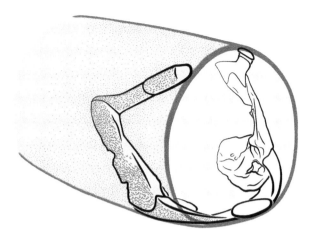

FIG. 4.8 Cutaway view of the pectoral girdle of *Tiktaalik roseae.*

the head while the body is in a fixed position, this feature lends support to the inferences drawn from the structure of the fins of *Tiktaalik* that the pectoral fins were used to support the body on a substrate (figs. 4.4 and 4.5). Comparison of *Tiktaalik* to limbed taxa reveals that the loss of the connection to the head happened prior to the complete reduction of the supracleithral series. The later shift involved freeing the dorsal shoulder from the body wall, a transition underway in *Acanthostega* and apparently complete in more derived Devonian forms such as *Ichthyostega, Tulerpeton,* and crown tetrapods.

Transitional Morphologies, Transitional Environments, and Transitional Taxa

The origin of the tetrapods, and their pectoral girdle structure, provides insights into the problem of major evolutionary transformations. At a macro level, comparing vertebrates that lived 60 million years apart, we find great morphological disparity between creatures specialized for life in water and those specialized for life on land. Fully terrestrial creatures of the Pennsylvanian are able to walk, sense, and feed in a completely different habitat than Devonian forms, which are fully or partially aquatic. As fossils from intervening time intervals are layered into the discussion, we find that the great transition is much more nuanced. Almost all the skeletal features that tetrapods use to walk on land arose within creatures that lived in an aquatic setting. Morphologies that are intermediate in structure between aquatic and terrestrial vertebrates, such as the combined dermal and endochondral shoulder girdle of *Tiktaalik* would be useful in benthic settings, where the animal supports itself on the substrate both with appendages and by the buoyancy of water. Benthic settings are part of a continuum that extends from deep water benthos, to shallow water, and ultimately to mudflats and dry land. Appendage support and head mobility are examples of traits that would be of use in each setting, and are exemplified in the different creatures that inhabit this range of environments.

Eusthenopteron, Elpistostege, Tiktaalik, and *Acanthostega* and other taxa are often described as transitional because they help link creatures at the correct part of the evolutionary tree to explain the origin of tetrapods. In a sense, the word "transitional" is a misno-

of caudal orientation such that the humerus likely projected both posteriorly, laterally, and ventrally.

The supracleithral series undergoes a reduction in the number and size of elements during the origin of tetrapods. *Eusthenopteron* reveals the general primitive condition consisting of four elements, anocleithrum, supracleithrum, posttemporal, and median extrascapular. The median extrascapular is part of the cranium. All elements of the dorsal series form broad dermal plates that have ornamented and unornamented facet surfaces. In *Tiktaalik,* there is no evidence of extrascapulars or the posttemporal. Furthermore, both the anocleithrum and supracleithrum are reduced in size relative to *Eusthenopteron*. Support for the dorsal part of the pectoral girdle is supplied by its insertion within the body wall and scales. The cranium, however, is not attached to the dorsal part of the pectoral girdle. In tetrapods, these elements are reduced even more than *Tiktaalik.* In the Late Devonian tetrapods *Acanthostega gunnari* (Coates and Clack 1991) and *Ventastega curonica* (Ahlberg et al. 2008), the only known supracleithral element is the anocleithrum. In these taxa, the anocleithrum is present, albeit highly reduced relative to finned forms. Most derived members of the tetrapod stem lineage have lost all remnants of the supracleithral series.

This transformation series of the pectoral girdle suggests two functional trends during the evolution of limbed forms. Head mobility, as inferred by the freeing of the cranium from the shoulder, predated the origin of digits. Since cranial mobility allows the animal to reorient mer when applied to taxa. Species are not themselves transitional, in any sense of the word: *Tiktaalik*, for example, was a successful organism in its own right. What is transitional is the morphology of individual organs; a fact that becomes apparent when these features are compared among species. *Tiktaalik*, like its close relatives on the tetrapodomorph tree, evolved intermediate structures of the skeleton to thrive in ecological settings that are themselves functionally transitional, or intermediate, between fully aquatic and terrestrial habitats.

* * *

Acknowledgments

Farish A. Jenkins Jr. passed away just as we started to put pen to paper for this contribution. We feel fortunate to have had the opportunity to have worked with him in the field for the past several decades. Indeed, one of the greatest joys in our collaboration lay in planning figures for the scientific papers on material we discovered. In our last meeting with him, we plotted the capstone "hoop" figures in figure 4.7. The illustrations were ultimately executed by John Westlund. Specimen preparation was by C F. Mullison and R. Masek. We thank the Nunavut Ministry of Culture, Languages, Elders and Youth, Grise Fiord Hamlet and HTA, and the Polar Continental Shelf Program. Comments by Jenny Clack and Michael Coates improved the manuscript. This research was supported by two anonymous donors; the Academy of Natural Sciences of Drexel University; the Putnam Expeditionary Fund (Harvard University); the University of Chicago; the National Science Foundation grants EAR 0207721 (EBD), EAR 0544093 (EBD), EAR 0208377 (NHS), and EAR 0544565 (NHS); and the National Geographic Society Committee for Research and Exploration grants 7223-02, 7665-04, 8040-06, and 8420-08.

References

Ahlberg, P. E. 1991. Tetrapod or near-tetrapod fossils from the Upper Devonian of Scotland. Nature 354:298–301.

Ahlberg, P. E. 1995. *Elginerpeton pancheni* and the earliest tetrapod clade. Nature 373:420–425.

Ahlberg, P. E., J. A. Clack, and E. Luksevics. 1996. Rapid braincase evolution between *Panderichthys* and the earliest tetrapods. Nature 381:61–64.

Ahlberg, P. E., J. A. Clack, E. Luksevics, H. Blom, and I. Zupins. 2008. *Ventastega curonica* and the origin of tetrapod morphology. Nature 453:1199–1204.

Ahlberg, P. E., E. Luksevics, and O. Lebedev. 1994. The first tetrapod finds from the Devonian (Upper Famennian) of Latvia. Phil. Trans. R. Soc. London B 343:303–328.

Ahlberg, P. E., E. Luksevics, and E. Mark-Kurik. 2000. A near tetrapod from the Baltic Middle Devonian. Palaeontology 43:533–548.

Andrews, S. M., and T. S. Westoll. 1970. The postcranial skeleton of *Eusthenopteron foordi* Whiteaves. Trans. R. Soc. Edinb. 68:207–329.

Boisvert, C. A. 2005. The pelvic fin and girdle of *Panderichthys* and the origin of tetrapod locomotion. Nature 438:1145–1148.

Boisvert, C. A., E. Mark-Kurik, and P. E. Ahlberg. 2008. The pectoral fin of *Panderichthys* and the origin of digits. Nature 456:636–638.

Brazeau, M. D., and P. E. Ahlberg. 2006. Tetrapod-like middle ear architecture in a Devonian fish. Nature 439:318–321.

Clack, J. A. 2012. Gaining Ground: The Origin and Evolution of Tetrapods. Bloomington: Indiana University Press.

Clément, G., P. E. Ahlberg, A. Blieck, H. Blom, J. A. Clack, E. Poty, J. Thorez, and P. Janvier. 2004. Palaeogeography: Devonian tetrapod from western Europe. Nature 427:412–413.

Coates, M. I. 1996. The Devonian tetrapod *Acanthostega gunnari* Jarvik: postcranial anatomy, basal tetrapod interrelationships and patterns of skeletal evolution. Trans. Royal Soc. Edinb., Earth Sci. 87:363–421.

Coates, M. I., and J. A. Clack. 1990. Polydactyly in the earliest known tetrapod limbs. Nature 347:66–69.

Coates, M. I., and J. A. Clack. 1991. Fish-like gills and breathing in the earliest known tetrapod. Nature 352:234–236.

Cressler, W. L., III, E. B. Daeschler, R. Slingerland, and D. A. Peterson. 2010. Terrestrialization in the Late Devonian: A Palaeoecological Overview of the Red Hill Site, Pennsylvania, USA. In Gaël Clement and Marco Vecoli, eds., The Terrestrialization Process: Modelling Complex Interactions at the Biosphere-Geosphere Interface. London: Geological Society.

Daeschler, E. B. 2000. Early tetrapod jaws from the Late Devonian of Pennsylvania, USA. J. Paleontol. 74(2):301–308.

Daeschler, E. B., N. H. Shubin, and F. A. Jenkins Jr. 2006. A Devonian tetrapod-like fish and the evolution of the tetrapod body plan. Nature 440:757–763.

Daeschler, E. B., N. H. Shubin, K. S. Thomson, and W. W. Amaral. 1994. A Devonian tetrapod from North America. Science 265:639–642.

Downs, J. P., E. B. Daeschler, F. A. Jenkins Jr., and N. H. Shubin. 2008. The cranial endoskeleton of *Tiktaalik roseae*. Nature 455:925–929.

Gibling, M. R., and N. S. Davies. 2012. Paleozoic landscapes shaped by plant evolution. Nature Geosci. 5:99–105.

Jarvik, E. 1980. Basic Structure and Evolution of Vertebrates. London: Academic Press.

Lebedev, O. A., and M. I. Coates. 1995. The postcranial skeleton of the Devonian tetrapod *Tulerpeton curtum* Lebedev. Zool. J. Linn. Soc. 114:307–348.

Lebedev, O. A., and J. A. Clack. 1993. Upper Devonian tetrapods from Andreyevka, Tula Region, Russia. Palaeontology 36(3):721–734.

Schultze, H.-P. 1996. The elpistostegid fish *Elpistostege*, the closest the Miguasha fauna comes to a tetrapod. Pp. 316–327 in H.-P. Schultze and R. Cloutier, eds., Devonian Fishes and Plants of Miguasha, Quebec, Canada. Munich: Verlag Dr. Friedrich Pfeil.

Schultze, H.-P., and M. Arsenault. 1985. The panderichthyid fish *Elpistostege*: a close relative of tetrapods? Palaeontology 28:292–309.

Shubin, N. H., E. B. Daeschler, and F. A. Jenkins Jr. 2006. The pectoral fin of *Tiktaalik roseae* and the origin of the tetrapod limb. Nature 440:764–771.

Vorobyeva, E. I. 1995. The shoulder girdle of *Panderichthys rhombolepis* (Gross) (Crossopterygii), upper Devonian, Latvia. GeoBios 19:285–288.

Vorobyeva, E. I. 2000. Morphology of the humerus in the Rhipidistian Crossopterygii and the origin of tetrapods. Paleontol. J. 34:632–641.

Vorobyeva, E. I., and H.-P. Schultze. 1991. Description and systematics of panderichthyid fishes with comments on their relationship to tetrapods. In H.-P. Schultze and L. Trueb, eds., Origins of the Higher Groups of Tetrapods. Ithaca: Cornell University Press.

Zhu, M., P. E. Ahlberg, W. Zhoa, and L. Jia. 2002. First Devonian tetrapod from Asia. Nature 420:760–761.

5

Origin of the Turtle Body Plan

Ann Campbell Burke*

Introduction

Turtles provide a classic example of many essential riddles of evolutionary biology: who are their ancestors, and how did their remarkable anatomy evolve from a more typical form? The literature abounds with reviews of this evolutionary conundrum. Rather than reiterating a comprehensive history of data and ideas, this chapter relies heavily on citations to extensive recent scholarship (e.g., Rieppel 2013; Nagashima et al. 2012). After a brief review of the unique anatomy of the shell, and the equally problematic characteristics of the skull, the chapter will focus on three main aspects of this major transition in vertebrate evolution: (1) the position of turtles among the amniotes, (2) critical fossil taxa, and (3) key developmental studies. It is usually logical to start the review of any taxonomic group with a discussion of the fossil and developmental characters used to infer their phylogeny. However, the departure of the order Testudines from the primitive amniote body plan, both cranial and postcranial as reviewed below, is such that the fossil record has not served to root the lineage with any degree of consensus. Likewise, fascinating developmental studies and new data on the molecular genetics of shell development have dramatically increased our understanding of the unique

* Department of Biology, Wesleyan University

chelonian anatomy, but these data also shed no direct light on specific ancestry. It therefore seems expedient to begin by describing the current phylogenetic debate first, allowing the reader to consider the recent fossil and developmental data in the context of the contemporary arguments.

This chapter is by no means a full review of all the data and ideas about turtle evolution. Instead, it will cover those issues that currently have the greatest significance, with full acknowledgment of and no apology for my own biases. As Farish Jenkins noted, it is the questions that are most important. Primarily, it is the author's intention to provide enough food for thought to inspire students of evolutionary morphology to follow up any of the persistent questions that define this fascinating group of animals.

Chelonian Anatomy

The Shell

Turtles are identified by the highly derived nature of their postcranial body. The association of the trunk vertebrae and ribs with a specialized dermis forms the carapace and brings about a novel relationship between axial and appendicular systems. The plastron, or bottom shell, contains the highly modified clavicles and interclavicle, as well as neomorphic bones. The nature of the ossifications in the shell was much discussed in the 19th century, and the debate continues (reviewed in

Rieppel 2013). Recently the character of shell ossification has been readdressed in arguments about the incipient stages of shell evolution (e.g., Lee 1997; Scheyer et al. 2008; Joyce et al. 2009; Lyson et al. 2010). There is persistent disagreement about appropriate histological terms and their phylogenetic implications. Here, the carapace and plastron are described in simple terms based on embryonic processes. The bones of the carapace and plastron are illustrated in figures 5.1 and 5.2.

The vertebrae and ribs are bones of the axial skeleton that undergo endochondral ossification and are thus referred to as endochondral bones. These elements are clearly visible in turtle embryos (fig. 5.2), though in most species they become largely obscured by ossifications of the surrounding loose mesenchyme of the dermis (fig. 5.1A). These latter elements are referred to as dermal bones, simply reflecting the fact that they are not preformed in cartilage. The carapace is thus a composite of endochondral and dermal bone, as described by 19th-century morphologists including Owen (1849). The plastron is comprised entirely of dermal ossifications. Apart from the modification of clavicles and interclavicles into epi- and entoplastra, respectively, there are no clear homologues of the hyo-, hypo-, or xiphiplastra. The carapace and plastron meet along the flank in what is called the bridge, with axillary and inguinal buttresses (fig. 5.1C). The nuchal bone in the anterior midline of the carapace is interesting in that it appears early, and seems to have two phases of ossification (Burke 1989a). This is also true of all the plastral

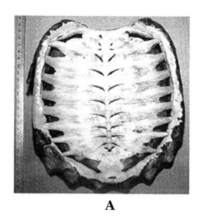
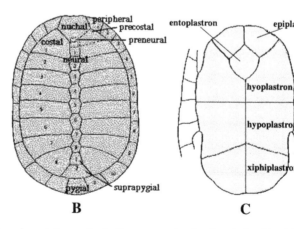

FIG. 5.1 Bones of the carapace and plastron. A, Ventral view of an adult snapping turtle carapace showing the relationship of the endochondral ribs and vertebrae to the dermal bones. B, Dorsal view of a schematized carapace showing the distribution of dermal bones (labeled). C, Ventral view showing the dermal bones of a schematized plastron (gray) and the bridge region where it meets the carapace (white). No epidermal scutes are shown in B. Reprinted with permission from Gilbert et al. (2001).

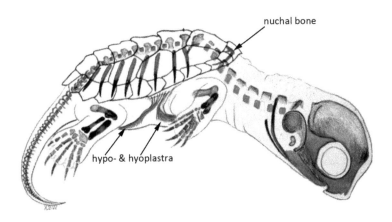

nuchal bone

hypo- & hyoplastra

FIG. 5.2 This drawing of a near-hatching snapping turtle embryo illustrates the underlying anatomy of the turtle shell and the "displacement" of the ribs within the dorsal dermis to form the carapace. The dermal bones present at this stage are the nuchal in the carapace, and all the bones of the plastron, though only hyo- and hypoplastra are visible in this view. Drawn by K. Wing-Brown in 1989 for Burke (1989a) and reprinted in Gilbert et al. (2001).

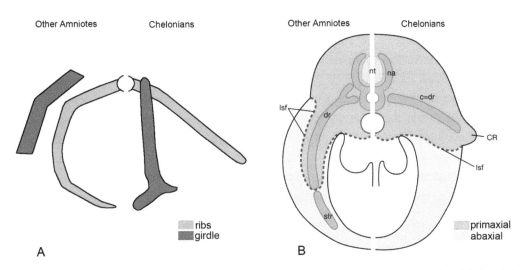

FIG. 5.3 These schematics illustrate the changes to the chelonian body plan in contrast to other amniotes. A, In the adult the relationship of axial and appendicular elements is polarized between turtles (right) and non-chelonians (left). B, In the embryo this relationship is generated by the primaxial (somitic only, green region) isolation of the ribs due to growth of the dermis at the CR. Reprinted from Shearman and Burke (2009), with permission from Nagashima et al. (2007) for panel B.

elements, where a discreet primary ossification center is eventually covered over by secondary ossification, except in species with highly reduced shells (reviewed in Gilbert et al. 2001). The degree of dermal ossification varies dramatically in turtles. Some authors see the polarity of this character as evidence of ancestry, and the nature and relevancy of shell ossification is still a matter of debate (see below).

There are many consequences of the association of the ribs with a specialized and often highly ossified dermis. For one, the absence of intercostal muscles requires alternative means of ventilating the lungs since the rib cage cannot expand. For another, the relationship between the pectoral girdle and the axial skeleton is different from the typical tetrapod body plan. This

later characteristic distinguishes the Chelonian body plan from the typical non-chelonian tetrapod (fig. 5.3A; and Burke 1989a, 1989b). The extremity of this distinctly chelonian variation can be considered an evolutionary novelty. The difficulty of imagining functional intermediates has been used to suggest that turtles appeared by saltation and can be classified as hopeful monsters (Rieppel 2001; Theissen 2009). Others argue that the "novelty" is exaggerated (Lyson and Joyce 2012) as the position of the scapula can be seen as primitive, not derived, and the body plan is the same as other tetrapods as regards the position of the primary body wall (Nagashima et al. 2012). In my opinion, these arguments pivot on semantic questions, that is, what is truly "novel"? How is a body plan defined? In fact everyone agrees there is an evolutionary puzzle here to be solved. What we want to understand are the morphogenetic steps that occurred to generate turtle anatomy from a typical tetrapod.

The Skull

In addition to the highly derived postcrania, the turtle skull presents an interesting array of primitive and autapomorphic characters, offering little definitive evidence of ancestry, but rich fodder for debate. The oldest turtle skulls in the fossil record (*Proganochelys* and *Odontochelys*; see below) have no temporal fenestration, an open basipterygoid articulation, and teeth along with general proportions characteristic of stem amniotes (Gaffney 1990; Li et al. 2008). The emargination of the skull roof and the dramatic modifications of the quadrate or pterygoid to act as a trochlea for the jaw adductor muscles appear later and mark the divergence of Cryptodires and Pleurodires (Gaffney 1979). The teeth are ultimately lost.

The traditional representation of turtles as unfenestrated anapsids is well entrenched in textbooks of vertebrate biology (e.g., Hyman 1942; Romer 1956; Romer and Parsons 1977; Kardong 2009). However, the primitive anapsid condition of the turtle skull has been challenged repeatedly in the 20th century (reviewed by Rieppel 2000; see also Werneburg 2012). There have been remarkably few recent studies directed at the ontogeny of cranial fenestration or its absence in turtles. Most recently, Werneburg (2012) reviewed the older literature on the temporal region in turtles, and presented an interesting discussion of both phylogenetic constraints and the functional forces that have influenced changes in the temporal region in all tetrapods. Fucik (1991) compared orbitotemporal development in turtle, lizard, and crocodile, and determined that significant independent changes occur in all three lineages. She concluded that this character suite is not suitable for phylogenetic inference (Fucik 1991). Müller (2003) explored the loss of the lower temporal arch across diapsids, and suggested the condition of the jugal and quadratojugal in *Proganochelys* can be interpreted as secondary closure of a temporal opening. Thus in general, recent developmental and fossil studies are skeptical at best about the anapsid nature of the chelonian skull.

The Phylogenetic Debate on Turtle Origins

In the past two centuries, both pre-Darwinian and evolutionary biologists debated the phylogenetic affinities of turtles. The phylogenetic position of turtles among amniotes is unresolved to this day and the subject of vigorous discussion. The debate can be simplified by three hypotheses with their obvious corollaries as listed in table 5.1 and represented by figure 5.4.

Morphological Data

Morphological data have been used to build phylogenies supporting all of these hypotheses, and many of these analyses are quite recent (see Rieppel 2000 and 2013 for excellent and extensive reviews of this history). The lack of cranial fenestration was used to support early consensus that turtles are a basal branch from the tetrapod tree, maintaining an anapsid skull. Amphibian origins have been proposed (Plagiosaurids: Nilsson 1945), but turtles are clearly amniotes. Using the absence of a Jacobson's organ in turtles, Gaffney (1980) placed them as the sister group to all other amniotes (fig. 5.4, 1a). Most often, however, they have been placed among so-called stem reptiles. Reptiles are refined as two major clades, the parareptiles and the eureptiles; the eureptiles in turn divide into the captorhinomorphs and the diapsids (fig. 5.4). Just about every named fossil group among them has been proposed as the source of turtle ancestry, as parareptiles (Pareiasaurs: Gregory 1946; Lee 1997; Procolophonids: Romer 1968; Laurin and

TABLE 5.1

1) Turtles are anapsids

   a) sister group to all other amniotes

   b) grouped among the parareptiles

      if so . . . which group of parareptiles is the sister group of turtles?

2) Turtles are a sister group of diapsids

   a) sister group to captorhinomorphs

   b) independent sister group to diapsids

3) Turtles are diapsids

   a) sister group of lepidosaurs

   b) sister group of archosaurs

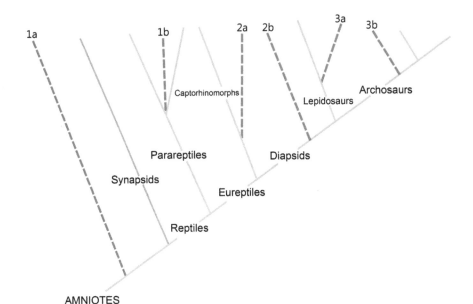

FIG. 5.4 The relationships shown in this cladogram of Amniotes are widely accepted except for the placement of turtles. Dotted red lines show the various relationships proposed for turtles that are discussed in this chapter. See text for explanations.

Reisz 1995), as captorhinomorphs (Gaffney 1980), as well as various diapsid groups (e.g., Placodonts: Broom 1924 and Jaekel 1902, cited in Gregory 1946).

As introduced above, the anapsid character of the turtle skull has been frequently questioned, and with it the basal position of turtles among living amniotes. Cranial characters have, however, been used to argue a sister group relationship with the captorhinomorphs (fig. 5.4, 2a) (Gaffney 1980). The inclusion of fossil taxa is still used to argue for the parareptile status of turtles (fig. 5.4, 1b) (Lyson et al. 2010). Comparison of characters based on the timing of organogenesis among crown amniotes has been interpreted to place turtles as the sister group to the Diapsida (fig. 5.4, 2b) (Werneburg and Sánchez-Villagra 2009). Goodrich argued in 1916 that the hooked fifth metatarsal and the disposition of the great vessels clearly establish turtles as diapsids ("sauropsidan branch"), and this idea has always had a following (Rieppel 2000, 2013). An extensive analysis of osteological characters from living taxa (Rieppel and deBraga 1996) placed turtles within crown group diapsids. Though several 20th-century authors used neontological characters to link turtles with Archosaurs (fig. 5.4, 3b), Rieppel reviews this literature and rejects the Archosaur alignment. Instead he argues that the morphological data place turtles as the sister group to Lepidosaurs (fig. 5.4, 3a) (Rieppel 2000), a conclusion upheld by inclusion of the recently discovered *Odontochelys* (Li et al. 2008).

The combination of potentially primitive *and* highly derived morphology in both cranial and postcranial anatomy of turtles confounds homology assessments. This situation has created something of a stalemate between the conflicting arguments (Rieppel and Kearney 2002; Harris et al. 2007).

Molecular Data

The 1990s and the molecular era brought the bright prospect of "independent," "quantifiable" sequence data, a potentially unlimited character set free of the pitfalls of morphological homology assessment. The molecular phylogenies published in the early 1990s not surprisingly suffered from small data sets and a poor range of taxa, and produced variable and contradictory results (cf. Hedges et al. 1990; Hedges 1994). The earliest molecular study that directly addressed the relationship of turtles was based on a single peptide that nested turtles within the diapsids (Platz and Conlon 1997). Mitochondrial sequences then placed turtles as the sister group of Archosaurs (Zardoya and Meyer 1998). This position was quickly corroborated by nuclear gene sequences (Hedges and Poling in 1999). To date, the weight of molecular data has been interpreted to support the hypothesis that turtles and Archosaurs are a monophyletic group (fig. 5.4, 3b). The data range from sequence including mtDNA (Kumazawa and Nishida 1999; Janke et al. 2000; Cao et al. 2000), nuclear protein-coding loci (NPCLs; Iwabe et al. 2005; Shen et al. 2011), Hox cluster sequence and structure (Liang et al. 2011), "ultraconserved elements" (UCEs; Crawford et al. 2012), the brain transcriptome (Tzika et al. 2011), and chromosomal structure and synteny (Matsuda et al. 2005). Some data sets further refine the node to one shared with crocodilians (Mannen and Li 1999; Hedges and Poling 1999; Cao et al. 2000).

Molecular data have also been found that support alternate hypotheses placing turtles as the sister group to all amniotes or as the diapsid sister group to Lepidosaurs. Mallatt and Winchell's (2007) comparisons of 18S and 28S ribosomal DNA across all deuterostomes support Gaffney's placement of turtles as the most basal branch of living amniotes (fig. 5.4, 1a). A comparison of a hormone precursor gene with a unique structure supported the Lepidosaur relationship (fig. 5.4, 3a) (Becker et al. 2011), and a study looking at microRNA families found unique overlap between turtles and lizards, and not between turtles and Archosaurs (Lyson et al. 2012b).

The molecular studies listed here vary on whether their data and methods definitively reject competing hypotheses, but there is no doubt that the current majority of molecular data has been interpreted to support turtle-Archosaur affinities. It is important to note Lee's discussion of the potential for a "systematic" bias in the literature that could result from a preference to publish data that conflict with older hypotheses, giving the skewed impression that molecules and morphology rarely agree (Lee et al. 2004). That said, a very recent study using 248 genes from 14 amniotes and a wide range of analytical techniques "unambiguously" supports the turtle-Archosaur clade (Hedges 2012; Chiari et al. 2012). Even more recently, draft genomes of the green sea turtle, *Chelonia mydas* and the soft shell *Pelodiscus sinensis* have been published (Wang et al. 2013). Along with very interesting observations of an expanded family of olfactory receptor genes, and transcriptome data supporting a phylotypic stage, the data firmly place turtles as the sister group to Archosaurs. Furthermore, the divergence time calculated between them clocked at between 248 and 267 mya, not inconsistent with the fossil record.

The placement of turtles among the amniotes has been a major battleground for molecular systematists and a case study of phylogenetic controversy (Harris et al. 2007). To the non-practitioner, this arena in which phylogenetic algorithms and sophisticated statistical methods of bioinformatics are pitched against one another is impressive but largely impenetrable. I look forward to a day when selection has simplified the field by eliminating meaningless data and faulty methods. At the time of writing the weight of molecular characters is tipping strongly toward the diapsid origin for turtles. In the meantime, the development and evolution of this remarkable body plan is still being revealed by new finds in the fossil record and new understanding of developmental mechanisms.

Turtles in the Fossil Record

The 2008 discovery of *Odontochelys* (Li et al. 2008) in late Triassic marine strata in China is one of the most exciting fossil finds of recent decades. Not only is the known history of turtles pushed back about 15 million years, the morphology of the shell and habitat of burial demanded the rethinking of early turtle evolution (Reisz and Head 2008; Lyson and Gilbert 2009; Burke 2009). Obviously the oldest fossil of a particular taxon determines our ideas about its original form and habits. *Proganochelys*, from Norian rocks (~204 my) in Germany, held the position of the "first turtle" for almost 100 years (Gaffney 1990). The skull shows the primitive characteristics mentioned above, but its identity as a turtle is unquestionable as the carapace and plastron of this animal were fully developed and very well ossified.

Despite its long-standing iconic position, *Proganochelys* was by no means temporally isolated, as rocks of the late Triassic have produced a number of fossil turtles. *Proterochersis*, also from Germany, is considered contemporaneous, and has the ilia fused to the carapace, which is one of the characters that define the Pleurodires. This indicates a significant history of diversification prior to 205 million years ago. There is also already significant geographic distribution in the late Triassic, with Proganochelid fossils from South Africa, fragments of shells from Thailand (Gaffney 1989), and similarly aged *Paleochersis* from Argentina (Rougier et al. 1995).

Kayentachelys merits mention here because of the interpretation that the dorsal process of the epiplastra (clavicles) seen in the Triassic turtles and *Kayentachelys* is a cleithrum (Joyce et al. 2006). These authors claim that the retention of a cleithrum in basal turtles is grounds for their placement outside of the Saurian (diapsid) clade. The position of *Kayentachelys* as the sister group to all other Cryptodires (Gaffney et al. 1987) is still debated in the literature (Gaffney and Jenkins 2010; Joyce and Sterli 2012), whereas the identification of a cleithrum in this taxon is considered quite uncertain (Rieppel 2013).

Eunotosaurus, a Permian "stem reptile" from South Africa was soundly rejected by Romer as a possible chelonian ancestor despite its expanded ribs and turtle-like vertebrae (Romer 1956). However, new material has recently been used to reposition this taxa as a stem chelonian (Lyson et al. 2010, 2012). This has interesting implications for the diapsid root suggested by the genomic data mentioned above and is likely to be a cause for lively debate.

Having replaced *Proganochelys* in the "first turtle" category, *Odontochelys* is now the focus for arguments about the paleoecology and morphological transition. The habitat that framed the selective environment in which turtle anatomy evolved is of obvious interest and arguments have been made for both terrestrial (Joyce and Gauthier 2004) and aquatic environments (Rieppel and Reisz 1999). The limb proportions of *Proganochelys* and *Palaeochersis* are argued to indicate a totally terrestrial ecology (Joyce and Gauthier 2004), though *Proganochelys* has also been described as semiaquatic based on taphonomy and deposition site (Gaffney 1990). The discovery of *Odontochelys* in marine sediments seems to support an aquatic origin, though the debate continues (e.g., Joyce et al. 2009).

The skull of *Odontochelys* shares some of the primitive characteristics of *Proganochelys* not found in younger turtles—an open basicranial articulation and palatal teeth, though the latter are more extensively distributed in the older species (Li et al. 2008). The real surprise, however, comes in the anatomy of the shell. The plastron is fully formed. The epiplastra are described with dorsal processes similar to those seen in *Proganochelys*, but otherwise this plastron is shockingly modern. The carapace, on the other hand, is lacking most of the dermal bone components so well developed in *Proganochelys*. Neural plates are present, but not fused to the neural spines; the ribs appear to have a flattened collar of periosteal bone, and are only slightly ventrally convex.

The authors are quick to assume that this fossil replaces *Proganochelys* as the definitive ancestor and come to the conclusion that "the plastron evolved before the carapace" (Li et al. 2008, 499). The interpretation that *Odontochelys* lacks a carapace is extreme and in my view an incorrect interpretation. Though the characteristic dermal bones are absent (nuchal, costals, pygals, and marginal; illustrated in fig. 5.1B) the disposition of the ribs indicates the presence of the specialized dermis in which these ossifications form. As pointed out by the authors and others, the morphology is very similar to that seen in the carapace of hatchling snapping turtles (Reisz and Head 2008; Burke 2009). This is also true of the border of the plastral bridge, which in the hatchling meets the unossified margin of the carapace as illustrated in figure 5.2. Overall, the morphology of *Odontochelys* provides a dramatic new window on the early history of turtles (fig. 5.5), and a morphology that is remarkably consistent with ideas generated by the developmental studies discussed below.

The Developmental Perspective

The growing interest in the development of the turtle shell has provided very rich embryonic context for trying to understand the sequence of morphological changes that could have lead from a typical tetrapod to the anatomy of *Odontochelys* and all subsequent turtles. Again, I will not cover all of this literature here, as it has been recently and extensively reviewed (Kuratani et al. 2011; Nagashima et al. 2012, 2013). Nagashima and coauthors beautifully illustrate a progression of morphogenetic models of shell development. These models begin with a 1929 descriptive study by Ruckes that in part inspired my thesis work (Burke 1989a). To my mind Ruckes's study of carapace and plastron development in embryos of *Chelydra serpentina* remains definitive, and forms a substantial basis for the experimental and molecular studies of the last 20 years.

Ruckes highlighted a number of factors that underlie the "encasement of the girdles by the shell." These can be paraphrased as (1) the association of a specialized dorsal dermis with the rib primordia, (2) the "displacement" of the ribs to a dorsal position caused by rapid growth of the dermis in a lateral dimension, followed by (3) their "horizontal displacement" resulting from radial expansion of the carapacial dermis. Finally,

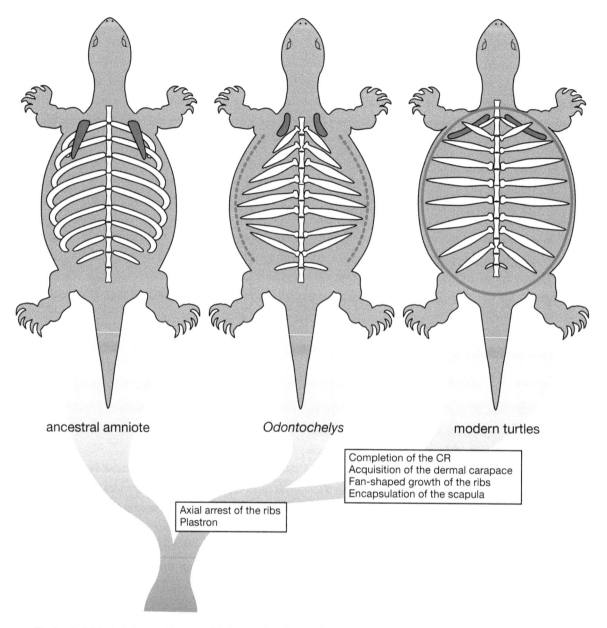

ancestral amniote   *Odontochelys*   modern turtles

Completion of the CR
Acquisition of the dermal carapace
Fan-shaped growth of the ribs
Encapsulation of the scapula

Axial arrest of the ribs
Plastron

FIG. 5.5 The fossil of *Odontochelys* provides a possible intermediate form in the early turtle lineage. This figure, reprinted with permission from Nagashima et al. (2008), represents *Odontochelys* as a sister taxon to "modern" turtles (including *Proganochelys*, etc.). This transitional form has a plastron (not shown), an early, incomplete CR, and primaxially arrested ribs.

(4) the unusual growth trajectory of the dermis causes the ribs to spread over the girdles and a consequent "involution of the underlying body wall" (Ruckes 1929, 86–87). These are the tissue-level events that divert the turtle embryo from a typical tetrapod trajectory and illustrate the fundamental novelty of the turtle bauplan (fig. 5.3). Turtles were hardly model systems for experimental developmental biology in the 20th century, but in the 1960s and 1970s Chester Yntema performed many types of experiments on embryos of the common snapping turtle, *Chelydra serpentina*. One of these studies confirmed the somitic origin of the carapace (Yntema 1970). Though not a particularly surprising finding, it did establish developmental homology with the typical vertebrate model systems and the basis for further comparative studies. For instance, the turtle body wall is segregated into primaxial and abaxial domains (*sensu* Burke and Nowicki 2003) at the carapacial ridge (Nagashima et al. 2007). These domains, based on the lineage of embryonic mesoderm are hypothesized

to act as both developmental and evolutionary modules that have influenced the evolution of vertebrate morphology (fig. 5.3B; reviewed in Shearman et al. 2009).

My own curiosity about turtles sprang from the lack of intermediate forms in the fossil record. Turning to embryology as a graduate student back at the Museum of Comparative Zoology, I set out to understand how a typical tetrapod ontogeny was diverted in turtle embryos. My extension of Ruckes's description was the identification of an epithelial-mesenchymal interaction (EMI) at the leading edge of the carapacial dermis, referred to as the carapacial ridge (CR; fig. 5.3B) (Burke 1989b). This provided the basis for a model of carapace evolution wherein the appearance of a novel inductive EMI in the early embryo was a pivotal step in the generation of the novel body plan (Burke 1989c and 1991).

The molecular nature of the CR has been the subject of a number of studies that confirmed the presence of morphogenetic molecules consistent with other EMI-derived structures (e.g., *Fgfs*: Loredo et al. 2001; *Msx* genes: Vincent et al. 2003; *Wnt* signaling members: Kuraku et al. 2005; *BMPs* and *shh*: Moustakas 2008), though there appears to be some differences between taxa (Kuraku et al. 2005). In comparison with the molecular development of somitic cell populations in chick and mouse, turtle somites also show distinctions. Turtle-specific differences in Hox gene expression have been reported (Ohya et al. 2005), and the *Myf-5* gene, a member of the muscle regulatory factors, has a turtle-specific deletion and an unusual expression pattern in the dermomyotome of *Pelodiscus sinensis* (Ohya et al. 2006). A novel distribution and role of hepatocyte growth factor (HGF), an important factor for migration of somitic myoblasts into the limb, has recently been proposed (Kawashima-Ohya et al. 2011).

These studies provide molecular-level details of the embryonic events surrounding shell formation. All of the molecular plays listed above are either transcription factors or cell-cell signaling molecules that affect cell behaviors, and thus morphogenesis, in many embryonic contexts. Of more interest to the morphologist perhaps, Nagashima and coauthors (2007, 2009, 2013) have presented a very satisfying model of the consequences of carapace growth on the muscles of the shoulder and body wall. These studies use immunohistochemistry and 3-D reconstructions to visualize the details of the folding of body wall and scapular muscles as the CR

expands the carapace to "encompass" (*sensu* Ruckes 1929) the limb girdles. They describe and illustrate how the body wall muscles are folded inward around the pectoral girdle as the carapace expands, and how attachments of the serratus muscle is altered to accommodate this process (Nagashima et al. 2007, 2009, and 2012; Kuratani et al. 2011).

The development of the plastron has also been the subject of study. Gilbert and colleagues have studied stages of shell development with immunohistochemistry using antibodies that label neural crest cells in vertebrate embryos. Based on this labeling, they propose that all the bones of the plastron, as well as the nuchal bone in the carapace, are derived from neural crest cells (Cebra-Thomas et al. 2007; Gilbert et al. 2007). Confirmation with cell-lineage mapping would increase the known repertoire of a remarkable cell population (Hall 1999; Donaghue et al. 2008) as well as illustrate another unusual developmental aspect of the turtle body plan.

Conclusions and Hopes for the Future

Olivier Rieppel has characterized the long-standing debate on turtle origins as divided by two different approaches to the problem (Rieppel and Kearney 2007; Rieppel 2013). The "transformational" approach views character states found in the fossil record as a transformational series. In the case of turtles, this approach is exemplified by current arguments that turtles arose by the gradual association of dermal osteoderms with the underlying endochondral axial elements (Lee 1993, 1997; Scheyer et al. 2008; Joyce et al. 2009; Lyson et al. 2010). The alternative view, which Rieppel calls "emergentist" or "generative," emphasizes the fact that any evolution of adult morphology is the consequence of changes in an ancestral ontogenetic process (Alberch 1980; Oster and Alberch 1982). For turtles, this approach is illustrated in the works cited here from Burke, Gilbert, Kuratani, and their collaborators. Studies of turtle ontogeny reveal developmental mechanisms and molecular genetic "programs" that can suggest early or intermediate stages of evolution. The inductive function of the carapacial ridge, the folding of the muscles of the body wall, and the necessitated changes in the attachments of shoulder muscles are integrated phenomena in a causal chain. The pace of these changes can be imagined as gradual (Nagashima et al. 2012) or as saltatory (Gilbert et al. 2001; Rieppel 2001).

The morphology of *Odontochelys* fits remarkably well with a generative view of morphological evolution (Nagashima et al. 2009; Burke 2009). On the other hand, *Odontochelys* brings a strict transformational view in line with the generative view. *Odontochelys* may be a true "missing link" potentially illustrating an early stage in the transformation of the turtle body plan as suggested by developmental data (fig. 5.5). If this fossil represents the most primitive form of turtle, arguments that the body plan evolved from a heavily armored ancestor are rendered moot, as the dermal carapace predates the ossified carapace. Furthermore, the unique aspects of the body plan would be seen to have evolved in an aquatic environment. However, new finds from older rocks could falsify these ideas by revealing that *Odontochelys* is not primitive, but instead evolved from an older and different turtle. It has been suggested that the lack of ossification may result from paedomorphosis perhaps in response to a newly adopted aquatic environment (Reisz and Head 2008). There is every reason to expect that these ideas will be tested by future discoveries in the fossil record.

An unshakable phylogeny of amniotes, should such a thing be possible, would provide greater context to our understanding of vertebrate evolution and the tempo and mode of morphological change. It is already clear, however, that a strong consensus of turtle roots will not "solve" all the riddles of their morphological evolution. This puzzle will continue to be explored on two fronts. First, the growing understanding of the molecular players that influence the behaviors of embryonic cells that give rise to morphology enable experimental approaches to test hypotheses regarding cellular dynamics in the development of the carapace and plastron. Further, comparison of the regulatory regions of key patterning genes between turtles and nonchelonian tetrapods will further illuminate the proximal targets of this dramatic evolutionary transition. And on the second front, there is great expectation that more intermediate forms will be revealed by new gems from the fossil record, which has illustrated so many of the great transformations in vertebrate history.

* * *

Acknowledgments

I am most grateful to the editors for providing the opportunity to make this contribution. Olivier Rieppel and Scott Gilbert provided excellent comments and corrections that substantially improved this chapter, and Hiroshi Nagashima kindly allowed me to reprint his figures. I am, as ever, indebted to the late Pere Alberch for early encouragement, and I am most grateful to Farish Jenkins who has always been an invaluable mentor. In the summer of 1978, I found myself in Bill Amaral's fossil preparation lab in the basement of the Museum of Comparative Zoology. Before me was a plaster jacket containing the postcrania of a turtle from the Kayenta formation in Arizona. Collected by Farish Jenkins and his field crew, this specimen was eventually named *Kayentachelys aprix* and identified as the oldest known Cryptodire (Gaffney et al. 1987). This chance introduction to the Order Chelonia started me on my intellectual path as an evolutionary morphologist. To paraphrase one of Farish's comments: "the answers are all well and good, but it is really our questions that are most important." I feel extraordinarily lucky to have found the joy of both discovery and teaching he referred to, and I am grateful to recall my first brush with such a remarkable man.

References

Alberch, P. 1980. Ontogenesis and morphological diversification. American Zoologist 20:653–667.

Becker, R. E., R. A. Valverde, and B. I. Crother. 2011. Proopiomelanocortin (POMC) and testing the phylogenetic position of turtles (Testudines). Journal of Zoological Systematics and Evolutionary Research 49:148–159.

Broom, R. 1924. On the classification of the reptiles. Bulletin of the American Museum of Natural History 51:39–65.

Burke, A. C. 1989a. Critical feature in chelonian development: the ontogeny and phylogeny of a unique tetrapod bauplan. Unpublished Ph.D. dissertation. Department of Organismic and Evolutionary Biology. Harvard University Cambridge, MA.

Burke, A. C. 1989b. Development of the turtle carapace: implications for the evolution of a novel bauplan. Journal of Morphology 199:363–378.

Burke, A. C. 1989c. Epithelial-mesenchymal interactions in the development of the chelonian Bauplan. In H. Splechtna and H. Hilgers, eds., Trends in Vertebrate Morphology, Progress in Zoology, vol. 35:206–209. Stuttgart: Fischer Verlag.

Burke, A. C. 1991. The development and evolution of the turtle body plan: inferring intrinsic aspects of the evolutionary process from experimental embryology. American Zoologist 31:616–627.

Burke, A. C. 2009. Turtles . . . again. Evolution & Development 11:622–624.

Burke, A. C., and J. L. Nowicki. 2003. A new view of patterning domains in the vertebrate mesoderm. Developmental Cell 4:159–165.

Cao, Y., M. D. Sorenson, Y. Kumazawa, D. P. Mindell, and M. Hasegawa. 2000. Phylogenetic position of turtles among amniotes: evidence from mitochondrial and nuclear genes. Gene 259:139–148.

Cebra-Thomas, J. A., E. Betters, M. Yin, C. Plafkin, K. Mcdow, and S. F. Gilbert. 2007. Evidence that a late-emerging population of trunk neural crest cells forms the plastron bones in the turtle Trachemys scripta. Evolution & Development 9:267–277.

Chiari, Y., V. Cahais, N. Galtier, and F. Delsuc. 2012. Phylogenomic analyses support the position of turtles as the sister group of birds and crocodiles (Archosauria). BMC Biology 10:65.

Crawford, N. G. B. C. Faircloth, J. E. McCormack, R. T. Brumfield, K. Winker, and T. C. Glenn. 2012. More than 1000 ultraconserved elements provide evidence that turtles are the sister group of archosaurs. Biology Letters 8:783–786.

Donoghue, P. C. J., A. Graham, and R. N. Kelsh. 2008. The origin and evolution of the neural crest. Bioessays 30:530–541.

Fucik, E. 1991. On the value of the orbitotemporal region for the reconstruction of reptilian phylogeny—ontogeny and adult skull analyses of the chelonian skull. Zoologischer Anzeiger 227:209–217.

Fujita, M. K., T. N. Engstrom, D. E. Starkey, and H. B. Shaffer. 2004. Turtle phylogeny: insights from a novel nuclear intron. Molecular Phylogenetics and Evolution 31:1031–1040.

Gaffney, E. S. 1979 Comparative cranial morphology of recent and fossil turtles. Bulletin of the American Museum of Natural History 164:65–375.

Gaffney, E. S. 1980 Phylogenetic relationships of the major groups of amniotes. Pp. 593–610 in The Terrestrial Environment and the Origin of Land Vertebrates. London: Academic Press.

Gaffney, E. S. 1989. Triassic and early Jurassic turtles. Pp. 183–187 in K. Padian, ed., The Beginning of the Age of Dinosaurs: Faunal Change across the Triassic-Jurassic Boundary. Cambridge: Cambridge University Press.

Gaffney, E. S. 1990 The comparative osteology of the Triassic turtle Proganochelys. Bulletin of the American Museum of Natural History 194:1–263.

Gaffney, E. S., J. H. Hutchison, F. A. Jenkins, and L. J. Meeker. 1987. Modern turtle origins—the oldest known Cryptodire. Science 237:289–291.

Gaffney, E. S., and F. A. Jenkins, Jr. 2010. The cranial morphology of Kayentachelys, an Early Jurassic cryptodire, and the early history of turtles. Acta Zoologica 91:335–368.

Gaffney, E. S., T. M. Scheyer, K. G. Johnson, J. Bocquentin, and O. A. Aguilera. 2008. Two new species of the side necked turtle genus, Bairdemys (Pleurodira, Podocnemididae), from the Miocene of Venezuela. Palaeontologische Zeitschrift 82:209–229.

Gilbert, S. F., G. Bender, E. Betters, M. Yin, and J. A. Cebra-Thomas. 2007. The contribution of neural crest cells to the nuchal bone and plastron of the turtle shell. Integrative and Comparative Biology 47:401–408.

Gilbert, S. F., G. A. Loredo, A. Brukman, and A. C. Burke. 2001. Morphogenesis of the turtle shell: the development of a novel structure in tetrapod evolution. Evolution & Development 3:47–58.

Goodrich, E. S. 1916 On the classification of the Reptilia. Proceedings of the Royal Society of London B 89:261–276.

Gregory, W. K. 1946 Pareiasaurs versus placodonts as near ancestors to turtles. Bulletin of the American Museum of Natural History 86:275–326.

Hall, B. K. 1999. The Neural Crest in Development and Evolution. New York: Springer.

Harris, S. R., D. Pisani, D. J. Gower, and M. Wilkinson. 2007. Investigating stagnation in morphological phylogenetics using consensus data. Systematic Biology 56:125–129.

Hedges, S. B. 1994. Molecular evidence for the origin of birds. Proceedings of the National Academy of Sciences USA 91:2621–2624.

Hedges, S. B. 2012. Amniote phylogeny and the position of turtles. BMC Biology 10:64.

Hedges, S. B., K. D. Moberg, and L. R. Maxson. 1990. Tetrapod phylogeny inferred from 18s-Ribosomal and 28s-ribosomal RNA sequences and a review of the evidence for amniote relationships. Molecular Biology and Evolution 7:607–633.

Hedges, S. B., and L. L. Poling. 1999. A molecular phylogeny of reptiles. Science 283:998–1001.

Hyman, L. H. 1942. Comparative Vertebrate Anatomy. Chicago: University of Chicago Press.

Iwabe N. H. Y., Y. Kumazawa, K. Shibamoto, Y. Saito, T. Miyata, and K. Katoh. 2005. Sister group relationship of turtles to the bird-crocodilian clade revealed by nuclear DNA-coded proteins. Molecular Biology and Evolution 22:810–813.

Jaekel, O. Ueber Placochelys n.g. und ihre Bedeutung für die Stammesgeschichte der Schildkrön. Neues Jahrb. Min., vol. 1:127–144.

Janke, A. E. D., D. Erpenbeck, M. Nilsson, and U. Arnason. 2000. The mitochondrial genomes of iguana (Iguana iguana) and the caiman (Caiman crocodylus): implications for amniote phylogeny. Proceedings of the Royal Society B: Biological Sciences 26B:623–631.

Joyce, W. G., and J. A. Gauthier. 2004. Palaeoecology of Triassic stem turtles sheds new light on turtle origins. Proceedings of the Royal Society B: Biological Sciences 271:1–5.

Joyce, W. G., F. A. Jenkins, T. Rowe. 2006. The presence of a cleithra in the basal turtle Kayentachelys. Pp. 93–103 in Fossil Turtle Research. Russian Journal of Herpetology (Suppl.), St. Petersburg.

Joyce, W. G., S. G. Lucas, T. M. Scheyer, A. B. Heckert, and A. P. Hunt. 2009. A thin-shelled reptile from the Late Triassic of

North America and the origin of the turtle shell. Proceedings of the Royal Society B: Biological Sciences 276:507–513.

Joyce, W. G., and J. Sterli. 2012. Congruence, non-homology, and the phylogeny of basal turtles. Acta Zoologica 93:149–159.

Kardong, K. V. 2009. Vertebrates: comparative anatomy, function, evolution. New York: McGraw Hill.

Kawashima-Ohya, Y., Y. Narita, H. Nagashima, R. Usuda, and S. Kuratani. 2011. Hepatocyte growth factor is crucial for development of the carapace in turtles. Evolution & Development 13:260–268.

Kumazawa, Y., and M. Nishida. 1999. Complete mitochondrial DNA sequences of the green turtle and blue-tailed mole skink: statistical evidence for Archosaurian affinity of turtles. Molecular Biology and Evolution 16:784–792.

Kuraku, S., R. Usuda, and S. Kuratani. 2005. Comprehensive survey of carapacial ridge-specific genes in turtle implies co-option of some regulatory genes in carapace evolution. Evolution & Development 7:3–17.

Kuratani, S., S. Kuraku, and H. Nagashima. 2011. Evolutionary developmental perspective for the origin of turtles: the folding theory for the shell based on the developmental nature of the carapacial ridge. Evolution & Development 13:1–14.

Laurin, M., and R. R. Reisz. 1995. A reevaluation of early amniote phylogeny. Zoological Journal of the Linnean Society 113:165–223.

Lee, M. 1993. The origin of the turtle body plan: bridging a famous morphological gap. Science 261:1716–1720.

Lee, M. S. Y. 1997. Pareiasaur phylogeny and the origin of turtles. Zoological Journal of the Linnean Society 120:197–280.

Lee, M. S. Y., T. W. Reeder, J. B. Slowinski, and R. Lawson. 2004. Resolving reptile relationships: molecular and morphological markers. Pp. 451–467 in J. C. M. J. Donoghue, ed., Assembling the Tree of Life. New York: Oxford University Press.

Li, C., X. C. Wu, O. Rieppel, L. T. Wang, and L. J. Zhao. 2008. An ancestral turtle from the Late Triassic of southwestern China. Nature 456:497–501.

Liang, D., R. G. Wu, J. Geng, C. L. Wang, and P. Zhang. 2011. A general scenario of Hox gene inventory variation among major sarcopterygian lineages. BMC Evolutionary Biology 11.

Loredo, G. A., A. Brukman, M. P. Harris, D. Kagle, E. E. Leclair, R. Gutman, E. Denney, E. Henkelman, B. P. Murray, J. F. Fallon, R. S. Tuan, and S. F. Gilbert. 2001. Development of an evolutionarily novel structure: fibroblast growth factor expression in the carapacial ridge of turtle embryos. Journal of Experimental Zoology 291:274–281.

Lyson, T., and S. F. Gilbert. 2009. Turtles all the way down: loggerheads at the root of the chelonian tree. Evolution & Development 11:133–135.

Lyson, T. R., G. S. Bever, B. A. S. Bhullar, W. G. Joyce, and J. A. Gauthier. 2010. Transitional fossils and the origin of turtles. Biology Letters 6:830–833.

Lyson T. R., E. A. Sperling, A. M. Heimberg, J. A. Gauthier, B. L. King, and K. J. Peterson. 2012. MicroRNAs support a turtle + lizard clade. Biology Letters 8:104–107.

Lyson, T. R., and W. G. Joyce. 2012. Evolution of the turtle bauplan: the topological relationship of the scapula relative to the ribcage. Biology Letters doi:10.1098/rsbl.2012.0462.

Mallatt, J., and C. J. Winchell. 2007. Ribosomal RNA genes and deuterostome phylogeny revisited: more cyclostomes, elasmobranchs, reptiles, and a brittle star. Molecular Phylogenetics and Evolution 43:1005–1022.

Mannen, H., and S. S. L. Li. 1999. Molecular evidence for a clade of turtles. Molecular Phylogenetics and Evolution 13:144–148.

Matsuda, Y., C. Nishida-Umehara, H. Tarui, A. Kuroiwa, K. Yamada, T. Isobe, J. Ando, A. Fujiwara, Y. Hirao, O. Nishimura, J. Ishijima, A. Hayashi, T. Saito, T. Murakami, Y. Murakami, S. Kuratani, and K. Agata. 2005. Highly conserved linkage homology between birds and turtles: bird and turtle chromosomes are precise counterparts of each other. Chromosome Research 13:601–615.

Meyer, A., and R. Zardoya. 2003. Recent advances in the (molecular) phylogeny of vertebrates. Annual Review of Ecology Evolution and Systematics 34:311–338.

Moustakas, J. E. 2008. Development of the carapacial ridge: implications for the evolution of genetic networks in turtle shell development. Evolution & Development 10:29–36.

Müller, J. 2003. Early loss and multiple return of the lower temporal arcade in diapsid reptiles. Naturwissenschaften 90:473–476.

Nagashima, H., S. Kuraku, K. Uchida, Y. K. Ohya, Y. Narita, and S. Kuratani. 2007. On the carapacial ridge in turtle embryos: its developmental origin, function and the chelonian body plan. Development 134:2219–2226.

Nagashima, H., S. Kuraku, K. Uchida, Y. Kawashima-Ohya, Y. Narita, and S. Kuratani. 2012. Body plan of turtles: an anatomical, developmental and evolutionary perspective. Anatomical Science International 87:1–13.

Nagashima, H., F. Sugahara, M. Takechi, R. Ericsson, Y. Kawashima-Ohya, Y. Narita, and S. Kuratani. 2009. Evolution of the turtle body plan by the folding and creation of new muscle connections. Science 325:193–196.

Nagashima H. K. S., K. Uchida, Y. Kawashima-Ohya, Y. Narita, and S. Kuratani. 2013. Origin of the turtle body plan—the folding theory to illustrate turtle-specific developmental repatterning. In D. B. Brinkman, P. A. Holroyd, and J. D. Gardner, eds., Morphology and Evolution of Turtles: Origin and Early Diversification. Dordrecht: Springer.

Nilsson, T. 1945. The structure of the cleithrum in Plagiosaurids and the descent of chelonia. Arkiv för Zoologie 37A:1–18.

Ohya, Y. K., S. Kuraku, and S. Kuratani. 2005. Hox code in embryos of Chinese soft-shelled turtle Pelodiscus sinensis correlates with the evolutionary innovation in the turtle. Journal of Experimental Zoology Part B: Molecular and Developmental Evolution 304B:107–118.

Ohya, Y. K., R. Usuda, S. Kuraku, H. Nagashima, and S. Kuratani. 2006. Unique features of Myf-5 in turtles: nucleotide deletion, alternative splicing, and unusual expression pattern. Evolution & Development 8:415–423.

Oster, G., and P. Alberch. 1982. Evolution and bifurcation of developmental programs. Evolution 36:444–459.

Owen, R. 1849. On the development and homologies of the carapace and plastron of the chelonian reptiles. Philosophical Transactions of the Royal Society of London. Series B: Biological Sciences 139:151–171.

Platz, J. E., and J. M. Conlon. 1997. Reptile relationships turn turtle . . . and turn back again. Nature 389:246.

Reisz, R. R., and J. J. Head. 2008. Turtle origins out to sea. Nature 456:450–451

Rest, J. S., J. C. Ast, C. C. Austin, P. J. Waddell, E. A. Tibbetts, J. M. Hay, and D. P. Mindell. 2003. Molecular systematics of primary reptilian lineages and the tuatara mitochondrial genome. Molecular Phylogenetics and Evolution 29:289–297.

Rieppel, O. 2000. Turtles as diapsid reptiles. Zoologica Scripta 29:199–212.

Rieppel, O. 2001. Turtles as hopeful monsters. Bioessays 23:987–991.

Rieppel, O. 2013. The evolution of the turtle shell. In D. B. Brinkman, P. A. Holroyd, and J. D. Gardner, eds., Morphology and Evolution of Turtles: Origin and Early Diversification. Dordrecht: Springer.

Rieppel, O., and M. Kearney. 2002. Similarity. Biological Journal of the Linnean Society 75:59–82.

Rieppel, O., and M. Kearney. 2007. The poverty of taxonomic characters. Biology & Philosophy 22:95–113.

Rieppel, O., and R. R. Reisz. 1999. The origin and early evolution of turtles. Annual Review of Ecology and Systematics 30:1–22.

Rieppel, O., and M. deBraga. 1996. Turtles as diapsid reptiles. Nature 384:453–455.

Romer, A. 1956. Osteology of the Reptile. Chicago: University of Chicago Press.

Romer, A. S. 1968. Notes and Comments on Vertebrate Paleontology. Chicago: University of Chicago Press.

Romer, A. S., and T. S. Parsons. 1977. The Vertebrate Body. Philadelphia: W. B. Saunders, Co.

Rougier, G. W., M. S. Delafuente, and A. B. Arcucci. 1995. Late Triassic turtles from South-America. Science 268:855–858.

Ruckes, H. 1929. Studies in Chelonian osteology, part II: the morphological relationships between the girdles, ribs and carapace. Annals of the New York Academy of Sciences 31:81—120.

Scheyer, T. M., B. Brullmann, and M. R. Sánchez-Villagra. 2008. The ontogeny of the shell in side-necked turtles, with emphasis on the homologies of costal and neural bones. Journal of Morphology 269:1008–1021.

Shearman, R. M., and A. C. Burke. 2009. The lateral somitic frontier in ontogeny and phylogeny. Journal of Experimental Zoology Part B, Molecular and Developmental Evolution 312:603–612.

Shen, X. X., D. Liang, J. Z. Wen, and P. Zhang. 2011. Multiple genome alignments facilitate development of NPCL markers: a case study of tetrapod phylogeny focusing on the position of turtles. Molecular Biology and Evolution 28:3237–3252.

Sterli, J. 2008. A new, nearly complete stem turtle from the Jurassic of South America with implications for turtle evolution. Biology Letters 4:286–289.

Sterli, J., and W. G. Joyce. 2007. The cranial anatomy of the Early Jurassic turtle Kayentachelys aprix. Acta Palaeontologica Polonica 52:675–694.

St. John, J. A., E. L. Braun, S. R. Isberg, L. G. Miles, A. Y. Chong, J. Gongora, P. Dalzell, C. Moran, B. Bed'Hom, A. Abzhanov, et al. 2012. Sequencing three crocodilian genomes to illuminate the evolution of archosaurs and amniotes. Genome Biology 13. 10.1186/gb-2012-13-1-415.

Theissen, G. 2009. Saltational evolution: hopeful monsters are here to stay. Theory in Biosciences 128:43–51.

Tzika, A. C., R. Helaers, G. Schramm, and M. C. Milinkovitch. 2011. Reptilian-transcriptome v1.0, a glimpse in the brain transcriptome of five divergent Sauropsida lineages and the phylogenetic position of turtles. EvoDevo 2:19.

Vincent, C., M. Bontoux, N. M. Le Douarin, C. Pieau, and A. H. Monsoro-Burq. 2003. Msx genes are expressed in the carapacial ridge of turtle shell: a study of the European pond turtle, Emys orbicularis. Development Genes and Evolution 213:464–469.

Wang, Z., J. Pascual-Anaya, A. Zadissa, W. Li, Y. Niimura, Z. Huang, C. Li, S. White, Z. Xiong, D. Fang, et al. 2013. The draft genomes of soft-shell turtle and green sea turtle yield insights into the development and evolution of the turtle-specific body plan. Nature Genetics 45:701–706.

Werneburg, I. 2012. Temporal bone arrangements in turtles: an overview. Journal of Experimental Zoology Part B: Molecular and Developmental Evolution 318:235–249.

Werneburg, I., and M. R. Sánchez-Villagra. 2009. Timing of organogenesis support basal position of turtles in the amniote tree of life. BMC Evolutionary Biology 9.

Wiens, J. J., C. A. Kuczynski, T. Townsend, T. W. Reeder, D. G. Mulcahy, and J. W. Sites. 2010. Combining phylogenomics and fossils in higher-level squamate reptile phylogeny: molecular data change the placement of fossil taxa. Systematic Biology 59:674–688.

Yntema, C. L. 1970. Extirpation experiments on the embryonic rudiments of the carapace of Chelydra serpentina. Journal of Morphology 132:235–244.

Zardoya, R., and A. Meyer. 1998. Complete mitochondrial genome suggests diapsid affinities of turtles. Proceedings of the National Academy of Sciences USA 95:14226–14231.

6

Anatomical Transformations and Respiratory Innovations of the Archosaur Trunk

Leon Claessens*

Introduction

The Archosauria, a group of tetrapods that dominated the terrestrial ecosystem for the majority of the Mesozoic era, provide a unique window into anatomical transformations of the trunk. Birds and crocodylians are the only living representatives of the Archosauria, a once diverse clade that included dinosaurs, pterosaurs, and other extinct groups encompassing an enormous range of anatomical disparity and functional diversity (fig. 6.1). Different archosaur groups successfully occupied a variety of terrestrial, aquatic, and aerial environments. Archosauria include, among others, the largest vertebrates that ever walked on land, the sauropods, as well as some of the smallest vertebrates that ever lived, such as hummingbirds. Pterosaurs and birds were the first two vertebrate clades to evolve active flapping flight, independently, in the Triassic and Jurassic periods, respectively. The clade Archosauria is also marked by the evolution of a diversity of respiratory mechanisms, including the hepatic piston pump of modern crocodylians and the lung-air sac system of modern birds. Because extant archosaurs offer examples of highly specialized respiratory systems serving both cold-blooded (ectothermic) and warm-blooded (endothermic) metabolic physiologies, the clade presents an

* Department of Biology, College of the Holy Cross

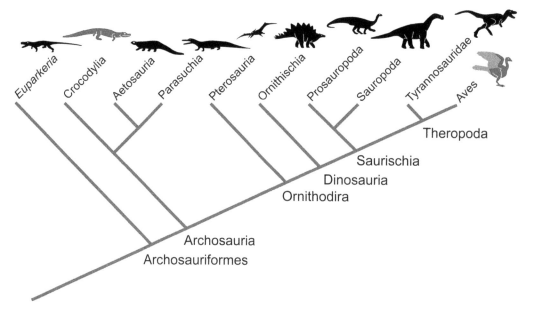

FIG. 6.1 Simplified phylogeny of the Archosauriformes, illustrating evolutionary relationships and the large degree of disparity in body size and form in this clade. Icons for the extant clades Crocodylia and Aves are shown in gray, and icons for extinct clades are shown in black.

unparalleled opportunity to explore and illustrate evolutionary pathways in respiratory design and function.

Similar to other tetrapods, the archosaur trunk houses the major organs of digestion, circulation, and respiration. The tetrapod trunk anchors the limbs, and assists with generating or withstanding forces associated with locomotion, feeding, reproduction, and other vital functions. The cranium is separated from the trunk due to the disassociation and subsequent caudad displacement of the pectoral girdle and the associated evolution of a neck, which allows the cranium a degree of mobility independent of the rest of the body. The presence of a tail, which often functions in locomotion, balance, and display, is plesiomorphic for Tetrapoda. It may be easy to underestimate the anatomical disparity in tetrapod trunk design when considering the overall similarity in the specific organs situated in the trunk and the number and location of appendages, neck, and tail that extend from the central core of the body. Indeed, the complexity in trunk design and function is underscored by centuries of research and debate on seemingly simple yet still not fully resolved questions such as the exact function of the ribs and intercostal musculature during respiration (see also Brainerd chapter in this volume) or the evolution and development of the axial skeleton (e.g., Gadow 1933; Burke et al. 1995; Mansfield and Abzhanov 2010).

Shifts in locomotor function as tetrapod clades diversified and expanded into new environments likely altered constraints on thoracic structure and function, which in turn may have allowed for changes in pulmonary organization and ventilatory mechanisms. Increased ossification and a reduction in the number of elements that make up the ribcage preceded the evolution of flight in both pterosaurs and birds, for instance, and would have increased the structural integrity of the trunk and likely aided in accommodating the increased stresses transmitted through the trunk during flight. In addition, these changes in trunk skeletal morphology placed constraints on the skeletal aspiration pump, and may have played an important role in the evolution of the respiratory system.

The examination of morphological and functional transformations of the archosaur trunk presented here is not intended to be comprehensive, but rather to provide a review of the fundamental organization of the musculoskeletal and pulmonary systems, to illustrate some lingering questions within the clade, and to explore exciting opportunities for further study. This chapter will hopefully serve as a starting point for future work on the evolution of the archosaur trunk and its many functions, and the general insights in vertebrate evolution that might be gleaned from this functionally and morphologically diverse clade.

Crocodylians and Birds: The Only Surviving Archosaurs

Crocodylians (alligators, caimans, crocodiles, and gharials) are the only living representatives of the clade Crocodylomorpha that originated in the early Mesozoic. Extant Crocodylia (a term used to indicate the crown group in this chapter) are all semiaquatic quadrupeds and exhibit only a modest degree of anatomical variation, mostly in body size and in the craniomandibular region of the body. The crocodylian trunk is elongate and near-cylindrical, and slightly wider than deep (fig. 6.2, top). Crocodylians possess the most complex multicameral lung of any reptile (i.e., non-avian sauropsid). Moreover, this group possesses a heart with fully separated atria and ventricles similar to the condition in extant birds and mammals (e.g., Perry 1990; Farmer 1999; Seymour et al. 2004).

By comparison with crocodylians, extant birds occupy a large diversity of habitats and exhibit an enormous degree of anatomical variation, particularly with regard to body size, and in the shape and function of both the appendicular and the craniomandibular systems. Birds are the second major vertebrate clade to have colonized the aerial environment (after the Pterosauria), and most extant birds engage in active flapping flight. The appendicular apparatus of birds is modular with fore and hind limb modules that can function

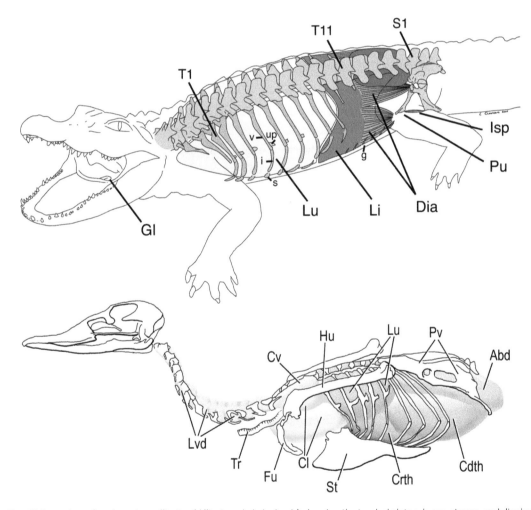

FIG. 6.2 Top, Oblique view of an American alligator (*Alligator mississippiensis*) showing the trunk skeleton, lungs, viscera, and diaphragmatic muscle; and bottom, the skeleton and pulmonary air sac system of a duck (*Anas crecca*), highlighting the extent of air sacs throughout the body. Abbreviations: Abd, abdominal air sac; Cdth, caudal thoracic air sac; Cl, clavicular air sac; Crth, cranial thoracic air sac; Cv, cervical air sac; Dia, diaphragmaticus muscle; Fu, furcula; Gl, glottis; Hu, humerus; Isp, ischiopubis muscle; Li, liver; Lu, lung; Lvd, lateral vertebral diverticula; Pu, pubis; Pv, pelvis; S1, first sacral vertebra, St, sternum; T1, first thoracic vertebra; T11, eleventh thoracic vertebra; Tr, trachea. Top, Modified from Claessens (2009a); bottom, from O'Connor and Claessens (2005) .

independently (Gatesy and Dial 1996); in most species the forelimbs are used primarily during flight, with the hind limbs used for bipedal terrestrial locomotion (but see the chapter by Dial et al. in this volume). A large diversity of additional wing and hind limb functions can be observed in different species (e.g., paddling, grasping, pushing off, combat). In birds, the trunk is less elongate than in crocodylians, and generally deeper than wide (fig. 6.2, bottom). Birds possess a fixed-volume, non-compliant lung that experiences a near-constant unidirectional flow of air during ventilation. Airflow patterns are maintained by the finely tuned interactions of an extensive network of bellows-like extrapulmonary air sacs and valving within the tubular portions of the respiratory tract (Hazelhoff 1951; Powell 2000; Maina 2005).

Extinct Archosaurs: Dinosaurs, Pterosaurs, and Many Others

Extinct archosaurs comprise a variety of groups with highly disparate anatomy and, by inference, paleobiology (figs. 6.1, 6.2). Basal archosauriforms include relatively small terrestrial carnivores such as *Euparkeria* and *Proterosuchus*, and large terrestrial carnivores such as *Erythrosuchus*. Basal crocodylomorphs display a much greater taxonomic and anatomical diversity than survives in crocodylians today, and include highly cursorial terrestrial forms such as *Terrestrisichus* as well as fully aquatic marine forms such as the Thalattosuchia. The clade Archosauria also contains the Sauropoda, the largest land vertebrates to have ever roamed the earth, and large barrel-shaped armored herbivores such as the aetosaurs and the ankylosaurs. The clade also includes the Pterosauria, the first vertebrate group that evolved active flapping flight, as well as the narrow, deep-bodied feathered theropod ancestors to birds, and a multitude of other taxa (figs. 6.1, 6.2).

Reconstructing the shape of the trunk in extinct taxa is often a complicated task. Many extinct tetrapods are known only from incomplete or deformed fossil skeletal material. Even for a specimen with relatively complete, three-dimensional preservation of the skeleton, crucial information regarding the shape of the ribcage is frequently missing: structures that are often cartilaginous such as the sternal ribs and the sternum are rarely preserved, except in forms where these structures are fully ossified (e.g., pterosaurs, derived non-avian theropods,

and birds), or under exceptional fossilization conditions, leading to uncertainty regarding the width and depth of the ventral ribcage. Gastralia (dermal rib-like ossifications located in the ventral abdominal region) are also often not included in museum mounts or other reconstructions of taxa that possess these abdominal ossifications.

Historically, many skeletal reconstructions consist of specimens in which the thoracic skeleton is poorly reconstructed. For example, the vertebral ribs on many theropod dinosaur museum mounts or reconstructions are not in articulation with the corresponding joint surfaces on the vertebral column, but instead project perpendicularly from the axial skeleton (fig. 6.3A, B). Obviously such extra girth would add significant weight to mass reconstructions and greatly influence scientific interpretations related to energetic and musculoskeletal function and constraints (Hutchinson et al. 2011).

Transformations of the Archosaur Ribcage

The trunk skeleton of extant crocodylians is differentiated into a rib-bearing thoracic and a rib-free lumbar region (fig 6.4A, B). The costal apparatus of crocodylians consists of metameric tripartite rib segments, which are composed of ossified vertebral ribs and cartilaginous intermediate and sternal ribs. Vertebral ribs are bicapitate and articulate with the diapophysis and parapophysis of each thoracic vertebra. The sternal ribs articulate directly with the sternum or, further caudad, with a xiphisternum via monocondylar joints. The cartilaginous intermediate ribs span between vertebral and sternal ribs, providing two additional monocondylar articulations per crocodylian rib segment (one articulation on each side of the body) in comparison to a bipartite metameric rib-segment arrangement. This increases the degrees of freedom for thoracic mobility along the ventral and lateral periphery of the trunk. In contrast to crocodylians, birds lack a rib-free lumbar region and exhibit a bipartite ribcage that consists of bicapitate vertebral ribs and sternal ribs, both of which are fully ossified. The avian thorax is also relatively short, and it is thought that the cranial portion of the avian synsacrum incorporates vertebrae that are homologous with the crocodylian lumbar vertebral series (Gegenbaur 1871; Burke et al. 1995). The sternum of extant birds is large, and flying birds generally possess a

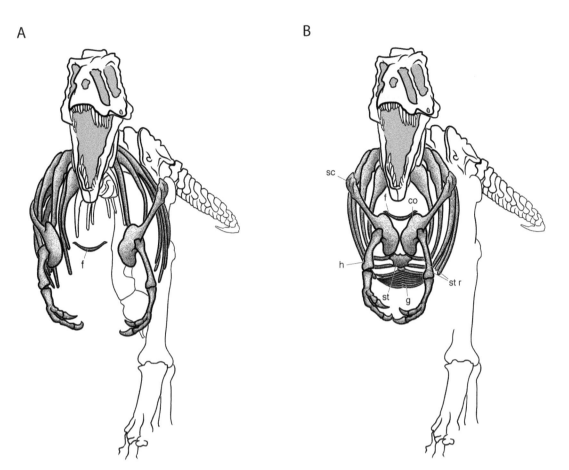

A B

FIG. 6.3 A, Example of an *Albertosaurus* skeletal mount with vertebral ribs that are nearly perpendicular to the vertebral column, and a pectoral girdle that is spaced too far apart in the midline. Gastralia are lacking in this reconstruction. The furcula has been drawn in place to show the extreme distance between the pectoral girdles. Incorrect positioning of the pectoral girdle may enhance the barrel-chested appearance of the reconstruction. Drawing based upon existing museum mount. B, Reconstruction of *Albertosaurus* with the vertebral ribs inclined caudally, resulting in a narrower profile. The scapulocoracoids have been repositioned closer to the midline, which enables contact of both arms of the furcula with the scapulocoracoids. Gastralia are positioned in the ventral abdominal area. In this reconstruction, a cartilaginous sternum and sternal ribs have also been added, although these structures are not known from the fossil record for *Albertosaurus*. Abbreviations: co, coracoid; f, furcula; g, gastralia; h, humerus; sc, scapula; st, sternum; st r, sternal ribs.

large median sternal keel (fig. 6.4C). The articulation between the vertebral and sternal ribs is monocondylar, but the articulation between the sternal ribs and the sternum is bicondylar, which further constrains costosternal movement by limiting rotation of the sternal rib (Claessens 2009b).

The tripartite construction of the crocodylian ribcage is probably plesiomorphic for archosaurs, since intermediate ribs are also present in the basal lepidosauromorph tuatara (*Sphenodon*; Günther 1867; Claessens 2009a), as well as several more derived lepidosauromorph clades (Seemann 1926). Interestingly, developmental evidence also indicates that there are three distinct parent tissues contributing to the development of the costal apparatus in chicks, with ventral-axial tissues being responsible for the formation of the proximal vertebral rib, surface ectoderm for the formation of the distal vertebral rib, and lateral plate mesoderm for the sternal rib (Aoyama et al. 2005). Moreover, whereas cartilaginous intermediate ribs or sternal ribs are not well known from the fossil record because they rarely fossilize, bipartite ribcages consisting of ossified vertebral and ossified sternal ribs are preserved in the fossil record and likely mark the transition from the plesiomorphic tripartite ribcage anatomy. The first occurrences of fully ossified bipartite archosaur ribcages are observed independently in Triassic pterosaurs and in some Cretaceous maniraptoriform theropods such as *Velociraptor*, which share common ancestry with birds. Ossified sternal ribs are not known in any other major archosaur clades.

A

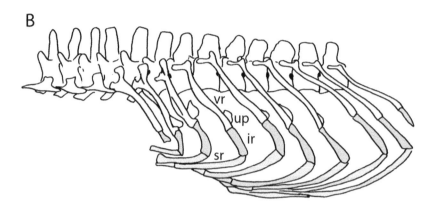

B

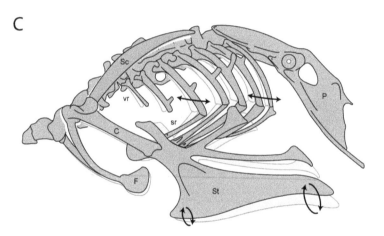

C

FIG. 6.4 A, Lateral view of the skeleton of the American alligator, showing the morphology of the trunk skeleton and the limited caudad extent of the shoulder girdle. B, Detailed view of the alligator thorax showing the vertebral ribs, cartilaginous intermediate and sternal ribs, and uncinate processes. C, Lateral view of the trunk skeleton of a galliform bird showing the vertebral ribs, sternal ribs, sternum, uncinate processes, and the elongate, caudally oriented scapula. Abbreviations: C, coracoid; F, furcula; ir, intermediate rib; P, pelvis; Sc, scapula; sr, sternal rib; St, sternum; vr, vertebral rib; up, uncinate process. A, Modified after Cong (1998); B, modified after Brühl (1862); C, from Claessens (2009b).

In extant crocodylians, the location of the parapophysis experiences an abrupt change in location from the cranioventral margin of the centrum on the second thoracic vertebra to the base of the transverse process in the third thoracic vertebra (fig. 6.4B; Claessens 2009a). A similar relatively abrupt transition in the location of the vertebral parapophysis can be observed in basal archosauriforms such as *Euparkeria* and phytosaurs, and therefore likely represents the basal archosaur condition (Claessens 2005). In crocodylians, the parapophysis is positioned increasingly further laterally on the transverse process throughout the thoracic vertebral column (Claessens 2009a; Schachner et al. 2011).

In birds, the parapophysis does not experience this abrupt change in position, and instead remains located on the centrum, immediately ventral to the diapophysis, throughout the thoracic vertebral column (fig 6.4C; Claessens 2009b).

Differences in the movements of the vertebral ribs during ventilation appear to reflect the differences in parapophyseal location between alligators and birds: in birds all vertebral ribs follow a "bucket-handle" motion, whereas in crocodylians only the cranial vertebral ribs of the thorax move according to this pattern. The caudal vertebral ribs of crocodylians have a distinct pump handle component to their movement (Claessens

2009a, 2009b). Birds thus exhibit a more uniform lateral widening of the ribcage upon inspiration in comparison to crocodylians (Claessens 2009a, 2009b). The crocodylian ribcage is capable of generating impressive volume changes of the trunk, in part due to the extra joint provided by the highly mobile cartilaginous intermediate rib. High compliance of the two cartilaginous distal rib elements likely also contributes to these volume dynamics (Claessens 2009a). The caudal sternal ribs of both crocodylians and extant birds are longer than the cranially positioned sternal ribs. For example, in extant birds, the caudal sternal ribs may be more than 10 times as long as the cranial sternal ribs (Claessens 2009b). Due to the longer moment arm of the caudal sternal ribs, the caudal region of the sternum experiences the greatest amount of displacement during lung ventilation, especially in birds (Claessens 2009a, 2009b).

The absence of the intermediate rib in the avian thorax provides fewer degrees of freedom for the movement of the avian thorax in comparison to the tripartite ribcage of crocodylians and, most likely, basal archosaurs. The ossification of the sternal ribs of birds (which provide a reduction in compliance in comparison to cartilaginous ribs), and the development of more constrained bicondylar distal sternal rib joints further decreases the relative degree of freedom of movement of the avian thorax. The reduction in the total degrees of freedom of movement of the avian thorax has been hypothesized to provide a much greater control over the timing, location, and amount of air sac expansion and compression (Claessens 2009b).

The avian ribcage is similar to that of pterosaurs in several structural aspects, including the absence of intermediate ribs and the ossification of the sternum and sternal ribs, which may be the result of similar selective pressures related to locomotor and respiratory evolution (Claessens et al. 2009).

Shifts in Locomotor Function, Constraints on Thoracic Design, and Respiratory Innovation

It seems likely that the lifting of existing constraints on thoracic structure and function, and the imposition of new constraints generated through novel locomotor mechanisms, played a major part in the morphological transformations of the avian thorax from the basal archosaur condition. The changes in the reduction of freedom of movement of the avian thorax, and changes in the parapophyseal arrangement in birds, may have evolved due to changes in locomotor system rather than selective pressures for improved gas exchange of the avian lung-air sac pulmonary apparatus.

In quadrupedal tetrapods with a sprawling, non-parasagittal gait, there is a conflict between the locomotor and respiratory functions of the hypaxial musculature (Carrier 1987). It has been argued that the evolution of bipedality in Theropoda, the ancestors to birds, freed the thorax from the design constraints inherent in its previous dual locomotor-respiratory function (Carrier 1987). In non-avian theropods, the release of the forelimbs from their function in locomotion would have enabled aspiratory movements of the theropod ribcage during locomotion.

Major changes in the stresses transmitted through the trunk associated with evolution of wing-powered aerial locomotion in the avian descendants of theropods may have, in turn, placed critical new constraints upon the morphology of the costovertebral joints of the ribcage. The ground impact force of the forelimb is transmitted onto, and dissipates over, relatively few cranially positioned vertebral ribs in crocodylians or basal archosauriforms such as *Euparkeria*. This is due to the limited caudad extent of the shoulder girdle. However, forces the trunks of birds are subjected to during flapping flight are dissipated over almost the entire length of the thoracic vertebral ribs due to the caudad elongation of the scapula (fig 6.4C). Furthermore, the placement of the parapophyseal articulation (for the costal capitulum) ventral to the diapophyseal facet of the transverse process (for the articulation of the costal tuberculum) provides a skeletal strut to help resist the forces transmitted onto the ribcage by the pectoral limb. The ventrally positioned parapophyses in the caudal thoracic region of birds support the caudally extended scapula in birds. The ossification of the sternal ribs, loss of the intermediate ribs, and bicondylar rib articulation with the sternum also confer greater structural integrity to the thorax, all of which would be a great asset in anchoring the well-developed pectoral musculature of flying birds and withstanding the compression of the ribcage during pectoral muscle contraction during the flight stroke.

In light of the potential importance of the greater structural integrity of the avian thorax in withstanding stress induced by flight, it is interesting to note that the Pterosauria, the first vertebrate group to evolve active flapping flight, is also the first group of vertebrates to display a bipartite, fully ossified ribcage (Claessens et al. 2009). The parallel evolution of bipartite, fully ossified ribcages in pterosaurs and birds may point to the greater degree of structural integrity required for the thoracic skeleton during flight. This interpretation is strengthened by the fact that the cranial vertebral ribs of the largest pterodactyloid taxa (e.g., *Pteranodon*) are completely fused to the vertebral column. This no doubt forms an impediment to respiratory movement of the ribcage but provides maximal structural integrity to withstand flight-induced stresses on the body skeleton. The biomechanical properties and function of the trunk skeleton in relationship to locomotor function and respiration is only poorly understood and continues to be a fertile ground for future research.

Interestingly, both elongation of the scapula toward the middle of the thorax and more ventrally positioned parapophyses throughout the middle thoracic vertebral column are already observed in non-avian theropod taxa. Bipartite ribcages and ossified sterna are also observed in derived non-avian maniraptoriform theropods, highlighting the possibility that the transformations of the avian ribcage may have evolved prior to the evolution of flight and the potential complexity of the function of the ribcage in extinct maniraptoriform theropods.

Lung Ventilation with Fused Vertebral Ribs

A few archosaur clades, including several large pterodactyloid pterosaurs (e.g., *Pteranodon*) as well as some ankylosaur taxa (e.g., *Ankylosaurus*), exhibit fusion of thoracic vertebral ribs to the vertebral column, rendering specific vertebral ribs immobile. The immobility of the proximal thorax has been used to argue for the presence of a diaphragmatic breathing system in Pterosauria (Carrier and Farmer 2000; Ruben et al. 2003), similar to crocodylians, and has been cited as precluding costal aspiration in ankylosaurs (e.g., Schachner et al. 2011). The reconstruction of extremely short, equal length sternal ribs in *Pteranodon* (e.g., Bennett 2001) seemed to indicate an immobile ribcage in this taxon; however, the discovery of sternal ribs that increase in length caudally in *Pteranodon* and other pterosaurs shows that even if the proximal ribcage may have been immobile in some taxa, the ventral ribcage would have still been capable of significant displacement (Claessens et al. 2009). Similarly, in ankylosaurs the ventral ribcage may have been capable of significant displacement, and fusion of vertebral ribs to vertebrae cannot be interpreted as evidence for thoracic immobility or the presence of accessory breathing systems similar to the crocodylian diaphragmaticus muscle (see below).

Accessory Breathing Mechanisms

Accessory breathing mechanisms, derived musculoskeletal systems that lower the energetic cost of lung ventilation or increase the potential for gas exchange, occur throughout Tetrapoda (e.g., Owerkowicz et al. 1999; Farmer and Carrier 2000; Claessens 2004a, 2004b; Codd et al. 2005; Brainerd and Owerkowicz 2006). Four of the anatomical systems that have evolved respiratory functions in specific archosaur clades are discussed below: costal processes, the pelvis, the diaphragmaticus muscle, and the gastralial apparatus.

Uncinate and Other Costal Processes

Both crocodylians and extant birds (except, apparently, the anseriform clade Anhimidae) have processes that project caudally off the vertebral ribs (fig. 6.4A–C). The uncinate processes of birds have long been hypothesized to provide a mechanical advantage for the thoracic musculature during inspiration based on geometric models (e.g., Sibson 1846; Zimmer 1935; Tickle et al. 2007), and recent electromyographic studies have confirmed active recruitment of the external oblique and appendicocostalis musculature in the Canada goose during lung ventilation (Codd et al. 2005). In extant birds, uncinate process length scales with resting metabolic rate, further highlighting the mechanical advantage of the processes for respiratory movements of the avian ribcage (Tickle et al. 2009). In addition to conferring a mechanical advantage for respiratory muscles, long uncinate processes that overlap one or multiple vertebral ribs provide additional support against compression of the ribcage (for instance, during the flight stroke or during dives). Ossified avian-like uncinate processes are present in at least some maniraptoriform theropods, and have been interpreted as evidence for the avian-style lung ventilation in this group (Tickle et al. 2012). The question whether the uncinate processes of theropods (including birds) are homologous with the vertebral processes of crocodylians or with the intercostal plates of ornithischian dinosaurs is a topic of debate. Developmental evidence indicates a different ossification mechanism of these structures in ornithischian dinosaurs, but the function of these processes in body support or respiration may have been analogous (e.g., Codd et al. 2008; Boyd et al. 2011).

Pterosaurs lack uncinate processes extending from the vertebral ribs, but in several rhamphorhynchoid and pterodactyloid pterosaurs, distinct processes extend from the sternal ribs that appear to align with intercostal muscle fiber directions (Claessens et al. 2009). Also, caudally projecting extensions of the intermediate ribs occur in large crocodylians, and large caudal and cranial extensions of the sternal ribs are present in the basal lepidosauromorph *Sphenodon*. It seems likely that these and other processes extending from the different rib segments of sauropsids functioned in providing a mechanical advantage for the intercostal musculature during respiration, similar to the uncinate processes of birds.

Pelvic Girdle Aspiration Breathing

The pelvic morphology of crocodylians is unique among extant tetrapods due to the presence of a mobile articulation between the pubis and the anterior process of the ischium (fig. 6.2, top; Claessens and Vickaryous 2012). Electromyographic (Farmer and Carrier 2000) and cineradiographic studies (Claessens 2004a, 2009a) have confirmed that the crocodylian pubis rotates caudoventrally during inspiration, thereby increasing trunk and thus pulmonary volume. The gradual exclusion of the pubis from the acetabulum and the evolution of the mobile ischial synchondrotic joint is well documented in the fossil record, and can be identified as a crocodyliform synapomorphy (Claessens 2004a; Claessens and Vickaryous 2012). The morphology of the prepubis of the Pterosauria strongly resembles the morphology of the crocodylian pubis. It seems likely that the pterosaurian prepubis could have functioned in changing trunk volume and visceral position similar to the crocodyliform pubis (Carrier and Farmer 2000; Claessens 2004a).

Involvement of the pelvis in aspiration breathing has also been suggested for birds, via a mechanism involving elevation of the pelvis through dorsiflexion at the joint between the caudalmost thoracic vertebra and the synsacrum (Baumel et al. 1990). Based upon the apparent existence of pelvic breathing mechanisms in both groups of extant archosaurs and a review of fossil archosaur pelvic morphology, Carrier and Farmer (2000) hypothesized that pelvic breathing mechanisms might be a basal archosaur characteristic. Since this time, cineradiographic study of ventilatory skeletal kinematics in emus, tinamous, and guinea fowl failed to record any vertebral movement between the thoracic vertebral column and the synsacrum (Claessens 2004a). Thus, based on both (1) the anatomical and functional disparity in the pelvic mechanisms hypothesized to be involved in aspiration breathing and (2) the apparent lack of pelvic dorsiflexion during inspiration in birds, it is most parsimonious to interpret the pelvic respiratory mechanisms of Archosauria as independently derived specializations (Claessens 2004a).

The Diaphragmatic Breathing Pump

Another unique component of the crocodylian aspiration pump is the diaphragmaticus, a muscle that originates from the pelvis and caudalmost gastralia, and inserts on a fascia surrounding the liver (Gadow 1882; Claessens 2009a). Contraction of the diaphragmaticus translates the liver caudad, thereby increasing the volume of the pleural cavity (e.g., Farmer and Carrier 2000; Claessens 2009a; Munns et al. 2012). This mechanism has been referred to as a "hepatic piston" (Gans and Clark 1976). The basal thoracic breathing pump of crocodylians is capable of supporting crocodylian metabolic demands at rest (Munns et al. 2012), but visceral translation through diaphragmatic muscle contraction can contribute significantly to overall tidal volume (Claessens 2009a).

There is no evidence to suggest that pubic mobility is a prerequisite for visceral translation in extant crocodylians (Claessens 2004a, 2009a). No osteological correlates for the diaphragmatic muscle have been identified (Claessens 2009a; Claessens and Vickaryous 2012), although evolution of a rib-free lumbar region in crocodylians may aid in accommodating the caudally displaced viscera during the non-ventilatory period (analogous to mammals). The conflict between the respiratory and the locomotor function of the intercostal muscles in extant crocodylians, and the ability of the diaphragmaticus to function during locomotion (Carrier 1987) as well as to sustain elevated metabolic demand (Munns et al. 2012) suggests a possible origin of this muscle among cursorial (terrestrial) basal crocodylomorphs (Carrier and Farmer 2000; Claessens and Vickaryous 2012). Alternatively, the use of the diaphragmaticus and pubic musculature during aquatic locomotion supports the possible origin of the hepatic piston system as an aquatic adaptation (Uriona and Farmer 2008; Claessens 2009a; Claessens and Vickaryous 2012), highlighting once again the possibility for locomotor and respiratory system based selection for changes in trunk anatomy. In absence of evidence to the contrary, Claessens and Vickaryous (2012) tentatively place the evolution of the hepatic piston breathing system at the point of origin of pubic mobility.

The possible existence of a diaphragmatic breathing pump has been hypothesized for other extinct archosaur clades, including theropods and pterosaurs (Ruben et al. 1997; Carrier and Farmer 2000; Ruben et al. 2003). Purported evidence for the sharp and abrupt subdivision of the theropod coelomic cavity into a caudal peritoneum bordered by the liver and a cranial pleural cavity (Ruben et al. 1997; Ruben et al. 1999) has since been proven to be an artifact of preservation (e.g., Currie and Chen 2001; Dal Sasso and Maganuco 2011). In the absence of conclusive osteological correlates for the diaphragmaticus muscle, evidence for the existence of this breathing system outside of Crocodylomorpha is not supported by the extant phylogenetic bracket nor by any direct fossil evidence, and remains equivocal in Theropoda, Pterosauria, or other basal archosaurs (Schachner et al. 2011; Benson et al. 2012; Claessens and Vickaryous 2012).

Gastralia

Gastralia, dermal ossifications located in the ventral abdominal region, are basal to tetrapods and plesiomorphic for Archosauria (Claessens 2004b), and have been interpreted as an accessory component of the breathing pump in theropod dinosaurs and basal birds (Carrier and Farmer 2000; Claessens 2004b). However, gastralia are only retained in extant crocodylians, and have been lost in extant birds, even though *Archaeopteryx* and other basal, non-ornithurine birds did possess gastralia. Today, gastralia are a rare component of the tetrapod

skeletal system. In addition to crocodylians, the only other extant vertebrates that retain gastralia are the tuataras (*Sphenodon*), endemic to New Zealand and sole surviving taxa within Sphenodontia, a sister group to modern lizards, amphisbaenians, and snakes. It is notable, however, that gastralia may also be incorporated in the ventral plastron of turtles (Gilbert et al. 2001; Gilbert et al. 2007). In the Archosauria, gastralia were also lost in ornithischian and possibly some sauropod dinosaurs (Claessens 2004b; Tschopp and Mateus 2013), but gastralia are known in representatives of all other major extinct archosauromorph lineages. Due to the rarity of gastralia in extant taxa, a central question bearing on the function of the gastralia is why so many tetrapod groups have lost these ventral abdominal skeletal elements (Perry 1983; Claessens 2004b). Perry (1983) hypothesized that the gastralia provided rigidity for the ventral abdominal wall, thereby preventing displacement of the viscera toward the pleural cavity during aspiration breathing, and that all tetrapod lineages that lost gastralia evolved other mechanisms to cope with visceral displacement during lung ventilation.

In theropods, imbrication of the medial gastralia and the development of distinct median joints likely enabled the hypaxial musculature to widen and narrow the gastralial apparatus, thus contributing to the total tidal volume generated during thoracic aspiration breathing, and possibly even ventilating an abdominal diverticulum (air sac) of the respiratory apparatus (Claessens 2004b). Basal birds retain a similar imbricating gastralial apparatus as non-avian theropods (e.g., Chiappe et al. 1999), and it has been hypothesized that the caudal enlargement of the sternum in ornithurine birds replaced the gastralia functionally and topographically (Perry 1983; Claessens 2004b).

In extant crocodylians, the gastralia may also serve a respiratory function. The gastralia of extant crocodylians are positioned in the superficial layers of the rectus abdominis muscle as separate metameric rows, which stiffen and provide a skeletal scaffold for the ventral abdominal wall. Deformation of the gastralial rows through hypaxial musculature contraction helps deform the abdominal wall and potentially store elastic energy during expiration. Also, the ventral scaffolding that is provided by the gastralial system may aid abdominal wall displacement during rotation of the pubis (Claessens 2009a).

The structure of the gastralial apparatus of pterosaurs is similar to that of crocodylians, and it seems likely that it would have functioned analogously in body wall stiffening and respiration. Displacement of the ventral abdominal wall would similarly be aided by the gastralial apparatus during rotation of the prepubis (Claessens et al. 2009).

Postcranial Skeletal Pneumaticity and the Structure of the Archosaur Pulmonary Apparatus

Soft tissues rarely preserve in the fossil record, and thus direct evidence of the shape and function of the lungs in extinct taxa is absent. However, indirect evidence for the structure of the pulmonary system in several archosaur clades exists in the form of pneumatic postcranial bones. In extant birds, diverticula of the lung–air sac apparatus invade, or "pneumatize," select postcranial skeletal elements (fig. 6.5a–d). The archosaur clades Pterosauria and Saurischia (including birds) exhibit postcranial skeletal pneumaticity. Often, the pneumatization of specific parts of the skeleton is associated with specific regions of the lung–air sac system (O'Connor 2006). Indeed, region-specific patterns of postcranial pneumaticity in extant birds have been used to reconstruct the pulmonary apparatus in extinct forms (Britt et al. 1998; Wedel 2003; O'Connor and Claessens 2005; O'Connor 2006; Claessens et al. 2009; Benson et al. 2012) (fig. 6.5e–g). Notably, pneumatization of skeletal elements in extinct taxa does not necessarily equate to the development of full-scale air sacs similar to that of living birds. Yet the most parsimonious interpretation for patterns of postcranial pneumaticity implies the incipient development of the pneumatizing tissue or structure at a minimum. By contrast, the absence of postcranial pneumaticity does not necessarily imply the absence of specific air sacs throughout the body in an extinct taxon; in certain extant bird taxa with reduced or completely absent postcranial pneumaticity (e.g., penguins) air sacs still extend throughout the body (O'Connor 2009). In fact, air sac development may have extended throughout the abdominal cavity and neck in many extinct archosaurs that lack or exhibit relatively limited postcranial pneumaticity. The developmental potential for lung diverticula is likely an archosaur characteristic (O'Connor 2009; Benson et al. 2012; Butler et al. 2012).

The distributional patterns of postcranial pneumaticity in archosaurs show that (incipient) compliant air sacs were positioned cranial and caudal to the gas exchange region of the lungs in sauropodomorphs, theropods, and pterosaurs (O'Connor and Claessens 2005; Claessens et al. 2009; Wedel 2009; Benson et al. 2012). This arrangement would have allowed flow-through ventilation of the lungs: air would have been drawn through the respiratory epithelium of the lung due to the bellows-like action of the air sacs in either a bidirectional or unidirectional flow pattern (O'Connor and Claessens 2005; Claessens et al. 2009; O'Connor 2009). Bidirectional ventilation of the lung epithelium through air sacs would have enabled a relatively low cost of breathing, double perfusion of the lungs with air, and a decreased pulmonary dead space (Wedel 2003; O'Connor and Claessens 2005; Benson et al. 2012). The recently demonstrated capacity for unidirectional airflow in the alligator (Farmer and Sanders 2010) underscores the possibility that the capacity for unidirectional airflow may have been a basal archosaur characteristic (O'Connor 2006, 2009; Benson et al. 2012; Butler et al. 2012).

Lung diverticula and pneumatic postcranial skeletal elements also result in a decrease in the specific gravity (density) of an animal by replacing heavier tissues such as bone, marrow, or muscle with air, and the evolution of postcranial pneumaticity in archosaurs is correlated to increases in body size in specific pterosaur and dinosaur (including birds) clades (O'Connor 2004, 2006; Claessens et al. 2009; O'Connor 2009; Benson et al. 2012; Butler et al. 2012). In addition, it has been hypothesized that air sacs may have functioned as a support system for the body (e.g., the sauropod neck; Schwarz and Frey 2010) or as a mechanism to modulate the mechanical properties of the body (e.g., the wings of pterosaurs or birds; Claessens et al. 2009; O'Connor 2009). However, the evolution of increased pneumaticity in derived, lighter-bodied theropods and pterosaurs appears to have been at least partially driven by increased metabolic demands (O'Connor and Claessens 2005; Claessens et al. 2009; Benson et al. 2012).

Are Crocodylians a Good Model for the Basal Archosaurian Condition?

Extant crocodylians have frequently been characterized as archaic in the popular arena, but many components

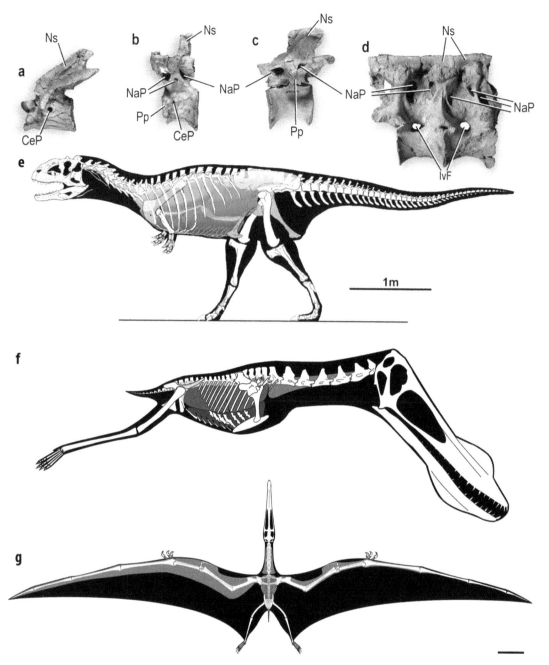

FIG. 6.5 a–d, Vertebrae of the theropod dinosaur *Majungasaurus atopus* illustrating pneumatic features that can be used to reconstruct pulmonary-air sac anatomy based on region-specific correlation with pneumatic topography in birds; a, second cervical vertebra of *Majungasaurus*, a region pneumatized by the cervical air sacs in birds; b, first thoracic vertebra of *Majungasaurus*, a region pneumatized by the cervical air sac or lungs in birds; c, ninth thoracic vertebra of *Majungasaurus* (reversed), a region pneumatized by the abdominal air sac in birds; d, sacral vertebrae of *Majungasaurus*, a region pneumatized by the abdominal air sac in birds; e, reconstruction of the pulmonary-air sac system and skeleton of *Majungasaurus atopus*, based on the region-specific pattern of pneumatization of the postcranial skeleton; and f and g, reconstruction of the pulmonary-air sac system and skeleton of the pterosaur *Anhanguera* in lateral (f) and dorsal (g) view, based on region-specific patterns of postcranial skeletal pneumaticity. In e–g, the cervical air sac system is shown in green, the lungs in orange, and the abdominal air sacs in blue, based upon direct fossil evidence. Clavicular and thoracic air sacs are also shown in gray, based upon their presence in modern birds, but this tertiary-level inference is not constrained by osteological evidence and has a higher degree of uncertainty. In the dorsal reconstruction of *Anhanguera* the subcutaneous pulmonary-air sac system is shown, which may have conferred greater biomechanical control over the wing. Abbreviations: CeP, central pneumatic foramen; IvF, intervertebral foramen; NaP, neural arch pneumatic foramen; Ns, neural spine; Pp, parapophysis. a–e, After O'Connor and Claessens (2005); and f–g, after Claessens et al. (2009).

of their anatomy are, in actuality, highly derived. Pubic mobility is a crocodyliform autapomorphy that is associated with aquatic locomotion and respiration, and diaphragmatic breathing is a unique mechanism that probably arose within the Crocodylomorpha. Even the relatively low metabolic lifestyle and the semiaquatic habit of crocodylians appear derived when considering their cursorial terrestrial ancestors and the apparent overcapacity of the cardiovascular and respiratory systems within living representatives (Carrier 1987; Seymour et al. 2004). However, the thoracic morphology of crocodylians, with intermediate ribs, largely cartilaginous distal thoracic skeletal elements, and "uncinate" processes, does likely resemble the general structure present in basal archosauriforms. Thus, depending on which anatomical system is being evaluated, crocodylians may or may not be an appropriate analogue for the basal archosaurian condition, and care should be taken in consistently evaluating the suitability of the group for the question under consideration.

Constraints and Limitations for the Study of Archosaur Trunk Evolution

One of the absolute limitations of the study of evolution of vertebrate anatomy and function is the incompleteness of the fossil record. The total number of extinct taxa currently known represents only a fraction of the diversity of life on earth, past and present, and of those species known from the fossil record, most are represented by incomplete skeletons. Specifically for the study of the evolution of the archosaur trunk, this means that soft tissue structures, in addition to often cartilaginous elements such as the intermediate and sternal ribs and sternum, are rarely preserved. Characterizing the morphological transformations of such elements by necessity relies on inferences related to other osteological correlates and extrapolations using approaches such as the extant phylogenetic bracket and phylogenetic inference. Vertebrae, vertebral ribs, gastralia, uncinate processes, and many other elements that are generally ossified have a relatively good chance of being preserved in the fossil record, but are often flattened or otherwise deformed. Finally, the skeleton or thorax is rarely preserved in its entirety.

A second major difficulty with the study of the function of the trunk skeleton of extinct (and even extant) forms is the extremely large number of degrees of freedom of movement of the bones that make up the trunk skeleton. Particularly in the ribcage, the theoretical range of movement of the serially linked multi-bar linkages confounds the interpretation of potential movements that the ribs experienced during lung ventilation or locomotion. The potential for slip or long axis rotation of skeletal elements at joints, or bending of thin or cartilaginous skeletal elements such as ribs, further complicates the potential to reconstruct skeletal movement or muscular function in extinct as well as extant taxa. We are only just starting to learn about the intricate details of form and function in only a select number of taxa (see, for instance, the chapters by Brainerd and by Gatesy and Baier in this volume).

Future Directions

New fossil discoveries, an ever-growing understanding of extant tetrapod biology, and new technology allow us to continuously formulate new questions or identify new approaches for addressing old questions. One of the exciting new methods that has been developed in the past decade is XROMM, offering the potential for characterizing high-resolution, three-dimensional detail of the functional anatomy of living taxa (see Brainerd and Gatesy and Baier chapters in this volume).

Advances in imaging (e.g., 3-D surface scanning, CT scanning, MRI) have also provided new pathways for qualitative and quantitative characterization of fossils, including the modeling and animation of biomechanical functions, the testing of structural integrity, or the examination of ontogenetic trajectories. These novel imaging and computer-based analytical methods are opening up new avenues of inquiry on existing fossil material, as well as newly discovered key taxa (see also Shubin et al. chapter in this volume). Digital anatomy data repositories (e.g. Digimorph, Paleoview 3D, Aves 3D, MorphoBank) are opening up possibilities for querying large data sets virtually instantaneously, and testing anatomical, functional or phylogenetic hypotheses over a large sample size or broad range of taxa.

Improved phylogenetic schemes, due to advances in systematic techniques as well as the growing number of fossil species known to science, have refined our ability to form and test functional evolutionary scenarios, which in turn may help elucidate broader principles underlying major anatomical and functional transformations. The resurgence of integrating fossil data with ontogenetic and developmental information thanks to novel techniques in evolutionary-developmental biology (see Abzhanov chapter in this volume) has led to a resurgence in focus on the evolutionary questions that stood at the base of the early paleontological and embryological investigations around the turn of the 19th century, and continue to fuel our understanding of the many facets of vertebrate evolution.

The sum of these new conceptual and methodological approaches allows scientists to address macroevolutionary questions, and integrate data from disparate sources, of an ever-expanding scope.

* * *

Acknowledgments

I would like to thank Farish A. Jenkins Jr. for the many years of expert guidance, for his uncompromising demand for the highest quality in research and technical writing, and for the exemplary role model of teacher-scholar that he provided. Much of the work described in this chapter is a part of, or developed out of, my thesis research at Harvard, and in addition to Farish I thank A. W. Crompton, Andrew Biewener, Michael Shapiro, Tomasz Owerkowicz, Corwin Sullivan, and all other members of the MCZ and OEB who provided assistance in many different ways. My earliest forays into vertebrate paleontology and archosaur respiration were guided by Albert van der Meulen (Utrecht University), Philip Currie (Royal Tyrrell Museum of Palaeontology, now University of Alberta), and Steven Perry (University of Bonn). I thank Patrick O'Connor, Mathew Wedel, Elizabeth Brainerd, Colleen Farmer, Matthew Vickaryous, Tatsuya Hirasawa, John Hutchinson, Steven Gatesy, David Unwin, and many others for stimulating discussions on archosaur evolution over the years. I thank John Westlund for drawing figure 6.3. Finally, I am also much indebted to Meinke Reindersma, Darren Naish, and Patrick O'Connor for their review of earlier drafts of this chapter.

References

Aoyama, H., Y. Mizutani-Koseki, and H. Koseki. 2005. Three developmental compartments involved in rib formation. International Journal of Developmental Biology 49:325–333.

Baumel, J. J., J. A. Wilson, and D. R. Bergren. 1990. The ventilatory movements of the avian pelvis and tail: function of the muscles of the tail region of the pigeon (Columba livia). Journal of Experimental Biology 151:263–277.

Bennett, S. C. 2001. The osteology and functional morphology of the Late Cretaceous pterosaur Pteranodon. Palaeontographica A 260:1–153.

Benson, R. B. J., R. J. Butler, M. T. Carrano, and P. M. O'Connor. 2012. Air-filled postcranial bones in theropod dinosaurs: physiological implications and the "reptile"-bird transition. Biological Reviews 86:168–193.

Boyd, C. A., T. P. Cleland, and F. Novas. 2011. Osteogenesis, homology, and function of the intercostal plates in ornithischian dinosaurs (Tetrapoda, Sauropsida). Zoomorphology 130:305–313.

Brainerd, E., and T. Owerkowicz. 2006. Functional morphology and evolution of aspiration breathing in tetrapods. Respiratory Physiology & Neurobiology 154:73–88.

Britt, B., P. J. Makovicky, J. Gauthier, and N. Bonde. 1998. Postcranial pneumatization in Archaeopteryx. Nature 395: 374–376.

Brühl, C. B. 1862. Skelet Der Krokodilinen. Vienna: Wilhelm Braumüller.

Burke, A. C., C. E. Nelson, B. A. Morgan, and C. Tabin. 1995. Hox genes and the evolution of vertebrate axial morphology. Development 121:333–346.

Butler, R. J., P. M. Barrett, and D. J. Gower. 2012. Reassessment of the evidence for postcranial skeletal pneumaticity in Triassic archosaurs, and the early evolution of the avian respiratory system. PLoS ONE 7:1–23.

Carrier, D. R. 1987. The evolution of locomotor stamina in tetrapods: circumventing a mechanical constraint. Paleobiology 13:326–341.

Carrier, D. R., and C. G. Farmer. 2000. The evolution of pelvic aspiration in archosaurs. Paleobiology 26:271–293.

Chiappe, L. M., S. Ji, Q. Ji, and M. A. Norell. 1999. Anatomy and systematics of the Confuciusornithidae (Theropoda, Aves) from the late Mesozoic of northeastern China. Bulletin of the American Museum of Natural History 242:1–89.

Claessens, L. P. A. M. 2004a. Archosaurian respiration and the pelvic girdle aspiration breathing of crocodyliforms. Proceedings of the Royal Society of London B: Biological Sciences 271:1461–1465.

Claessens, L. P. A. M. 2004b. Dinosaur gastralia; origin, morphology, and function. Journal of Vertebrate Paleontology 24:89–106.

Claessens, L. P. A. M. 2005. The evolution of breathing mechanisms in the Archosauria. 258 pp. Cambridge, MA:

Department of Organismic and Evolutionary Biology, Harvard University.

Claessens, L. P. A. M. 2009a. A cineradiographic study of lung ventilation in *Alligator mississippiensis*. Journal of Experimental Zoology 311A:563–585.

Claessens, L. P. A. M. 2009b. The skeletal kinematics of lung ventilation in three basal bird taxa (emu, tinamou, and guinea fowl). Journal of Experimental Zoology 311A:586–599.

Claessens, L. P. A. M., P. M. O'Connor, and D. M. Unwin. 2009. Respiratory evolution facilitated the origin of pterosaur flight and aerial gigantism. PLoS ONE 4:1–8.

Claessens, L. P. A. M., and M. K. Vickaryous. 2012. The evolution, development and skeletal identity of the crocodylian pelvis: revisiting a forgotten scientific debate. Journal of Morphology 273:1185–1198.

Codd, J. R., D. F. Boggs, S. F. Perry, and D. R. Carrier. 2005. Activity of three muscles associated with the uncinate processes of the giant Canada goose *Branta canadensis maximus*. Journal of Experimental Biology 208:849–857.

Codd, J. R., P. L. Manning, M. A. Norell, and S. F. Perry. 2008. Avian-like breathing mechanics in maniraptoran dinosaurs. Proceedings of the Royal Society of London B: Biological Sciences 275:157–161.

Cong, L. 1998. The Gross Anatomy of *Alligator sinensis* Fauvel. Ke xue chu ban she, Beijing.

Currie, P. J., and P. Chen. 2001. Anatomy of *Sinosauropteryx prima* from Liaoning, northeastern China. Canadian Journal of Earth Sciences 38:1705–1727.

Dal Sasso, C., and S. Maganuco. 2011. *Scipionyx samniticus* (Theropoda: Compsognathidae) from the Lower Cretaceous of Italy: osteology, ontogenetic assessment, phylogeny, soft tissue anatomy, taphonomy and palaeobiology. Memorie della Società Italiana di Scienze Naturali e del Museo Civico di Storia Naturales di Milano 37:1–281.

Farmer, C. G. 1999. Evolution of the vertebrate cardio-pulmonary system. Annual Review of Physiology 61:573–592.

Farmer, C. G., and D. R. Carrier. 2000. Pelvic aspiration in the American alligator (*Alligator mississippiensis*). Journal of Experimental Biology 203:1679–1687.

Farmer, C. G., and K. Sanders. 2010. Unidirectional airflow in the lungs of alligators. Science 327:338–340.

Gadow, H. 1882. Untersuchungen über die Bauchmuskeln der Krokodile, Eidechsen und Schildkröten. Morphologisches Jahrbuch 7:57–100.

Gadow, H. 1933. The Evolution of the Vertebral Column. Cambridge: Cambridge University Press.

Gans, C., and B. Clark. 1976. Studies on ventilation of *Caiman crocodilus* (Crocodilia; Reptilia). Respiration Physiology 26:285–301.

Gatesy, S. M., and K. P. Dial. 1996. Locomotor modules and the evolution of avian flight. Evolution 50:331–340.

Gegenbaur, C. 1871. Beiträge zur Kenntniss des Beckens der Vögel. Jenaische Zeitschrift für Medicin und Naturwissenschaft 6:157–220.

Gilbert, S. F., G. Bender, E. Betters, M. Yin, and J. A. Cebra-Thomas. 2007. The contribution of neural crest cells to the nuchal bone and plastron of the turtle shell. Integrative and Compartive Biology Advance Access:1–8.

Gilbert, S. F., G. A. Loredo, A. Brukman, and A. C. Burke. 2001. Morphogenesis of the turtle shell: the development of a novel structure in tetrapod evolution. Evolution & Development 3:47–58.

Günther, A. 1867. Contribution to the anatomy of *Hatteria* (*Rhynchocephalus* Owen). Philosophical Transactions of the Royal Society of London 157:595–629.

Hazelhoff, E. H. 1951. Structure and function of the lung of birds. Poultry Science 30:3–10.

Hutchinson, J. R., K. T. Bates, J. Molnar, V. Allen, and P. J. Makovicky. 2011. A computational analysis of limb and body dimensions in *Tyrannosaurus rex* with implications for locomotion, ontogeny, and growth. PLoS ONE 6:e26037.

Maina, J. N. 2005. The Lung-Air Sac System of Birds: Development, Structure, and Function. Berlin: Springer.

Mansfield, J. H., and A. Abzhanov. 2010. Hox expression in the American alligator and the evolution of archosaurian axial patterning. Journal of Experimental Zoology, B: Molecular and Developmental Evolution 314B:629–644.

Munns, S. L., T. Owerkowicz, S. J. Andrewartha, and P. B. Frappell. 2012. The accessory role of the diaphragmaticus muscle in lung ventilation in the estuarine crocodile *Crocodylus porosus*. Journal of Experimental Biology 215:845–852.

O'Connor, P. M. 2004. Pulmonary pneumaticity in the postcranial skeleton of extant Aves: a case study examining Anseriformes. Journal of Morphology 261:141–161.

O'Connor, P. M. 2006. Postcranial pneumaticity: an evaluation of soft-tissue influences on the postcranial skeleton and the reconstruction of pulmonary anatomy in archosaurs. Journal of Morphology 267:1199–1226.

O'Connor, P. M. 2009. Evolution of archosaurian body plans: skeletal adaptations of an air-sac-based breathing apparatus in birds and other archosaurs. Journal of Experimental Zoology 311A:629–646.

O'Connor, P. M., and L. P. A. M. Claessens. 2005. Basic avian pulmonary design and flow-through ventilation in nonavian theropod dinosaurs. Nature 436:253–256.

Owerkowicz, T., C. G. Farmer, J. W. Hicks, and E. L. Brainerd. 1999. Contribution of gular pumping to lung ventilation in monitor lizards. Science 284:1661–1663.

Perry, S. F. 1983. Reptilian lungs: functional anatomy and evolution. Advances in Anatomy, Embryology, and Cell Biology 79:1–81.

Perry, S. F. 1990. Gas exchange strategy in the Nile crocodile: a morphometric study. Journal of Comparative Physiology B 159:761–769.

Powell, F. L. 2000. Respiration. Pp. 233–264 in G. Causey Whittow, ed., Sturkie's Avian Physiology. San Diego: Academic Press.

Ruben, J. A., C. Dal Sasso, N. R. Geist, W. J. Hillenius, T. D. Jones, and M. Signore. 1999. Pulmonary function and metabolic physiology of theropod dinosaurs. Science 283:514–516.

Ruben, J. A., T. D. Jones, and N. Geist. 2003. Respiratory and reproductive paleophysiology of dinosaurs and early birds. Physiological and Biochemical Zoology 76:141–164.

Ruben, J. A., T. Jones, N. Geist, and W. J. Hillenius. 1997. Lung structure and ventilation in theropod dinosaurs and early birds. Science 278:1267–1270.

Schachner, E. R., C. G. Farmer, A. T. McDonald, and P. Dodson. 2011. Evolution of the dinosauriform respiratory apparatus: new evidence from the postcranial axial skeleton. Anatomical Record 294:1532–1547.

Schwarz, D., and E. Frey. 2010. Is there an option for a pneumatic stabilization of sauropod necks? an experimental and anatomical approach. Palaeontologia Electronica 11:17A:11–26.

Seemann, G. 1926. Die Gliederung der Rippen bei den Reptilien. Gegenbaurs Morphologisches Jahrbuch 56:105–135.

Seymour, R. S., C. L. Bennett-Stamper, S. D. Johnston, D. R. Carrier, and G. C. Grigg. 2004. Evidence for endothermic ancestors of crocodiles at the stem of archosaur evolution. Physiological and Biochemical Zoology 77:1051–1067.

Sibson, F. 1846. On the mechanism of respiration. Philosophical Transactions of the Royal Society of London 136:501–550.

Tickle, P., R. Nudds, and J. Codd. 2009. Uncinate process length in birds scales with resting metabolic rate. PLoS ONE 4: 1–6.

Tickle, P. G., R. A. Ennos, L. E. Lennox, S. F. Perry, and J. R. Codd. 2007. Functional significance of uncinate processes in birds. Journal of Experimental Biology 210:3955–3961.

Tickle, P. G., M. A. Norell, and J. Codd. 2012. Ventilatory mechanics from maniraptoran theropods to extant birds. Journal of Evolutionary Biology 25:740–747.

Tschopp, E., and O. Mateus. 2013. Clavicles, interclavicles, gastralia, and sternal ribs in sauropod dinosaurs: new reports from Diplodocidae and their morphological, functional and evolutionary implications. Journal of Anatomy 222:321–340.

Uriona, T. J., and C. G. Farmer. 2008. Recruitment of the diaphragmaticus, ischiopubis and other respiratory muscles to control pitch and roll in the American alligator (*Alligator mississippiensis*). Journal of Experimental Biology 211:1141–1147.

Wedel, M. J. 2003. Vertebral peumaticity, air sacs, and the physiology of sauropod dinosaurs. Paleobiology 29:243–255.

Wedel, M. J. 2009. Evidence for bird-like air sacs in saurischian dinosaurs. Journal of Experimental Zoology 311A:611–628.

Zimmer, K. 1935. Beiträge zur Mechanik der Atmung bei den Vögeln in Stand und Flug auf Grund anatomisch-physiologischer und experimenteller Studien. Zoologica (Stuttgart) 33:1–69.

7

Evolution of Hind Limb Posture in Triassic Archosauriforms

Corwin Sullivan*

Introduction

From the Middle Triassic until the end of the Cretaceous, almost all aerial and large terrestrial vertebrates were members of the Archosauriformes. This reptilian clade includes an exotic array of Triassic forms as well as such familiar groups as the dinosaurs, pterosaurs, and crocodylomorphs (fig. 7.1). The diversity and ecological predominance of archosauriforms during the Mesozoic is a remarkable chapter in vertebrate history, and the highly modified theropod dinosaurs known as birds remain a salient part of the global biota today. Does some suite of advantageous anatomical and/or physiological attributes shared by archosauriforms explain their spectacular diversity and persistence?

Charig (1972) was a strong advocate of the idea that the evolution of an upright hind limb posture within archosauriforms was a crucial driver of their success. In an influential discussion of hind limb evolution, Charig (1972) described the basal-most archosauriforms as lizard-like "sprawlers," modern crocodilians and many Triassic archosauriforms as "semi-improved" with respect to hind limb posture, and dinosaurs as "fully improved." The "improvement" in question was a change in hind limb kinematics in which the femur gradually came to protract and retract

* Key Laboratory of Vertebrate Evolution and Human Origins, Institute of Vertebrate Paleontology and Paleoanthropology, Chinese Academy of Sciences

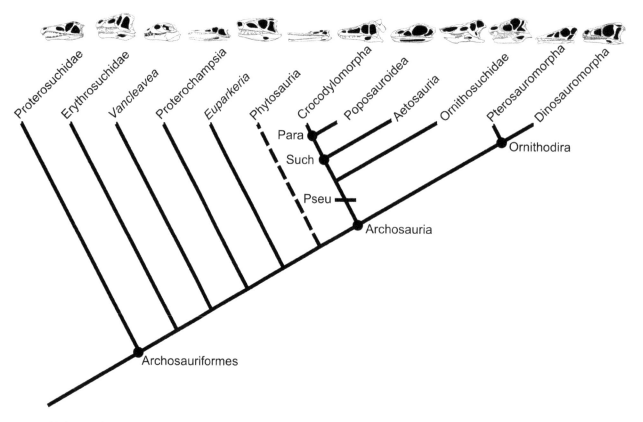

FIG. 7.1 Phylogeny of Archosauriformes, simplified from Nesbitt (2011). Circles indicate node-based taxa, bar indicates stem-based taxon. Dashed line emphasizes unorthodox, but analytically well-supported, placement of phytosaurs in this phylogeny. Doswelliidae does not appear on the cladogram because this taxon was not included in the Nesbitt (2011) analysis and has appeared in varying positions in studies by other authors (Dilkes and Sues 2009; Desojo et al. 2011; Dilkes and Arcucci 2012). Skull illustrations not to scale. Para, Paracrocodylomorpha; Pseu, Pseudosuchia; Such, Suchia.

in an approximately parasagittal plane, rather than an approximately horizontal one as in sprawlers (fig. 7.2). In quantitative terms, this may be described as an increase in the maximum angle of femoral adduction (below a horizontal plane passing through both hip joints) that is reached during the stance phase of locomotion. This angle is low in sprawling gaits, high in "fully improved" gaits, and intermediate in "semi-improved" gaits such as the "high walk" of extant crocodilians, although there is no consensus as to the precise numerical ranges that should be attached to these categories.

Subsequent workers (e.g., Reilly and Elias 1998) have followed Bakker (1971) in using the more neutral terms "sprawling," "semierect," and "fully erect" (or simply "erect") to describe postural categories, and Gatesy (1991) argued persuasively that limb posture should be regarded as a continuum in any case. Perhaps most importantly, advances in scientific understanding of both archosauriform phylogeny and the locomotor function of particu-

lar taxa have made it clear that the orderly progression from sprawling to erect hind limb posture envisioned by Charig (1972) is an oversimplification, and that patterns of postural evolution were more complex and to some degree homoplastic. Nevertheless, it is clear that a broad trend toward increasingly erect hind limb orientation existed among Triassic archosauriforms, affecting numerous aspects of pelvic and hind limb morphology and potentially having profound evolutionary and ecological consequences (e.g., Charig 1972; Parrish 1986a).

Determining even the approximate position of a given fossil archosauriform on the continuum from sprawling to erect posture is far from a trivial problem, and accordingly provides an intriguing challenge to scientists concerned with functional anatomy and paleontology. This chapter briefly reviews the most important available osteological and ichnological evidence as it pertains to various archosauriforms, and the implications of this evidence for reconstructing the postural shift in the context of archosauriform phylogeny. The consequences of the shift for archosauriform evolution are also considered, and promising avenues for further research into this major functional transition are suggested.

Archosauriform Phylogeny

For simplicity and clarity, this chapter largely adopts the results and terminology of the recent and highly comprehensive analysis of archosauriform relationships by Nesbitt (2011) as a basic framework for discussing archosauriform phylogeny (fig. 7.1). The name Archosauria is used for the crown-group defined by extant crocodilians and birds, whereas Archosauriformes denotes a more inclusive group that corresponds to the "traditional" Archosauria of some authors (e.g., Gower 2003; Kubo and Benton 2007; Sullivan 2010). Crocodilian-line and avian-line archosaurs are referred to as Pseudosuchia and Ornithodira (strictly speaking, a nodal taxon based on the common ancestor of dinosaurs and pterosaurs), respectively. Previous phylogenetic analyses (e.g., Benton and Clark 1988; Sereno 1991; Parrish 1993; Brusatte, Benton, et al. 2010) obtained topologies differing from that adopted here in various respects, but most notably in consistently finding the superficially crocodilian-like phytosaurs to be basal pseudosuchians rather than

archosauriforms falling just outside Archosauria as accepted here. Archosauriforms outside the clade containing phytosaurs and archosaurs are termed "basal archosauriforms" in this chapter.

Institutional Abbreviations

NM, National Museum, Bloemfontein, South Africa; SAM, Iziko South African Museum, Cape Town, South Africa; UCMZ, University Museum of Zoology, Cambridge, England; USNM, National Museum of Natural History, Washington, DC.

Anatomical Abbreviations

aa, articular surface for astragalus; af, articular surface for fibula; at, articular surface for tibia; ce, centrale; cp, calcaneal process; fh, femoral head; il, ilium; is, ischium; it, internal trochanter; lp, lateral process; mp, medial process; pl, *M. peroneus longus*; pu, pubis; sc, supracetabular crest; tr4, fourth trochanter.

Hind Limb Posture in Extinct Archosauriforms

Studying the postural transition in archosauriforms entails drawing conclusions about the placement of

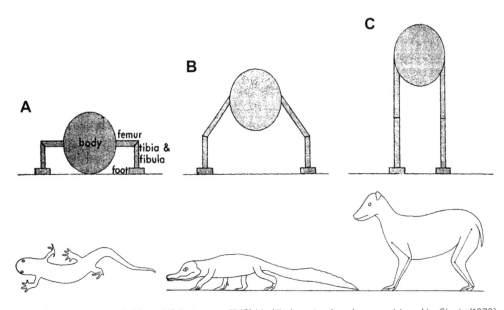

FIG. 7.2 Sprawling (A), "semi-improved" (B), and "fully improved" (C) hind limb postural grades as envisioned by Charig (1972), respectively exemplified by a salamander, a crocodilian performing the high walk, and a dog. Upper illustrations show hind limb posture considered characteristic of each grade, in schematic anterior or posterior view. After Charig (1972).

particular taxa on the continuum from sprawling to erect posture. Such conclusions have traditionally been based on qualitative assessment of the osteology of the hind limb and pelvis (e.g., Charig 1972), and Parrish (1986a) detailed numerous morphological features that he considered to be characteristic of different postural grades. The most persuasive of these relate to the configuration of the hip joint, which sets geometric limits on how far the femur can be adducted during the stance phase. In modern lizards, for example, the acetabulum is shallow and laterally directed, and the proximal end of the femur is not deflected medially to form a distinct head. The femur cannot be adducted far below the horizontal without compromising the articular contact between the femur and the acetabulum. In birds, by contrast, the acetabulum is deep and the femur forms a medial head, so that the femur can be adducted into a nearly sagittal plane. In crocodilians, the proximal end of the femur has a slight medial deflection and the acetabulum is relatively shallow but partly roofed by a supracetabular crest that engages the head of the femur, reflecting the ability of crocodilians to adduct the femur to intermediate angles during the "high walk." There have been virtually no attempts to quantify the range of motion of the hip joint in Triassic archosauriforms. This chapter uses the terms "sprawling," "semierect," and "fully erect" to suggest 0°–30°, 30°–60°, and 60°–90° degrees of femoral adduction below the horizontal, respectively, but the approximate nature of these boundaries must be emphasized.

The structure of the calcaneum is used as a secondary indication of limb posture, partly following Parrish (1986a). In most Triassic archosauriforms, the calcaneum bears a flange-like or tuber-like process, which may be posteriorly directed as in aetosaurs, posterolaterally directed as in phytosaurs, or laterally directed as in proterosuchids. An analogous lateral or posterolateral process is present in some lizards, and Sullivan (2010) studied the role of this structure in the savannah monitor *Varanus exanthematicus* using the technique of scientific rotoscoping (Gatesy et al. 2010). Retraction of the femur in a sprawling position tends to rotate the crural bones so that their anatomically anterior surfaces become directed laterally, and long-axis rotation of the femur simultaneously tends to force the tibia into an inclined rather than upright orientation. The resulting limb position is shown in figure 7.3.

An important consequence of lateral rotation and inclination of the crus is that the calcaneal process is brought into an approximately dorsal orientation, with functional implications that were explored by Sullivan (2010). *M. peroneus longus* wraps around the end of the calcaneal process, which in the position illustrated in figure 7.3 can operate mechanically to increase the moment arm of the muscle with respect to propulsive extension of the ankle. This varanid mechanism demonstrates how a posterolaterally directed calcaneal process can contribute to generating propulsive torque, and the function of the calcaneum was probably analogous in extinct archosauriforms with a similarly oriented process. Because this mechanism depends on the outward rotation of the crus that is intrinsic to sprawling kinematics, a posterolaterally to laterally oriented calcaneal process is diagnostic of sprawling in fossil taxa.

In extant crocodilians, by contrast, the calcaneum contributes to propulsion in a different manner. The calcaneal process is only slightly laterally deflected, and during the semierect high walk there is little rotation of the distal part of the limb because the femur is retracting in an oblique rather than a horizontal orientation. Like the mammalian heel, the crocodilian calcaneal process increases the moment arm of *Mm. gastrocnemii* rather than *M. peroneus longus*, and remains close to a parasagittal plane throughout the stance phase rather than being carried into one by crural rotation (Brinkman 1980; Sullivan 2007). Accordingly, a posterior calcaneal process indicates semierect or erect posture, just as a lateral to posterolateral one indicates sprawling.

The following discussion of hind limb posture in particular archosauriform clades is based primarily on these features of the hip and ankle, in addition to the limited evidence available from ichnology (Padian et al. 2010) and a study of femoral stresses (Kubo and Benton 2007).

Proterosuchids

The well-known species *Proterosuchus fergusi* from the Lower Triassic of South Africa, a sturdy carnivorous quadruped with a maximum body length of about 1.5 m (Cruickshank 1972; Charig and Sues 1976), is among the most basal archosauriforms and exemplifies the locomotor apparatus in a proterosuchid clade (Gower and Sennikov 1997) or grade (Ezcurra et al. 2010). The

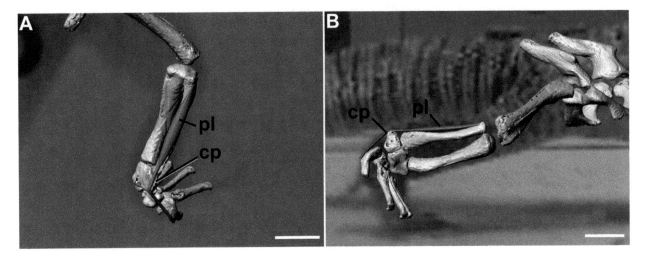

FIG. 7.3 Rotoscopic figure of the right hind limb of the savannah monitor *Varanus exanthematicus* in the late part of the stance phase, showing a skeletal model superimposed on frames from (A) X-ray and (B) light footage of an individual walking on a treadmill. A, Dorsal view; B, lateral view. Muscle belly and distal tendon of *M. peroneus longus* have been approximated as an addition to the rotoscopic model. Scale bars equal 1 cm. After Sullivan (2010).

acetabulum is shallow and laterally directed as in primitive diapsids (fig. 7.4A), and the supracetabular crest is weak. The femur (fig. 7.4B) lacks a medially deflected head, and flares proximally into a prominent internal trochanter for muscle insertion. The calcaneum bears a large, laterally directed process (fig. 7.4C).

P. fergusi has always been interpreted as a sprawler (Charig and Sues 1976; Parrish 1986a). The lateral orientation of the acetabulum, coupled with the lack of a medially deflected femoral head and the presence of a well-developed internal trochanter, would have made it impossible to significantly adduct the femur without dislocating the hip. In non-parasagittal (sprawling to semierect) tetrapods the femur undergoes long-axis rotation during retraction, so that the flexor surface comes to face posteriorly rather than ventrally (Kubo and Benton 2007). In the position expected at midstance (fig. 7.4D), hip adduction would result in a collision between the internal trochanter and the ventral rim of the acetabulum. The same constraint exists in the extant lizard *Varanus*, in which the hip configuration is grossly similar to that of *P. fergusi* (fig. 7.5).

The lateral orientation of the calcaneal process indicates that rotation of the crus would have been necessary in order for the process to contribute mechanically to propulsion. Because such rotation is characteristic only of sprawling hind limb kinematics, this represents corroborating evidence that *P. fergusi* was a sprawler.

Other Basal Archosauriforms

The hind limb and pelvic morphology of other basal archosauriforms resembles that of *P. fergusi* to varying degrees. The two major basal archosauriform clades, the large (up to about 4 m in total body length) erythrosuchids (Young 1964; Cruickshank 1978; Gower 2003) and the smaller, crocodilian-like proterochampsians (Romer 1971, 1972; Trotteyn and Haro 2011; Dilkes and Arcucci 2012; Trotteyn et al. 2012), share most of the indicators of sprawled hind limb posture seen in *Proterosuchus* (fig. 7.6A–C).

However, important differences from *Proterosuchus* also exist in both groups. In the well-studied erythrosuchid *Erythrosuchus*, the internal trochanter is reduced and displaced somewhat distally relative to the condition in *Proterosuchus* (Gower 2003). Proterochampsians possess a fourth trochanter that is probably homologous to the primitive internal trochanter (Nesbitt 2011) but is situated still more distally. Distal displacement of the internal trochanter presumably allowed slightly greater femoral adduction during locomotion. Nevertheless, the shallowness of the acetabulum and the lack of medial deflection of the proximal end of the femur imply that the amount of adduction needed to perform the "high walk" would have been impossible. The suggestion that sprawling was widely characteristic of basal archosauriforms agrees with the conclusions

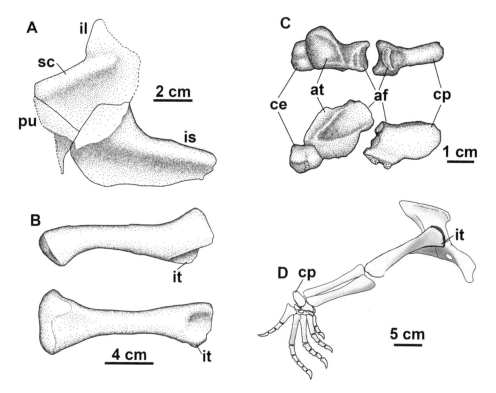

FIG. 7.4 A, Partial left pelvis of *Proterosuchus fergusi* (NM QR 1484) in lateral view; B, left femur of *P. fergusi* (SAM K140) in lateral view, above, and flexor view, below; C, left astragalus, centrale and calcaneum of *P. fergusi* (NM QR 1484) in proximal view (above) and extensor view (below); D, reconstructed sprawling hind limb and pelvis of *P. fergusi* near midpoint of stance phase, in anterolateral view.

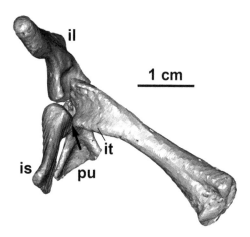

FIG. 7.5 Right hip joint of the extant savannah monitor *Varanus exanthematicus* in orthographic posterior and slightly lateral view, showing that even moderate adduction with the femur in a retracted and rotated position causes the internal trochanter to collide with the rim of the acetabulum (black arrow).

of Parrish (1986a) and is supported by the orientation of the calcaneal process, which is lateral in *Erythrosuchus* (Cruickshank 1978; Gower 1996). In the proterochampsian *Tropidosuchus* the calcaneal process is deflected somewhat posteriorly (Sullivan 2010), but not to the degree seen in extant crocodilians.

Kubo and Benton (2007) attempted to infer hind limb posture in *Erythrosuchus* and various other taxa by estimating the mid-stance femoral stresses associ-

ated with various possible angles between the ground reaction force and the femoral shaft (α), with lower angles corresponding to more upright postures. For each angle considered, Kubo and Benton (2007) generated both minimum and maximum stress estimates, corresponding to different kinematic assumptions. Maximum stress estimates for *Erythrosuchus* simply decreased at lower α values, indicating that the hind limb may have been held nearly upright in order to minimize stress.

Minimum estimates were lowest for α = 25° and α = 35°, angles implying semierect hind limb posture. Taken together, the results contradict the interpretation that *Erythrosaurus* was a sprawler.

However, the analysis relied on a number of estimated parameters, particularly body mass, muscle moment arms, and the moment arms of the ground reaction force about the knee and ankle. These estimates were made on the basis of relatively simple measurements and assumptions, introducing potential error. Furthermore, Kubo and Benton (2007) acknowledged that their results for large-bodied animals such as *Erythrosuchus* were particularly uncertain. The maximum stress estimates calculated for *Erythrosuchus* were extremely high, ranging up to a magnitude of 203 MPa for compressive stress and 180 MPa for tensile stress. By comparison, peak stresses in the bones of large modern mammals are on the order of 50 to 150 MPa (Kubo and Benton 2007), and Erickson et al. (2002) reported that structural failure occurred at a bending stress of 188 MPa in a femur of *Varanus exanthematicus*. The very high

stress estimates for *Erythrosuchus* may be a sign that the analysis was not realistic enough to reliably indicate the posture of the hind limb. Kubo and Benton (2007) opened a promising avenue of investigation into archosauriform locomotion, but their results were based on too many uncertain assumptions to refute the osteological evidence that *Erythrosuchus* was close to the sprawling end of the postural continuum.

A few other basal archosauriforms are sufficiently well known for inferences of hind limb posture to be possible. *Euparkeria*, a small (body length well under 1 m) and gracile form from the Middle Triassic of South Africa (Ewer 1965; Botha-Brink and Smith 2011), is similar to proterochampsians in most functionally significant pelvic and hind limb features. However, the proximal end of the femur is slightly medially deflected in *Euparkeria*. This implies that *Euparkeria* may have been able to adduct the hip to a slightly greater degree than in most other basal archosauriforms (fig. 7.6E), although probably not nearly enough to achieve an erect posture as suggested by Ewer (1965).

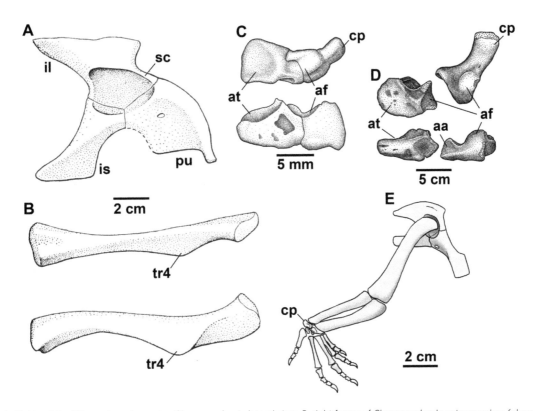

FIG. 7.6 A, Right pelvis of the proterochampsian *Chanaresuchus* in lateral view; B, right femur of *Chanaresuchus* in extensor view (above) and medial view (below); C, left astragalus and calcaneum of the proterochampsian *Tropidosuchus*; D, left astragalus and calcaneum of a phytosaur tentatively identified as *Leptosuchus* (USNM 18313); E, reconstructed near-sprawling hind limb and pelvis of basal archosauriform *Euparkeria* near midpoint of stance phase, in anterolateral view. C–D, Proximal views above, extensor views below. A–B after Romer (1972); C–D after Sullivan (2010).

Vancleavea (Nesbitt, Stocker, et al. 2009; Nesbitt 2011) and the clade Doswelliidae (Sues et al. 2013) are similar in being heavily armored, possibly semierect basal archosauriforms, but are not thought to be closely related. In *Vancleavea* the limbs are extremely small relative to the body, and the hind limb lacks a femoral internal trochanter but has the other key plesiomorphic indicators of sprawling posture present in *P. fergusi*. Doswelliids resemble *Vancleavea* and other basal archosauriforms in having a shallow, laterally directed acetabulum, although the dorsal part of the iliac blade is everted (Dilkes and Sues 2009; Desojo et al. 2011; Sues et al. 2013). In the doswelliid *Doswellia* (USNM 186989) the proximal end of the femur is medially deflected to about the same degree as in *Euparkeria*, suggesting a similar capacity for limited adduction.

Phytosaurs

The pelvic and hind limb morphology of the large, superficially crocodilian-like phytosaurs (Chatterjee 1978; Long and Murry 1995; Lucas et al. 2002) is similar in many respects to that of proterochampsians and *Euparkeria*. The femur has a small fourth trochanter and lacks an inturned proximal head. In combination with the shallowness and lateral orientation of the acetabulum, this suggests that phytosaurs remained close to the sprawling end of the postural continuum (Parrish 1986a, 1986b).

This assessment is consistent with the posterolateral orientation of the calcaneal process (fig. 7.6D), which resembles the condition in proterochampsids and *Euparkeria*. A novelty in phytosaurs, however, is that the calcaneum bears a convex joint facet for the fibula and a shallow concave one for the astragalus (Parrish 1986b). The calcaneum would have twisted upon the matching convexity formed by the astragalus, increasing the range of plantarflexion. However, manipulation of scanned models of a phytosaur ankle (Sullivan 2007) suggests that the range of motion for this "crurotarsal" movement of the calcaneum upon the astragalus and fibula was small.

Based on their attribution of the ichnotaxon *Apatopus* to a phytosaur, Padian et al. (2010) argued that members of this group were capable of adopting a gait similar to the modern crocodilian "high walk." However,

this proposal conflicts with the osteological evidence for a near-sprawling hind limb posture in phytosaurs and is also questionable on its own terms. Attribution of *Apatopus* to a phytosaur depends on the assumption that the fourth digit is the longest in the phytosaur foot, but phytosaur pedal morphology has not been described in sufficient detail for this to be certain.

Ornithodirans

Ornithodira includes the Pterosauromorpha and Dinosauromorpha (Nesbitt 2011). Dinosauromorphs such as *Dromomeron* and *Marasuchus* (fig. 7.7) display such correlates of erect posture as a large supracetabular crest, an inturned femoral head, and a hinge-like ankle joint between the proximal and distal tarsals (Sereno and Arcucci 1994a, 1994b; Dzik 2003; Nesbitt, Irmis, et al. 2009; Nesbitt et al. 2010; Brusatte, Nesbitt, et al. 2010; Langer et al. 2010). The calcaneal process is reduced or absent.

Pterosaur terrestrial locomotion is more controversial. A distinct femoral head is present in basal pterosaurs, but varies in orientation from medial to proximomedial (Padian 1983a, 1983b; Jenkins et al. 2001). With the acetabulum facing approximately laterally, as was probably the case (Padian 1983a, 1983b), a medially directed femoral head would allow erect posture whereas a proximomedial one might limit adduction to semierect angles. As noted by Padian (1983a, 1983b), the distal part of the pterosaur hind limb appears suited to parasagittal movement. However, the stance-phase movements of the crus and pes are approximately parasagittal in the high walk of extant crocodilians (Sullivan 2007), just as in erect walking. Some basal pterosaurs evidently resembled basal dinosauromorphs in being capable of erect hind limb posture (Padian 1983a, 1983b, 2008), but this was not necessarily the typical and/or primitive condition.

Pseudosuchians

In Triassic pseudosuchians (fig. 7.8) the ankle forms a more mobile crurotarsal joint than is present in phytosaurs (Parrish 1986a; Sereno and Arcucci 1990; Sereno 1991; Sullivan 2007). Typical pseudosuchians have a "crocodile-normal" ankle (fig. 7.8D) in which a prominence on the astragalus contacts a socket and adjacent

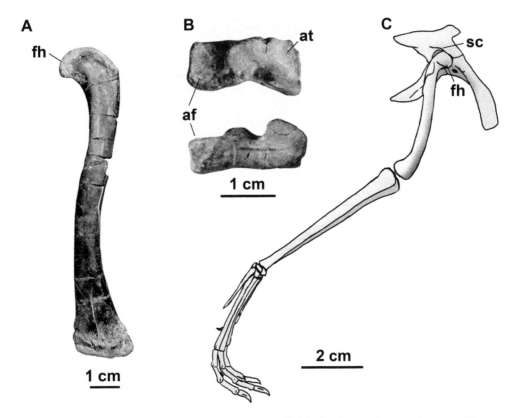

FIG. 7.7 A, Left femur of basal dinosauromorph *Dromomeron* in anterior view; B, right fused astragalus and calcaneum of *Dromomeron* in proximal view (above) and extensor view (below); C, reconstructed buttress-erect hind limb and pelvis of basal dinosauromorph *Marasuchus* near midpoint of stance phase, in anterolateral view. A and B after Nesbitt et al. (2009b).

flange on the calcaneum (Chatterjee 1978; Cruickshank 1979; Parrish 1986a; Sereno 1991). The calcaneum can rotate upon the astragalar "peg," carrying with it the distal tarsals and pes. Ornithosuchids, large terrestrial predators of which the best known are *Ornithosuchus* from Scotland (Walker 1964) and *Riojasuchus* from Argentina (Bonaparte 1972), possess a functionally similar but anatomically distinct "crocodile-reversed" ankle (fig. 7.8E) in which a prominence on the calcaneum rotates within a recess on the astragalus (Baczko and Ezcurra 2013). In both ankle types the calcaneal process is posteriorly directed and plantarflexion and dorsiflexion would have taken place about a transverse axis, implying that the distal part of the limb functioned parasagittally.

The evident lack of distal hind limb rotation is compatible with both erect and semierect postures, but hip structure helps distinguish between these possibilities. In ornithosuchids (Walker 1964; Bonaparte 1972), the relatively basal pseudosuchian *Gracilisuchus* (Lecuona and Desojo 2011) and the armored aetosaurs (e.g.,

Sawin 1947; Bonaparte 1972; Long and Murry 1995; Desojo and Báez 2005), the proximal end of the femur may be modestly deflected medially but lacks a distinct head. This is also generally true of "rauisuchians" (e.g., Gower and Schoch 2009), a paraphyletic assemblage (Nesbitt 2011; Nesbitt, Brusatte et al. 2013; but see Brusatte, Benton, et al. 2010) comprising Poposauroidea and various other forms on the crocodylomorph stem. The most derived poposauroids are gracile bipeds, but the quadrupedal non-poposauroid *Batrachotomus* (Gower and Schoch 2009) is a more typical "rauisuchian."

Hind limb posture was clearly variable in pseudosuchians. In ornithosuchids and *Gracilisuchus*, the acetabulum is shallow and laterally directed (fig. 7.8A, B). Combined with the lack of a medial femoral head (fig. 7.8C), this feature suggests semierect rather than erect posture. In some "rauisuchians," however, the ilium is tilted so that the acetabulum is directed somewhat ventrally (Bonaparte 1984), allowing the femur to be fully adducted even though a medial head is absent

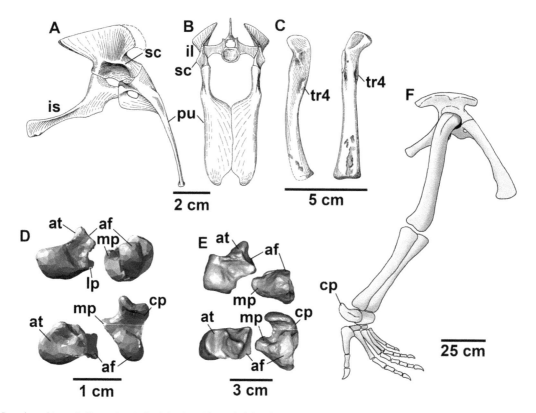

FIG. 7.8 Pseudosuchians. A, Reconstructed pelvis of ornithosuchid *Ornithosuchus* in right lateral view; B, same in anterior view; C, femur of *Ornithosuchus*, represented by lateral view of left femur (left) and extensor view of right femur (right); D, left astragalus and calcaneum of crocodylomorph *Alligator* in extensor view (above) and proximal view (below); E, left astragalus and calcaneum of ornithosuchid *Riojasuchus* in extensor view (above) and proximal view (below); F, reconstructed pillar-erect hind limb and pelvis of "rauisuchian" *Batrachotomus* near midpoint of stance phase, in anterolateral view. A–C after Walker (1964), D–E after Sullivan (2007).

(fig. 7.8F). This "pillar-erect" condition (fig. 7.9A) contrasts strikingly with the "buttress-erect" condition (fig. 7.9B) of dinosauromorphs (Benton and Clark 1988), and represents an alternative mechanism for allowing the femur to assume a near-vertical orientation. In derived poposauroids, however, the ilia are more vertical (Nesbitt 2011), and erect posture appears to have been achieved by expanding the supracetabular crest to form a ventrally directed cup for the head of the femur. Pillar-erect inclination of the ilium is also broadly characteristic of aetosaurs (Benton 1986a; Parrish 1986a, 1987).

The small, gracile and seemingly terrestrial crocodylomorphs of the Triassic and Jurassic, the "sphenosuchians," are the only buttress-erect pseudosuchians. Early crocodylomorphs such as *Terrestrisuchus* are characterized by a prominent supracetabular crest, a laterally directed acetabulum, and a femur with a distinct medial head (Colbert 1952; Walker 1970; Crush

1984; Parrish 1991; Wu and Chatterjee 1993; Sues et al. 2003; Göhlich et al. 2005), a combination indicating strong femoral adduction. The sprawling to semi-erect hind limb posture and semiaquatic habitus seen in modern crocodilians are reversals from a primitive pattern of erect terrestriality (Parrish 1987).

Discussion

Osteological indicators point unambiguously to a broad shift toward more erect hind limb postures in Triassic archosauriforms, in agreement with the general conclusions of Charig (1972) and Parrish (1986a). The morphology of the pelvis and hind limb leave little doubt that *Proterosuchus* was a sprawler. In other basal archosauriforms the conversion of the primitive internal trochanter of the femur into a more distally placed fourth trochanter, combined in *Euparkeria* and the doswelliids with slight deflection of the proximal end of the femur, may have allowed slightly more femoral adduction. However, basal archosauriforms and indeed phytosaurs uniformly retain key indicators of sprawling hind limb posture, including a shallow, laterally directed acetabulum, a femur lacking strong medial deflection of the proximal part of the shaft, and a lateral or posterolateral calcaneal process.

The phylogenetic framework adopted in this chapter, in which phytosaurs are considered to fall outside Archosauria (Nesbitt 2011), implies that a single transition from sprawling to semierect or possibly even erect posture coincided with the origin of archosaurs. This view is consistent with the fact that living crocodilians are capable of walking semierect, and living birds of walking fully erect (Gauthier et al. 2011). However, mapping probable locomotor transitions onto a simplified but current archosauriform phylogeny (fig. 7.10) reveals unexpectedly complex patterns of postural evolution within Archosauria. Basal dinosauromorphs appear to be uniformly buttress-erect, but it is less certain that this was also true of basal pterosaurs, raising the possibility that early ornithodirans may have been only semierect. A semierect hind limb posture seems even more likely for the ornithosuchids, which were recovered by Nesbitt (2011) as the basalmost pseudosuchians

(fig. 7.1). This implies that the common ancestor of Archosauria may have been only semierect, as indicated in figure 7.10, and that the pillar-erect posture of aetosaurs and pseudosuchians evolved independently from the buttress-erect posture of dinosauromorphs and at least some pterosaurs. The buttress-erect hip of early crocodylomorphs seems to have evolved from a pillar-erect predecessor, and crocodilians subsequently reverted to a semierect hind limb posture (fig. 7.10). Postural evolution in archosaurs, and particularly in pseudosuchians, must have involved a somewhat homoplastic pattern of shifts among the semierect, buttress-erect, and pillar-erect conditions, in contrast to the simpler scenario of linear postural "improvement" proposed by Charig (1972).

When Did Erect Hind Limb Posture Evolve?

The question of which archosauriform clades were erect is linked to, but distinct from, the question of when erect posture appeared in geological time. A compelling source of evidence with respect to the latter issue is the ichnological record. Kubo and Benton (2009) suggested on the basis of evidence from extant tetrapods that pace angulation, defined as the angle between the lines

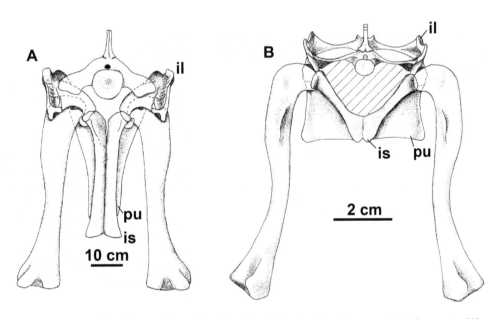

FIG. 7.9 Schematic contrast between the pillar-erect hip configuration of typical "rauisuchians," represented by *Saurosuchus* (A), and the buttress-erect hip configuration of typical dinosauromorphs, represented by a taxon cf. *Lagerpeton* (B). Both views posterior. After Bonaparte (1984).

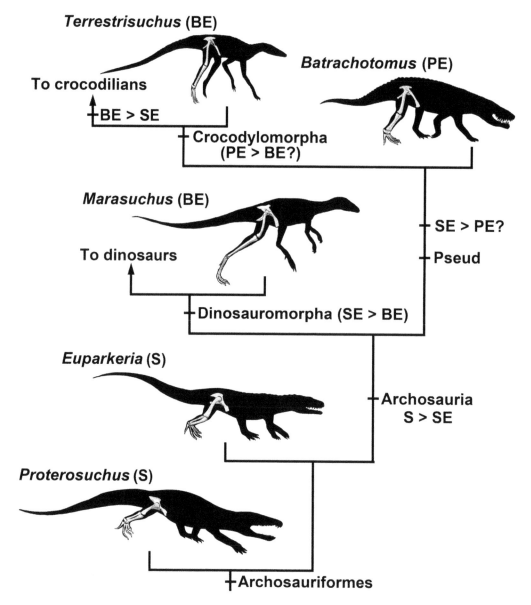

FIG. 7.10 Postural evolution in Triassic archosauriforms. Simplified cladogram shows silhouettes of forms representing various postural grades, all depicted in anterolateral view with the hind limb in a mid-stance position, and inferred positions of shifts between sprawling (S), semierect (SE), buttress-erect (BE), and pillar-erect (PE) grades. Pseud, Pseudosuchia. Silhouettes not to scale.

connecting the tips of the middle toes of three successive pes or manus prints (fig. 7.11), is typically higher in more erect walkers. Their data set showed that pace angulation values for tetrapod tracks in general increased abruptly across the Permo-Triassic boundary. The median value for the Late Permian trackways included in their study (88°) was well within the range expected for sprawling tetrapods based on data from extant species, whereas that for Early Triassic trackways (140°) was high enough to indicate semierect to erect posture. The median pace angulation for Early Triassic trackways

that could be attributed to archosauromorphs (a clade broader than Archosauriformes, and presumably including more sprawlers) was 156°, equivalent to the median value obtained for modern birds. Individual archosauromorph trackways with pace angulation values below the median for extant lizards (110°) did exist in the Triassic, but were rare.

The functional inferences presented above suggest that the sprawling archosauromorph tracks in the Kubo and Benton (2009) sample should represent phytosaurs, basal archosauriforms, and non-archosauriform archosauromorphs, whereas the semierect to erect tracks should primarily represent archosaurs. The paucity of trackways with low pace angulation values throughout the Triassic is then surprising, given the prevalence of phytosaurs and basal archosauriforms in the body fossil record. However, the data for modern forms suggest that pace angulation values are not always reliable indicators of posture, as considerable overlap exists between the ranges reported by Kubo and Benton (2009) for lizards on the one hand and birds and mammals on the other. Lateral undulation, for example, can increase pace angulation in a sprawler by bringing the feet closer to the trackway midline (Kubo and Benton 2009). Furthermore, it is possible that the trace fossil and body fossil records sample different faunas, particularly given that some sprawling archosauriforms were probably at least semiaquatic.

However, the ichnological data of Kubo and Benton (2009) do suggest that erect archosauriforms were present as early as the Early Triassic. This is corroborated by the occurrence of the poposauroid *Ctenosauriscus* and the fragmentary possible paracrocodylomorph *Vytshegdosuchus* in Lower Triassic (late Olenekian) strata in Germany and Russia, respectively (Butler et al. 2011). *Xilousuchus* from the Heshanggou Formation of China may be a second Olenekian poposauroid (Nesbitt et al. 2011), although the age of this taxon is uncertain (Butler et al. 2011). Further support for the early appearance of erect posture comes from the fact that some Early Triassic

tracks can be plausibly attributed to dinosauromorphs (Brusatte, Niedzwiedzki, et al. 2010). These discoveries indicate that many of the major clades within Archosauria, including forms with erect hind limb posture, must have appeared before the end of the Early Triassic.

The Postural Shift and the Rise of Archosauriforms

Charig (1972, 121) believed that erect hind limb posture was an advantageous condition that ultimately permitted archosauriforms to achieve a "dominant position in the land faunas of the Mesozoic." This scenario implies that erect hind limb posture offered some general functional advantage, and that more erect taxa gradually increased in diversity and/or abundance at the expense of less erect ones. In principle, the first of these predictions is testable through biomechanical analysis and modeling, whereas the second is testable through analysis of patterns of abundance and diversity in the Triassic fossil record.

However, the functional benefits and disadvantages of erect as opposed to sprawling posture remain surprisingly poorly understood. Charig (1972) suggested that an erect walker would require less muscular effort to hold its body clear of the ground, would be less likely to slip upon the substrate, and would benefit from an increase in the "potential efficiency of the limb as a locomotor organ" (Charig 1972, 129) because the limb could

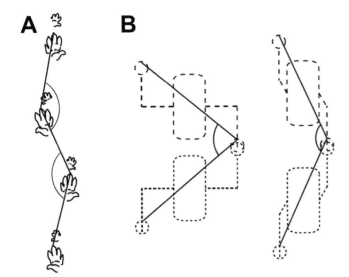

FIG. 7.11 Relationship between posture and pace angulation. A, Pace angulation angles measured on trackway; B, diagram showing that pace angulation is lower in more sprawling tetrapods (left) than more erect ones (right). Note, however, the lack in (A) of lateral undulation of the body, which could bring the tracks closer to the trackway midline and increase the measured pace angulation. After Kubo and Benton (2009).

be used as an extensible strut rather than only as a lever. Bakker (1974, 500) asserted that erect posture resulted in "much greater maneuverability at high speed on level ground." Kemp (1980) suggested that erect locomotion might be characterized by greater efficiency, speed, and maneuverability, and perhaps by a less "complex" pattern of stresses at the ankle joint. These various ideas are intuitively plausible and deserve further investigation, but have not yet been rigorously formulated and tested in the context of archosauriform evolution. As a result, the hypothesis that erect posture conferred a competitive advantage suffers from the shortcoming of being unable to spell out the nature of the advantage in clear and specific terms.

Similarly, patterns in the fossil record provide little support for the hypothesis of competitive replacement of less erect forms by more erect ones. While there was clearly a shift in the Late Triassic from faunas dominated by rhynchosaurs and synapsids to faunas dominated by archosauriforms, evidence that this process was competitive is elusive. Benton (1986b, 1994, 2006), for example, defended an opportunistic replacement model in which an extinction at the end of the Carnian age of the Late Triassic decimated rhynchosaurs and synapsids, allowing archosauriforms and particularly dinosaurs to radiate during the succeeding Norian. Olsen and Sues (1986) identified the Carnian itself as a time of rapid turnover in tetrapod faunas, an interpretation that also posits a diversification of archosauriforms in the Late Triassic. As noted above, however, strong evidence exists that erect hind limb posture was present in some archosauriforms by the end of the Early Triassic, well before the emergence of archosauriform-dominated faunas. The competitive model may have to be abandoned if it cannot be elaborated to explain why, if erect posture was highly advantageous, the rise of archosauriforms to ecological dominance did not occur soon after the appearance of erect posture in this group.

The possibility remains that erect posture might still provide a partial explanation for why dinosaurs and other archosauriforms were so successful in radiating into whatever empty ecological space existed following the Carnian-Norian boundary. However, at least some archosauriforms would also have benefited from the physiological advantages of rapid ontogenetic growth and a unidirectional lung (Sookias et al. 2012).

It may never be clear which features were most helpful to the archosauriform groups that flourished during the Triassic, but biomechanical comparisons between the buttress-erect and pillar-erect conditions and among different postural grades would at least be helpful in assessing the likelihood that postural evolution played a key role in the rise of archosauriforms.

The Postural Shift and Bipedality

The attainment of erect posture paved the way for the evolution of bipedality in derived poposauroids (Nesbitt 2007; Gauthier et al. 2011), dinosauromorphs (Sereno and Arcucci 1994a, 1994b; Langer et al. 2010), and possibly pterosaurs (Padian 1983a, 1983b, 2008). Whereas some lizards can run bipedally despite their sprawled hind limb posture (Gauthier et al. 2011), sprawled tetrapods are rarely if ever capable of bipedal standing or walking. Erect hind limb posture was therefore an essential prerequisite or corequisite of bipedality in archosauriform evolution (Charig 1972).

Bipedality, in turn, was unquestionably an important innovation in the history of the ornithodiran lineage. This adaptation freed the forelimb for such functions as prey capture, possibly climbing in some derived theropods (e.g., Zhang et al. 2002), and ultimately aerial locomotion in birds and perhaps some of their close non-avialan relatives. Among pseudosuchians, by contrast, the occurrence of bipedality in a few derived poposauroids did not lead to extensive taxonomic diversification or morphological experimentation.

Future Directions

To date, the locomotion of Triassic archosauriforms has been investigated primarily by the qualitative and semi-quantitative methods of traditional functional morphology, an approach that has provided many insights but nevertheless was aptly criticized by Hutchinson (2006) as too intuitive and too reliant on imperfect analogies with modern animals. However, some recent studies, particularly that by Bates and Schachner (2012), have adopted a greater degree of quantitative rigor. Other authors have contributed to an improved understanding of the evolution of locomotion in Triassic archosaurs by employing explicit methods of phylogenetic inference (e.g., Gauthier et al. 2011) or by using clear criteria to infer posture from ichnological data (Kubo and Benton 2009; Padian et al. 2010).

The central challenge in the coming years will be to use methods of scanning, computer modeling, and phylogenetic inference to test hypotheses about locomotion in basal archosauriforms, particularly by arriving at quantitative estimates of such parameters as ranges of joint motion and muscle moment arms. Examining basal archosauriforms and phytosaurs in order to confirm or refute the inference that they were sprawlers would be a particularly worthy endeavor, as would further analysis of terrestrial locomotion in basal pterosaurs. The work of Bates and Schachner (2012) offers an excellent case study in the application of quantitative methods to the hind limb biomechanics of a Triassic archosauriform. Extending this approach to a larger set of judiciously chosen taxa, in the context of rapidly improving knowledge of archosauriform phylogeny (Dilkes and Sues 2009; Brusatte, Benton, et al. 2010; Ezcurra et al. 2010; Desojo et al. 2011; Nesbitt 2011; Dilkes and Arcucci 2012) and limb function in extant archosauriforms (Gatesy et al. 2010) will make it possible to rigorously test old ideas about evolutionary shifts in hind limb posture and no doubt also generate some new ones. The more theoretical problem of determining whether there are systematic differences among postural grades in such parameters as stability, maneuverability, and locomotor efficiency is also ripe for renewed attack, ideally by a combination of abstract modeling and experiments on living animals representing various points along the continuum from sprawling to erect posture.

***

Acknowledgments

Many of the ideas expressed in this chapter were formulated during my PhD research on the evolution of ankle structure and function in archosauriforms, carried out under the invaluable but never overbearing guidance of Farish A. Jenkins Jr. Some of the firsthand observations and illustrations presented in this chapter resulted from museum visits I made as a graduate student, during which I was welcomed and assisted by individuals too numerous to mention but particularly including Jennifer Botha-Brink, Matthew Carrano, Sandra Chapman, Paul Barrett, Sheena Kaal, and Jaime Powell. I am indebted to David Baier, Michael Benton, Leon Claessens, Alfred Crompton, Stephen Gatesy, David Gower, Agustina Lecuona, Russell Main, Molly Markey, Charles Marshall, Kevin Padian, and Hans-Dieter Sues for much valuable discussion, and to Brian Choo for drafting many of the illustrations in this chapter. The manuscript also benefited greatly from reviews by Julia Desojo and Stephen Gatesy, and editorial guidance from Kenneth Dial. The component of this work that dates from my PhD thesis was supported by a Chapman Fellowship and by the Department of Organismic and Evolutionary Biology, Harvard University. Since relocating to the Institute of Vertebrate Paleontology and Paleoanthropology, I have been supported by the National Natural Science Foundation of China and the Chinese Academy of Sciences.

References

Baczko, M. B. von, and M. D. Ezcurra. 2013. Ornithosuchidae: a group of Triassic archosaurs with a unique ankle joint. Geological Society, London, Special Publications 379:187–202.

Bakker, R. T. 1971. Dinosaur physiology and the origin of mammals. Evolution 25:636–658.

Bakker, R. T. 1974. Dinosaur bioenergetics—a reply to Bennett and Dalzell, and Feduccia. Evolution 28:497–503.

Bates, K. T., and E. R. Schachner. 2012. Disparity and convergence in bipedal archosaur locomotion. Journal of the Royal Society Interface 9:1339–1353.

Benton, M. J. 1986a. The Late Triassic reptile *Teratosaurus*— a rauisuchian, not a dinosaur. Palaeontology 29:293–301.

Benton, M. J. 1986b. The Late Triassic tetrapod extinction events. Pp. 303–320 in K. Padian, ed., The Beginning of the Age of Dinosaurs. New York: Cambridge University Press.

Benton, M. J. 1994. Late Triassic to Middle Jurassic extinctions among continental tetrapods: testing the pattern. Pp. 366–397 in N. C. Fraser and H.-D. Sues, eds., In the Shadow of the Dinosaurs. Cambridge: Cambridge University Press.

Benton, M. J. 2006. The origin of the Dinosaurs. Pp. 11–19 in Colectivo Arqueológico-Paleontológico Salense, ed., Actas de las III jornadas sobre dinosaurios y su entorno. Burgos, Spain: Salas de los Infantes.

Benton, M. J., and J. M. Clark. 1988. Archosaur phylogeny and the relationships of the Crocodylia. Pp. 295–338 in M. J. Benton, ed., The Phylogeny and Classification of the Tetrapods, Volume 1: Amphibians, Reptiles, Birds. Systematics Association Special Volume No. 35A. Oxford: Clarendon Press.

Blob, R. W. 2001. Evolution of hindlimb posture in nonmammalian therapsids: biomechanical tests of paleontological hypotheses. Paleobiology 27:14–38.

Bonaparte, J. F. 1972. Los tetrápodos del sector superior de la Formación Los Colorados, La Rioja, Argentina (Triásico Superior). I Parte. Opera Lilloana 22:1–183.

Bonaparte, J. F. 1984. Locomotion in rauisuchid thecodonts. Journal of Vertebrate Paleontology 3:210–218.

Botha-Brink, J., and R. M. H. Smith. 2011. Osteohistology of the Triassic archosauromorphs Prolacerta, Proterosuchus, Euparkeria, and Erythrosuchus from the Karoo Basin of South Africa. Journal of Vertebrate Paleontology 31:1238–1254.

Brinkman, D. 1980. The hind limb step cycle of Caiman sclerops and the mechanics of the crocodile tarsus and metatarsus. Canadian Journal of Zoology 58:2187–2200.

Brusatte, S. L., M. J. Benton, J. B. Desojo, and M. C. Langer. 2010. The higher-level phylogeny of Archosauria (Tetrapoda: Diapsida). Journal of Systematic Palaeontology 8:3–47.

Brusatte, S. L., S. J. Nesbitt, R. B. Irmis, R. J. Butler, M. J. Benton, and M. A. Norell. 2010. The origin and early radiation of dinosaurs. Earth-Science Reviews 101:68–100.

Brusatte, S. L., G. Niedzwiedzki, and R. J. Butler. 2010. Footprints pull origin and diversification of dinosaur stem lineage deep into Early Triassic. Proceedings of the Royal Society B 278:1107–1113.

Butler, R. J., S. L. Brusatte, M. Reich, S. J. Nesbitt, R. R. Schoch, and J. J. Hornung. 2011. The sail-backed reptile Ctenosauriscus from the latest Early Triassic of Germany and the timing and biogeography of the early archosaur radiation. PLoS One 6:e25693.

Charig, A. J. 1972. The evolution of the archosaur pelvis and hind-limb: an explanation in functional terms. Pp. 121–155 in K. A. Joysey and T. S. Kemp, eds., Studies in Vertebrate Evolution. Edinburgh: Oliver & Boyd.

Charig, A. J., and H.-D. Sues. 1976. Proterosuchia. Pp. 11–39 in O. Kuhn, ed., Handbuch der paläoherpetologie. Stuttgart: Gustav Fischer.

Chatterjee, S. 1978. A primitive parasuchid (phytosaur) reptile from the Upper Triassic Maleri Formation of India. Palaeontology 21:83–127.

Colbert, E. H. 1952. A pseudosuchian reptile from Arizona. Bulletin of the American Museum of Natural History 99:561–592.

Cruickshank, A. R. I. 1972. The proterosuchian thecodonts. Pp. 89–119 in K. A. Joysey and T. S. Kemp, eds., Studies in Vertebrate Evolution. Edinburgh: Oliver & Boyd.

Cruickshank, A. R. I. 1978. The pes of Erythrosuchus africanus Broom. Zoological Journal of the Linnean Society 62:161–177.

Cruickshank, A. R. I. 1979. The ankle joint in some early archosaurs. South African Journal of Science 75:168–178.

Crush, P. J. 1984. A late Upper Triassic sphenosuchid crocodilian from Wales. Palaeontology 27:131–157.

Desojo, J. B., and A. M. Báez. 2005. El esqueleto postcraneano de Neoaetosauroides (Archosauria: Aetosauria) del Triásico Superior del centro-oeste de Argentina. Ameghiniana 42:115–126.

Desojo, J. B., M. D. Ezcurra, and C. L. Schultz. 2011. An unusual new archosauriform from the Middle-Late Triassic of southern Brazil and the monophyly of Doswelliidae. Zoological Journal of the Linnean Society 161:839–871.

Dilkes, D., and A. Arcucci. 2012. Proterochampsa barrionuevoi (Archosauriformes: Proterochampsia) from the Late Triassic (Carnian) of Argentina and a phylogenetic analysis of Proterochampsia. Palaeontology 55:853–885.

Dilkes, D., and H.-D. Sues. 2009. Redescription and phylogenetic relationships of Doswellia kaltenbachi (Diapsida: Archosauriformes) from the Upper Triassic of Virginia. Journal of Vertebrate Paleontology 29:58–79.

Dzik, J. 2003. A beaked herbivorous archosaur with dinosaur affinities from the early Late Triassic of Poland. Journal of Vertebrate Paleontology 23:556–574.

Erickson, G. M., J. Catanese III, and T. M. Keaveny. 2002. Evolution of the biomechanical material properties of the femur. Anatomical Record 268:115–124.

Ewer, R. F. 1965. The anatomy of the thecodont reptile Euparkeria capensis Broom. Philosophical Transactions of the Royal Society of London B 248:379–435.

Ezcurra, M. D., A. Lecuona, and A. Martinelli. 2010. A new basal archosauriform diapsid from the Lower Triassic of Argentina. Journal of Vertebrate Paleontology 30:1433–1450.

Gatesy, S. M. 1991. Hind limb movements of the American alligator (Alligator mississippiensis) and postural grades. Journal of Zoology, London 224:577–588.

Gatesy, S. M., D. B. Baier, F. A. Jenkins, and K. Dial. 2010. Scientific rotoscoping: a morphology-based method of 3-D motion analysis and visualization. Journal of Experimental Zoology 313A:244–261.

Gauthier, J. A., S. J. Nesbitt, E. R. Schachner, G. S. Bever, and W. G. Joyce. 2011. The bipedal stem crocodilian Poposaurus gracilis: inferring function in fossils and innovation in archosaur locomotion. Bulletin of the Peabody Museum of Natural History 52:107–126.

Göhlich, U. B., L. M. Chiappe, J. M. Clark, and H.-D. Sues. 2005. The systematic position of the Late Jurassic alleged dinosaur Macelognathus (Crocodylomorpha: Sphenosuchia). Canadian Journal of Earth Sciences 42:307–321.

Gower, D. J. 1996. The tarsus of erythrosuchid archosaurs, and implications for early diapsid phylogeny. Zoological Journal of the Linnean Society 116:347–375.

Gower, D. J. 2003. Osteology of the early archosaurian reptile Erythrosuchus africanus Broom. Annals of the South African Museum 110:1–84.

Gower, D. J., and R. Schoch. 2009. Postcranial anatomy of the rauisuchian archosaur Batrachotomus kupferzellensis. Journal of Vertebrate Paleontology 29:103–122.

Gower, D. J., and A. G. Sennikov. 1997. Sarmatosuchus and the early history of the Archosauria. Journal of Vertebrate Paleontology 17:60–73.

Hutchinson, J. R. 2001. The evolution of femoral osteology and soft tissues on the line to extant birds (Neornithes). Zoological Journal of the Linnean Society 131:169–197.

Hutchinson, J. R. 2006. The evolution of locomotion in archosaurs. Comptes Rendus Palevol 5:519–530.

Jenkins, F. A., Jr., N. H. Shubin, S. M. Gatesy, and K. Padian. 2001. A diminutive pterosaur (Pterosauria: Eudimorphodontidae) from the Greenlandic Triassic. Bulletin of the Museum of Comparative Zoology 156:151–170.

Kemp, T. S. 1980. The primitive cynodont Procynosuchus: structure, function and evolution of the postcranial skeleton.

Philosophical Transactions of the Royal Society of London B 288:217–258.

Kubo, T., and M. J. Benton. 2007. Evolution of hind limb posture in archosaurs: limb stresses in extinct vertebrates. Palaeontology 50:1519–1529.

Kubo, T., and M. J. Benton. 2009. Tetrapod postural shift estimated from Permian and Triassic trackways. Palaeontology 52:1029–1037.

Langer, M. C., M. D. Ezcurra, J. S. Bittencourt, and F. E. Novas. 2010. The origin and early evolution of dinosaurs. Biological Reviews 85:55–110.

Lecuona, A., and J. B. Desojo. 2011. Hind limb osteology of *Gracilisuchus stipanicicorum* (Archosauria: Pseudosuchia). Earth and Environmental Science Transactions of the Royal Society of Edinburgh 102:105–128.

Long, R. A., and P. A. Murry. 1995. Late Triassic (Carnian and Norian) tetrapods from the southwestern United States. New Mexico Museum of Natural History and Science Bulletin 4:1–254.

Lucas, S. G., A. B. Heckert, and R. Kahle. 2002. Postcranial anatomy of *Angistorhinus*, a Late Triassic phytosaur from west Texas. New Mexico Museum of Natural History and Science Bulletin 21:157–164.

Nesbitt, S. 2007. The anatomy of *Effigia okeeffeae* (Archosauria: Suchia), theropod-like convergence, and the distribution of related taxa. Bulletin of the American Museum of Natural History 302:1–84.

Nesbitt, S. J. 2011. The early evolution of archosaurs: relationships and the origin of major clades. Bulletin of the American Museum of Natural History 352:1–292.

Nesbitt, S. J., S. L. Brusatte, J. B. Desojo, A. Liparini, M. A. G. de França, J. C. Weinbaum, and D. J. Gower. 2013. Rauisuchia. Geological Society, London, Special Publications 379:241–274.

Nesbitt, S. J., J. Liu, and C. Li. 2011. A sail-backed suchian from the Heshanggou Formation (Lower Triassic: Olenekian) of China. Earth and Environmental Science Transactions of the Royal Society of Edinburgh 101:271–284.

Nesbitt, S. J., R. B. Irmis, W. G. Parker, N. D. Smith, A. H. Turner, and T. Rowe. 2009. Hindlimb osteology and distribution of basal dinosauromorphs from the Late Triassic of North America. Journal of Vertebrate Paleontology 29:498–516.

Nesbitt, S. J., C. A. Sidor, R. B. Irmis, K. D. Angielczyk, R. M. H. Smith, and L. A. Tsuji. 2010. Ecologically distinct dinosaurian sister group shows early diversification of Ornithodira. Nature 464:95–98.

Nesbitt, S. J., M. R. Stocker, B. J. Small, and A. Downs. 2009. The osteology and relationships of *Vancleavea campi* (Reptilia: Archosauriformes). Zoological Journal of the Linnean Society 157:814–864.

Olsen, P. E., and H.-D. Sues. 1986. Correlation of continental Late Triassic and Early Jurassic sediments, and patterns of the Triassic-Jurassic tetrapod transition. Pp. 321–351 in K. Padian, ed., The Beginning of the Age of Dinosaurs. New York: Cambridge University Press.

Padian, K. 1983a. A functional analysis of flying and walking in pterosaurs. Paleobiology 9:218–239.

Padian, K. 1983b. Osteology and functional morphology of *Dimorphodon macronyx* (Buckland) (Pterosauria: Rhamphorhynchoidea) based on new material in the Yale Peabody Museum. Postilla 189:1–44.

Padian, K. 2008. Were pterosaur ancestors bipedal or quadrupedal? morphometric, functional and phylogenetic considerations. Zitteliana B 28:21–33.

Padian, K., C. Li, and J. Pchelnikova. 2010. The trackmaker of *Apatopus* (Late Triassic, North America): implications for the evolution of archosaur stance and gait. Palaeontology 53:175–189.

Parrish, J. M. 1986a. Locomotor adaptations in the hindlimb and pelvis of the Thecodontia. Hunteria 1(2):1–35.

Parrish, J. M. 1986b. Structure and function of the tarsus in the phytosaurs (Reptilia: Archosauria). Pp. 35–43 in K. Padian, ed., The Beginning of the Age of Dinosaurs. New York: Cambridge University Press.

Parrish, J. M. 1987. The origin of crocodilian locomotion. Paleobiology 13:396–414.

Parrish, J. M. 1991. A new specimen of an early crocodylomorph (cf. *Sphenosuchus* sp.) from the Upper Triassic Chinle Formation of Petrified Forest National Park, Arizona. Journal of Vertebrate Paleontology 11:198–212.

Parrish, J. M. 1993. Phylogeny of the Crocodylotarsi, with reference to archosaurian and crurotarsan monophyly. Journal of Vertebrate Paleontology 13:287–308.

Reilly, S. M., and J. A. Elias. 1998. Locomotion in *Alligator mississippiensis*: kinematic effects of speed and posture and their relevance to the sprawling-to-erect paradigm. Journal of Experimental Biology 201:2559–2574.

Rewcastle, S. C. 1983. Functional adaptations in the lacertilian hind limb: a partial analysis of the sprawling posture and gait. Copeia 1983:476–487.

Romer, A. S. 1971. The Chañares (Argentina) Triassic reptile fauna. XI. Two new long-snouted thecodonts, *Chanaresuchus* and *Gualosuchus*. Breviora 379:1–22.

Romer, A. S. 1972. The Chañares (Argentina) Triassic reptile fauna. XII. The postcranial skeleton of the thecodont *Chanaresuchus*. Breviora 385:1–21.

Sawin, H. J. 1947. The pseudosuchian reptile *Typothorax meadei*. Journal of Paleontology 21:201–238.

Sereno, P. C. 1991. Basal archosaurs: phylogenetic relationships and functional implications. Society of Vertebrate Paleontology Memoir 2:1–53.

Sereno, P. C., and A. B. Arcucci. 1990. The monophyly of crurotarsal archosaurs and the origin of bird and crocodile ankle joints. Neues Jahrbuch für Geologie und Paläontologie, Abhandlung 180:21–52.

Sereno, P. C., and A. B. Arcucci. 1994a. Dinosaurian precursors from the Middle Triassic of Argentina: *Lagerpeton chanarensis*. Journal of Vertebrate Paleontology 13:385–399.

Sereno, P. C., and A. B. Arcucci. 1994b. Dinosaurian precursors from the Middle Triassic of Argentina: *Marasuchus lilloensis*, gen. nov. Journal of Vertebrate Paleontology 14:53–73.

Sookias, R. B., R. J. Butler, and R. B. J. Benson. 2012. Rise of dinosaurs reveals major body-size transitions are driven by passive processes of trait evolution. Proceedings of the Royal Society B 279:2180–2187.

Sues, H.-D., J. B. Desojo, and M. D. Ezcurra. 2013. Doswelliidae: a clade of unusual armoured archosauriforms from the Middle and Late Triassic. Geological Society, London, Special Publications 379:49–58.

Sues, H.-D., P. E. Olsen, J. G. Carter, and D. M. Scott. 2003. A new crocodylomorph archosaur from the Upper Triassic of North Carolina. Journal of Vertebrate Paleontology 23:329–343.

Sullivan, C. 2007. Function and evolution of the hind limb in Triassic archosaurian reptiles. PhD thesis, Harvard University.

Sullivan, C. 2010. The role of the calcaneal "heel" as a propulsive lever in basal archosaurs and extant monitor lizards. Journal of Vertebrate Paleontology 30:1422–1432.

Trotteyn, M. J. 2011. Material postcraneano de *Proterochampsa barrionuevi* Reig 1959 (Diapsida: Archosauriformes) del Triásico Superior del centro-oeste de Argentina. Ameghiniana 48:424–446.

Trotteyn, M. J., and J. A. Haro. 2011. The braincase of a specimen of *Proterochampsa* Reig (Archosauriformes: Proterochampsidae) from the Late Triassic of Argentina. Paläontologische Zeitschrift 85:1–17.

Trotteyn, M. J., R. N. Martínez, and O. A. Alcober. 2012. A new proterochampsid *Chanaresuchus ischigualastensis* (Diapsida, Archosauriformes) in the early Late Triassic Ischigualasto Formation, Argentina. Journal of Vertebrate Paleontology 32:485–489.

Walker, A. D. 1964. Triassic reptiles from the Elgin area: *Ornithosuchus* and the origin of carnosaurs. Philosophical Transactions of the Royal Society of London B 248:53–134.

Walker, A. D. 1970. A revision of the Jurassic reptile *Hallopus victor* (Marsh), with remarks on the classification of crocodiles. Philosophical Transactions of the Royal Society of London B 257:323–372.

Wu, X.-C., and S. Chatterjee. 1993. *Dibothrosuchus elaphros*, a crocodylomorph from the Lower Jurassic of China and the phylogeny of the Sphenosuchia. Journal of Vertebrate Paleontology 13:58–89.

Young, C. C. 1964. The pseudosuchians in China. Palaeontologia Sinica 151:1–205.

Zhang, F., Z. Zhou, X. Xu, and X. Wang. 2002. A juvenile coelurosaurian theropod from China indicates arboreal habits. Naturwissenschaften 89:394–398.

8

Fossils, Trackways, and Transitions in Locomotion: *A Case Study of Dimetrodon*

James A. Hopson*

The limb girdles and proximal limb bones of basal synapsids of the Late Paleozoic strongly suggest a sprawling posture in which the humerus and femur were held out horizontally from the body. The shoulder joint of "pelycosaurs" and other basal amniotes (Jenkins 1971, 1973; Holmes 1977) is formed by an elongate screw-shaped articular surface that severely constrained movement of the humerus. The hip joint of "pelycosaurs" was less constrained (Jenkins 1971), but the femur nonetheless lacked extensive freedom of movement. Both proximal limb bones appear to have had greater freedom of long-axis rotation than of adduction/abduction or fore-aft movement (Jenkins 1971, 1973). My approach to locomotor behavior in basal synapsids involves biomechanical analysis of skeletal structure, but also evidence on footfall parameters provided by trackways (see Gatesy et al. 1999). I have developed a protocol for estimating step length from skeletal structure and testing the results against evidence derived from trackways. An important outcome of this analysis is the demonstration of how important lateral bending of the trunk was to the locomotor repertoire of basal synapsids.

Extinct sprawlers with constrained limb movements have long been expected to have relatively short step lengths and slow locomotor speeds, as is entertainingly described for pelycosaurian locomotion by Romer and Price (1940, 172): "The

* Department of Organismal Biology and Anatomy, University of Chicago

legs were widely sprawled out from the body, the track-way wide, the stride short, and at the best they must have been exceedingly slow and essentially turtle-like in locomotion." This improbable image would indeed be the case unless, as I propose to show, lateral bending of the trunk was used to augment the distance the limbs moved with each step. Thus, side-to-side bending of the trunk through a fairly large angle would have been essential for even moderately rapid locomotion in basal synapsids, as it is in living lizards (Sukhanov 1974; Jenkins and Goslow 1983).

Lateral bending of the trunk in primitive synapsids may have set severe constraints on their ability both to move at high speed and to supply their locomotor muscles with oxygen. Carrier (1987a, 1987b, 1989, 1991; Wang et al. 1997), in a series of insightful studies, showed that in fast-moving lizards side-to-side bending of the trunk, resulting from *alternate* contractions of hypaxial muscles, precluded the bilateral compression and expansion of the lungs, which requires synchronous *bilateral* contraction of these same muscles. Thus, during high-speed running that includes extensive lateral axial bending, fast-moving sprawlers suspend breathing and rely on anaerobic glycolysis to fuel muscle contraction. This suggests that the use of the trunk in alternate lateral bending for increased locomotor speed in basal synapsids and other early amniotes imposed a reliance on anaerobic activity metabolism in which the buildup of the metabolic byproduct lactic acid caused the locomotor muscles to rapidly fatigue. This constrained high-speed locomotion to a short quick dash and forced these primitive amniotes to adopt a low energy sit-and-wait ambush mode of prey capture (Carrier 1987a).

The evidence of skeletal structure permits stance and gait to be reconstructed within somewhat broad limits, but it cannot demonstrate the relative contributions of the limbs and trunk to stride length. In part, this is because evidence of trunk movement cannot be determined from the skeleton alone. An independent form of evidence is required to assess how the feet were placed during locomotion in extinct species, permitting not only stride length and transverse distance between left and right feet to be measured, but also permitting the relative contribution of axial bending to be determined. Such information is available only from fossil trackways.

The analysis presented here combines reconstruction of limb movement through functional analysis of early synapsid skeletal morphology and comparison of the reconstructed placement of the feet with the actual placement of feet in the trackway of an individual of the same or a closely related taxon. The degree of mismatch between the step pattern of the skeletally reconstructed "trackway" and the step pattern made by the (once-) living and moving animal, as represented by its preserved trackway, permits an assessment of how much compensatory lateral bending of the trunk in the skeletally based animal is required to approximate the features of the trackway.

Two essential aspects of such an analysis are that (1) the skeletal elements must possess features (e.g., articular surfaces) that place recognizable limits on possible movements of the limbs and (2) some dimension related to body size be determinable for both the skeleton and the trackmaker. The forelimbs of basal synapsids are ideally suited to such analysis because of the constrained nature of both the shoulder and elbow joints (Jenkins 1971, 1973). The hind limb of "pelycosaurs" is less constrained, as I have determined in a preliminary study, and so is less suitable for such an analysis.

With respect to a common measurement between skeleton and trackway maker, a reasonably accurate measure of glenoacetabular length (i.e., the distance between shoulder and hip joints) can be determined for both (see Baird 1952; Padian and Olsen 1984). With this information, the fossil skeleton can be scaled to the same glenoacetabular length as the trackmaker and thus act as a proxy for the actual trackmaker. I have carried out such an analysis on the Early Permian sphenacodontid "pelycosaur" *Dimetrodon* and a trackway of the supposed sphenacodontid ichnogenus *Dimetropus*.

Before discussing the fossil material used in this study, I shall explain the protocol I have developed for (1) determining step length from skeletal anatomy; (2) converting measurements derived from fossil skeletons to those of the maker of a trackway, thus producing a proxy trackmaker for comparing skeletal structure and trackway parameters; (3) measuring discrepancies between step lengths determined from trackways and those determined from limb structure, the shortfall of the latter being the proportion of the step to be compensated for by axial bending of the trunk; and, finally, (4) determining the amount of lateral bending required to yield the step lengths determined from the trackway. An independent test of the accuracy of the reconstructed locomotor pattern is to compare the reconstructed transverse distance between left and right feet with that determined from the trackway. This distance should be commensurately reduced as step length increases, so, provided that the functional reconstruction of the trackmaker surrogate is correct, this inverse relationship should exist. Where reconstructed trackway width varies widely from that determined from the actual trackway, the functional analysis must be based on incorrect assumptions of skeletal functional morphology.

Trackway Analysis of the Sphenacodontid Ichnogenus *Dimetropus*

Trackways of Lower Permian "pelycosaurs" that possess proportions of the manus and pes resembling the skeletal elements of *Dimetrodon* are well known from Europe (Haubold 1971; Voigt and Ganzelewski 2010) and the western United States (Hunt et al. 1993; MacDonald 1994; Haubold et al. 1995). They are placed in the ichnogenus *Dimetropus* (Romer and Price 1940, 336) and assigned to several species. Abundant trackways attributed to *Dimetropus*, collected from the southern Robledo Mountains, Doña Ana County, New Mexico, are housed in the New Mexico Museum of Natural History and Science, Albuquerque, and several other institutions. A long series of contiguous slabs bearing *Dimetropus* trackways from the New Mexico Museum of Natural History and Science, Albuquerque (NMMNHS) locality 846, illustrated by Hunt et al. (1993, fig. 7), is housed in the Carnegie Museum of Natural History, Pittsburgh (CMNH), where it is cataloged as CMNH 87684 and tentatively identified as *D. leisnerianus*. I have photographed and measured a sequence of 10 manus and pes prints from this set of trackways (fig. 8.1), chosen because they are well preserved and clearly pertain to a single individual. I chose as a common point preserved in all prints the base of the notch between digits 2 and 3.

The principal features measured in the *Dimetropus* trackway are those described for *Varanus komodoensis*

by Padian and Olsen (1984, fig. 1 and p. 665). They are measured for both manus and pes and listed in appendix table 8.1, although only the manus is included in the subsequent analysis of limb and trunk movements. Measurements utilized in conjunction with the skeletal analysis are forelimb pace (step) length (the longitudinal distance between sequential left and right manus prints), stride length (the longitudinal distance between sequential prints of the same manus), and transverse distance between left and right manus tracks. Of particular importance for the present analysis is glenoacetabular length (defined, respectively, for trackway measurement in fig. 8.1, and for skeletal measurement in fig. 8.2). This assumes that all four feet were in contact with the substrate at once, however briefly. Padian and Olsen (1984) determined the glenoacetabular length from the trackway of a Komodo monitor (*Varanus komodoensis*) at the San Diego Zoo (best estimate from their data: 68.3 cm) and demonstrated its reliability by comparing it with a measurement (71 cm) made from the living animal. Glenoacetabular length permits comparisons of body size, and, therefore, skeletal proportions between a fossil skeleton and a presumably related trackmaker known only from footprints (see below).

Measurements of the *Dimetropus* trackway, both manus and pes, are presented in appendix table 8.1. The important measurements for this analysis, as noted above, are pace (step) and stride length of the manus, transverse distance between left and right manus, and glenoacetabular length.

Skeletal Analysis of *Dimetrodon*

For the skeletal portion of the analysis, I chose a small skeleton (fig. 8.2A) identified by Romer and Price (1940) as a female individual of *D. limbatus* (MCZ 1347) because these authors have published measurements of most limb bones of this specimen, as well as a restored body length. Glenoacetabular length (fig. 8.2A) was determined from their reconstruction of this specimen (Romer and Price 1940, fig. 61). The ratio of glenoacetabular length of the *Dimetropus* trackmaker to that of the *Dimetrodon limbatus* skeleton permits the measured elements of the skeleton to be scaled to these elements in the trackmaker. These measurements are provided in appendix table 8.2.

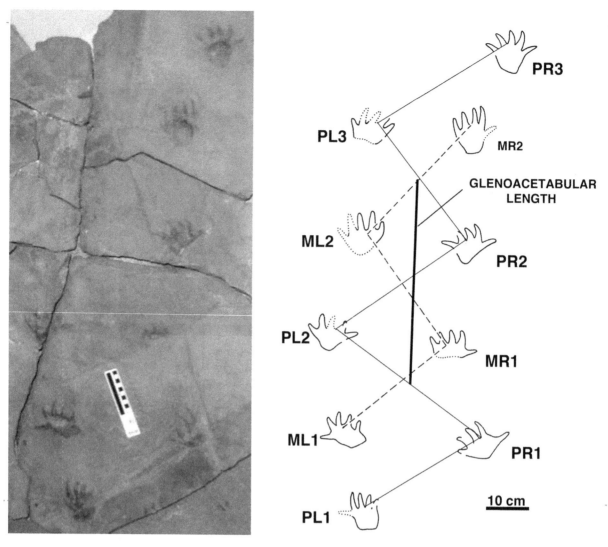

FIG. 8.1 Photograph of 10 consecutive footprints of a single individual from a set of *Dimetropus* trackways (CMNH 87684) in the Carnegie Museum of Natural History, Pittsburgh, and a tracing of the trackway with diagonal lines connecting footprints of manus (broken lines) and pes (solid lines). Prints are labeled manus (M) or pes (P), left (L) or right (R), and numbered in sequence (1–3). Thick vertical line represents glenoacetabular length, which is measured along the line connecting the midpoint of the diagonal line between consecutive manus prints (herein ML2 and MR2) and the midpoint of the diagonal between the next more posterior consecutive pes prints (PR1 and PL2). This estimate of glenoacetabular length on trackway assumes that all four feet contact substrate simultaneously, however briefly. Scale in both figures is 10 cm.

In order to reconstruct the step and stride lengths of the forelimb and also investigate the possible contribution of lateral bending of the trunk to locomotion in this undoubted sprawler (see below), the interglenoid width of the shoulder girdle (figs. 8.2B, 8.4) and length of the humerus must be determined, inasmuch as it is their alternate side-to-side rotation in the horizontal plane that determines the contribution of axial bending to stride length. As in other early tetrapods, the massive shoulder girdle of *Dimetrodon*, with its long interclavicle stem on the ventral midline, was almost certainly incapable of independent movement on the rib cage. Interglenoid width across the *Dimetrodon* shoulder girdle was determined from a drawing of the scapulocoracoid in lateral and ventral views (Romer and Price 1940, fig. 18) scaled to measurements of the *D. limbatus* scapula (Romer and Price 1940, table 4). The reconstructed horizontal span of the shoulder girdle and humeri is given in appendix table 8.2.

The contribution of the appendicular skeleton to step and stride length requires determination of total long axis rotation of the humerus and functional length of the lower limb. The latter is defined here as that portion of the limb, primarily the radius but also the radial condyle of the humerus and the proximal carpals (as discussed below), that rotate with the humerus during the step cycle. I have assumed that the manus of *Dimetrodon* distal to the proximal carpals was primarily plantigrade (see Romer and Price 1940), so not contributing to the functional step length. Ventroflexion of the normally horizontal distal carpals and metacarpals contributed to step length only at the end of the step, so the length of their lever arm is separately added to the step length determined by humeral rotation and retraction of the functional lower limb.

Determination of degree of axial rotation of the humerus.—For the functional analysis of the glenohumeral joint, particularly the amount of long-axis rotation of the humerus, two well-preserved specimens of an associated scapulocoracoid and humerus (FMNH UC816 [Field Museum of Natural History, Chicago]; and MCZ 1542 [Museum of Comparative Zoology, Harvard University, Cambridge]; the latter was previously described and analyzed by Jenkins 1971) permitted precise articulation of the articular head of the humerus in the glenoid and determination of minimum and maximum angles of rotation of the humerus. The possibility of abduction-adduction (elevation-depression)

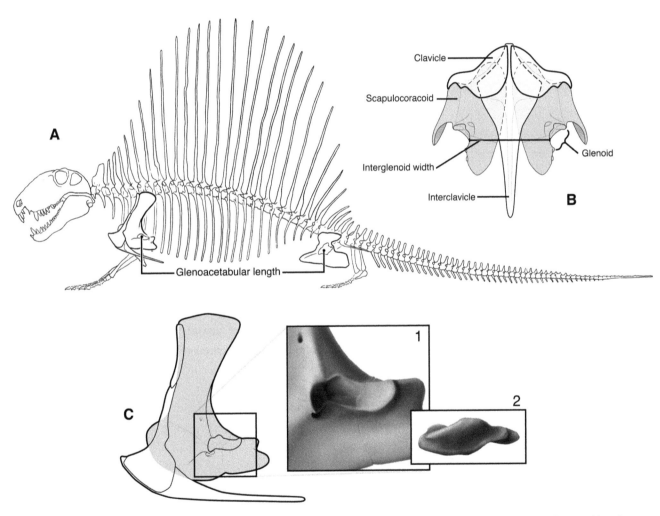

FIG. 8.2 A, Skeleton of *Dimetrodon limbatus* "female" (MCZ 1347) redrawn from Romer and Price (1940, fig. 61), with a total restored length of 256 cm. Left limbs removed in order to show the articular surfaces of the limb girdles. Heavy line indicates glenoacetabular length, distance between glenoid cavity of scapulocoracoid and acetabulum of pelvis. B, Shoulder girdle of *Dimetrodon* in ventral view, based on *D. milleri*, from Romer and Price (1940, fig. 18). Scapulocoracoid shaded and clavicles and interclavicle drawn as semitransparent. Heavy line represents interglenoid width. C, Shoulder girdle in lateral view with scapulocoracoid shaded. Box around glenoid represents area enlarged to the right (indicated by "1"), which shows detailed morphology of the screw-shaped glenoid. Box "2" shows the screw-shaped head of the left humerus, with anterior to the right. Rotation of the convex humeral head 180 degrees to the left would fit it into the concave glenoid. Not to scale.

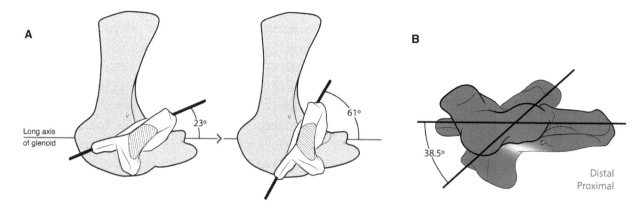

FIG. 8.3 A, Lateral view of scapulocoracoid and humerus, based on FMNH UC1134. The distal end of the humerus is incomplete (see text), and its preserved portion is represented by the area of parallel lines. The positions of the head of the humerus in relation to the glenoid are based on photographs. On the left, the humerus is rotated to the extreme clockwise position, which orients the long axis of the expanded proximal end of the humerus at 23 degrees posterodorsal to the long axis of the elongated glenoid (indicated by the horizontal line). On the right, the humerus is rotated to the extreme counterclockwise position, which orients the long axis of the proximal end of the humerus at 61 degrees posterodorsal to the long axis of the glenoid. B, Left humerus, based on MCZ 1542, as seen looking proximally down the long axis of its shaft to show the torsion of the expanded distal end (blue) on the expanded proximal end (red). The angle between the expanded ends is 38.5 degrees, although it varies in other specimens. Not to scale.

and protraction-retraction of the humerus was also examined.

Humeral rotation to the glenoid was determined by the best fit of the humeral head to the glenoid for the FMNH specimen (fig. 8.3). I found one stable and best fit of the humeral head and the glenoid for the posteriorly rotated position (23 degrees: fig 8.3A, left) and two stable fits (56.5 and 61.0 degrees: fig. 8.3A, right) in the anteriorly rotated position. The mean of the latter two angles (58.75 degrees) was used in the full analysis. The difference in angle between the most posteriorly and anteriorly rotated positions represents the total amount of long axis rotation of the humerus contributing to the step cycle (see appendix table 8.3).

As noted by Romer and Price (1940, 138), the humerus of basal synapsids is greatly expanded at both ends, and the "twist of the distal upon the proximal plane varies from about 60 to 35 degrees." The distal end of FMNH UC1134 is incomplete. Nonetheless, the angle of the transversely expanded distal end relative to the proximal end is estimated to be 38.5 degrees (fig. 8.3B), partly based on comparison to better-preserved landmarks of more complete specimens, notably MCZ 1542. The angle of torsion does not effect the amount of rotation the humerus undergoes, but it changes the direction in which the expanded distal end is oriented relative to the horizontal and thus would

affect the angle at which the lower limb meets the ground at the beginning of a step and the angle at which the foot is lifted from the ground at the end of the step.

Determination of functional length of the lower limb.—The functional length of the lower limb, as determined here, consists of the added lengths of the radial condyle below the rotational axis of the humerus, the radius, and the proximal carpal elements (radiale and medial centrale) that were elevated above the ground during locomotion. Inclusion of the proximal carpals as extensions of the radius is based on the observation by Romer and Price (1940, 162) that the radius "was butted firmly on the radiale, so that radius and carpus had little play on one another." These authors also note that "the distal carpals turned forward, in average articulation, so that the [presumably horizontal] metacarpals extended anteriorly at about a 45 degree angle from the plane of the proximal carpals." As there was no markedly apparent intracarpal joint, "each element could apparently move slightly on its neighbors, allowing the structure as a whole to alter its shape by increase or decrease of its vertical and transverse curvatures." In light of these observations, I include the proximal half of the carpus as an extension of the antebrachium and the distal carpals and metacarpals as an additional (distal) lever that pushed off from the ground at the end of the step cycle. Therefore, I have added the length of the fourth distal carpal plus metacarpal at the end of the step related to humeral rotation, assuming this part of the manus became vertical when the rest of the lower limb was maximally retracted. The lengths of the carpal and metacarpal elements are based on the lengths of these elements in *Dimetrodon milleri* (MCZ 1365) as figured by Romer and Price (1940, fig. 40H), in which manus width is close to that of the *Dimetropus* trackmaker. Measurements of the functional length of the lower limb are presented in appendix table 8.4.

Step and stride length determinations of the trackmaker were made using angles of rotation of the humerus derived from analysis of FMNH UC1134. Step length values and the method for estimating them are presented in appendix table 8.4.

Determination of contribution of lateral trunk rotation to step length.—To determine the contribution of lateral bending of the trunk to step length, it is necessary to compare the step lengths determined from the trackway, which combine both limb and trunk movements of the trackmaker, with those determined from limb movement only, as reconstructed from the *Dimetrodon* skeleton scaled to that of the trackmaker (see appendix table 8.4). If the latter step lengths are subtracted from the former, the remaining portion of trackway step length must be a function of lateral rotation of the trunk and shoulder girdle. If the transverse distance between the distal ends of left and right humeri also equals the transverse distance between left and right manus, which would be the case if the lower limb were oriented perpendicular to the long axes of the horizontal humeri, then horizontal bending of the trunk would move the manus a distance equal to that of the distal end of the humerus. Because of the variation in observed step (pace) length of the *Dimetropus* trackway (appendix table 8.1), two determinations of the contribution of lateral bending were made. The first utilized *mean* step length (22.8 cm), and the second used the *maximum* recorded step length (25.5 cm). The same contribution to step length (due to extension and retraction of the functional lower limb and foot ventroflexion) was subtracted from each total value, yielding a lower and higher value for degree of trunk bending required to produce each step length (appendix table 8.5). The contribution of lateral bending of the trunk was

determined by treating the resulting forward movement of each manus due *only* to trunk bending as the equivalent of a step. A series of right triangles were constructed, this time with the length of the hypotenuse equal to the distance between the ends of the humeri, as viewed from above, and the length of the triangle's base being that portion of step length due to lateral bending. This permitted the angle between the hypotenuse and the vertical side to be determined. This angle represents the degree of lateral bending of the trunk. The values (angles and lengths) associated with lateral bending of the trunk are given in appendix table 8.5.

Measurements from the analysis of the *Dimetrodon limbatus* skeleton are presented in appendix tables 8.2, 8.4, and 8.5. Figure 8.4 shows graphic results of the analysis of axial bending, and figure 8.5 shows a reconstruction of the step cycle in *Dimetrodon*.

A common measurement between skeleton and trackway is required to scale measured skeletal elements to the animal that made the trackway. As noted above, glenoacetabular length, the distance between the centers of the shoulder and hip joints, was determined in the skeleton of *Dimetrodon limbatus* (MCZ 1347; fig. 8.2A) and reconstructed from the *Dimetropus* trackway (CMNH 87684; fig. 8.1). The *Dimetropus* trackmaker has a glenoacetabular length of 163 cm, which scales as 0.637 to the reconstructed glenoacetabular length of 256 cm for the skeleton of *D. limbatus*. Appendix table 8.2 presents the skeletal measurements of *D. limbatus* and the scaled-down measurements (by 0.637) of the trackmaker. The latter are used in the determination of functional limb parameters from which are determined step and stride length in the trackmaker.

An essential part of the analysis is the determination of total rotation of the humerus from the beginning to the end of the step cycle. Measurements of humeral rotation angles and methods for obtaining these measurements are presented in appendix table 8.3. The composition and contribution of the functional lower limb and manus to step length and methods of their calculation are presented in appendix table 8.4. Details of forelimb function are presented in the step cycle section, below.

The main objective of this study has been to determine whether basal synapsids, notably the sphenacodontid *Dimetrodon*, (1) had a fully sprawling posture and (2) used alternate lateral bending of the trunk as

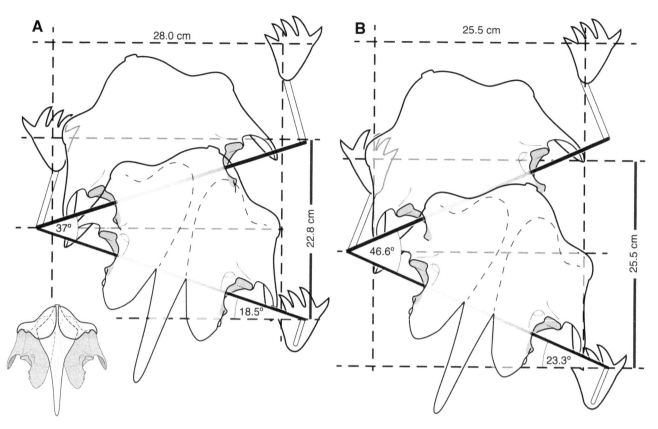

FIG. 8.4 Ventral view of scapulocoracoid of *Dimetrodon* at furthest right and left bending of trunk and girdle in the horizontal plane (bone lengths scaled down by 50% from Romer and Price 1940, fig. 18). Angles represent degree of bending of trunk and protraction from the transverse axis of the humeri (for step length on right and stride length on the left). Heavy diagonal lines contacting the glenoid cavity of the shoulder girdle represent left and right humeri; light area enclosed by thin lines extending diagonally forward from the distal ends of the humeri represent the lower limb; manus shown diagramatically. A, Degree of bending of trunk required to compensate for the discrepancy between forelimb step length as determined from functional analysis of *Dimetrodon* skeleton and *mean* step length derived from trackway. B, Same determined for *maximum* step length derived from trackway. Numbers on side of drawings indicate step lengths and at top indicate transverse distances between left and right manus for the different degrees of lateral bending. See text for further explanation.

an important component of locomotion. The data on lateral bending are presented in appendix table 8.5 and figure 8.4.

Figure 8.4 represents ventral views of the outline of the scapulocoracoid at the beginning and end of the propulsive phase of a stride. Figure 8.4A represents trunk and girdle rotation to produce *mean* step length of the trackway; figure 8.4B represents *maximum* step length. In both, the smaller angle on the lower right side represents the deviation (i.e., horizontal rotation) from the longitudinal midline of the trunk and, therefore, of the firmly attached shoulder girdle, during a single step. The larger angle on the left represents the full horizontal rotation of the shoulder girdle from the farthest right position to an equal position on the left; this represents the total angular movement of the girdle, and,

therefore, of the trunk, during a full stride. Because the humeri (represented by the heavy lines extending diagonally out from the glenoid of the scapulocoracoid) maintain their perpendicular orientation to the glenoid, rotation of the shoulder girdle rotates the humeri horizontally by the same angle. The horizontal distance moved by the distal ends of the humeri during a step or a stride is proportional to the interglenoid width plus the lengths of the two humeri (34.6 cm for the trackmaker; see appendix table 8.2).

The functional lower limb and distal carpals + metacarpals in figure 8.4 are portrayed diagrammatically so as to convey their total contribution to step length. The functional lower limbs are represented by the unshaded space bounded by thin diagonal lines that extend antero- or posteromedially from the distal ends of the humeri. They are perpendicular to the humeri because of the presumed fixed nature of the elbow joint (Jenkins 1973). Their contribution to longitudinal step length is indicated by the fore-aft distance from the elbow joint to the anteromedial end of the thin lines representing the functional forelimb. The outline of the manus minus the digits represents the combined lengths of the distal carpals and metacarpals, which add their length to total step length at the end of the propulsive phase. Trackways indicate that the manus and pes are oriented parallel to the direction of travel. The slender, elongated digits are interpreted as not contributing significant force to the end of the step cycle, so their lengths are not included in the determination of step length.

The full length of a step is thus indicated as the distance between the horizontal broken line intersecting the distal end of the right humerus to the base of the digits of the left forelimb representing the left distal carpals + metacarpals. As measured from the half-scale diagrams, this length is 22.8 cm for mean step length and 25.5 for maximum step length. The first step extends from the distal end of the right humerus (lower right) to the tip of the left metacarpals and the second step from the distal end of the left humerus to the tip of the right metacarpals. The left lower limb is not included in the determination of the second step because it will have rotated back on the humerus to lie under its distal end (as is the case on the lower right).

Figure 8.4A, representing the lateral bending angle derived from using the *mean* step length (22.8 cm) of the trackmaker yields an angle of axial bending of 18.5 degrees for augmenting the step and of 37 degrees for augmenting the stride. Compared with angles of lateral trunk bending in a varanid lizard (Jenkins and Goslow 1983, 205), this undoubtedly represents a slowly walking individual. Therefore, I took the *maximum* step length recorded from the trackway (25.5 cm) and ran the same analysis. The results are depicted in figure 8.4B. The angle of axial bending contributing to step length is now 23.3 degrees and to stride length 46.6 degrees. These values fall well within the range of a varanid walking at moderate speed (Jenkins and Goslow 1983).

Dimetropus trackways tend to be narrower than expected in a fully sprawling animal, leading MacDonald (1994) and Hunt and Lucas (1998) to argue that the narrow trackways of presumed sphenacodontids like *Dimetrodon* indicate that they had adducted limbs and

carried their bodies well off the ground. In the studied trackway (fig. 8.1), mean distance between left and right manus is 20.25 cm and maximum distance is 23.5 cm (appendix table 8.1). The transverse distance between reconstructed left and right manus was determined from the half-scale reconstruction that serves as the basis for figure 8.4. For mean step length, this distance is 28.0 cm, and for maximum step length it is 25.5 cm. Although less than the intermanual distances of the trackway, it contrasts greatly with the maximum possible distance between manus prints of 34.6 cm (i.e., the transverse distance between the distal ends of the humeri), which would pertain in the absence of lateral bending of the trunk. Although the medial components of the reconstructed step lengths are less than those of the trackway, it is clear that as stride length increases with increased lateral bending of the trunk, so intermanual distance concomitantly decreases. Therefore, narrow *Dimetropus* trackways are not an indication of adducted limbs, as proposed by MacDonald (1994) and Hunt and Lucas (1998), but of sprawling limbs combined with extensive lateral bending of the trunk.

Step Cycle of the "Pelycosaur" *Dimetrodon*

Figure 8.5 illustrates three stages in the step cycle of *Dimetrodon* as determined by the present analysis. It is modeled on a figure of pelycosaur limb movement by Jenkins (1973, fig. 5) that uses the convention of depicting the proximal and distal expansions of the humerus as triangular plates that are rotated on one another where they meet at their apices and the long axis of the humerus as a circular rod. In addition, thin oblique lines indicate the degree of inclination of the proximal expansion during the step cycle. The torsion angle of 35 degrees between the proximal and distal plates is based on humerus FMNH UC1134 (table 8.3). The slight elevation of the humerus in the right-hand drawing and slight depression in the middle and left-hand drawings are shown in photographs of the humerus articulated in the glenoid of this specimen. In the middle and left-hand drawings, the orientation of the limb in the next drawing to the right is shown in gray. The limb elements and inferred step length are not scaled to the reconstructed trackmaker.

At the beginning of the propulsive phase of the step cycle (right drawing), the humerus is maximally rotated

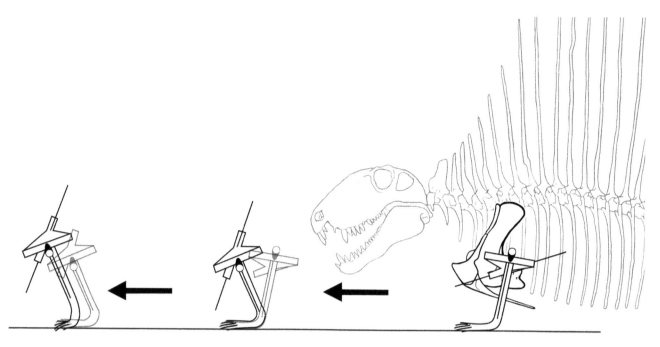

FIG. 8.5 Diagrammatic representation of forelimb step cycle of *Dimetrodon*. Features of humerus based on conventions used by Jenkins (1973): triangular plates joined at their apices represent proximal and distal expanded portions of humerus; circular rod represents long axis of humerus; thin line passes through expanded proximal end of humerus. In center and left figures, the position of the limb in the preceding figure is shown in light gray. Contribution of lateral bending of trunk not included. Figure at right represents limb at beginning of step, with proximodorsal surface of humerus rotated to most posterior (clockwise) position and forelimb and manus in most anterior position. In center, proximodorsal surface of humerus rotated to most anterior (counterclockwise) position and lower limb rotated to most posterior position. Manus remains planted in same position, and humerus and trunk move forward over it. In left figure, manus ventroflexed so proximal part is lifted vertically upward, causing upper limb and trunk to move further forward. Not to scale.

posteriorly (clockwise), so that the plane of the anteroproximal end lies at 23 degrees below the horizontal long axis of the glenoid and the plane of the anterodistal end lies at 12 degrees above the horizontal, its articular surface facing anteroventrally. The radius, which is perpendicular to the distal humeral plane, extends anteroventrally at 12 degrees from the vertical. At the end of humeral rotation (middle drawing), the humerus is maximally rotated anteriorly (counterclockwise), and the plane of the anteroproximal end now lies at 58.75 degrees below the horizontal and that of the anterodistal end at 24 degrees below the horizontal. The radius and proximal carpals, the functional lower limb, now extend posteroventrally at an angle of 24 degrees from the vertical. The axis ("axle") of the humerus has thus rotated 36 degrees and the lower limb ("spoke") has moved its upper end forward of the planted forefoot a distance of 7.6 cm. At full retraction of the functional lower limb (left drawing), the distal carpals and metacarpals are ventroflexed approximately 90 degrees so that they are

lifted from the substrate to a vertical orientation, adding a length of 4.25 cm to the step. The slender digits will also ventroflex as they lift off the ground at the beginning of the recovery phase, but I have assumed that they add little to the length of the step. Although some abduction and adduction of the distal end of the humerus occurs, no evident protraction or retraction of the humerus adds to step length. My interpretation of limb function in *Dimetrodon* is similar to that of Jenkins (1971, 1973) in that we both consider the humerus to have been incapable of significant adduction, let alone the strong adduction suggested by several recent workers (MacDonald 1994; Hunt and Lucas 1998).

Contribution of Lateral Trunk Bending to Locomotion in *Dimetrodon*

Step length in a *Dimetrodon* individual the size of the trackmaker, as estimated from analysis of humeral long axis rotation and length of the functional lower limb, appears to fall well short of the step length of the trackmaker (by 46%–48% for mean step length and 52%–54% for maximum step length). Thus, lateral bending of the trunk contributes about half of total stride length during what were probably, given the sharpness of the prints, moderate speeds for the trackmaker. Assuming that maximum step length is constrained by fixed maximal rotational movements of the humerus, then higher speeds could only be achieved either by increasing the amount of trunk bending, increasing step frequency, or extending the forelimb at the elbow joint as the limb approached maximal protraction (or a combination of the three). It is likely that extension of the forelimb at the elbow was very limited (Jenkins 1973, 283; this study), so that the extended lower limb would have remained essentially perpendicular to the humeral long axis (in the horizontal plane) throughout the step (fig. 8.4). With increased lateral bending of the trunk, the distal end of the humerus would extend increasingly forward, thus increasing the medial swing of the forelimb as well as step length. As a consequence, medial placement of the manus would gradually increase with bending and the contribution of the forelimb to longitudinal step length would gradually decrease, both being equal at 45 degrees of axial bending. Thus, there is a limit on how much lateral bending of the trunk can increase step length in an animal with limited extension of the forearm at the elbow (also see Pridmore 1992). In my analyses of step length I have not taken into account the *relative* decrease in step length with increased medial inflection of the forelimb because, at the small angles involved, an increase in lateral bending by a few degrees increases forward movement of the manus much more than it does medial movement.

The amount of lateral bending of the trunk in *Dimetrodon* during what is probably slow to moderate walking (as indicated by the CMNH trackway) falls within the range of bending reported by Jenkins and Goslow (1983) for slow to moderate walking (0.7–2.2 km/h) in the Savannah monitor (*Varanus exanthematicus*). Axial bending during a stride is about 40–60 degrees in the lizard (Jenkins and Goslow 1983, 205); in the reconstructed *Dimetrodon* it ranges from about 35 to 47 degrees (appendix table 8.5). In the absence of body size information on the lizards studied, relative speeds of *V. exanthematicus* and *Dimetrodon* cannot be

meaningfully compared, although it is most likely that the latter was walking slowly when it made the trackway. A significant increase in speed in order to capture a prey item was probably achieved mainly by increasing lateral bending; this would not only increase step length but would also further reduce the transverse distance between left and right manus. It would be worthwhile to analyze a variety of *Dimetropus* trackways showing different patterns of footfall, especially those with greater step lengths than studied here.

Kemp (1982, 61) has argued on the basis of zygapophysis orientation that "pelycosaurs" show "a marked tendency to reduce the extent of lateral undulation of the trunk during locomotion" and that in *Dimetrodon* it was essentially absent (Kemp 2005, 101). This is unlikely because zygopophyseal angles in *Dimetrodon* (Romer and Price 1940) are similar to those in crocodylians (Hoffstetter and Gasc 1969), in which lateral bending of the trunk occurs. In the more mammal-like gorgonopsian therapsids, Kemp argues, it was also essentially absent and "played no significant part in the locomotory mechanism" (Kemp 1982, 124). Pridmore (1992) studied lateral bending of the trunk in the opossum *Monodelphis* and reviewed its occurrence in other mammals. He took exception to Kemp's (1982) suggestion that lateral bending was lost in early therapsids, pointing out that its presence in living mammals "argues that neither mammals nor their immediate ancestors ever lost lateral bending movements" (Pridmore 1992, 145). Inasmuch as I believe lateral trunk bending was essential to the locomotion of synapsids, including early therapsids (see below), I concur with Pridmore's conclusion.

Evidence of lateral bending in therapsids is provided by Smith (1993, 345–346, fig. 11), who described a large trackway of a presumed dinocephalian from the Middle Permian of South Africa in which the left and right tracks show that the medial sides of the footprints (especially digit I) lie on the longitudinal midline. Smith noted that "the tracemaker had an elbows-out, 'rolling' gait . . . and a certain amount of lateral flexure of the vertebral column." Placement of the feet nearly on the axial midline could occur in a sprawling tetrapod only if the trunk is significantly flexed from side to side, as Smith's reconstruction of the walking trackmaker indicates. Smith's observations demonstrate that primitive therapsids retained some lateral flexion of the trunk as part of their locomotor activity, even though the limbs were probably held in a somewhat more upright position than in primitive synapsids.

The retention of lateral bending of the trunk well into the radiation of therapsids indicates that the notion of "semi-upright" therapsids with reduced lateral bending is not supported. Rather, at least some therapsids appear to have retained about the same degree of trunk bending as described here for more basal synapsids.

A consequence of the need for lateral bending in basal synapsids to increase stride length during pursuit of prey was that, due to cooptation of the hypaxial musculature for bending the trunk, breathing had to be suspended during high speed locomotor activity (Carrier 1987a). This could imply that therapsids, like "pelycosaurs," were restricted to short, anaerobically fueled, bursts of speed to capture prey. If so, it would also imply that therapsids retained low levels of aerobically mediated locomotor activity and, thus, lacked the physiological means for supporting the widely foraging activity that I (Hopson 2012) have proposed was the key to the great increase in aerobic capacity and stamina that ultimately led to the achievement of mammalian endothermy.

Owerkowicz et al. (1999) have demonstrated that varanid lizards, which use lateral bending of the trunk during rapid locomotion, have overcome the bending constraint on respiration by using a gular pump that forces air bilaterally into the lungs. By using this breathing mechanism, varanids can maintain higher speeds for longer periods than can lizards without the accessory pump (Owerkowicz et al. 1999). In mammals, a muscular diaphragm that actively draws air into the lungs, serves the function of an accessory oxygen pump. Therefore, I (Hopson 2012) have suggested that the diaphragm evolved during or shortly after the transition from sphenacodontids to basal therapsids. Probably at low to moderate foraging speeds, when bending would be moderate, diaphragmatic ventilation would compensate for the alternate compression of each lung, as apparently occurs in the widely foraging varanids. At high speeds, therapsids would have used some anaerobic muscle metabolism to sustain a short dash, as do living mammals. The difference from typical ectothermic sprawlers would be that their presumed greater aerobic capacity would cause relatively rapid recovery from the incurred oxygen debt.

* * *

Acknowledgments

For permission to study the Robledo *Dimetropus* trackways in the Carnegie Museum of Natural History, Pittsburgh, I thank Dr. David Berman, and for their help in facilitating the trackway research, I thank Amy Henrici, Norman Wuerthele, and Dr. John Wible. I am also grateful to Drs. Christian Sidor and Zhe-Xi Luo for providing very helpful reviews of the manuscript, to Dr. Elizabeth Brainerd and Neil Shubin for editorial advice, and to Claire Vanderslice and April Isch for the illustrations.

To all who organized the symposium and other activities to honor the life and work of Dr. Farish A. Jenkins Jr., I offer my sincere thanks for a wonderful celebration. A special acknowledgment goes to Dr. Ken Dial, the moving spirit behind the event. My thanks also to Farish and Eleanor Jenkins for their hospitality following the event. This celebration was a fitting tribute to a great scientist, teacher, and friend.

Appendix

TABLE 8.1 Measurements (in cm) of *Dimetropus* sp. trackway (CMNH 87684)

Datum	Measurements	Mean
A. Pace length forelimb		
ML1 → MR1	21.0	
MR1 → ML2	22.0	
ML2 → MR2	25.5	22.8
B. Pace length hind limb		
PL1 → PR1	14.0	
PR1 → PL2	25.0	
PL2 → PR2	23.0	
PR2 → PL3	23.0	
PL3 → PR3	22.0	23.25
		(PL1 → PR1 omitted)
C. Stride length forelimb		
ML1 → ML2	42.5	
MR1 → MR2	48.0	45.25
D. Stride length hind limb		
PL1 → PL2	40.0	
PR1 → PR2	45.0	
PL2 → PL3	48.0	
PR2 → PR3	44.5	44.4
E. Diagonal distance between consecutive prints of L & R manus		
ML1 → MR1	30.5	
MR1 → ML2	29.0	
ML2 → MR2	32.0	30.5
F. Diagonal distance between consecutive prints of L & R pes		
PL1 → PR1	28.0	
PR1 → PL2	40.5	
PL2 → PR2	35.5	
PR2 → PL3	33.0	
PL3 → PR3	33.0	34.0
G. Distance between L & R manual tracks		
ML1 and (MR1 & MR2)	23.5	
ML2 and (MR1 & MR2)	20.5	
MR1 and (ML1 & ML2)	19.0	
MR2 and (ML1 & ML2)	18.0	20.25

(continued)

TABLE 8.1 (continued)

Datum	Measurements	Mean
H. Distance between L & R pedal tracks		
PL1 and (PR1 & PR2)	23.5	
PL2 and (PR1 & PR2)	29.5	
PL3 and (PR2 & PR3)	24.0	
PR2 and (PL2 & PL3)	24.0	
PR3 and (PL2 & PL3)	24.0	25.0
I. Glenoacetabular length (midline distance between diagonal lines connecting ML2 & MR2 and PR1 & PL2)	46.5	46.5

TABLE 8.2 Dimensions (in cm) of skeleton of *Dimetrodon limbatus* (MCZ 1347) and maker of trackway (CMNH 87684), provisionally identified as *Dimetropus leisnerianus*

Measurements of *D. limbatus* skeleton from Romer and Price (1940) or calculated from those measurements; measurements of trackmaker, except for glenoacetabular length, which is calculated from trackway, derived from those of *D. limbatus* scaled down by 0.637. Measurements of manual elements of trackmaker based on *Dimetrodon milleri* (MCZ 1365).

	Dimetrodon limbatus (MCZ 1347)	*Dimetropus leisnerianus* (CMNH 87684)
Glenoacetabular length	73.0	46.5
Reconstructed total body length	256.0	163.0
Interglenoid width of scapulocoracoid	20.7	13.2
Humerus length	16.8	10.7
Height of radial condyle of humerus	1.6	1.0
Radius length	15.0	9.5
Ulna length (to base of sigmoid notch)	15.5	9.9
Length proximal carpals (radiale + medial centrale)	—	2.5
Length 4th distal carpal + metacarpal	—	4.25
Functional transverse diameter of forelimb (interglenoid width + 2 x length of humerus)	54.3	34.6

TABLE 8.3 Humeral rotation in the glenoid during the step cycle in *Dimetrodon*

Angles below horizontal of the humeral head in the horizontally oriented glenoid and of the distal end of the humerus at the beginning and end of the step cycle[1,2].

	FMNH UC1134	MCZ 3357
Angle (in degrees) of proximal articulation of humerus below horizontal		
Beginning of step	23.0	38.5
End of step	58.75	77.0
	(mean of 56.5 & 61.0)	
Total rotation	35.8	38.5
Torsion angle (between proximal and distal expansions of humerus)	35.0	38.5
Angle of distal articulation of humerus		
Beginning of step	12.0 anterovent.	0.0
End of step	24.0 posterovent.	38.5 posterovent.

[1] Angles of humeral rotation to the glenoid of the scapulocoracoid were established by measurement on photographs of the rearticulated humerus and scapulocoracoid. The scapulocoracoid was placed in a sandbox with the glenoid facing vertically upward. To represent the beginning of the step cycle, the humerus was placed vertically in the glenoid with its posterior surface in the most posteriorly rotated position (i.e., for a left humerus the furthest clockwise position, representing the beginning of the step cycle) in which it fit snugly (one could almost say "locked") in the glenoid (fig. 8.3A, left). It was photographed from directly above, and the angle between the long axis of the glenoid (considered to represent the horizontal) and the transversely expanded proximal head of the humerus were measured on the photographic print. The same was done for the humerus oriented in the most stable anteriorly rotated position (furthest counterclockwise rotation, representing the end of the step cycle) (fig. 8.3A, right).
[2] Torsional angle between transversely expanded proximal and distal ends of humerus was estimated by measurement on photographs of two humeri in this study. The humeri were oriented in the sandbox with their long axes vertical and photographed first with the proximal and then with the distal end upward. The angle between the expanded ends was measured from both photos and the mean value recorded. The distal end of FMNH UC1134 is incomplete, but an estimate by photograph of the angle of transverse expansion relative to the proximal end is 35.0 degrees, as compared to better preserved landmarks of MCZ 1542 (fig. 8.3B).

TABLE 8.4 Reconstruction of trackmaker step length[1]

Functional lower limb length (in cm) for calculating portion of step length caused by humeral rotation = radial condyle + radius + proximal carpals (ulnare + medial central) = 1.0 + 9.5 + 2.5 = 13. Length of functional part of manus added at end of step = distal carpal IV + metacarpal IV = 4.25.

	From FMNH UC1134	From MCZ 3357
Total humeral rotation (in degrees):	35.8	38.5
Contribution of humeral rotation + functional lower limb (= 13 cm) to step length	7.6	8.1
Contribution of distal carpals and metacarpals to end of step	4.25	4.25
Total step length	11.85	12.35

[1] Step length is constructed from a series of right triangles, using the program Carbide Depot Trigonometry Calculator (carbide depot.com/formulas-trigright.asp), with the length of the functional lower limb (as seen in lateral view: fig. 8.5) entered as the hypotenuse and the angle of humeral rotation entered as the angle between the hypotenuse and the vertical side (the second required angle being 90 degrees); the length of the base of the triangle equals the step length.

TABLE 8.5 Contribution of lateral bending of the trunk to step length (in cm)

Trackway step length minus step length from forelimb = step length from trunk bending.

	FMNH UC1134	MCZ 3357
Mean trackway step length	22.8	22.8
Contribution of step length from forelimb	11.85	12.35
Contribution of step length from lateral bending	10.95	10.45
Angle of axial bending of trunk determined by:		
Distance between ends of humeri	34.6	34.6
Length of base (step length)	11.0	10.5
Length of base x 2 (stride length)	22.0	21.0
Angle of axial bending for step (degrees)	18.5	17.7
Angle of axial bending for stride (degrees)	37.0	35.4
Maximum trackway step length	25.5	25.5
Contribution of step length from forelimb	11.85	12.35
Contribution of step length from lateral bending	13.65	13.15
Angle of axial bending of trunk determined by:		
Distance between ends of humeri	34.6	34.6
Length of base (step length)	13.7	13.2
Length of base x 2 (stride length)	27.4	26.4
Angle of axial bending for step (degrees)	23.3	22.4
Angle of axial bending for stride (degrees)	46.6	44.8

References

Baird, D. 1952. Revision of the Pennsylvanian and Permian footprints of *Limnopus, Allopus* and *Baropus*. J. Paleont. 26:832–840.

Carrier, D. R. 1987a. The evolution of locomotor stamina in tetrapods: circumventing a mechanical constraint. Paleobiology 13:326–341.

Carrier, D. R. 1987b. Lung ventilation during walking and running in four species of lizards. Exper. Biol. 47:33–42.

Carrier, D. R. 1989. Ventilatory action of the hypaxial muscles of the lizard *Iguana iguana*: a function of slow muscle. J. Exper. Biol. 143:435–457.

Carrier, D. R. 1991. Conflict in the hypaxial musculo-skeletal system documenting an evolutionary constraint. Am. Zool. 31:644–654.

Gatesy, S. M., K. M. Middleton, F. A. Jenkins Jr., and N. H. Shubin. 1999. Three-dimensional preservation of foot movements in Triassic theropod dinosaurs. Nature 399:141–144.

Haubold, H. 1971. Ichnia amphibiorum et reptiliorum fossilium. Encyclopedia of paleoherpetology 18:1–124.

Haubold, H., A. P. Hunt, S. G. Lucas, and M. G. Lockley. 1995. Wolfcampian (Early Permian) vertebrate tracks from Arizona and New Mexico. New Mexico Mus. Nat. Hist. Sci. 6:135–165.

Hoffstetter, R., and Gasc, J.-P. 1969. Vertebrae and ribs of modern reptiles. Pp. 201–310 in C. Gans, A. d'A. Bellairs, and T. S. Parsons, eds., Biology of the Reptilia, vol. 1, Morphology A. London and New York: Academic Press.

Holmes, R. 1977. The osteology and musculature of the pectoral limb of small captorhinids. J. Morph. 152:101–140.

Hopson, J. A. 2012. The role of foraging mode in the origin of therapsids: implications for the origin of mammalian endothermy. Fieldiana: Life & Earth Sci. 5:126–148.

Hunt, A. P., M. G. Lockley, S. G. Lucas, J. P. MacDonald, N. Hotton III, and J. Kramer. 1993. Early Permian tracksites in the Robledo Mountains, South-Central New Mexico. Pp. 23–31 in S. G. Lucas and J. Zidek, eds., Vertebrate Paleontology in New Mexico. Bulletin 2. Albuquerque: New Mexico Museum of Natural History and Science.

Hunt, A. P., and S. G. Lucas. 1998. Vertebrate tracks and the myth of the belly-dragging, tail-dragging tetrapods of the Late Paleozoic. Pp. 67–70 in S. G. Lucas, J. W. Estep, and J. M. Hoffer, eds., Permian Stratigraphy and Paleontology of the Robledo Mountains, New Mexico. Bulletin 12. Albuquerque: New Mexico Museum of Natural History and Science.

Jenkins, F. A., Jr. 1971. The postcranial skeleton of African cynodonts. Bull. Peabody Mus. Nat. Hist. 36:1–216.

Jenkins, F. A., Jr. 1973. The functional anatomy and evolution of the mammalian humero-ulnar articulation. Am. J. Anat. 137:281–298.

Jenkins, F. A., Jr., and G. E. Goslow. 1983. The functional anatomy of the shoulder of the Savannah monitor (*Varanus exanthematicus*). J. Morph. 175:195–216.

Kemp, T. S. 1982 Mammal-Like Reptiles and the Origin of Mammals. London: Academic Press.

Kemp, T. S. 2005. The Origin and Evolution of Mammals. Oxford: Oxford University Press.

MacDonald, J. 1994. Earth's First Steps: Tracking Life before the Dinosaurs. Boulder, CO: Johnson Printing. 290 pp.

Owerkowicz, T., C. G. Farmer, J. W. Hicks, and E. L. Brainerd. 1999. Contribution of gular pumping to lung ventilation in monitor lizards. Science 284:1661–1663.

Padian, K., and P. E. Olsen. 1984. Footprints of the Komodo monitor and the trackways of fossil reptiles. Copeia 1984:662–671.

Pridmore, P. A. 1992. Trunk movements during locomotion in the marsupial *Monodelphis domestica* (Didelphidae). J. Morph. 211:137–146.

Romer, A. S., and L. I. Price. 1940. Review of the Pelycosauria. Spec. Pap. Geol. Soc. Am. 28:1–538.

Smith, R. M. H. 1993. Sedimentology and ichnology of floodplain paleosurfaces in the Beaufort Group (Late Permian), Karoo Sequence, South Africa. Palaios 8:339–357.

Sukhanov, V. B. 1974. General system of symmetrical locomotion of terrestrial vertebrates and some features of movement of lower tetrapods. Acad. Sciences., U.S.S.R. (English translation of 1968 publication). 274 pp.

Voigt, S., and M. Ganzelewski. 2010. Toward the origin of amniotes: diadectomorph and synapsid footprints from the early Late Carboniferous of Germany. Acta Palaeontol. Pol. 55:57–72.

Wang, T., D. R. Carrier, and J. W. Hicks. 1997. Ventilation and gas exchange in lizards during treadmill exercise. J. Exper. Biol. 200:2629–2639.

9

Respiratory Turbinates and the Evolution of Endothermy in Mammals and Birds

Tomasz Owerkowicz,* Catherine Musinsky,† Kevin M. Middleton,‡ and A. W. Crompton§

Summary

For almost thirty years, respiratory turbinates have been hailed as a bona fide hallmark of endothermy in mammals and birds. Located in the nasal cavity, respiratory turbinates act as a temporal countercurrent exchanger to reduce heat and water loss in expired air. Surface area of respiratory turbinates scales to (body mass)$^{0.73}$ in both mammals and birds, but is three times greater in the former clade. Interspecific variation in respiratory turbinate surface area shows positive correlation with field metabolic rates in mammals, but not birds. Loss or reduction of respiratory turbinates in various bird lineages suggests that turbinates are not required for avian endothermy, if heat and water savings can occur in the trachea. Tracheal surface area also scales to (body mass)$^{0.73}$ in both mammals and birds, but is 3.5 times greater in the latter clade. In vivo temperature measurements and surgical ablation of respiratory turbinates in emus offer experimental evidence that temporal countercurrent exchange operates in the avian trachea. We propose that craniocervical organization of cynodonts—specifically, short necks

* Department of Biology, California State University, San Bernardino
† Department of Organismic and Evolutionary Biology, Harvard University
‡ Department of Pathology and Anatomical Sciences, University of Missouri
§ Museum of Comparative Zoology, and Department of Organismic and Evolutionary Biology, Harvard University

and massive heads—constrained their countercurrent exchanger to the nasal cavity. In contrast, a trend of neck elongation among theropod dinosaurs, enabled by the flow-through lung design and driven by lightening of the cranial skeleton, allowed their trachea to become a major site of respiratory heat and water conservation. Presence of a prominent nasal preconcha in extant crocodilians suggests that respiratory turbinates and ectothermy are not mutually exclusive. Instead of heat conservation, the original selection pressure for respiratory turbinates may have been heat dissipation by evaporative heat loss. We conclude that respiratory turbinates may have been exapted for heat and water conservation from their original function in selective brain cooling.

Endothermy and Its Indicators

Endothermy evolved independently in the ancestors of mammals and birds, and possibly in other extinct vertebrate taxa (e.g., pterosaurs). Endothermy is considered one of the great evolutionary transformations in the vertebrates (Lane 2010), as it affords the organism a large degree of thermal independence from its environment and obviates the need for acclimation to changing ambient temperatures. Thermal independence, in turn, allows fine-tuning of numerous enzymatic reactions to the preferred body temperature range and, presumably, optimization of physiologic performance (Angiletta et al. 2010).

Numerous scenarios for the acquisition of endothermy have been proposed, each arguing a different proximate reason for chronic elevation of metabolic rate: facilitation of diurnal niche invasion (Crompton et al. 1978; Taylor 1980), consequence of body miniaturization (McNab 1978, 2009), selection for greater aerobic capacity (Bennett and Ruben 1979; Bennett 1991), support of improved parental care (Farmer 2000), corollary of increased foraging and assimilation costs (Koteja 2000; Hopson 2012), or enhancement of metabolic power (Clarke and Pörtner 2010). Rarely pointed out is the possibility that the original metabolic up-regulation might have served disparate adaptive purposes in the mammalian and avian lineages, and only later converged on a similar thermoregulatory function. This hypothesis is not implausible, considering that non-shivering thermogenesis in extant mammals and birds is supported by

distinct mitochondrial uncoupling mechanisms (Cannon and Nedergaard 2004; Jastroch et al. 2008; Walter and Seebacher 2009).

Presence or absence of endothermy as a thermoregulatory strategy in extinct vertebrate taxa has been hotly debated by biologists for decades, and various anatomic "indicators" of high metabolic rate have been proposed.

(i) Posture

The argument dates back to 1968, when Heath suggested that upright, parasagittal posture is associated with higher levels of heat production by locomotor muscles. This idea was invoked by Bakker (1971), who argued that the gradual transition from an abducted- to more adducted-limb posture be used to infer the gradual acquisition of endothermy in cynodonts and dinosaurs. It was ultimately disproved by Bakker's own metabolic measurements, which showed that lizards with a sprawling gait have the same cost of transport as similarly sized but cursorially adapted mammals (Bakker 1972). Although later disavowed by Bakker (1974), the general idea of upright posture being associated with endothermy keeps reappearing under various guises (e.g., Pontzer et al. 2009). Never directly involved in the controversy, Jenkins provided irrefutable anatomic evidence to counter this hypothesis. With his masterful cineradiographic technique and insightful functional analysis, Jenkins visualized the abducted position of proximal limb elements (humerus and femur) in numerous endothermic mammals (echidna, opossum, and diverse placentals) with non-cursorial postures (Jenkins 1970b, 1971a; Jenkins and Camazine 1977). Jenkins's illustration of the diversity of limb orientation during mammalian locomotion, coupled with subsequent evidence of sprawling-to-erect postural spectrum in crocodilians (Gatesy 1991a), and documentation of significant hind limb abduction and rotation in birds (Gatesy 1995; Rubenson et al. 2007), underscores the argument that it is wrong to conflate posture with metabolic rate.

(ii) Pelage and Plumage

It has always been tacitly assumed that the telltale sign of endothermy is a coat of insulation. After all, ectotherms rely on infrared radiation for warmth, so any insulating layer may prevent them from gaining heat from the environment. Endotherms, on the other hand, strive to minimize heat loss to the environment in order to reduce their energy expenditure. Many scenarios of the origins of endothermy (see above) recognize that sparse insulation would not have been able to reduce heat loss to the environment and, as such, would not have been adaptive to animals with slightly elevated metabolic rates (often referred to as "incipient endothermy"; Ruben 1996; Kemp 2006). Relatively few scenarios, however, have considered the possibility that proto-hair and proto-feathers could initially have played a non-thermoregulatory role, as they do in some "primitive" mammals (Gould and Eisenberg 1966). Only recently have arguments been advanced that such early ectodermal derivatives might have had a functional role in intra- and interspecific visual display or camouflage (Caro 2005; Norell and Xu 2005) and acoustic communication (Bostwick and Prum 2005; Clark et al. 2011). The fact that either hair or feathers might have evolved to improve the animal's tactile perception of the environment remains surprisingly unappreciated by the paleobiologist and physiologist community (Maderson 1972; Rowe et al. 2011), yet both excel at tactile signal transduction in extant mammals (Zelená 1994; Sarko et al. 2011) and birds (Küster 1905; Cunningham et al. 2011). Pressure sensing by feathers (Gewecke and Woike 1978; Brown and Fedde 1993) would have been a great adaptation for aerodynamic control, and perhaps a permissive step in the origin of flight among theropods (Ruben and Jones 2000). At the same time, an increase in sensory input from hair- and feather-like structures would have required enlargement of the sensorimotor cortex in synapsids (Rowe et al. 2011) and archosaurs (Witmer et al. 2008), and contributed to the increase in resting metabolic rate. Discoveries of fossil imprints of hair (Meng and Wyss 1997; Ji et al. 2006) and feathers (Chen et al. 1998; Zhang and Zhou 2000; Ji et al. 2001; Clarke 2013) continue to add fuel to the debate on the origins of endothermy.

(iii) Respiratory Turbinates and Other Characters

Persisting disagreement on when and how endothermy had been acquired by the therapsid ancestors of mammals and theropod ancestors of birds (Hillenius and Ruben 2004; Seymour et al. 2004; Chinsamy and Hillenius 2004; Padian and Horner 2004; Kemp 2006; Nespolo et al. 2011; Ruben et al. 2012) is perhaps a reflection on the lack of consensus by the paleobiologist community of what constitutes a verifiable osteologic correlate of endothermy. Yet, as long ago as 1986, Bennett and Ruben (p. 208) provided a comprehensive review of evidence for endothermy among the ancestors of mammals, in which they postulated:

> In order for an anatomical structure or other criterion to be diagnostic of a high metabolic rate or of endothermy, it must satisfy at least the following three conditions:
> - *it should have a clear and logical association with high rates of metabolism;*
> - *it should not be explicable in terms unrelated to high levels of metabolism; and*
> - *it should be absent in ectotherms.*

Of the osteologic features (limb posture, ribcage morphology, bony secondary palate, parietal foramen), which had previously been suggested as diagnostic of endotherms, most were found "untenable" or "inconclusive" at best (Bennett and Ruben 1986). Bone histology was deemed "suggestive" because certain microstructural characters of bone seemed to indirectly correlate with the thermoregulatory status of vertebrates (de Ricqlès 1974, and references therein). Other confounding factors—animal size, growth rate, phylogeny, biomechanic loading history, environmental conditions—have made interpretation of bone microstructure difficult, especially in terms of determining the metabolic physiology of extinct vertebrates. Paleohistology remains a vibrant yet highly contentious field of research (Chinsamy-Turan 2005; Cubo et al. 2012; Padian and Lamm 2013), with physiologic conclusions dependent on individual interpretation of bone microstructural data. The only gross-anatomic character found "very suggestive" of endothermy was the respiratory turbinates ("nasal turbinals" of Bennett and Ruben 1986). These fine scrolls of bone (in mammals) or cartilage (in birds) protrude into the stream of respiratory airflow, augment the surface area of the nasal cavity, and can substantially reduce evaporative heat and water loss from expired air.

Since then, respiratory turbinates (RTs) have been hailed as unequivocal indicators of high resting metabolic

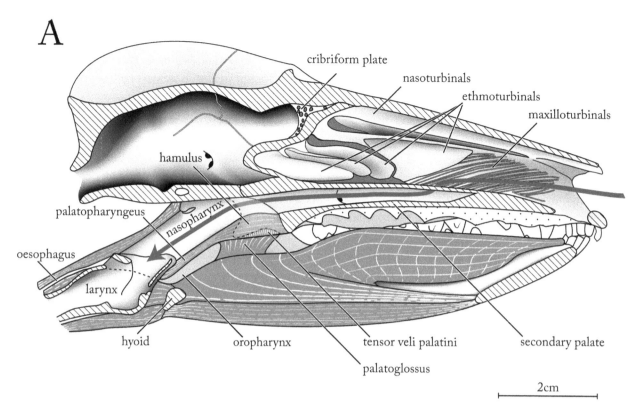

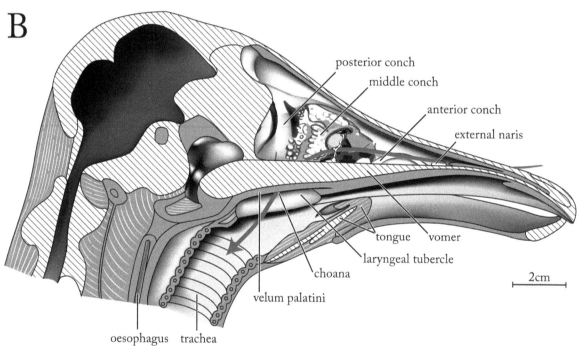

FIG. 9.1 Comparison of the nasal cavity morphologies in a mammal (A) and a bird (B) with similar respiratory turbinate surface areas. A, Sagittal section of the head of a opossum (*Didelphis virginiana*; Mb = 3.43 kg, RTSA = 11.9 cm²). B, Sagittal section of the head of an ostrich (*Struthio camelus*; Mb = 100 kg, RTSA = 11.5 cm²). Hatched surfaces indicate plane of section through bones of the skull. Scale bar = 20 mm.

rates in extinct amniotes. Ruben et al. (1997) referred to RTs as the "Rosetta stone" of endothermy. Putative presence (or absence) of RTs in nasal cavities of fossil taxa has been used to argue for endothermy among nonmammalian therapsids (Hillenius 1992, 1994) and against endothermy among non-avian theropods (Ruben et al. 1996). Respiratory turbinates as indicators of endothermy have already entered college-level textbooks in comparative anatomy (Liem et al. 2000; Kardong 2011), paleontology (Benton 2004), and zoology (Pough et al. 2012). However, no quantitative relationship between RT morphology and metabolic rate has been established.

Our chapter aims to

(i) test the correlation of respiratory turbinate structure and field metabolic rates in a diverse assemblage of extant mammals and birds;

(ii) discuss the role of the trachea as an alternative site of heat and water conservation;

(iii) consider how clade-specific constraints might have determined the relative roles of turbinates and trachea in thermoregulation;

(iv) show that respiratory turbinates are not, after all, unique to endotherms; and

(v) speculate how incipient turbinate structures may have originally served a different function in the ancestors of mammals and birds.

Mechanism of Respiratory Heat and Water Conservation

The main cost of endothermy is its high energetic expenditure. Basal metabolic rate of endotherms is 6–10 times higher than the standard metabolic rate of similarly sized ectotherms at equivalent body temperature (Dawson and Hulbert 1970; Bennett and Dawson 1976). Average body temperature of endotherms is generally higher than that of ectotherms, and most sustainable metabolic processes are sensitive to temperature effects, with the Q_{10} coefficient determined to vary between 2 and 3 among tetrapods (White et al. 2006). Daily energetic expenditures, therefore, are expected to be exponentially greater in endotherms. Indeed, field metabolic rates (FMRs) of endotherms exceed those of similarly sized ectotherms by a factor of 12–20 (Nagy 2005). As

all sustainable metabolic activity must be matched by appropriate oxygen supply, endothermy necessitates relatively high rates of lung ventilation in air-breathing vertebrates. The higher the minute ventilation (= breathing frequency x tidal volume), the greater the amount of heat and water lost *per unit time* from the respiratory tract. In addition, the higher the animal's body core temperature, the greater the heat and water content *per unit volume* of pulmonary air. Taken together, endotherms are at risk of substantial rates of heat and water loss in expired air. It is, therefore, likely that any structural modification of the upper respiratory tract, which could improve heat and water conservation, would have been adaptive in the early stages of evolution of endothermy among mammal and bird ancestors.

Evaporative loss can be reduced in the upper respiratory tract by temporal countercurrent exchange (TCCE). During inspiration, the relatively cool (and dry) ambient air absorbs heat and moisture from the respiratory tract mucosa. During expiration, the cool mucosa reclaims most of the heat and moisture from the warm saturated air leaving the lungs, and air is exhaled below body core temperature. The TCCE mechanism, therefore, relies on transfer of heat between a stationary structure (respiratory tract mucosa) and a convective current (airflow), the direction of which reverses with inspiration and expiration. The stationary mucosa acts as a heat source (during inspiration) or heat sink (during expiration) at different times of every ventilatory cycle. TCCE thus differs from spatial countercurrent exchange (Schmidt-Nielsen 1981), which is found in the vascular supply of vertebrate limbs (Scholander and Schevill 1955; Scholander and Krog 1957) and other extremities (Rommel et al. 1992). Therein heat transfer occurs between two antiparallel convective currents in close proximity, with the heat source (arteries) and heat sink (veins) anatomically distinct and constant.

Operation of the TCCE mechanism depends on the animal's breathing pattern. The heat and water contents of expired air are a function of temperature at the exit from the respiratory tract, that is, the nasal cavity. The TCCE is most efficient when temperature of the nasal mucosa remains as low as that of ambient air prior to expiration. This occurs when inspiration is immediately followed by expiration (I/E). Any delay between inspiration and subsequent expiration allows gradual warming of the mucosa by capillary perfusion, which

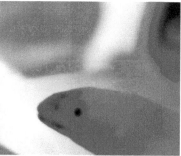

FIG. 9.2 Thermographic images of the head of a juvenile savannah monitor (*Varanus exanthematicus*, Mb = 0.303 kg) during expiration (left), and inspiration (right). During expiration, nostril temperature increases and approximates body core temperature. This suggests that temporal countercurrent exchange is not functional in the nasal cavity. Images recorded with a FLIR thermographic camera, with animal body core temperature at 30°C and room air temperature at 23°C.

decreases the efficiency of TCCE. Most mammals and birds breathe continuously (i.e., without apnea), which allows them to establish a standing temperature gradient along their nasal cavities (e.g., Adams 1972; Murrish 1973). A short end-expiratory pause is observed during quiet breathing in resting neonate (Mortola 2001; Simpson et al. 2012) and adult terrestrial mammals (Frappell and Baudinette 1995; Mortola and Seguin 2009), but is unlikely to interfere with TCCE because it occurs after expiration. In contrast, all extant ectothermic amniotes (lepidosaurs, turtles, and crocodilians) exhibit a reversed breathing pattern (E/I), whereby expiration precedes inspiration (Gans and Clark 1978; Brainerd and Owerkowicz 2006). In this case, the nasal mucosa remains at body core temperature during expiration (fig. 9.2), which precludes condensation of water vapor from expired air, effectively reducing the capacity of the nasal cavity to conserve heat and water.

Breathing pattern in resting amniotes is dictated primarily by the combined compliance of lungs and thoracic wall. Total respiratory compliance is a function of animal size (Stahl 1967), age (Mortola 2001), lung architecture (Perry and Duncker 1978; 1980), coelomic septation (Klein and Owerkowicz 2006), body armor (Frappell et al. 1998), and thoracoabdominal morphology (Brainerd, this volume; Claessens, this volume). Some reorganization of cranial, postcranial, and visceral morphology would have been required to alter the breathing pattern from E/I to I/E, prior to the TCCE mechanism becoming operational in the upper respiratory tract of mammalian and avian ancestors.

Conservation of heat and water in the upper respiratory tract of man was first reported in late 19th century (Goodale 1896), but its operating mechanism remained poorly understood until mid-20th century (Cole 1953;

Ingelstedt 1956). The principle of countercurrent exchange, with temporal separation of flow, was coined by Jackson and Schmidt-Nielsen (1964). Theirs was the first study to use a nontraditional research model (the kangaroo rat) and underscore the role of nasal morphology in TCCE. Operation of TCCE has since been established in the nasal cavities of various xeric and mesic mammal and bird species.

The TCCE mechanism can lead to significant savings in daily energy and water expenditure. Cold- or desert-adapted endotherms can recover as much as 83–88% of respiratory heat and water from the expired air when breathing through their nose (Schmidt-Nielsen, Hainsworth, and Murrish 1970; Murrish 1973; Langman et al. 1979). Mammals (Hillenius 1992) and birds (Geist 2000), whose external nares have been experimentally blocked, are forced to resort to oral breathing and incur significant heat and water losses. Lizards, whose nasal cavities lack respiratory turbinates and which exhibit the E/I breathing pattern with variable periods of apnea, are less efficient at respiratory heat exchange than various endotherms (Murrish and Schmidt-Nielsen 1970). These observations support the argument that the presence of respiratory turbinates is a key benefit to modern endotherms, as the efficiency of TCCE in heat and water conservation depends on the strategic location and complex architecture of respiratory turbinates.

Respiratory Turbinates

Morphology and Histology

Efficient countercurrent exchange demands intimate contact between the airstream and the epithelium. Respiratory turbinates—primarily the maxilloturbinal in mammals and the middle concha in birds—are strategically positioned and suitably shaped to do just that. Protruding from the lateral walls of the nasal capsule, respiratory turbinates lie between the external and internal nares in the path of breathed air (fig. 9.1) (Negus 1958; Schmidt-Nielsen, Hainsworth, and Murrish 1970; Bang 1971). The scrolled structure of turbinates combines a large surface area per unit volume of nasal cavity with the narrow passages between individual turbinate laminae, in a way that maximizes the epithelial surface available for exchange and minimizes the distance between the epithelial surface and the center of airstream (Collins et al. 1971; Schroter and Watkins 1989). Autonomic nervous regulation of blood flow (Lung and Wang 1989) in the dense vascular network underlying the turbinate epithelium (Dawes and Prichard 1953) allows fine control of the temperature gradient along the nasal cavity (Murrish 1973; Johnsen et al. 1985). This allows the turbinate mucosa to engage in heat conservation or dissipation (Negus 1958; Schmidt-Nielsen, Bretz, and Taylor 1970), depending on the heat balance of the animal.

Although convergent in function, respiratory turbinates of mammals and birds have distinct morphologies (fig. 9.3), which reflect their independent development and evolution (Witmer 1995). Mammalian maxilloturbinals are composed of delicate, often perforate osseous sheets with a scroll pattern of varying complexity (fig. 9.3A–C; Negus 1958; Moore 1981; Van Valkenburgh et al. 2004; Rowe et al. 2005; Smith et al. 2011). Each maxilloturbinal, and in some species parts of the nasoturbinal and the anterior-most ethmoturbinal, is evenly covered by a mucociliary respiratory epithelium with a rich capillary network, which suggests their participation in evaporative heat exchange. The medial surface of the nasoturbinal lies outside of the main respiratory airstream in most mammals and instead serves to provide a high-speed conduit for olfactory airflow directed at ethmoturbinals (Morgan et al. 1991; Craven et al. 2007, 2010).

In contrast, the relatively sturdy, dome-shaped conchae of birds are supported by cartilage, which may, in rare cases, ossify (e.g., in woodpeckers). Conchae are usually of a single-scroll configuration (fig. 9.3D), but may be more (fig 9.3E) or less (fig 9.3F) elaborate (Negus 1958; Bang 1971). In some taxa (fig 9.3G), the conchae have been secondarily lost due to permanent covering of external nares with the rhamphotheca (Ew-

art 1881; Pycraft 1899; MacDonald 1960). Vascular and mucous epithelium is restricted largely to the convex surface of middle concha and the laterodorsal surface of anterior concha. In many species, the concave surface of the middle concha supports only a thin avascular epithelium, which potentially limits its role in evaporative heat exchange. The anterior concha can be quite elaborate on its vestibular (medioventral) aspect, especially in passerine birds (Bang 1971), but it is unlikely to participate in countercurrent exchange because (i) its medioventral surface is keratinized, which precludes supply of moisture from the underlying vascular bed, and (ii) the inferior vestibule is a cul-de-sac and thus most likely bypassed by the excurrent airstream.

The distinct morphologies of mammalian and avian respiratory turbinates suggest that their surface area may scale with different allometric relationships to body mass.

Scaling

When measured in a diverse assemblage of mammals and birds, from a wide variety of mesic and xeric environments, spanning over five orders of magnitude in body mass—from a hummingbird to an ostrich, from a shrew to an elephant seal—the respiratory turbinate surface area (RTSA) scales with slight positive allometry: $Mb^{0.72}$ for mammals ($n = 67$), and $Mb^{0.74}$ for birds ($n = 67$). On average, RTSA is almost threefold greater in mammals than in similarly sized birds (fig. 9.4A), despite the latter having 1.2–2 times higher FMRs (Nagy et al. 1999). This scaling exponent is similar to the recent results for carnivoran mammals based on measurements obtained from CT scans of skeletonized skulls (Van Valkenburgh et al. 2011; Green et al. 2012).

RTSA is especially high in mammals adapted to extremely cold or dry habitats (e.g., camel, harp seal, reindeer), whose survival may depend on heat and water conservation. On the other hand, RTSA is low in those mammalian taxa (e.g., primates) evolved and adapted to the tropics, where high temperature and humidity reduce the need for stringent heat and water conservation. In addition, lower-than-predicted RTSA is found in various mammal species (e.g., echidna, wombat, naked mole rat) with relatively low body temperatures and metabolic rates (Dawson 1973; Frappell et al. 2002). Respiratory turbinates have been lost only in

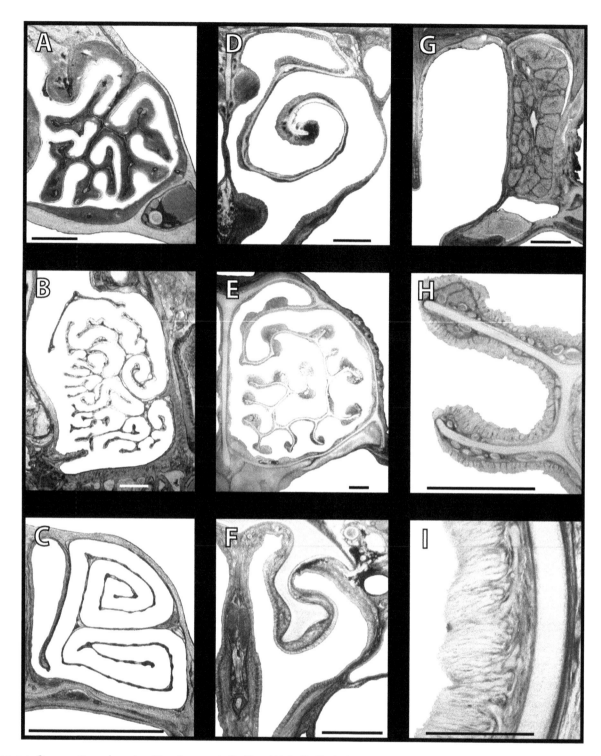

FIG. 9.3 Cross-sections of nasal cavities of mammals (A-C) and birds (D-G): A, short-beaked echidna (*Tachyglossus aculeatus*); B, Virginia opossum (*Didelphis virginiana*); C, star-nosed mole (*Parascalops breweri*); D, Canada goose (*Branta canadensis*); E, emu (*Dromaius novaehollandiae*); F, jackass penguin (*Spheniscus demersus*); G, double-crested cormorant (*Phalacrocorax auritus*). Magnified view of the vascular epithelium lining the respiratory turbinate (H) and trachea (I) of an emu. Scale bar = 2 mm.

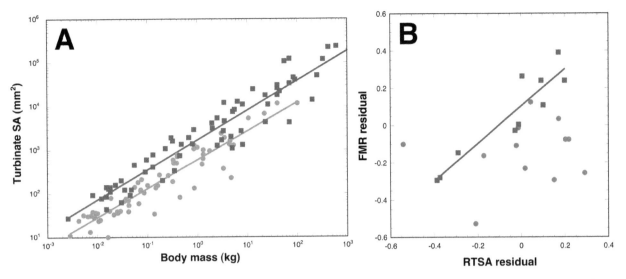

FIG. 9.4 (A) Scaling relationship of the respiratory turbinate surface area (RTSA) on body mass (Mb) in mammals (blue squares) and birds (red circles). In both clades, RTSA scales with slight positive allometry (common slope $Mb^{0.73}$), but is approximately three times greater in mammals than in similarly sized birds (ANCOVA, $n = 131$, $r^2 = 0.93$, $p < 0.0001$). (B) Regression residuals of field metabolic rate (FMR) and respiratory turbinate surface area (RTSA) on body mass in mammals (blue squares) and birds (red circles). RTSA is a strong predictor of FMR for mammals ($r^2 = 0.86$, $p < 0.01$), but not birds ($r^2 < 0.01$, $p = 0.93$). For mammals, FMR residuals show a direct linear correspondence (slope = 1) with RTSA residuals. FMR data from Nagy et al. (1999).

those mammalian taxa, which are obligatorily aquatic (e.g., cetaceans) and rely on episodic breathing, which precludes operation of the TCCE. In contrast, extensive turbinates with higher-than-predicted RTSA are found in many semiaquatic marine mammals, which spend considerable periods of time on land. Some mammalian orders are yet to be sampled for RT measurements—for instance, the snouters, a particularly macrosmatic taxon with great diversity in nasal morphology (Stümpke 1967).

No such pattern of environmental or metabolic influence on RTSA is discernible among birds. Many species adapted to cold or dry habitats (e.g., penguins, pigeons) show severely reduced respiratory turbinates. Other species with relatively low metabolic requirements (e.g., some procellariiform species) have larger-than-predicted respiratory turbinates. Complete absence of respiratory turbinates occurs in some diving birds (e.g., pelecaniform species; fig. 9.3G), despite the fact that they engage in apnea only during diving and breathe fairly continuously otherwise (Bartholomew et al. 1968).

Correspondence of mass-scaling exponents of morphologic and physiologic variables does not necessarily indicate a functional relationship between them. A single regression line for the cross-sectional area of nasal cavities of endotherms (e.g., Ruben et al. 1996) can camouflage clear morphologic differences between mammalian turbinals and avian conchae. Furthermore, cross-sectional or surface area of many biologic structures approximates isometric scaling (i.e., $Mb^{2/3}$; Calder 1984; Schmidt-Nielsen 1984) and may bear only *accidental* similarity to the scaling exponent of basal or resting metabolic rate (Frappell et al. 2001; White and Seymour 2003). The crucial step in verifying respiratory turbinates as the osteologic indicator of endothermy in air-breathing vertebrates is to show strong positive correlation of interspecific variation in RTSA and metabolic rate (*sensu* Garland and Huey 1987). A residuals analysis of our sample (10 mammal and 12 bird species) uncovered a highly significant correlation of RTSA and FMR in mammals, but none in birds (fig. 9.4B). This suggests that RTSA is a good proxy of FMR of extant mammals, but not of birds.

The threefold greater RTSA of mammals implies that birds either tolerate higher evaporative heat and water loss than mammals, or exhibit higher efficiency of heat and water exchange per unit turbinate area. Measurements of water content and temperature of expired air in mammals and birds indicate their respiratory heat and water losses are comparable (Schmidt-Nielsen, Hainsworth, et al. 1970; Langman et al. 1979; Johnsen et al. 1985). The efficiency of TCCE in the nasal cavity of birds is unlikely to be higher than that of mammals given the wider spacing of avian conchae. Further, two independent lines of evidence suggest that respiratory turbinates may not be the only site where conditioning of breathed air occurs. First, in vivo temperature measurements in penguins show that cold inspired air does not approximate body core temperature by the time it reaches the choana (Murrish 1973), as it must if the delicate lung parenchyma is to be protected from damage. Second, respiratory evaporative water loss in desert-adapted larks is not increased significantly by experimental occlusion of external nares (Tieleman et al. 1999), as it should be if heat/water conservation was confined to the nasal cavity (*contra* Geist 2000). Altogether, this strongly suggests that another region of the respiratory tract participates in TCCE—the trachea.

Trachea

Histology, Scaling, and Topography

The inner surface of the trachea is covered by a vascular pseudostratified columnar epithelium, similar to that of the respiratory turbinates (fig. 9.3H, I). The elongate, cylindrical morphology, coupled with a regular breathing pattern, should allow the trachea to participate in conditioning of respiratory airflow. The rate of flux in a countercurrent exchanger is proportional to its surface area, and inversely related to the distance between the center of airstream and the respiratory epithelium (Collins et al. 1971). For an approximately cylindrical exchanger, such as the trachea, an increase in the radial dimension brings a trade-off—as the surface area increases, so does the airstream-epithelium distance. In contrast, elongation of the trachea does not carry a similar trade-off—the longer the trachea, the longer the time air spends in contact with its respiratory epithelium, the more complete the exchange of heat and water. TCCE should, therefore, become more efficient with increasing tracheal length. Trachea of birds is approximately 2.7 times longer than of similarly sized mammals (Tenney and Bartlett 1967; Hinds and Calder

1971). A longer-than-predicted trachea compensates for the lack of turbinates in pelecaniform birds, whereas several penguin species (with reduced turbinates) possess a paired trachea (Zeek 1951; McLelland 1989) with half the effective air-epithelium distance and double the surface area. Although other factors (e.g., vocalization; Fitch 1999) may account for extreme tracheal elongation in some species, there is likely to be strong positive selection for a longer trachea in those taxa, whose survival depends on heat conservation, but whose RT morphology is limited by nasal geometry, phylogeny, or feeding behavior.

Just like RTSA, the tracheal surface area (TrSA) scales with positive allometry: $Mb^{0.71}$ for mammals ($n = 62$), and $Mb^{0.77}$ for birds ($n = 88$). On average, TrSA is almost 3.5 times greater in birds than in similarly sized mammals (fig. 9.5A). A residuals analysis, based admittedly on a limited sample size, shows no significant correlation of TrSA and FMR in either mammals ($n = 15$), or birds ($n = 9$; fig. 9.5B). There is no apparent association between TrSA with ecologic habitat or body temperature in either clade. Instead, phylogenetic factors appear to play a role in shaping tracheal morphology—carnivorans and chiropterans have relatively greater TrSA than other mammals, whereas falcons and owls have relatively smaller TrSA than other birds.

The idea that the trachea is a site of TCCE has been criticized by Geist (2000) and Ruben et al. (2012) on the premise that a temperature gradient established in the trachea would adversely affect the temperature of arterial blood supply to the brain. Proximity of the common carotid arteries to the trachea would make this argument plausible, although fluctuations in carotid blood temperature when breathing cold air have not been demonstrated experimentally. In contrast to the cervical topography in mammals, where carotid arteries lie alongside the trachea, carotid arteries of birds are located in a groove along the ventral surface of cervical vertebrae, and are separated from the trachea by a thick layer of cervical flexor musculature, the longus colli (Baumel 1964, 1993; Nickel et al. 1977; West et al. 1981). Such great distance between the trachea and carotid arteries effectively precludes heat transfer between them, so the temperature of arterial blood supply to the head is unlikely to be affected by TCCE in the trachea.

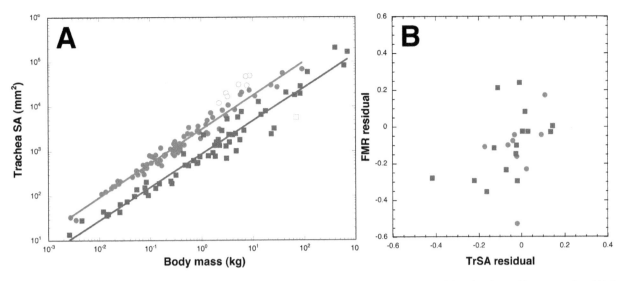

FIG. 9.5 (A) Scaling relationship of the tracheal surface area (TrSA) on body mass (Mb) in mammals (closed and open blue squares) and birds (closed and open red circles). Filled symbols represent species upon which the regression is based; open symbols represent species whose vocal adaptation (elongated trachea of cranes and swans; shortened trachea of humans) disqualifies them from the regression. In both clades, TrSA scales with positive allometry ($Mb^{0.70}$ for mammals, $Mb^{0.76}$ for birds), but is approximately 3.5 times greater in birds than in similarly sized mammals (ANCOVA, $n = 142$, $r^2 = 0.96$, $p < 0.01$). (B) Regression residuals of field metabolic rate (FMR) and trachea surface area (TrSA) on body mass in mammals (blue squares) and birds (red circles). TrSA is not a predictor of FMR for either clade. FMR data from Nagy et al. (1999).

Plasticity of Tracheal TCCE

That the trachea of birds actively participates in TCCE is revealed by direct temperature measurements in the upper respiratory tract of adult emus, which were raised with or without respiratory turbinates under seminatural conditions of the Australian outback. Direct temperature measurements of the tracheal epithelium showed a persistent temperature gradient from the glottis toward the body core (fig. 9.6A). A steeper temperature gradient in the distal trachea of emus without turbinates suggests a greater efficiency of their tracheal TCCE mechanism, likely due to plasticity of vascular networks in the tracheal epithelium. In addition, expired air temperature measurements at the external nares and the glottis indicate heat conservation was significantly greater in emus with intact turbinates and the bulk (75%) of respiratory heat conservation occurred in their nasal cavity (fig. 9.6B). In contrast, emus without turbinates relied primarily on their trachea as the site of most (60%) heat conservation (fig. 9.6B). A similar scenario might have been present among theropod ancestors of birds, which would have relied exclusively on

their trachea for heat and water conservation prior to the origin of respiratory turbinates.

Constraints on TCCE Location

Despite mammals and birds converging on a similar metabolic strategy, these two taxa appear to emphasize different sites of their respiratory tract for temporal countercurrent exchange. Why? We posit that their divergent strategies derive from taxon-specific developmental, biomechanic, and respiratory constraints.

Mammals are almost invariably constrained in development to seven cervical vertebrae (Galis 1999; Narita and Kuratani 2005). Among extant mammals, only three species are known to circumvent the cervical count constraint (two sloth species and the manatee). Whether this morphology is dictated by cancer incidence (Galis et al. 2006; Varela-Lasheras et al. 2011) or determined by diaphragmatic development (Buchholtz et al. 2012) remains to be determined. Regardless, this vertebral Bauplan appears to have been established early among the cynodont ancestors of mammals (Jenkins 1971b; Müller

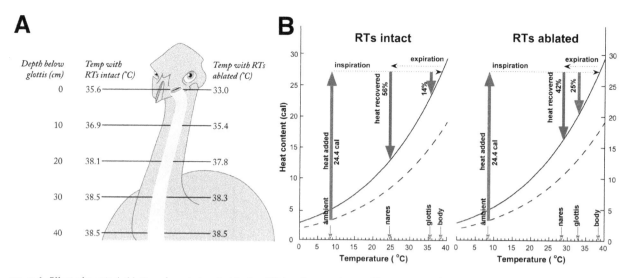

A

Depth below glottis (cm)	Temp with RTs intact (°C)	Temp with RTs ablated (°C)
0	35.6	33.0
10	36.9	35.4
20	38.1	37.8
30	38.5	38.3
40	38.5	38.5

B

FIG. 9.6 Effect of surgical ablation of respiratory turbinates (RTs) on temperature and heat content of expired air in lightly anesthetized emus (*Dromaius novaehollandiae*). Body core temperature 38.5°C; ambient temperature 8.5°C. (A) Expired air temperature at the glottis is significantly lower in emus with ablated RTs than in sham-operated controls ($n = 4$, t-test, $p < 0.001$). The temperature gradient in the trachea is steeper in emus with ablated RTs. (B) Expired air temperatures recorded at the nostrils of emus with intact RTs were significantly lower than those found in emus with surgically ablated RTs ($n = 4$, t-test, $p < 0.02$). Curves show heat capacity of 1 L of air at 100% relative humidity (solid line), and at 65% relative humidity (dashed line). Color arrows show heat gained by inspired air (red), and heat lost in expired air at the glottis (pink) and external nares (purple). These results suggest the avian trachea becomes more competent at heat and water conservation in the absence of functional RTs.

et al. 2010) and might have been influenced by the biomechanic need to support a massive head.

Mass of the cynodont head was determined primarily by their complex masticatory morphology. Selection for improved hearing acuity likely drove increased jaw adductor muscle mass, which allowed for balancing of adductor torques around the jaw joint and unloading of the original articular-quadrate joint (Crompton 1963, 1972; Crompton and Parker 1978; Crompton and Hylander 1986). This, in turn, required a commensurately robust skull, the mass of which may have constrained neck elongation and thus limited tracheal length. In fact, when skulls of extant mammals and birds are compared, the skull mass of mammals is twice that of similarly sized birds (fig. 9.7). This difference is presumably a result of feeding biomechanics—high forces engendered by chewing in mammals, and lack thereof in birds. Evolution of a bony secondary palate, via fusion of medial maxillary and palatal shelves, may have been driven by the need to brace the snout against high masticatory forces (Thomason and Russell 1986). At the same time, the secondary palate permitted enlargement of the nasal cavity and offered protection to the

incipient respiratory turbinates in more derived cynodonts (Hillenius 1992; Crompton et al., this volume).

In stark contrast to mammals, the cervical vertebral count of extant avian taxa varies from 13 to 25 (Burke et al. 1995). The ancestral cervical segment number for archosaurs is nine (Mansfield and Abzhanov 2010), as still seen in extant crocodilians. Although the trachea is rarely preserved in fossils, its length is likely to be determined by the length and number of cervical vertebrae. Measurements of relative neck length of select extinct and extant archosaurs (fig. 9.8) reveal a general trend in the archosaur phylogeny—necks of birds are longer than those of most non-avian theropods, which in turn are longer than those of basal archosaurs. *Archaeopteryx* is found at an intermediate position between non-avian and avian theropods, with a neck length of 40% precaudal vertebral length. Necks of crocodilians and tyrannosaurids are relatively short, perhaps as an adaptation to supporting a massive head (by shortening of the cervical cantilever arm). Cranial lightening apparent in other theropod taxa—achieved by loss of skeletal elements and teeth, reduced cranial robustness, (Gauthier 1986; Weishampel et al. 2004), and

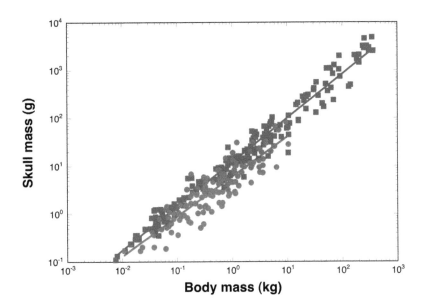

FIG. 9.7 Scaling relationship of skull mass on body mass (Mb) in mammals (blue squares) and birds (red circles). Reduced-major axis regression reveals that skull mass scales with isometry in mammals ($Mb^{0.98}$) and birds ($Mb^{1.02}$), but skulls of mammals are twice as heavy as those of similarly sized birds (ANCOVA, $n = 310$, $r^2 = 0.95$, $p < 0.0001$). Skull mass in mammals may be related to dietary specialization, as species with relatively light skulls include the echidna, armadillo, and pangolin. Notable exceptions of birds with relatively massive skulls include a frogmouth and a hornbill.

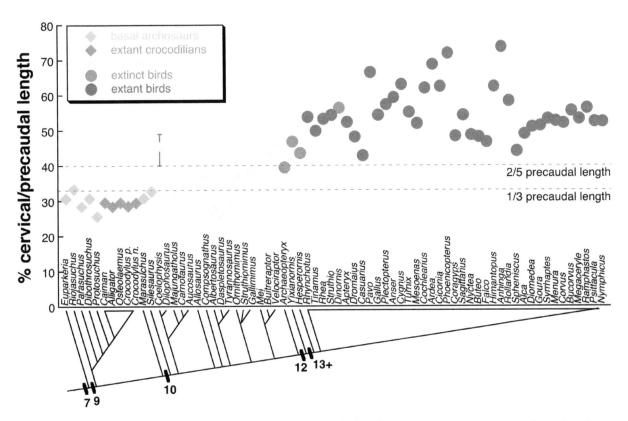

FIG. 9.8 Relative neck length in diverse extinct and extant archosaurs. Length of cervical vertebrae is expressed in proportion to that of precaudal vertebrae (a proxy for body size). Linear measurements along the ventral vertebral border were obtained from published skeletal reconstructions or photographs of extinct taxa, and from select complete skeletons of extant taxa (from the MCZ collection). Number of cervical vertebrae is indicated on the cladogram below the graph. Note a general trend in neck elongation from basal archosaurs to modern birds, which appears associated with gradual lightening of the cranial skeleton. Species with relatively massive skulls (crocodilians, tyrannosaurid dinosaurs) have relatively short necks.

extensive pneumatization of the skull (Witmer 1997)—may have lifted a biomechanic constraint on their neck elongation.

Progressive tracheal elongation among theropod dinosaurs would considerably have augmented the anatomic dead space of their respiratory system, and could not have occurred in the absence of an avian-style flow-through lung design. Unidirectional airflow through the rigid parabronchial lungs of birds allows for efficient gas exchange despite large tracheal volume, 4.5 times greater than that in similarly sized mammals (Hinds and Calder 1971). In contrast, tidal pool ventilation of the alveolar lungs of mammals is sensitive to dead space (Crosfill and Widdicombe 1961), and any increase in tracheal length (and volume) would likely have incurred significant costs associated with work of breathing. Fossil evidence of extensive pneumaticity in theropod dinosaurs suggests presence of a lung–air sac system similar to that of modern birds (O'Connor and Claessens 2005). Recent reports of unidirectional lung ventilation in modern crocodilians indicate that this ventilatory design probably arose early in the evolutionary history of archosaurs (Farmer and Sanders 2010; Schachner et al. 2013). If so, the respiratory constraint on tracheal dead space, and neck elongation, would have been effectively circumvented in avian ancestors.

A longer trachea with an efficient TCCE would have reduced respiratory heat loss, and thus reduced the cost of maintaining a high metabolic rate and a high body temperature. Among extinct dinosaurs, small theropods were likely to benefit the most from acquisition of endothermy, as their size could not afford them homeothermy via mass inertia (Seebacher 2003). This scenario dovetails with fossil evidence of gradual acquisition of integumental insulation (Chen et al. 1998), although theropods were not the only ornithodiran clade known to evolve feather-like structures (Clarke 2013). In contrast to ornithischian dinosaurs, whose nasal cavities were probably large enough to house respiratory turbinates (Maryańska 1977; Miyashita et al. 2011), the relatively short palates and narrow nasal cavities of theropods might have precluded presence of extensive respiratory turbinates (Ruben et al. 1996). Contribution of the nasal passages to conditioning of the respiratory airstream would have been dependent on the location of external nares (Witmer 2001) and nasal topography (Witmer and Ridgely 2008). Theropods

might have had to rely mainly on the TCCE in the trachea to control respiratory heat and water loss. Small theropods are likely to have benefited the most from an improved heat and water economy, because of higher mass-specific metabolic rate and greater surface area-to-volume ratio.

Admittedly, selection for improved thermoregulation alone might not have driven the increase in tracheal dimensions. A longer trachea of birds may simply be a function of gradual neck elongation, which itself would have been under various selection pressures, such as increased head mobility for preening or predation, especially if coupled with relative hind limb elongation (Gatesy 1991b; Gatesy and Middleton 1997) and functional (locomotor) constraints on the range of forelimb movement (Jenkins 1993; Gatesy and Dial 1996; Middleton and Gatesy 2000; Baier et al. 2007).

If the trachea did play a major role in TCCE, was there a selective pressure for "respiratory turbinates" in archosaurs at all?

Turbinates in Ectotherms

The idea of respiratory turbinates as the "Rosetta stone" of endothermy falls apart when nasal cavity morphology in extant ectotherms is revisited. Although no turbinate-like extensions are found protruding from the nasal capsule of lepidosaurs and chelonians, the singular concha of squamates is situated within the respiratory airstream and exhibits greater morphologic variation (Stebbins 1948; Bellairs 1949; Parsons 1967) than has been acknowledged in the turbinate debate. In addition, a turbinate-like structure is prominently present in crocodilians (Parker 1883; Reese 1901; Meek 1911; Parsons 1967; Witmer 1995). Constrained by the narrow geometry of the crocodilian nasal cavity, a slightly curled cartilaginous preconcha extends from the dorsolateral wall of the nasal capsule into the lumen and runs the length of the nasal passage between the external and internal nares (fig. 9.9). Its location and morphology suggests it could potentially participate in the TCCE, although the E/I breathing pattern of crocodilians likely precludes it. The crocodilian preconcha is partly covered with a chemosensory epithelium (Saint Girons 1976; Hansen 2007), appears to play a role in olfaction (Weldon and Ferguson 1993), and may contribute to the chemofensor (chemosensory defense) system of the upper

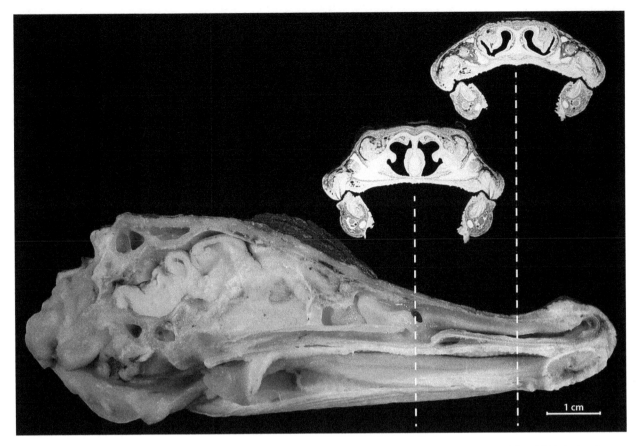

FIG. 9.9 Longitudinal and transverse sections of the head of a juvenile American alligator (*Alligator mississippiensis*, Mb = 1.36 kg), showing the prominent preconcha protruding into the nasal passage between the external and internal nares. The preconcha is covered with a chemosensory vascular epithelium, but does not appear to participate in temporal countercurrent exchange. Location of this "proto-turbinate" in the nasal passage of an extant ectotherm suggests that the original role of turbinates may not have been related to heat or water conservation.

airways (Tizzano and Finger 2013). Epithelium of the preconcha may also house CO_2-sensitive chemoreceptors, which modulate the animal's instantaneous breathing pattern (Tattersall et al. 2006).

Presence of the preconcha in extant crocodilians suggests that turbinate-like structures and ectothermy are not mutually exclusive. Its lack of homology with any turbinate-like structure of extant mammals and birds (Witmer 1995) suggests that it must have arisen independently in crocodilians. The preconcha may be a vestigial RT left over from the endothermic ancestors of crocodilians (Seymour et al. 2004), or evolved de novo for improved olfactory acuity. Neither scenario weakens the argument that turbinate-like structures protruding into the respiratory airstream can perform nonthermoregulatory roles in ectotherms (*contra* Bennett and Ruben 1986). This supports the contention that the origin of "proto-turbinates" does not correlate uniquely

with endothermy, but rather with presence of a complete secondary palate in various amniote lineages.

A similar combination of characters seems to have been present in nonmammalian cynodonts, whose oral and nasal cavities became separated by a secondary palate and whose nasal cavities likely housed cartilaginous turbinates (Crompton et al., this volume). If turbinates were present in ectothermic cynodonts, what was their function?

Heat Conservation or Heat Dissipation?

In extant mammals, respiratory turbinates participate not only in heat and water conservation, but also play a key role in heat dissipation and selective brain cooling (SBC) during environment- or exercise-induced heat stress. Under conditions of net heat gain, many mammals will switch off the TCCE by inspiring via the nasal passage, and expiring via the mouth (Schmidt-Nielsen, Bretz, and Taylor 1970). This mechanism can facilitate heat loss to the environment and uncouple temperature control of the brain and the body core (Taylor and Lyman 1972; Johnsen et al. 1987; Kuhnen and Jessen 1991; Mitchell et al. 2002; Caputa 2004). Venous blood from the RTs, cooled by inspiration, is used as a heat sink at the cavernous sinus/carotid artery interface, lowering the temperature of arterial blood supply to the brain (Daniel et al. 1953; Baker 1982). Close anatomic association of lateral nasal glands with evaporative surfaces of the nose in mammals (Blatt et al. 1972) suggests that demand for an effective cooling mechanism is an important factor in shaping the internal morphology of the nasal passage. Whether greater RTSA correlates with superior SBC ability in mammals has not been investigated.

A similar brain cooling mechanism operates in birds, but heat exchange occurs primarily via the well-developed *rete mirabile ophthalmicum*. The rete receives venous drainage primarily from the eyes (Kilgore et al. 1976; Bech and Midtgård 1981), but also from the beak, palate, and nasal cavity (Baumel 1975; Midtgård 1984). Bird eyes are significantly larger than those of similarly sized mammals (Kiltie 2000; Howland et al. 2004). The surface area of the ophthalmic rete scales isometrically ($Mb^{0.65}$) in 40 species of birds sampled by Midtgård (1983), which suggests that SBC in birds is a function of evaporative heat loss from eyes. This is corroborated by experimental evidence of elevation in brain temperature in birds whose eyes are covered during heat exposure (Bernstein et al. 1979) or whose blood flow to the rete is surgically obstructed (Kilgore et al. 1979). In addition, other richly vascular surfaces in the head and neck region (Crowe and Crowe 1979; Baumel et al. 1983; Arad et al. 1989; Tattersall et al. 2009) may participate in control of brain temperature in birds. Considering the relatively small surface area of RTs, and the existence of many other avenues for heat exchange, we predict that the nasal cavity plays a minor role in SBC in birds. To date, however, contribution of the nasal passages to heat dissipation in birds has not been quantified.

Nonmammalian cynodonts, with relatively small eyes and short necks, might initially have evolved respiratory turbinates for excess heat loss and some degree of SBC (Crompton et al., this volume), as their activity levels and/or ambient temperatures were increasing. On the other hand, non-avian theropods would not have had to rely on morphologic modifications to the nasal cavity, if they could dissipate sufficient heat from the relatively large surface area of their eyes, oropharynx, and trachea. Only once heat generation potential of theropods increased substantially, perhaps in association with flight ability, did the nasal cavities expand and accommodate additional cooling surfaces of RTs for better brain temperature homeostasis. The accuracy of this prediction will be tested as more data on orbit morphology in extinct cynodont and theropod taxa become available (K. Angielczyk and L. Schmitz, personal communication). Our scenario envisions the evolution of RTs *before* acquisition of endothermy by cynodont ancestors of mammals (Crompton et al., this volume), but *after* acquisition of endothermy by theropod ancestors of birds.

Conclusions

Respiratory turbinates (RTs) have been hailed as the "Rosetta stone" of endothermy. If proven correct, evidence of RTs in the fossil record would allow paleobiologists to infer the evolutionary pattern of this costly metabolic strategy in various tetrapod lineages. Our work supports the functional association of RT morphology and metabolic rate in extant mammals, but not birds.

Further, we present evidence of a turbinate-like structure in the anterior nasal cavity of extant crocodilians. Its presence counters the original assertion that RTs are an anatomic correlate of endothermy. Instead, we propose that heat and water conservation at RTs may be an exaptation of their original role in heat dissipation and selective brain cooling.

In birds, the relatively small RTs play an accessory role in control of brain temperature, given the large surface area of eyes, oropharynx, and other craniofacial regions available for evaporative heat loss. In mammals, in contrast, large RTs are an integral part of temperature control of arterial blood supply to the brain. This supports our contention that the original selective pressure for RT evolution may have been heat dissipation, not heat conservation.

Endothermy was a milestone in the evolutionary history of vertebrates. Independently acquired in the mammalian, avian, and possibly other lineages, endothermy freed the animals from the vagaries of ambient temperature and boosted their metabolic power. It did not, however, instigate the appearance of turbinate-like structures.

Avenues of Future Investigation

Structure of the upper respiratory tract of vertebrates has recently received lots of attention with regard to endothermy, yet some of the original literature has languished forgotten during the debate (vide crocodilian preconcha; Parker 1883). Much work remains to be done for a fuller understanding of RT form and function. We identify a few as-yet unexplored directions, which may yield exciting new information on the thermoregulatory flexibility of the upper respiratory tract.

(i) Morphology

A quantitative database of respiratory (and olfactory) turbinate morphology is needed to identify patterns of variation with species behavior, ecology, and geographic distribution. Considering their potential effects on metabolic rate and lung ventilation, any of these factors may affect the morphology of RTs. Sampling of extant carnivoran mammals suggests a significant effect of diet and habitat, as well as phylogenetic affiliation (Van Valkenburgh et al. 2011; Green et al. 2012). Similar studies on birds can elucidate factors determining the size and complexity of the avian conchae.

Investigation of RT development and growth in mammals and birds can inform our understanding of selective pressure behind their disparate evolutionary origins. Although ontogenetic development of mammalian turbinates has been illustrated long ago (Parker 1885), no quantitative studies of RT morphology and their correlation to thermoregulation have been undertaken to date. Studies of metabolic competency in neonatal mammals reveal that endothermy is acquired gradually with age and body size (e.g., Frappell 2008). Nasal cavity resistance to airflow increases with age in placental mammals (Mortola and Fisher 1981), possibly

because of RT growth. Whether RT development correlates with ontogenetic changes in metabolic rate has not been ascertained. Ontogeny of RT function in birds, in relation to their metabolic maturation, is not known at all.

(ii) Physiology

In order to understand how RTs function, it is necessary to quantify heat transfer during inspiration and expiration. This can be particularly informative for the few endotherms, which show an episodic breathing pattern with a post-inspiratory pause (Frappell 2003; Mortola and Seguin 2009). Recent advances in computational fluid dynamics (CFD) have allowed pioneering work on mammalian turbinates by Craven et al. (2007, 2010). To date, however, little is known about airflow and heat transfer in nasal cavities of birds. After its validation in extant species of ecto- and endotherms, the CFD technique may help us interpret heat transfer dynamics in fossil taxa (J. Bourke and L. Witmer, personal communication).

Heat and water transfer are always conflated in analyses of TCCE in the upper respiratory tract. Water conservation is likely more important than heat conservation to desert-adapted species (Mitchell et al. 2002; Williams and Tieleman 2005), and this may have been the case for ectothermic ancestors of mammals and birds. Some mammals (Schmidt-Nielsen et al. 1981) and birds (Withers and Siegfried 1981) are capable of desaturating expired air, but whether respiratory turbinates are involved in this mechanism has not been investigated. Future research should dissect out the role played by the nasal passages in heat versus water exchange.

(iii) Environment

Ambient conditions are likely to play a significant role in shaping RT morphology. An early report suggested that high ambient temperatures suppress development of RTs in laboratory rats (Riesenfeld 1973), but subsequent investigation has failed to corroborate the original findings (Hill 1982). No such experimental work has been undertaken on birds.

A recent scenario proposed that the origin of RTs may have been precipitated by the bout of global hypoxia, rather than driven by elevation in resting metabolic rate. Huey and Ward (2005) hypothesized that the drop in atmospheric oxygen level at the end of the Triassic (Berner et al. 2003) would have resulted in elevated ventilation rates. Global hypoxia, thus, might have increased selection pressure for heat and water conserving structures, such as RTs. This testable hypothesis (Powell 2010) is yet to be confirmed in the fossil record (Angielczyk and Walsh 2008) and under laboratory conditions.

* * *

Acknowledgments

We wish to thank everyone who donated specimens to this study, but in particular we recognize the contributions of Farish A. Jenkins Jr. of Harvard University. Farish's secret collection of specimens hidden in the electrical closets of MCZ Labs was a veritable treasure trove for this project. We also extend our gratitude to institutions: Museum of Comparative Zoology, Concord Field Station, New England Aquarium, South Australian Museum, Cleland Wildlife Park, Adelaide Zoo, and Monarto Zoological Park—special places that still recognize the value of destructive specimen analysis. Special thanks are due to several Biology 121 students at Harvard College, who provided preliminary data for this project. We appreciate the constructive critiques offered by J. B. Williams and L. M. Witmer of our draft manuscript. This work was supported by a grant from the Frank M. Chapman Memorial Fund (American Museum of Natural History) and the postdoctoral Discovery fellowship from the Australian Research Council to T.O.

References

Adams, D. R. 1972. Olfactory and non-olfactory epithelia in the nasal cavity of the mouse, *Peromyscus*. American Journal of Anatomy 133:37–50.

Angielczyk, K., and M. Walsh. 2008. Patterns in the evolution of nares size and secondary palate length in anomodont therapsids (Synapsida): implications for hypoxia as a cause of end-Permian tetrapod extinction. Journal of Paleontology 82:528–542.

Angiletta, M. J., Jr., B. S. Cooper, M. S. Schuler, and J. G. Boyles. 2010. The evolution of thermal physiology in endotherms. Frontiers in Bioscience E2:861–881.

Arad, Z., E. Midtgård, and M. H. Bernstein. 1989. Thermoregulation in turkey vultures: vascular anatomy, arteriovenous heat exchange, and behavior. Condor 91:505–514.

Baier, D. B., S. M. Gatesy, and F. A. Jenkins Jr. 2007. A critical ligamentous mechanism in the evolution of avian flight. Nature 445:307–310.

Baker, M. A. 1982. Brain cooling in endotherms in heat and exercise. Annual Review of Physiology 44:85–96.

Bakker, R. T. 1971. Dinosaur physiology and the origin of mammals. Evolution 25:636–658.

Bakker, R. 1972. Locomotor energetics of lizards and mammals compared. Physiologist 15:278.

Bakker, R. T. 1974. Dinosaur bioenergetics—a reply to Bennett and Dalzell, and Feduccia. Evolution 28:497–503.

Bang, B. G. 1971. Functional anatomy of the olfactory system in 23 orders of birds. Acta Anatomica 79 (Suppl. 58):1–76.

Bartholomew, G. A., R. C. Lasiewski, and E. C. Crawford Jr. 1968. Patterns of panting and gular flutter in cormorants, pelicans, owls, and doves. Condor 70:31–34.

Baumel, J. J. 1964. Vertebral-dorsal carotid artery interrelationships in the pigeon and other birds. Anatomischer Anzeiger 114:113–130.

Baumel, J. J. 1975. Heart and blood vessels. In R. Getty, ed., Sisson and Grossman's The Anatomy of the Domestic Animals, vol. 2. 5th ed. Philadelphia: W. B. Saunders.

Baumel, J. J. 1993. Handbook of Avian Anatomy: Nomina Anatomica Avium. 2nd ed. Cambridge: Nuttall Ornithological Club.

Baumel, J. J., A. F. Daley, and T. H. Quinn. 1983. The collar plexus of subcutaneous thermoregulatory veins in the pigeon, *Columba livia*; its association with esophageal pulsation and gular flutter. Zoomorphology 102:215–239.

Bech, C., and U. Midtgård. 1981. Brain temperature and the *rete mirabile ophthalmicum* in the zebra finch (*Poephila guttata*). Journal of Comparative Physiology 145:89–93.

Bellairs, A. A. 1949. Observations on the snout of *Varanus* and a comparison with that of other lizards and snakes. Journal of Anatomy 83:116–146.

Bennett, A. F. 1991. The evolution of activity capacity. Journal of Experimental Biology 160:1–23.

Bennett, A. F., and W. R. Dawson. 1976. Metabolism. In C. Gans and W. R. Dawson, eds., Biology of the Reptilia, vol. 5. London: Academic Press.

Bennett, A. F., and J. A. Ruben. 1979. Endothermy and activity in vertebrates. Science 206:649–654.

Bennett, A. F., and J. A. Ruben. 1986. The metabolic and thermoregulatory status of therapsids. In N. Hotton III, P. D. MacLean, J. J. Roth, and E. C. Roth, eds., The Ecology and Biology of Mammal-Like Reptiles. Washington, DC: Smithsonian Institution Press.

Benton, M. J. 2004. Vertebrate Palaeontology. 3rd ed. Oxford: Blackwell Science.

Berner, R. A., D. J. Beerling, R. Dudley, J. M. Robinson, and R. A. Wildman Jr. 2003. Phanerozoic atmospheric oxygen. Annual Review of Earth & Planetary Sciences 31:105–134.

Bernstein, M. H., I. Sandoval, M. B. Curtis, and D. M. Hudson. 1979. Brain temperatures in pigeons: effects of anterior respiratory bypass. Journal of Comparative Physiology 129:115–118.

Blatt, C. M., C. R. Taylor, and M. B. Habal. 1972. Thermal panting in dogs: the lateral nasal gland, a source of water for evaporative cooling. Science 177:804–805.

Blix, A. S., H. K. Johnsen, and J. B. Mercer. 1983. On nasal heat exchange and the structural basis for its regulation in reindeer. Journal of Physiology (London) 343:108P–109P.

Bostwick, K. S., and R. O. Prum. 2005. Courting bird sings with stridulating feathers. Science 309:736.

Brainerd, E. L., and T. Owerkowicz. 2006. Functional morphology and evolution of aspiration breathing in tetrapods. Respiratory Physiology & Neurobiology 154:73–88.

Brown, R. E., and M. R. Fedde. 1993. Airflow sensors in the avian wing. Journal of Experimental Biology 179:13–30.

Buchholtz, E. A., H. G. Ballin, S. A. Laves, J. T. Yang, M.-Y. Chan, and L. E. Drozd. 2012. Fixed cervical count and the origin of the mammalian diaphragm. Evolution & Development 14:399–411.

Burke, A. C., C. E. Nelson, B. A. Morgan, and C. Tabin. 1995. Hox genes and the evolution of vertebrate axial morphology. Development 121:333–346.

Calder, W. A. III 1984. Size, Function, and Life History. Cambridge, MA: Harvard University Press.

Cannon, B., and J. Nedergaard. 2004. Brown adipose tissue: function and physiological significance. Physiological Reviews 84:277–359.

Caputa, M. 2004. Selective brain cooling: a multiple regulatory mechanism. Journal of Thermal Biology 29:691–702.

Caro, T. 2005. The adaptive significance of coloration in mammals. BioScience 55:125–136.

Chen, P.-J., Z.-M. Dong, and S.-N. Zhen. 1998. An exceptionally well-preserved theropod dinosaur from the Yixian Formation of China. Nature 391:147–152.

Chinsamy, A., and W. J. Hillenius. 2004. Physiology of nonavian dinosaurs. In D. B. Weishampel, P. Dodson, and H. Osmólska, eds., The Dinosauria. 2nd ed. Berkeley: University of California Press.

Chinsamy-Turan, A. 2005. The Microstructure of Dinosaur Bone: Deciphering Biology with Fine-Scale Techniques. Baltimore: Johns Hopkins University Press.

Clark, C. J., D. O. Elias, and R. O. Prum. 2011. Aeroelastic flutter produces hummingbird feather song. Science 333:1430–1433.

Clarke, A., and H.-O. Pörtner. 2010. Temperature, metabolic power and the evolution of endothermy. Biological Reviews 85:703–727.

Clarke, J. 2013. Feathers before flight. Science 340:690–692.

Cole, P. 1953. Temperature, moisture and heat relationships in the upper respiratory tract. Journal of Laryngology 67:449–456.

Collins, J. C., T. C. Pilkington, and K. Schmidt-Nielsen. 1971. A model of respiratory heat transfer in a small mammal. Biophysical Journal 11:886–914.

Craven, B. A., T. Neuberger, E. G. Paterson, A. G. Webb, E. M. Josephson, E. E. Morrison, and G. S Settles. 2007. Reconstruction and morphometric analysis of the nasal airway of the dog (Canis familiaris) and implications regarding olfactory airflow. Anatomical Record 290:1325–1340.

Craven, B. A., E. G. Paterson, and G. S. Settles. 2010. The fluid dynamics of canine olfaction: unique nasal airflow patterns as an explanation of macrosmia. Journal of the Royal Society Interface 7:933–943.

Crompton, A. W. 1963. On the lower jaw of Diarthrognathus and the origin of the mammalian lower jaw. Proceedings of the Zoological Society of London 140:697–750.

Crompton, A. W. 1972. The evolution of the jaw articulation in cynodonts. Pp. 231–251 in K. A. Joysey and T. S. Kemp, eds., Studies in Vertebrate Evolution. Edinburgh: Oliver & Boyd.

Crompton, A. W., and W. L. Hylander. 1986. Changes in mandibular function following the acquisition of a dentary-squamosal jaw articulation. In N. Hotton III, P. D. MacLean, J. J. Roth, and E. C. Roth, eds., The Ecology and Biology of Mammal-Like Reptiles. Washington, DC: Smithsonian Institution Press.

Crompton, A. W., and P. Parker. 1978. Evolution of mammalian masticatory apparatus. American Scientist 66:192–201.

Crompton, A. W., C. R. Taylor, and J. A. Jagger. 1978. Evolution of homeothermy in mammals. Nature 272:333–336.

Crossfill, M. L., and J. G. Widdicombe. 1961. Physical characteristics of the chest and lungs and the work of breathing in different mammalian species. Journal of Physiology 158:1–14.

Crowe, T. M., and A. A. Crowe. 1979. Anatomy of the vascular system of the head and neck of the helmeted guineafowl, Numida meleagris. Journal of Zoology (London) 188:221–233.

Cubo, J., N. Le Roy, C. Martinez-Maza, and L. Montes. 2012. Paleohistological estimation of bone growth rate in extinct archosaurs. Paleobiology 38:335–349.

Cunnigham, S. J., M. R. Alley, and I. Castro. 2011. Facial bristle feather histology and morphology in New Zealand birds: implications for function. Journal of Morphology 272:118–128.

Damiani, R., S. Modesto, A. Yates, and J. Neveling. 2003. Earliest evidence of cynodont burrowing. Proceedings of the Royal Society B 270:1747–1751.

Daniel, P. M., J. D. K. Dawes, and M. M. L. Prichard. 1953. Studies on the carotid rete and its associated arteries. Philosophical Transactions of the Royal Society of London B 237:173–208.

Dawes, J. D. K., and M. M. L. Prichard. 1953. Studies of the vascular arrangements of the nose. Journal of Anatomy 87:311–323.

Dawson, T. J. 1973. "Primitive" mammals. In G. C. Whittow, ed., Comparative Physiology of Thermoregulation III: Special Aspects of Thermoregulation. New York: Academic Press.

Dawson, T. J., and A. J. Hulbert. 1970. Standard metabolism, body temperature, and surface areas of Australian marsupials. American Journal of Physiology 218:1233–1238.

de Ricqlès, A., 1974. Evolution of endothermy: histological evidence. Evolutionary Theory 1:51–80.

Ewart, J. C. 1881. On the nostrils of the cormorant. Journal of the Linnean Society 15:455–456.

Farmer, C. G. 2000. Parental care: the key to understanding endothermy and other convergent features in birds and mammals. American Naturalist 155:326–334.

Farmer, C. G., and K. Sanders. 2010. Unidirectional airflow in the lungs of alligators. Science 327:338–340.

Fitch, W. T. 1999. Acoustic exaggeration of size in birds via tracheal elongation: comparative and theoretical analyses. Journal of Zoology (London) 248:31–48.

Frappell, P. B. 2003. Ventilation and metabolic rate in the platypus: insights into the evolution of the mammalian breathing pattern. Comparative Biochemistry & Physiology A 136:943–955.

Frappell, P. B. 2008. Ontogeny and allometry of metabolic rate and ventilation in the marsupial: matching supply and demand from ectothermy to endothermy. Comparative Biochemistry & Physiology A 150:181–188.

Frappell, P. B., and R. V. Baudinette. 1995. Scaling of respiratory variables and the breathing pattern in adult marsupials. Respiration Physiology 100:83–90.

Frappell, P. B., R. V. Baudinette, P. M. MacFarlane, P. R. Wiggins, and G. Shimmin. 2002. Ventilation and metabolism in a large semifossorial marsupial: the effect of graded hypoxia and hypercapnia. Physiological & Biochemical Zoology 75:77–82.

Frappell, P. B., D. F. Boggs, and D. L. Kilgore Jr. 1998. How stiff is the armadillo? A comparison with the allometrics of mammalian respiratory system. Respiration Physiology 113:111–122.

Frappell, P. B., D. S. Hinds, and D. F. Boggs. 2001. Scaling of respiratory variables and the breathing patterns in birds: an allometric and phylogenetic approach. Physiological & Biochemical Zoology 74:75–89.

Galis, F. 1999. Why do almost all mammals have seven cervical vertebrae? Developmental constraints, Hox genes, and cancer. Journal of Experimental Zoology (Molecular Development & Evolution) 285:19–26.

Galis, F., T. J. M. Van Dooren, J. D. Feuth, J. A. J. Metz, A. Witkam, S. Ruinard, M. J. Steigenga, and L. C. D. Wunaendts. 2006. Extreme selection in humans against homeotic transformations of cervical vertebrae. Evolution 60:2643–2654.

Gans, C., and B. D. Clark. 1978. Air flow in reptilian ventilation. Comparative Biochemistry & Physiology 60A:453–457.

Garland, T., Jr., P. H. Harvey, and A. R. Ives. 1992. Procedures for the analysis of comparative data using phylogenetically independent contrasts. Systematic Biology 41:18–32.

Garland, T., Jr., and R. B. Huey. 1987. Testing symmorphosis: does structure match functional requirements? Evolution 41:1404–1409.

Gatesy, S. M. 1991a. Hind limb movements of the American alligator (Alligator mississippiensis) and postural grades. Journal of Zoology (London) 224:577–588.

Gatesy, S. M. 1991b. Hind limb scaling in birds and other theropods: implications for terrestrial locomotion. Journal of Morphology 209:83–96.

Gatesy, S. M. 1995. Functional evolution of the hind limb and tail from basal theropods to birds. In J. J. Thomason, ed., Functional Morphology in Vertebrate Paleontology. Cambridge: Cambridge University Press.

Gatesy, S. M., and K. P. Dial. 1996. Locomotor modules and the evolution of avian flight. Evolution 50:331–340.

Gatesy, S. M., and K. M. Middleton. 1997. Bipedalism, flight, and the evolution of theropod locomotor diversity. Journal of Vertebrate Paleontology 17:308–329.

Gauthier, J. 1986. Saurischian monophyly and the origin of birds. Memoirs of the California Academy of Sciences 8:1–55.

Geist, N. R. 2000. Nasal respiratory turbinate function in birds. Physiological & Biochemical Zoology 73:581–589.

Gewecke, M., and M. Woike. 1978. Breast feathers as an air-current sense organ for the control of flight behaviour in a songbird (Carduelis spinus). Zeitschrift für Tierpsychologie 47:293–298.

Goodale, J. L. 1896. An experimental study of the respiratory functions of the nose. New England Journal of Medicine 135:457–460.

Gould, E., and J. F. Eisenberg. 1966. Notes on the biology of the Tenrecidae. Journal of Mammalogy 47:660–686.

Green, P. A., B. Van Valkenburgh, B. Pang, D. Bird, T. Rowe, and A. Curtis. 2012. Respiratory and olfactory turbinal size in canid and arctoid carnivorans. Journal of Anatomy 221:609–621.

Hansen, A. 2007. Olfactory and solitary chemosensory cells: two different chemosensory systems in the nasal cavity of the American alligator, Alligator mississippiensis. BMC Neuroscience 8:64.

Heath, J. E. 1968. The origins of thermoregulation. In Evolution and Environment. New Haven: Yale University Press.

Hill, R. W. 1982. A reexamination of the effects of high rearing temperature on the nonolfactory turbinates in rats. American Journal of Physical Anthropology 58:357–362.

Hillenius, W. J. 1992. The evolution of nasal turbinates and mammalian endothermy. Paleobiology 18:17–29.

Hillenius, W. J. 1994. Turbinates in therapsids: evidence for Late Permian origins of mammalian endothermy. Evolution 48:207–229.

Hillenius, W. J., and J. A. Ruben. 2004. The evolution of endothermy in terrestrial vertebrates: who? when? why? Physiological & Biochemical Zoology 77:1019–1042.

Hinds, D. S., and W. A. Calder. 1971. Tracheal dead space in the respiration of birds. Evolution 25:429–440.

Hopson, J. A. 2012. The role of foraging mode in the origin of therapsids: implications for the origin of mammalian endothermy. Fieldiana: Life and Earth Sciences 5:126–148.

Howland, H. C., S. Mereola, and J. R. Basarab. 2004. The allometry and scaling of the size of vertebrate eyes. Vision Research 44:2043–2065.

Huey, R. B., and P. D. Ward. 2005. Hypoxia, global warming and terrestrial Late Permian extinctions. Science 308:398–401.

Ingelstedt, S. 1956. Studies on the conditioning of air in the respiratory tract. Acta Oto-laryngologica 131:1–180.

Jackson, D. C., and K. Schmidt-Nielsen. 1964. Countercurrent heat exchange in the respiratory passages. Proceedings of the National Academy of Sciences 51:1192–1197.

Jastroch, M., K. W. Withers, S. Taudien, P. B. Frappell, M. Helwig, T. Fromme, V. Hirschberg, G. Helmaier, B. M. McAllan, B. T. Firth, T. Burmester, M. Platzer, and M. Klingenspor. 2008. Marsupial uncoupling protein 1 sheds light on the evolution of mammalian nonshivering thermogenesis. Physiological Genomics 32:161–169.

Jenkins, F. A., Jr. 1970a. Cynodont postcranial anatomy and the "prototherian" level of mammalian organization. Evolution 24:230–252.

Jenkins, F. A., Jr. 1970b. Limb movements in a monotreme (Tachyglossus aculeatus): a cineradiographic analysis. Science 168:1473–1475.

Jenkins, F. A., Jr. 1971a. Limb posture and locomotion in the Virginia opossum (Didelphis marsupialis) and in other non-cursorial mammals. Journal of Zoology (London) 165:303–315.

Jenkins, F. A., Jr. 1971b. The postcranial skeleton of African cynodonts: problems in the early evolution of the mammalian postcranial skeleton. Bulletin of the Peabody Museum of Natural History (Yale University) 36:1–216.

Jenkins, F. A., Jr. 1993. The evolution of the avian shoulder joint. American Journal of Science 293A:253–267.

Jenkins, F. A., Jr., and S. M. Camazine. 1977. Hip structure and locomotion in ambulatory and cursorial carnivores. Journal of Zoology (London) 181:351–370.

Ji, Q., Z.-X. Luo, C.-Xi. Yuan, and A. R. Tabrum. 2006. A swimming mammaliaform from the Middle Jurassic and ecomorphological diversification of early mammals. Science 311:1123–1127.

Ji, Q., M. A. Norell, K.-Q. Gao, S.-A. Ji, and R. Dong. 2001. The distribution of integumentary structures in a feathered dinosaur. Nature 410:1084–1088.

Johnsen, H. K., A. S. Blix, L. Jørgensen, and J. B. Mercer. 1985. Vascular basis for regulation of nasal heat exchange in reindeer. American Journal of Physiology 249:R617–R623.

Johnsen, H. K., A. S. Blix, J. B. Mercer, and K. D. Bolz. 1987. Selective cooling of the brain in reindeer. American Journal of Physiology 253:R848–R853.

Kardong, K. V. 2009. Vertebrates: Comparative Anatomy, Function, Evolution. 5th ed. Boston: McGraw Hill.

Kemp, T. S. 2006. The origin of mammalian endothermy: a paradigm for the evolution of complex biological structure. Zoological Journal of the Linnean Society 147:473–488.

Kilgore, D. L., Jr., M. H. Bernstein, and D. M. Hudson. 1976. Brain temperatures in birds. Journal of Comparative Physiology 110:209–215.

Kilgore, D. L., Jr., D. F. Boggs, and G. F. Birchard. 1979. Role of the rete mirabile ophthalmicum in maintaining the body-to-brain temperature difference in pigeons. Journal of Comparative Physiology 129:119–122.

Kiltie, R. A. 2000. Scaling of visual acuity with body size in mammals and birds. Functional Ecology 14:226–234.

Klein, W., and T. Owerkowicz. 2006. Function of intracoelomic septa in lung ventilation of amniotes: lessons from lizards. Physiological & Biochemical Zoology 79:1019–1032.

Koteja, P. 2000. Energy assimilation, parental care and the evolution of endothermy. Proceedings of the Royal Society B 267:479–484.

Kuhnen, G., and C. Jessen. 1991. Threshold and slope of selective brain cooling. Pflügers Archiv 418:176–183.

Küster, E. 1905. Die Innervation und Entwicklung der Tastfeder. Morphologisches Jahrbuch 34:126–148.

Lane, N. 2010. Life Ascending: The Ten Great Inventions of Evolution. New York: W. W. Norton. 344 pp.

Langman, V. A., G. M. O. Maloiy, K. Schmidt-Nielsen, and R. C. Schroter. 1979. Nasal heat exchange in the giraffe and other large mammals. Respiration Physiology 37:325–333.

Liem, K. F., W. Bemis, W. F. Walker Jr., and L. Grande. 2000. Functional Anatomy of the Vertebrates: An Evolutionary Perspective. 3rd ed. Belmont: Thomson/Brooks Cole.

Lung, M. A., and J. C. C. Wang. 1989. Autonomic nervous control of nasal vasculature and airflow resistance in the anaesthetized dog. Journal of Physiology (London) 419:121–139.

MacDonald, J. D. 1960. Secondary external nares of the gannet. Proceedings of the Zoological Society of London 135:357–363.

Maderson, P. F. A. 1972. When? Why? and How?: Some speculations on the evolution of the vertebrate integument. American Zoologist 12:159–171.

Mansfield, J. F., and A. Abzhanov. 2010. Hox expression in the American alligator and evolution of archosaur axial patterning. Journal of Experimental Zoology (Molecular and Developmental Evolution) 314B:629–644.

Maryańska, T. 1977. Ankylosauridae (Dinosauria) from Mongolia. Palaeontologia Polonica 37:85–151.

McNab, B. K. 1978. The evolution of endothermy in the phylogeny of mammals. American Naturalist 112:1–21.

McNab, B. K. 2009. Resources and energetics determined dinosaur maximal size. Proceedings of the National Academy of Sciences 106:12184–12188.

McLelland, J. 1989. Larynx and trachea. In A. S. King, and J. McLelland, eds., Form and Function in Birds, vol. 4. New York: Academic Press.

Meek, A. 1911. On the morphogenesis of the head of the crocodile. Journal of Anatomy & Physiology 45:357–377.

Meng, J., and A. R. Wyss. 1997. Multituberculate and other mammal hair recovered from Paleogene excreta. Nature 385:712–714.

Middleton, K. M., and S. M. Gatesy. 2000. Theropod forelimb design and evolution. Zoological Journal of the Linnean Society 128:149–187.

Midtgård, U. 1983. Scaling of the brain and the eye cooling system in birds: a morphometric analysis of the rete ophthalmicum. Journal of Experimental Zoology 225:197–207.

Midtgård, U. 1984. Blood vessels and the occurrence of arteriovenous anastomoses in cephalic heat loss areas of mallards, Anas platyrhynchos (Aves). Zoomorphology 104:323–335.

Mitchell, D., S. K. Maloney, C. Jessen, H. P. Laburn, P. R. Kamerman, G. Mitchell, and A. Fukler. 2002. Adaptive heterothermy and selective brain cooling in arid-zone mammals. Comparative Biochemistry & Physiology B 131:571–585.

Miyashita, T., V. M. Arbour, L. M. Witmer, and P. J. Currie. 2011. The internal cranial morphology of an armoured dinosaur Euoplocephalus corroborated by X-ray computed tomographic reconstruction. Journal of Anatomy 219:661–675.

Moore, W. J. 1981. The Mammalian Skull. Cambridge: Cambridge University Press.

Morgan, K. T., J. S. Kimbell, T. M. Monticello, A. L. Patra, and A. Fleishman. 1991. Studies of inspiratory airflow patterns in the nasal passages of the F344 rat and rhesus monkey using nasal molds: relevance to formaldehyde toxicity. Toxicology and Applied Pharmacology 110:223–240.

Mortola, J. P. 2001. Respiratory Physiology of Newborn Mammals: A Comparative Perspective. Baltimore: Johns Hopkins University Press.

Mortola, J. P., and J. T. Fisher. 1981. Mouth and nose resistance in newborn kittens and puppies. Journal of Applied Physiology 51:641–645.

Mortola, J. P., and J. Seguin. 2009. End-tidal CO_2 in some aquatic mammals of large size. Zoology 112:77–85.

Müller, J., T. M. Scheyer, J. J. Head, P. M. Barrett, I. Werneburg, P. G. Ericson, D. Pol, and M. Sánchez-Villagra. 2010. Homeotic effects, somitogenesis and the evolution of vertebral numbers in recent and fossil amniotes. Proceedings of the National Academy of Sciences 107:2118–2123.

Murrish, D. E. 1973. Respiratory heat and water exchange in penguins. Respiration Physiology 19:262–270.

Murrish, D. E., and K. Schmidt-Nielsen. 1970. Exhaled air temperature and water conservation in lizards. Respiration Physiology 10:151–158.

Nagy, K. A. 2005. Field metabolic rate and body size. Journal of Experimental Biology 208:1621–1625.

Nagy, K. A., I. A. Girard, and T. K. Brown. 1999. Energetics of free-ranging mammals, reptiles and birds. Annual Review of Nutrition 19:247–277.

Narita, Y., and S. Kuratani. 2005. Evolution of the vertebral formulae in mammals: a perspective on developmental constraints. Journal of Experimental Zoology 304B:91–106.

Negus, V. E. 1958. Comparative Anatomy and Physiology of the Nose and Paranasal Sinuses. Edinburgh: E.& S. Livingstone.

Nespolo, R. F., L. D. Bacigalupe, C. C. Figueroa, P. Koteja, and J. C. Opazo. 2011. Using new tools to solve an old problem: the evolution of endothermy in vertebrates. Trends in Ecology and Evolution 26:414–423.

Nickel, R., A. Schummer, E. Seiferle, W. G. Siller, and P. A. L. Wright. 1977. Anatomy of the Domestic Birds. Berlin: Paul Parey.

Norell, M. A., and X. Xu. 2005. Feathered dinosaurs. Annual Review of Earth and Planetary Sciences 33:277–299.

O'Connor, P. M., and L. P. A. M. Claessens. 2005. Basic avian pulmonary design and flow-through ventilation in non-avian theropod dinosaurs. Nature 436:253–256.

Padian, K., and J. R. Horner. 2004. Dinosaur physiology. In D. B. Weishampel, P. Dodson, and H. Osmólska, eds., The Dinosauria. 2nd ed. Berkeley: University of California Press.

Padian, K., and E.-T. Lamm. 2013. Bone Histology of Fossil Tetrapods: Advancing Methods, Analysis, and Interpretations. Berkeley: University of California Press.

Parker, W. K. 1883. On the structure and development of the skull in Crocodilia. Transactions of the Zoological Society of London 11:263–310.

Parker, W. K. 1885. On the structure and development of the skull in the Mammalia. Philosophical Transactions of the Royal Society of London 176:1–275.

Parsons, T. S. 1967. Evolution of nasal structure in the lower tetrapods. American Zoologist 7:397–413.

Perry, S. F., and H. R. Duncker. 1978. Lung architecture, volume and static mechanics in five species of lizards. Respiration Physiology 34:61–81.

Perry, S. F., and H. R. Duncker. 1980. Interrelationships of static mechanical factors and anatomical structure. Journal of Comparative Physiology 138:321–334.

Pontzer, H., V. Allen, and J. R. Hutchinson. 2009. Biomechanics of running indicates endothermy in bipedal dinosaurs. PLoS One 4:e7783.

Pough, F. H., C. M. Janis, and J. B. Heiser. 2008. Vertebrate Life. 8th ed. San Francisco: Benjamin Cummings.

Powell, F. L. 2010. Studying biological responses to global change in atmospheric oxygen. Respiration Physiology & Neurobiology 173S:S6–S12.

Pycraft, W. P. 1899. Note on the external nares of the cormorant. Journal of the Linnean Society 27:207–209.

Reese, A. M. 1901. The nasal passages of the Florida alligator. American Naturalist 53:457–464.

Riesenfeld, A. 1973. The effects of extreme temperatures and starvation on the body proportions of the rat. American Journal of Physical Anthropology 39:427–460.

Rommel, S. A., D. A. Pabst, W. A. McLellan, J. G. Mead, and C. W. Potter. 1992. Anatomical evidence for a countercurrent heat exchange associated with dolphin testes. Anatomical Record 232:150–156.

Rowe, T. B., T. P. Eiting, T. E. Macrini, and R. A. Ketcham. 2005. Organization of the olfactory and respiratory skeleton in the nose of the gray short-tailed opossum Monodelphis domestica. Journal of Mammalian Evolution 12:303–336.

Rowe, T. B., T. E. Macrini, and Z.-X. Luo. 2011. Fossil evidence on origin of the mammalian brain. Science 332:955–957.

Ruben, J. A. 1995. The evolution of endothermy in mammals and birds: from physiology to fossils. Annual Review of Physiology 57:69–95.

Ruben, J. A. 1996. Evolution of endothermy in mammal, birds and their ancestors. In I. A. Johnston and A. F. Bennett, eds., Animal and Temperature: Phenotypic and Evolutionary Adaptation. Society for Experimental Biology Seminar Series 59. Cambridge: Cambridge University Press.

Ruben, J. A., W. J. Hillenius, N. R. Geist, A. Leitch, T. D. Jones, P. J. Currie, J. R. Horner, and G. Espe III. 1996. The metabolic status of some Late Cretaceous dinosaurs. Science 273: 1204–1207.

Ruben, J. A., W. J. Hillenius, T. S. Kemp, and D. E. Quick. 2012. The evolution of mammalian endothermy. In A. Chinsamy-Turan, ed., Forerunners of Mammals: Radiation, Histology, Biology. Bloomington: Indiana University Press,.

Ruben, J. A., and T. D. Jones. 2000. Selective factors associated with the origin of fur and feathers. American Zoologist 40:585–596.

Ruben, J. A., A. Leitch, W. J. Hillenius, N. R. Geist, and T. Jones. 1997. New insights into the metabolic physiology of dinosaurs. In J. O. Farlow and M. K. Brett-Surman, eds., The Complete Dinosaur. Bloomington: Indiana University Press.

Rubenson, J., D. G. Lloyd, T. F. Besier, D. B. Heliams, and P. A. Fournier. 2007. Running in ostriches (Struthio camelus): three-dimensional joint axes alignment and joint kinematics. Journal of Experimental Biology 210:2548–2562.

Saint Girons, H. 1976. Données histologiques sur les fosses nasales et leurs annexes chez Crocodylus niloticus Laurenti et Caiman crocodilus (Linnaeus) (Reptilia, Crocodylidae). Zoomorphology 84:301–318.

Sarko, D. K., F. L. Rice, and R. L. Reep. 2011. Mammalian tactile hair: divergence from a limited distribution. Annals of the New York Academy of Sciences 1225:90–100.

Schachner, E., J. R. Hutchinson, and C. G. Farmer. 2013. Pulmonary anatomy in the Nile crocodile and the evolution of unidirectional airflow in Archosauria. PeerJ 1:e60.

Schmidt-Nielsen, K. 1981. Countercurrent systems in animals. Scientific American 244:118–128.

Schmidt-Nielsen, K. 1984. Scaling: Why Is Animal Size So Important? Cambridge: Cambridge University Press.

Schmidt-Nielsen, K., W. L. Bretz, and C. R. Taylor. 1970. Panting in dogs: unidirectional air flow over evaporative surfaces. Science 11:1102–1104.

Schmidt-Nielsen, K., F. R. Hainsworth, and D. E. Murrish. 1970. Counter-current heat exchange in the respiratory passages: effect on water and heat balance. Respiration Physiology 9:263–276.

Schmidt-Nielsen, K., R. C. Schroter, and A. Shkolnik. 1981. Desaturation of exhaled air in camels. Proceedings of the Royal Society of London B 211:305–319.

Scholander, P. F., and J. Krog. 1957. Countercurrent heat exchange and vascular bundles in sloths. Journal of Applied Physiology 10:405–411.

Scholander, P. F., and W. E. Schevill. 1955. Counter-current vascular heat exchange in the fins of whales. Journal of Applied Physiology 8:279–282.

Schroter, R. C., and N. V. Watkins. 1989. Respiratory heat exchange in mammals. Respiration Physiology 78:357–368.

Seebacher, F. 2003. Dinosaur body temperatures: the occurrence of endothermy and ectothermy. Paleobiology 29:105–122.

Seymour, R. S., C. L. Bennett-Stamper, S. D. Johnston, D. R. Carrier, and G. C. Grigg. 2004. Evidence for endothermic ancestors of crocodiles at the stem of archosaur evolution. Physiological & Biochemical Zoology 77:1051–1067.

Simpson, S. J., A. Y. Fong, K. J. Cummings, and P. B. Frappell. 2012. The ventilatory response to hypoxia and hypercapnia is absent in the neonatal fat-tailed dunnart. Journal of Experimental Biology 215:4242–4247.

Smith, T. D., T. P. Eiting, and J. B. Rossie. 2011. Distribution of olfactory and nonolfactory surface area in the nasal fossa of *Microcebus murinus*: implications for microcomputed tomography and airflow studies. Anatomical Record 294:1217–1225.

Stahl, W. R. 1967. Scaling of respiratory variables in mammals. Journal of Applied Physiology 22:453–460.

Stebbins, R. C. 1948. Nasal structures in lizards with reference to olfaction and conditioning of the inspired air. American Journal of Anatomy 83:183–221.

Stümpke, H. 1967. The Snouters: Form and Life of the Rhinogrades. Garden City: Natural History Press.

Tattersall, G. J., D. V. Andrade, and A. S. Abe. 2009. Heat exchange from the toucan bill reveals a controllable vascular thermal radiator. Science 325:468–470.

Tattersall, G. J., D. V. de Andrade, S. P. Brito, A. S. Abe, and W. K. Milsom. 2006. Regulation of ventilation in the caiman (*Caiman latirostris*): effects of inspired CO_2 on pulmonary and upper airway chemoreceptors. Journal of Comparative Physiology B 176:125–138.

Taylor, C. R. 1980. Evolution of mammalian homeothermy: a two-step process? In K. Schmidt-Nielsen, L. Bolis, and C. R. Taylor, eds., Comparative Physiology: Primitive Mammals. Cambridge: Cambridge University Press.

Taylor, C. R., and C. Lyman. 1972. Heat storage in running antelopes: independence of brain and body temperatures. American Journal of Physiology 222:114–117.

Tenney, S. M., and D. Bartlett. 1967. Comparative quantitative morphology of the mammalian lung: trachea. Respiration Physiology 3:130–135.

Thomason, J. J., and A. P. Russell. 1986. Mechanical factors in the evolution of the mammalian secondary palate: a theoretical analysis. Journal of Morphology 189:199–213.

Tieleman, B. I., J. B. Williams, G. Michaeli, and B. Pinshow. 1999. The role of the nasal passages in the water economy of crested larks and desert larks. Physiological & Biochemical Zoology 72:219–226.

Tizzano, M., and T. E. Finger. 2013. Chemosensors in the nose: guardians of the airways. Physiology 28:51–60.

Van Valkenburgh, B., A. Curtis, J. Z Samuels, D. Bird, B. Fulkerson, J. Meachen-Samuels, and G. J. Slater. 2011. Aquatic adaptations in the nose of carnivorans: evidence from the turbinates. Journal of Anatomy 218:298–310.

Van Valkenburgh, B., J. Theodor, A. Friscia, A. Pollack, and T. Rowe. 2004. Respiratory turbinates of canids and felids: a quantitative comparison. Journal of Zoology (London) 264:281–293.

Varela-Lasheras, I., A. J. Bakker, S. D. van der Mije, J. A. J. Metz, J. van Alpern, and F. Galis. 2011. Breaking evolutionary and pleiotropic constraints in mammals: on sloths, manatees and homeotic mutations. EvoDevo 2:11.

Walter, I., and F. Seebacher. 2009. Endothermy in birds: underlying molecular mechanisms. Journal of Experimental Biology 212:2329–2336.

Weishampel, D. B., P. Dodson, and H. Osmòlska. 2004. The Dinosauria. 2nd ed. Berkeley: University of California Press.

Weldon, P. J., and M. W. J. Ferguson. 1993. Chemoreception in crocodilians: anatomy, natural history, and empirical results. Brain, Behavior and Evolution 41:239–245.

West, N. H., B. L. Langille, and D. R. Jones. 1981. Cardiovascular system. In A. S. King and J. McLelland, eds., Form and Function in Birds, vol. 2. London: Academic Press.

White, C. R., N. F. Phillips, and R. S. Seymour. 2006. The scaling and temperature dependence of vertebrate metabolism. Biology Letters 2:125–127.

White, C. R., and R. S. Seymour. 2003. Mammalian basal metabolic rate is proportional to body mass 2/3. Proceedings of the National Academy of Sciences 100:4046–4049.

Williams, J. B., and B. I. Tieleman. 2005. Physiological adaptations in desert birds. BioScience 55:416–425.

Withers, P. C., and W. R. Siegfried. 1981. Desert ostrich exhales unsaturated air. South African Journal of Science 77:569–570.

Witmer, L. M. 1995. Homology of facial structures in extant archosaurs (birds and crocodilians), with special reference to paranasal pneumaticity and nasal conchae. Journal of Morphology 225:269–327.

Witmer, L. M. 1997. The evolution of the antorbital cavity of archosaurs: a study in soft-tissue reconstruction in the fossil record with an analysis of the function of pneumaticity. Journal of Vertebrate Paleontology 17, Suppl. (1):1–73.

Witmer, L. M. 2001. Nostril position in dinosaurs and other vertebrates and its significance for nasal function. Science 293:850–853.

Witmer, L. M., and R. C. Ridgely. 2008. The paranasal air sinuses of predatory and armoured dinosaurs (Archosauria: Theropoda and Ankylosauria) and their contribution to cephalic structure. Anatomical Record 291:1362–1388.

Witmer, L. M., R. C. Ridgely, D. L. Dufeau, and M. C. Semones. 2008. Using CT to peer into the past: 3D visualization of the brain and ear regions of birds, crocodiles, and nonavian dinosaurs. In H. Endo and R. Frey, eds., Anatomical Imaging: Towards a New Morphology. Tokyo: Springer.

Zeek, P. M. 1951. Double trachea in penguins and sea lions. Anatomical Record 111:327–343.

Zelená, J. 1994. Nerves and Mechanoreceptors: The Role of Innervation in the Development and Maintenance of Mammalian Mechanoreceptors. London: Chapman & Hall.

Zhang, F., and Z. Zhou. 2000. A primitive enantiornithine bird and the origin of feathers. Science 290:1955–1959.

10

Origin of the Mammalian Shoulder

Zhe-Xi Luo*

Introduction

The shoulder girdle connects the forelimb and the axial skeleton (fig. 10.1). Its structure has undergone significant changes through the origins of extant mammals from the pre-mammalian cynodonts, accompanied by evolution of locomotor function (Jenkins 1970a, 1971a, 1971b; Jenkins and Weijs 1979). Thanks to the newly discovered fossils and more extensive comparative studies in the last two decades, the shoulder girdle structure has become better known in a wide range of pre-mammalian cynodonts, stem mammaliaforms that are extinct relatives to modern mammals, and Mesozoic clades of the crown Mammalia defined by the common ancestor of monotremes, marsupials, and placentals, plus fossil mammals among these living groups (Krebs 1991; Rougier 1993; Sereno and McKenna 1995; Hu et al. 1997; Ji et al. 1999; Gow 2001; Luo and Wible 2005; Martin 2005; Hu 2006; Sereno 2006; Sues and Jenkins 2006; Luo, Chen, et al. 2007; Hurum and Kielan-Jaworowska 2008; Chen and Luo 2013). The better-preserved fossils have revealed more complex patterns of evolution of the shoulder girdles through the great transition from cynodonts to mammals.

Monotremes are different from marsupials and placentals, collectively known as the crown therians (or Theria *sensu* Rowe 1988), in patterns of embryogenesis, and

* Department of Organismal Biology and Anatomy, University of Chicago

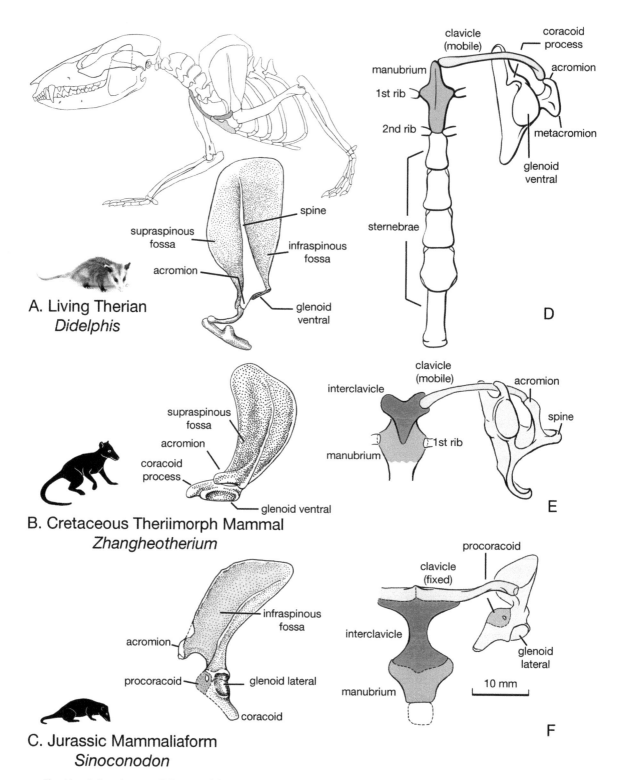

A. Living Therian
Didelphis

supraspinous fossa
spine
acromion
infraspinous fossa
glenoid ventral

B. Cretaceous Theriimorph Mammal
Zhangheotherium

supraspinous fossa
acromion
coracoid process
glenoid ventral

C. Jurassic Mammaliaform
Sinoconodon

acromion
infraspinous fossa
procoracoid
glenoid lateral
coracoid

clavicle (mobile)
coracoid process
manubrium
1st rib
acromion
2nd rib
metacromion
sternebrae
glenoid ventral

D

clavicle (mobile)
interclavicle
acromion
manubrium
spine
1st rib

E

procoracoid
clavicle (fixed)
interclavicle
glenoid lateral
manubrium
10 mm

F

FIG. 10.1 Shoulder girdles of mammaliaforms and therian mammals. Top, Extant marsupial *Didelphis virginiana*: shoulder girdle in lateral view (A) and ventral view (D). Middle, Early Cretaceous theriimorph (spalacotheroid) *Zhangheotherium quinquecuspedens*: scapulocoracoid in lateral view (B) and restoration of shoulder girdle in ventral view (E). Bottom, Early Jurassic mammaliaform *Sinoconodon rigneyi*: scapulocoracoid in lateral view (C), restoration of shoulder girdle in ventral view (F). Stylistic figures not on the same scale.

in adult structure and function of the shoulder girdle and forelimb (Lessertisseur and Saban 1967a; Jenkins 1970b; Klima 1973, 1985; Walter 1988). Among therians, marsupials differ from placentals in timing of embryonic development of the shoulder girdle (Sears 2004, 2005; Hübler et al. 2010), a difference that is attributable to gene patterning (Keyte and Smith 2010; Hübler et al. 2013). Marsupials have a more conserved pattern of variation of the scapula than placentals, which is related to an accelerated development of the shoulder girdle and forelimb (Sánchez-Villagra and Maier 2002; Sears 2004; Hübler et al. 2010; Keyte and Smith 2010). This ontogenetic feature is required for fetuses to climb to maternal tits after their premature birth, a life history constraint of all marsupials (Lillegraven 1975; Gemmell et al. 2002). Placental mammals have diverse morphologies, related to the fact that their shoulder girdle is unconstrained (relative to marsupials) and to their versatile locomotor adaptations (Kardong 1998; Hildebrand and Goslow 2001; Polly 2007).

Recent advances in developmental genetics of the mouse, a model organism in laboratory studies, have made it possible to trace the morphogenesis of some adult features of the shoulder girdle to gene networks and signaling pathways. Sternal, shoulder girdle, and forelimb characters of mice are now attributable, in an increasingly precise way, to gene patterning (Timmons et al. 1994; Matsuoka et al. 2005; McIntyre et al 2007; Capellini et al. 2010). The understanding of the genetic control of morphogenesis is a good step toward the deciphering of the evolution, as it can offer a more explicit genetic underpinning for embryogenesis and on the mechanism of macroevolutionary transformation (Vickaryous and Hall 2006; Sears et al. 2013).

Shoulder Girdle and Musculature of Extant Mammals

The transformation in the shoulder girdle in early mammal evolution is evident from the prominent differences between extant therians and stem mammaliaforms (fig. 10.1), and from the distinctive patterns among Mesozoic mammal groups phylogenetically intermediate between therians and monotremes (figs. 10.2–10.4). Extant therians have fewer bony elements in the shoulder girdle and the sternum than stem mammaliaforms, through an evolutionary reduction and eventual loss of the interclavicle bone, a membranous element in the sternal series. With exception of those of flying and burrowing adaptations, therians have proportionally smaller sternal elements than those of monotremes and mammaliaforms (fig. 10.4). This is especially evident in the manubrium, the anterior-most sternal element developed embryonically from the endochondral sternebrae. Therians and Mesozoic theriimorph ("therian-like") mammals lost the procoracoid that is a primitive character of some mammaliaforms and monotremes. Therians and their theriimorph relatives show a simplified scapulocoracoid, with the much-diminished embryonic coracoid (also known as metacoracoid *sensu* Vickaryous and Hall 2006) integrated into the glenoid of the adult scapula.

Extant therians (figs. 10.1–10.3)—The scapulas of marsupials and placentals have two topographical areas: a ventral region of the coracoid process, the glenoid and the acromion, and a dorsal region of the plate-like blade that has the septal part of the scapular spine, a crest-like septum dividing the lateral surface of the scapular blade into the supraspinous muscle fossa and the infraspinous muscle fossa.

The ventral part of the spine bears the acromion (thus called here the acromiospine). The acromion is projected ventrally or anteroventrally to articulate with the lateral (distal) end of the clavicle, forming the acromioclavicular joint. The acromial part of the spine can have a metacromion in some (although not all) therians, and it marks the separation of the spino-deltoid part from the acromio-clavicular part of the deltoid muscle (Jenkins and Weijs 1979; Evans 1993; Großmann et al. 2002). The part of the spine dorsal to the acromion is a simple crest, herein called the septal part of the spine, or the septospine, which separates the supraspinous fossa and the infraspinous fossa.

The acromiospine and septospine have distinctive embryonic development: the former developed endochondrally, while the latter developed as appositional bone without cartilaginous precursor (Sánchez-Villagra and Maier 2002, 2003) (fig. 10.5). More recently, genetic studies show that the coraco-gleno-acromial area has different gene patterning, from the plate-like blade and the septospine in mouse (Timmons et al. 1994; Dietrich and Gruss 1995; Pellegrini et al. 2001; Capellini et al. 2010, 2011). The relative independence of the

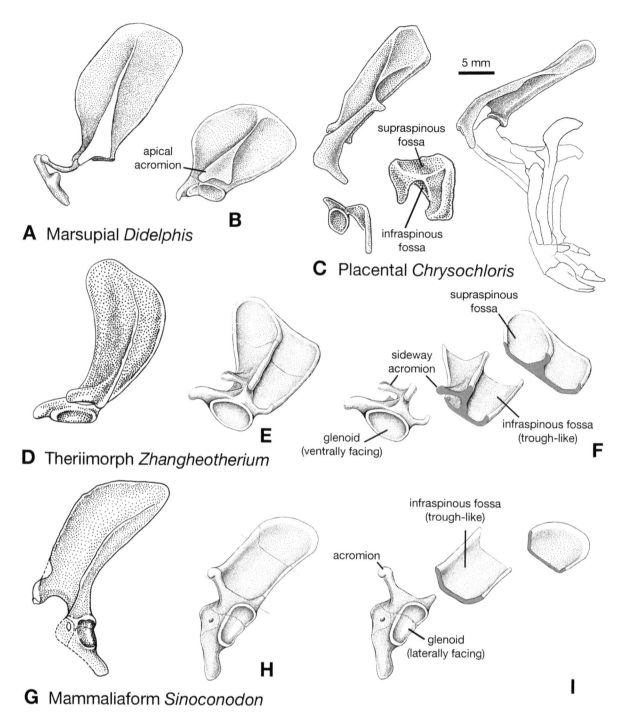

A Marsupial *Didelphis*

B

apical acromion

5 mm

supraspinous fossa

infraspinous fossa

C Placental *Chrysochloris*

D Theriimorph *Zhangheotherium*

E

supraspinous fossa

sideway acromion

glenoid (ventrally facing)

infraspinous fossa (trough-like)

F

G Mammaliaform *Sinoconodon*

H

infraspinous fossa (trough-like)

acromion

glenoid (laterally facing)

I

FIG. 10.2 Scapulocoracoid of mammaliaforms and mammals: offset acromion of stem therians and apical acromion of extant therians. A and B, Marsupial *Didelphis* left scapula in (A) lateral and (B) oblique (ventrolateral) views showing the acromion stemming from the top of the spine. C, Placental *Chrysochloris asiaticus* (African golden mole, a fossorial mammal), left scapula in lateral, ventral, and dorso-posterior views, and in articulation with the forelimb. D, E, and F, Early Cretaceous spalacotheroid *Zhangheotherium*, (a theriimorph) left scapula in lateral view (D), and oblique (ventrolateral) view (E), showing the acromion arising from the sidewall of the scapular spine, and the trough-like infraspinous fossa, and (F), reconstruction in cross-sections. G, H, and I, Early Jurassic mammaliaform *Sinoconodon rigneyi*, left scapula in lateral view (G), oblique (ventrolateral) view (H), and cross-sectional reconstruction (I). Sources: *Chrysochloris* redrawn from Lessertisseur and Saban (1967a), Asher et al. (2007), Wible et al. (2009), and public domain image from www.digimorph.org (accessed in June 2012). Stylistic figures not on the same scale.

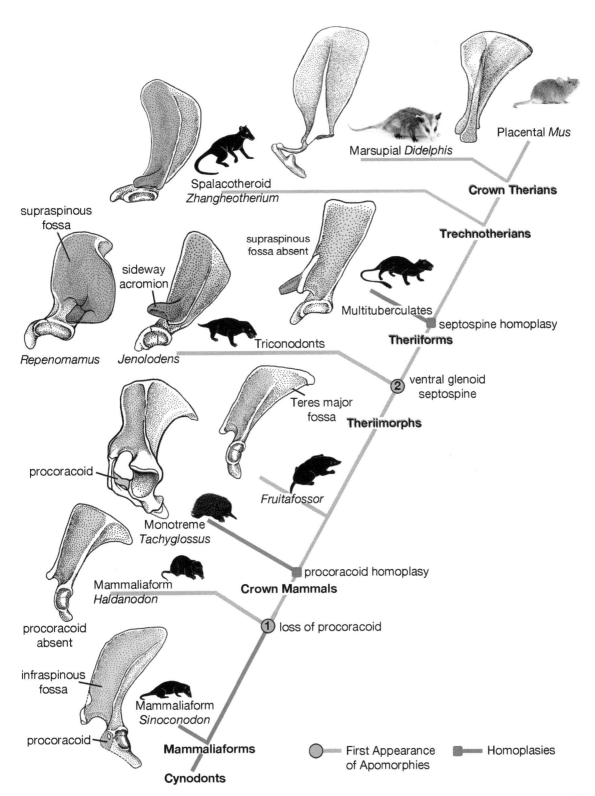

FIG. 10.3 Evolutionary patterns of scapulocoracoids through the cynodont-mammal transition. Two major evolutionary apomorphies: (1) loss of the procoracoid in derived mammaliaforms starting with *Haldanodon* (although homoplastic in monotremes and unknown in *Pseudotribos*), and (2) reorientation of scapular glenoid, and appearance of the supraspinous fossa with triconodonts and spalacotheroids, although lost in multituberculates by homoplasy. A general evolutionary trend: the size of the sternal elements, especially the interclavicle, is greatly reduced in theriimorph mammals. The schematic illustrations standardized to the length of the clavicle, not on the same scale. Sources: *Haldanodon* redrawn from Martin (2005), *Tachyglossus* redrawn from Jenkins and Parrington (1976), *Fruitafossor* modified from Luo and Wible (2005), *Repenomamus* redrawn from Hu (2006) with personal observation of other specimens, multituberculate a composite restoration from *Kryptobaatar* (Sereno 2006), *Catopsbaatar* from Hurum and Kielan-Jaworowska (2008) and personal observation of several multituberculates, *Zhangheotherium* based on Chen and Luo (2013) and observation of other specimens.

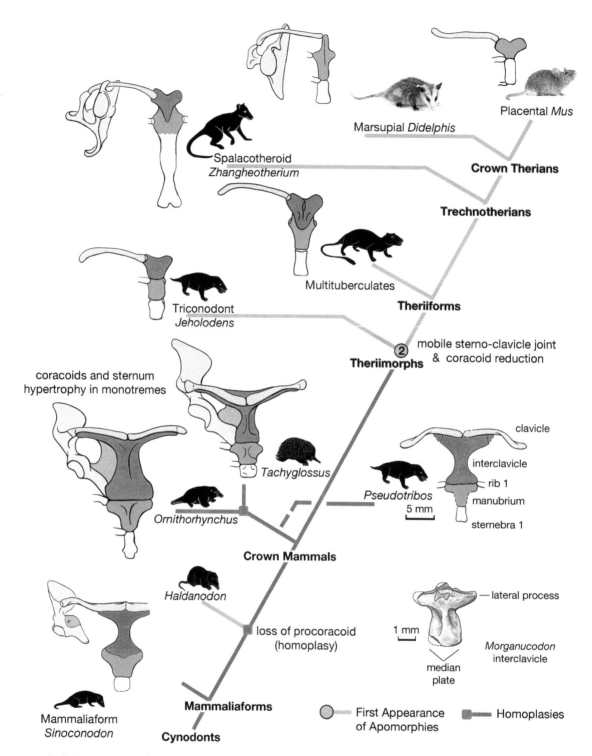

Placental *Mus*

Marsupial *Didelphis*

Crown Therians

Spalacotheroid
Zhangheotherium

Trechnotherians

Multituberculates

Theriiforms

Triconodont
Jeholodens

② mobile sterno-clavicle joint
& coracoid reduction

Theriimorphs

coracoids and sternum
hypertrophy in monotremes

clavicle

interclavicle

Tachyglossus

Pseudotribos

rib 1

manubrium

5 mm

sternebra 1

Ornithorhynchus

Crown Mammals

Haldanodon

lateral process

1 mm

Morganucodon
interclavicle

loss of procoracoid
(homoplasy)

median
plate

Mammaliaform
Sinoconodon

Mammaliaforms

First Appearance
of Apomorphies

Homoplasies

Cynodonts

FIG. 10.4 Evolutionary patterns of the clavicular and sternal structures through the cynodont-mammal transition. A major evolutionary apomorphy is the mobile claviculo-interclavicular joint through reduction of the interclavicle and loss of the procoracoid, which first occurred in mammal evolution with *Fruitafossor* and theriimorphs, although loss of the procoracoid likely occurred independently in *Haldanodon*. Sources: *Morganucodon* based on a specimen courtesy of Prof. Susan Evans.

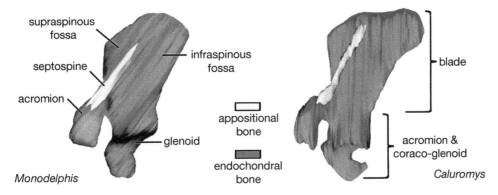

Independent Embryonic Components of Scapula in Opossums

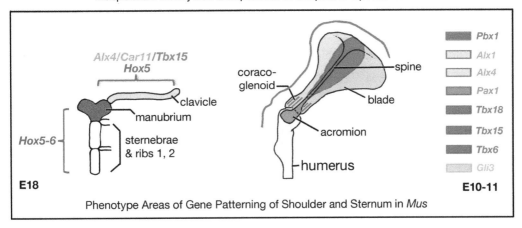

Phenotype Areas of Gene Patterning of Shoulder and Sternum in *Mus*

FIG. 10.5 Phenotype areas of gene patterning of the shoulder girdle and sternal structures and modular development of embryonic components of scapula. Top, Modular development of the scapular spine in didelphid marsupials; adapted with permission from Sánchez-Villagra and
Maier (2003). Bottom (Box), Functional phenotype areas of major genes in morphogenesis of the mouse shoulder girdle and sternal structure:
scapular patterns about embryonic days 10–11 adapted with modification from Capellini et al. (2010) with additional information from Wehn
and Chapman (2010), and Hübler et al (2013); sternal patterns of mouse about embryonic day 18 after McIntyre et al. (2007). The gene
expression areas may overlap, and are based on stylistic outline, not on the same scale.

coraco-gleno-acromial region is also manifest in the gross
morphological features during the ontogenetic growth,
and in morphometric patterns of variation (Sears et al.
2013). In marsupials, the coracoid process in the coraco-
gleno-acromial region shows a distinctly slower developmental rate, and has a different growth trajectory from
the rest of the scapula, likely by different gene patterning
in marsupials (Hübler et al. 2010, 2013).

The glenoid of the scapula in therians is a shallow,
uniformly concave fossa and has a nearly oval outline in
ventral view (fig. 10.1D, E). The concave glenoid and the
hemispherical head of the humerus form a spheroidal
(ball-in-socket) or an ovoidal joint (Williams et al. 1989;
Hildebrand and Goslow 2001). The glenohumeral joint
of therians is capable of movement of the humeral head
to the scapula in any direction with three degrees of
freedom in flexion-extension, adduction-abduction, and
circumduction and mediolateral rotation around the

humeral long-axis, or any combination of these (Williams et al. 1989). The concave glenoid with oval outline
is the most common, presumably a general condition of
therians (Lessertisseur and Saban 1967a; Evans 1993).
However, the shape of the glenoid can vary with different habits or locomotor specializations among closely
related species of marsupials and placentals (Taylor
1974; Argot 2001; Sargis 2002).

The glenoid of therians is oriented nearly perpendicularly to the scapular blade; its functional synovial
surface is faced ventrally in life. The scapular blade is
dorsal to the glenohumeral joint and the humerus shaft.
This perpendicular orientation of a ventrally facing glenoid, or nearly so, is important for the parasagittal or
nearly parasagittal posture of the forelimb in therians
(Jenkins 1971a; Sereno 2006), in contrast to the more
sprawling forelimb posture of the majority of the nontherian amniotes including monotremes (Gambaryan

and Kielan-Jaworowska 1997). The glenoid is shallow, partly incongruent with, and smaller than the much larger humeral head. This kind of joint enables a great freedom of movement, but it is also weak due to the lack of restraining. The highly mobile but simultaneously weak glenohumeral joint is stabilized compensatorily by the tension of ligamentous capsule of the glenohumeral joint and reinforced by the large supraspinous muscle, an evolutionary apomorphy of therians (Jenkins and Goslow 1983; Williams et al. 1989; Evans 1993; Fischer 2001).

The coracoid process (= "metacoracoid") is a small projection anterior to the margin of the glenoid (fig. 10.2B, C). It shows different relationships to the scapular glenoid and blade among therians, monotremes, and stem mammaliaforms. Also within different therian groups, the coracoid can vary in size, curvature, and shape, related to functional differences of the biceps and coracobrachialis muscles among species of different habits within families or genera (Taylor 1974; Argot 2001; Sargis 2002). The coracoid process is known for its different growth rates in different phases of ontogeny in marsupials (Hübler et al. 2013).

The clavicle is present in the most extant therians although some have lost this bone. The aclaviculate mammals, by losing the clavicle, can achieve a wider range of movement of the scapula relative to the axial skeleton, such as seen in therians of cursorial (running) adaptation. In claviculate therians, the medial end of the clavicle articulates with the manubrium by a mobile joint (fig. 10.1). The clavicle serves as a spoke to guide the arcuate movement of the glenohumeral joint, and also as a strut to support the shoulder joint (Jenkins 1974). The dual function of strut and spoke of clavicle is crucial for a wide range of locomotor functions, especially among placentals (Jenkins 1974; Hildebrand and Goslow 2001; Polly 2007).

Monotremes (figs. 10.3 and 10.4)—Extant monotremes differ significantly from therians in having an interclavicle. The large interclavicle has a broad median plate and long lateral processes, developed, respectively, from the unpaired chondral element and the paired desmal elements in embryogenesis (Klima 1973, 1985). The lateral process forms a tongue-in-groove contact with the clavicle along the entire length of the latter. Bounded by ligament, the clavicle is immobile relative to the in-

terclavicle. The lateral process of interclavicle and the clavicle both reach the acromio-clavicle joint, which is mobile (Klima 1973; Jenkins and Parrington 1976; Augee et al. 2006). The interclavicle may have highly variable distal epiphyses at the joint with the manubrium (Cave 1970). The interclavicle and the manubrium are not mobile, but their junction has a synovial joint for the coracoid process of the scapulocoracoid (Cave 1970; Sereno 2006).

Monotremes have a hypertrophied coracoid process, and it is homologous to the metacoracoid of non-mammalian amniotes (reviewed by Vickaryous and Hall [2006]). The coracoid is so massive that it reaches the sternal series and articulates with the manubrium-interclavicle joint. Monotremes differ further from therians and are unique among amniotes in having a large, plate-like procoracoid (Lessertisseur and Saban 1967a; Vickaryous and Hall 2006), which is developed in ontogeny from the medial part of the embryonic coraco-scapular plate (Klima 1973). Posteriorly, the procoracoid abuts tightly the coracoid process (metacoracoid) of the scapulocoracoid. Anteriorly and medially it can either overlap, or attach loosely to the interclavicle (fig. 10.3). Besides a variable relationship to the interclavicle, the procoracoid can be variable in shape and often asymmetric between two sides (Cave 1970).

With mobile joints of the acromion and clavicle and between the coracoid process and the interclavicle-manubrium, the scapulocoracoid is capable of limited dorso-ventral movement to the sternum. The clavicle and the hypertrophied coracoid process cannot function as a spoke in a therian fashion. Instead, these collectively serve as a strong strut for the glenohumeral joint, for powerful forelimb movement for burrowing of monotremes (Jenkins 1970b; Augee et al. 2006), and secondarily for rowing in swimming by the platypus of semiaquatic habits.

The glenoid of scapulocoracoid in monotremes has a half-saddle shape, also known as hemi-sellar joint (*sensu* Jenkins 1993) (fig. 10.3). By comparison to the shallow ovoidal glenoid of therians capable of movement in any direction, the hemi-sellar joint is more restrictive for elevation and depression of the humerus to the glenohumeral joint, although not for protraction and retraction (Jenkins 1993). The half-saddle joint permits translational movement along the short axis of the glenoid surface, and long-axis rotation of the humerus that is held more or less perpendicular to the sagittal plane (Jenkins 1970b).

The glenoid of monotremes is oriented obliquely to the main plane of the scapular blade. With the scapula oriented anterodorsally in life, the glenoid is facing laterally, in contrast to the ventrally facing glenoid of therians (Lessertisseur and Saban 1967a; Jenkins 1971b; Sues and Jenkins 2006). The glenoid is more similar to those of stem mammaliaforms than therians, and plesiomorphic in its characteristics. The coracoid and the scapula contribute equally to the synovial surface of the glenoid joint in monotremes, different from the therian condition of the glenoid formed mostly by the scapula, almost to the exclusion of the coracoid. The glenoid surface of monotremes is relatively larger for the humeral head than in therians. The scapula-coracoid junction is reinforced by the greater bone thickness, likely for bracing the forelimb for burrowing.

The scapulas of monotremes have an extensive fossa for the infraspinous muscle (*sensu* Diogo and Abdala 2010), also known as subscapularis fossa (Jouffroy and Lessertisseur 1971). This is an ancestral feature of pre-mammalian cynodonts and stem mammaliaforms (Jenkins 1971b; Jenkins and Parrington 1976; Sun and Li 1985; Sues and Jenkins 2006). The scapula lacks a fully developed supraspinous fossa as seen in therians. A small muscle, recognized either as a small supraspinous muscle by some (Coues 1871; Walter 1988; Diogo and Abdala 2010) or as the "pars suprascapularis" of the supracoracoideus by others (Jouffroy and Lessertisseur 1971; Jenkins 1971b) is attached to the base of the acromion (Jouffroy and Lessertisseur 1971; Walter 1988). These alternative terms of this muscle are both consistent because the supraspinous muscle of therians is partly homologous to the supracoracoideus of non-mammalian amniotes by ontogeny (Cheng 1955; Romer 1956; Diogo and Abdala 2010). However, in monotremes, the supraspinous muscle on the scapula is miniscule (Walter 1988), and much smaller than those of therians. It coexists with the much larger supracoracoideus muscle that originates from the plate-like procoracoid bone and inserts on the great tubercle of the humerus, to pull the humerus forward and simultaneously rotate it along its long axis (Jouffroy and Lessertisseur 1971, figures 627, 631; Jenkins 1971a; Walter 1988).

The scapula has a well-developed dorso-posterior angle and a prominent teres major muscle fossa marked by a distinctive crest from the infraspinous fossa, for a hypertrophied teres major muscle for the burrowing movement of the forelimb. A similar, large teres major fossa is present in some specialized fossorial therians (Hildebrand 1985; Hildebrand and Goslow 2001), although this is not the case for the majority of therians with generalized forelimb function.

A major difference in musculature and muscle function between monotremes and therians is related to how the glenohumeral joint is stabilized. The supraspinous muscle is an evolutionary apomorphy of therians, and it is a major muscle in stabilizing the glenohumeral joint that has a considerable laxity due to a shallow glenoid smaller than the large humeral head (Fischer 2001; Sánchez-Villagra and Maier 2002). The supraspinous fossa and the infraspinous fossa are both well developed on the scapular blade; the supraspinous and the infraspinous muscles insert on the greater tubercle of the humerus, with equitable muscle fiber contributions to stabilizing the glenohumeral joint during movement (Jenkins and Goslow 1983; Evans 1993; Fischer 2001).

By contrast, the supraspinous muscle (or equivalent to the therian supraspinous) is miniscule in monotremes (Howell 1937; Jouffroy and Lessertisseur 1971; Walter 1988). The glenohumeral joint is stabilized by the infraspinous muscle and by the large supracoracoideus muscle originating from the procoracoid to insert on the greater tubercle of the humerus (Jouffroy and Lessertisseur 1971; Walter 1988), analogous to the supracoracoideus function to stabilize the glenohumeral joint, as seen in lizards (Jenkins and Goslow 1983). In contrast to the lax joint of therians that maximizes mobility at the expense of stability, the monotreme glenohumeral joint is more massive and strongly reinforced, suggesting the stabilization by a miniscule supraspinous muscle, if needed at all, is not as critical for monotremes as for therians.

Shoulder Girdles of Mammaliaforms

Mammaliaforms are defined by the common ancestor of *Sinoconodon*, *Morganucodon*, and living mammals, and can be diagnosed by the key skull features (Rowe 1988, as modified by Luo 2007). Stem mammaliaforms are Mesozoic relatives to living mammals. Many of their features are intermediate for the cynodont-mammal transition, more derived than cynodonts, but primitive for crown Mammalia.

Morganucodon is an insectivorous mammaliaform from the Late Triassic to Early Jurassic. Its shoulder girdle provides important information on the ancestral condition from which the girdles of mammals have likely evolved (Jenkins and Parrington 1976; Jenkins and Weijs 1979; Sues and Jenkins 2006). The scapulocoracoid of the Late Jurassic docodont *Haldanodon* is also represented by extensive fossils and has been reconstructed (Martin 2005). However, the clavicle and interclavicle are unknown for these mammaliaforms (Evans 1981; Martin 2005).

Sinoconodon from the Early Jurassic (fig. 10.1C) has relatively well-preserved interclavicle and clavicle, which made it possible to reconstruct their relationship to the rest of the shoulder girdle (Luo, Ji, and Yuan 2007). The interclavicle has a dumbbell shape. Its median plate has a constricted middle portion, a widened posterior part abutting the manubrium, and two short lateral processes in the anterior part overlapping with clavicles. The interclavicle of *Sinoconodon* has nearly the same shape as an isolated interclavicle of *Morganucodon* (specimen courtesy of Prof. Susan Evans). The broad interclavicle with a constricted waist and an expanded posterior end is monotreme-like, more derived features than the ancestral cynodont condition of a narrow interclavicle with a tapering posterior end, as seen in several cynodonts (Jenkins 1970a, 1971b; Sun and Li 1985; Sues and Jenkins 2006). Most cynodonts lack the lateral processes (Jenkins 1970a, 1971b; Romer and Lewis 1973), although these are present in tritylodontids (Sun and Li 1985; Sues and Jenkins 2006).

The interclavicle, manubrium, and procoracoid of *Sinoconodon* provide an expanded area for muscles, presumably for the pectoralis and supracoracoideus muscles as in monotremes (Jouffroy and Lessertisseur 1971; Walter 1988). The clavicle is curved in boomerang shape, as those of tritylodontids (Sues and Jenkins 2006) with its lateral end articulating with the acromion process of the scapula. The acromio-clavicle joint appears to be mobile.

The procoracoid is integrated into the scapulocoracoid. It is interpreted here to have a coracoid foramen, and is excluded from the glenoid (fig. 10.1C). The coracoid process (= metacoracoid) has a pointed apex, directed posteromedially toward, but is short of

contacting the sternum. The short coracoid process without sternal contact is obviously different from the monotreme condition of a long and massive coracoid process in synovial articulation to the sternum. The hemi-sellar glenoid is oriented to face ventrolaterally (fig. 10.1C). The infraspinous fossa is a deep, trough-like structure, and occupies the entire lateral aspect of the scapular blade (figs. 10.1C and 10.2G). The acromion is a short protuberance on the scapular spine at the anterior (cranial) margin of the scapula. If the supraspinous muscle is present, it would be attached to a small surface on the anterodorsal aspect of the spine, and its muscle fibers would pass below the clavicle and acromio-clavicle joint to insert on the humerus, as previously interpreted for *Morganucodon* (Jenkins and Weijs 1979). The posterior (or posteroventral) margin of the infraspinous fossa is curved.

There is not an expanded fossa for the teres major muscle although this muscle could be attached to the posterodorsal angle of the scapula. It would be small if present (fig. 10.2G). The large teres major muscle fossa is present in some pre-mammalian cynodonts and mammaliaforms, inferred to have fossorial habits by limb features for digging (Martin 2005; Luo and Wible 2005; Sues and Jenkins 2006), although the fossorial adaptation is not a uniform condition of mammaliaforms, as evident from *Sinoconodon*.

The scapulocoracoid of the mammaliaform *Haldanodon* is well represented by extensive fossils (Martin 2005), and is similar to *Morganucodon* and *Sinoconodon* in many features (fig. 10.4). However, it differs from and is more derived than *Morganucodon* and *Sinoconodon* in that the scapulocoracoid lacks a distinctive procoracoid that typically bears a coracoid foramen (Martin 2005). The coracoid process has a blunt apex. It is smaller and shorter than the glenoid, in contrast to *Sinoconodon* and *Morganucodon*, both of which have a longer coracoid process than the length of the glenoid. Because the coracoid process is so short it is unlikely to have contacted the sternal series as in monotremes. This needs to be verified by sternal elements to be discovered for *Haldanodon*. A prominent fossa is present on the posterodorsal part of the scapula (= postscapular fossa of Martin 2005) for the teres major muscle. Its large size indicates a massive muscle, for powerful flexion and retraction of the humerus to the scapula, related to *Haldanodon*'s fossorial and semiaquatic habits, as can be inferred from other skeletal features (Martin 2005). The difference of *Haldanodon* from other mammaliaforms suggests that basal mammaliaforms already have developed different locomotor functions, related to different niches (Martin 2005; Ji et al. 2006).

Shoulder Girdles of Mesozoic Mammalia

Mesozoic lineages of the Mammalia have diverse forelimb features, indicating locomotor functions for swimming, gliding, and fossorial adaptations (Luo and Wible 2005; Ji et al. 2006; Meng et al. 2006; Luo 2007). Their shoulder girdles show a wide range of the primitive to derived characteristics. The partially preserved shoulder girdle of the Middle Jurassic mammal *Pseudotribos* has a monotreme-like interclavicle. It has an expanded anterior end with lateral processes, and a broader posterior end of the interclavicle in juxtaposition with a broad manubrium, also present in *Sinoconodon* (fig. 10.4). The shoulder features of *Pseudotribos* are likely related to extensive pectoral muscles and the sternocleidomastoid muscle for fossorial adaptation (Luo, Ji, and Yuan 2007).

Fruitafossor is a Late Jurassic mammal with convergent dental features to extant armadillos and aardvarks specialized for tongue feeding (Luo and Wible 2005). It appears to be fossorial, from the distinctive forelimb features otherwise known only in the burrowing mammals (Hildebrand 1985), and convergently in *Haldanodon* and in the tritylodontid *Kayentatherium* of fossorial adaptation (Martin 2005; Sues and Jenkins 2006). Its scapula has a prominent teres major fossa in addition to the infraspinous fossa, and a hemi-sellar glenoid that is oriented obliquely to the scapular blade with a laterally facing surface (fig. 10.3). However, there is no procoracoid bone, and the coracoid process is much smaller than those of monotremes and mammaliaforms, and evolutionarily more derived than monotremes.

Theriimorph mammals—Triconodonts, multituberculates, and spalacotheroids (also known as "symmetrodonts") are three mammal clades with an extensive fossil record in the Mesozoic (Kielan-Jaworowska et al. 2004). They are closer to extant therians than to monotremes, and belong to the theriimorph clade defined by the common ancestor of triconodonts through living therians (Rowe 1988).

Triconodonts, such as *Jeholodens* and gobiconodontids (fig. 10.3), are characterized by an ovoid to spheroidal glenoid, oriented perpendicularly (or nearly so) to the scapular blade (Jenkins and Schaff 1988; Ji et al 1999; Hu 2006; Luo, Chen, et al. 2007). The coracoid process is pointed anteroventrally. The scapular blade has a fully developed supraspinous fossa separated by the spine from the infraspinous fossa. The supraspinous fossa and the perpendicular glenoid suggest that triconodonts have acquired a therian-like glenohumeral joint, stabilized and reinforced by the supraspinous muscle, in addition to the infraspinous. Theriimorph mammals including triconodonts differ from crown therians in a prominent feature—the acromion is projected from the side of the scapular spine and has complex morphology (Hu 2006; Chen and Luo 2013), not from the apex of the spine as seen in extant therians (figs. 10.2 and 10.3). Among triconodonts known so far, there are significant variations in shape of the spine, the acromion, and the size of the supraspinous fossa (Jenkins and Schaff 1988; Ji et al. 1999; Hu 2006; Luo, Chen, et al. 2007), suggesting size difference in muscles attached to these bony features, and likely also functional differences for shoulder girdle.

Triconodonts have retained the interclavicle (Ji et al. 1999; Hu 2006; Luo, Chen, et al. 2007). In extant monotremes, the interclavicle is distinguished from the sternal manubrium by embryogenesis: the former is primarily a membranous bone while the latter is primarily an endochondral bone consisting of a pair of ossifications (Lessertisseur and Saban 1967b; Cave 1970; Klima 1973). Also the interclavicle can be distinguished from the manubrium by contacting relationship with the first thoracic rib (Lessertisseur and Saban 1967b; Hu et al. 1997). For extinct mammals that cannot be examined for embryogenesis, the distinction of the interclavicle and the manubrium can be established only on their topographic relationship to the first costal ribs (fig. 10.1D). In general, the manubrium is the anteriormost sternal element with direct contact with the first costal rib while the interclavicle has no exclusive contact with this rib. In the case of the manubrium preserved as paired elements, as has been recognized in tritylodontids, it can be distinguished from the single interclavicle (Sun and Li 1985; Sues and Jenkins 2006). These topographical characters for distinguishing the interclavicle from the manubrium in fossils are consistent with the embryological observation that manubrium is developed from the paired sternebral bands connected to anlagen of the first costal ribs, while the median chondral element (the "pars chondralis interclaviculae") of the presumptive interclavicle has no such connection (Klima 1973, 1985).

The interclavicle of triconodonts lacks the lateral process (Ji et al. 1999; Hu 2006; Luo, Chen, et al. 2007). It has a point contact to the clavicle. The clavicle is a fully mobile spoke to the interclavicle for the arcuate swing of the shoulder girdle. Triconodonts are the basal-most group among Mesozoic mammals to show a mobile shoulder girdle (Ji et al. 1999; Hu 2006; Sereno 2006) and have acquired the therian-like function of the shoulder and forelimb, as seen in living therians (Jenkins 1971a, 1974).

Multituberculates are omnivorous to herbivorous mammals with a diverse range of feeding adaptations (Wilson et al. 2012). They are the most abundant group in the mammal faunas from the Late Jurassic to Cretaceous (Kielan-Jaworowska et al. 2004) and can have diverse habits: some inferred to be terrestrial (Kielan-Jaworowska and Gambaryan 1994; Sereno and McKenna 1995; Hu and Wang 2002), fossorial (Kielan-Jaworowska and Qi 1990), or arboreal (Jenkins and Krause 1983; Krause and Jenkins 1983). There are also alternative interpretations of the forelimb posture of multituberculates (Kielan-Jaworowska and Gambaryan 1994; Sereno and McKenna 1995; Gambaryan and Kielan-Jaworowska 1997; Sereno 2006; Kielan-Jaworowska and Hurum 2006). But the scapular morphology is relatively uniform across a range of multituberculates, and so is the rest of the shoulder girdle (Jenkins and Weijs 1979; Sereno 2006; Hurum and Kielan-Jaworowska 2008).

The multituberculates' shoulder girdle has a mobile clavicle-interclavicle joint, and is fully capable of the arcuate shoulder movement of modern therians (Sereno and McKenna 1995; Sereno 2006). The scapular glenoid is oriented perpendicularly to the scapular blade, and facing ventrally. In these features, multituberculates are unquestionably therian-like and derived (Krause and Jenkins 1983; Kielan-Jaworowska and Gambaryan 1994; Sereno 2006; Kielan-Jaworowska and Hurum 2006). Besides a well-developed acromion, the scapular spine also has a prominent metacromion (Sereno 2006), a derived feature of some therians (Lessertisseur and Saban 1967a; Großmann et al. 2002), not present in monotremes and

mammaliaforms. Among the primitive features, the scapular spine is truncated dorsal to the metacromion (Jenkins and Weijs 1979; Sereno 2006). The short multituberculate spine appears to be equivalent to the acromial (ventral-most) part of the therian spine, but lacks the septal part of the therian spine (Sánchez-Villagra and Maier 2003). The position of the spine along the anterior scapular margin and the absence of a supraspinous fossa are primitive conditions of monotremes and stem mammaliaforms.

Spalacotheroids are characterized by molars with a symmetrical triangle of cusps (thus known as the "symmetrodonts") for insectivory. The scapulas of spalacotheroids and more derived therian relatives, such as *Henkelotherium* and *Vincelestes*, already have the derived condition of living therians, such as the full supraspinous fossa, and the perpendicular glenoid (Krebs 1991; Rougier 1993; Hu et al. 1997; Rougier et al. 2003; Li and Luo 2006; Chen and Luo 2013).

In *Zhangheotherium* and *Akidolestes*, the acromion process and its topographical relationship to the spine are quite different from those of living therians (Chen and Luo 2013). In *Didelphis*, the acromion is directly arising from the spine, and is apical with the spine in the majority of living therians (fig. 10.2). By contrast, the acromion of spalacotheroids arises from the anterior side of the spine, and the base of the acromion is offset from the apex of the spine (fig. 10.2D–F). This "sideway" attachment of the acromion is well documented in extensive fossils of the gobiconodontid *Repenomamus* (Hu 2006, figure 3-19; Luo, personal observation) and other triconodonts (Ji et al. 1999; Luo, Chen, et al. 2007; Chen and Luo 2013). Multituberculates have a similar condition of the acromion, although lacking the full supraspinous fossa of triconodonts (Sereno 2006; Hurum and Kielan-Jaworowska 2008). The offset acromion is not present in basal eutherians and metatherians (Argot 2001; Horovitz 2003; Wible et al. 2009). In *Vincelestes*, a Cretaceous mammal closer to crown therians than spalacotheroids (Rougier 1993), the acromion is apical to the crest of the spine as in crown therians. Another prominent, but primitive feature is the trough-like infraspinous fossa with its posterior margin curled and elevated (fig. 10.2), now documented in many stem mammaliaforms, triconodonts, and multituberculates (Martin 2005; Hu 2006; Sereno 2006; Hurum and Kielan-Jaworowska 2008; Chen and Luo 2013).

Mesozoic crown theria and kin—The scapula of the Early Cretaceous eutherian *Eomaia* (Ji et al. 2002) appears to have a plate-like scapular blade, and its infraspinous fossa is flat. The Late Cretaceous eutherians *Maelestes* and *Ukhaatherium*, however, differ from *Eomaia* in that the crest of the scapular spine is curved and overhangs the infraspinous fossa. The infraspinous fossa is trough-like (Horovitz 2003; Wible et al. 2009). A similar configuration is also present in *Vincelestes* (Rougier 1993, figs. 81 and 82), indicating that this may be phylogenetically primitive for eutherians (Wible et al. 2009). The trough-like infraspinous fossa suggests that some stem eutherians are either terrestrial as indicated by their generalized forelimbs as in the case of *Ukhaatherium* (Horovitz 2003), or could be fossorial, as indicated by the deep trough-like infraspinous fossa best developed in fossorial marsupial mole *Notoryctes* and the placental golden mole *Chrysochloris* (Lessertisseur and Saban 1967a; Asher et al. 2007; Wible et al. 2009) (fig. 10.2C). In this trough-like infraspinous fossa, *Ukhaatherium* and *Maelestes* differ from the shallow and flat infraspinous fossa of several other stem eutherians and metatherians interpreted to be scansorial or even arboreal (Szalay 1994; Argot 2001; Ji et al. 2002; Luo et al. 2003). Given the diversity of scapular features, likely the Cretaceous eutherians had already evolved differences in locomotor functions and habits (Goswami et al. 2011; Luo et al. 2011).

Major Features of Phylogeny

Coracoids and sternal series—In basal synapsids, the procoracoid is a large element of the scapulocoracoid complex and a major component of the glenoid (Romer 1956; Jenkins 1971b). The procoracoid appears to be a discrete bone in some subadults but can merge with the scapula in skeletally mature specimens of some stem synapsids (Vickaryous and Hall 2006). More derived cynodonts differ from stem synapsids in that the procoracoid is excluded from the glenoid (Sues and Jenkins 2006), although it remains integrated in the scapula-coracoid in cynodonts and some stem mammaliaforms (fig. 10.1C). The mammaliaform *Haldanodon* has no procoracoid (Martin 2005), and it is phylogenetically intermediate between other mammaliaforms with an integrated procoracoid and monotremes that have an unfused procoracoid (fig. 10.3). In monotremes, the procoracoid is developed from the medial part of the pre-cartilaginous coracoid-scapular plate of the embryonic and fetal stages (Klima 1973; Vickaryous and Hall 2006) and becomes an independent bony element in adults, loosely attached to the rest of the scapulocoracoid, but not sutured or fused to the latter like in several mammaliaforms, except *Haldanodon*.

The homoplasy of the procoracoid in *Haldanodon* and monotremes can be interpreted, alternatively, either as an independent loss of the procoracoid in *Haldanodon* among mammaliaforms (fig. 10.3), or as a convergent acquisition of this bone by monotremes. It is not possible to discriminate between these alternative interpretations because the procoracoid is unknown in the successive sister taxa of the monotreme lineage. For example, it is not known in the Jurassic *Pseudotribos*, a putative relative of monotremes (Luo, Ji, and Yuan 2007) (fig. 10.4). If it can be ruled out that this is not an artifact of preservation, then it can augment the likelihood that the procoracoid is a separate acquisition of monotremes. The loss of the procoracoid enhances the mobility of the shoulder girdle, accompanied by concomitant transformation of the glenohumeral joint.

Glenohumeral joint—The half-saddle or semi-sellar joint, an ancestral mammaliaform condition, is retained in monotremes and in *Fruitafossor*, the most derived Mesozoic mammal known to have this feature (Luo and Wible 2005) (fig. 10.3). Through the rise of crown therians and their theriimorph kin, the glenoid underwent changes from the half-saddle shape to the ovoidal or spheroidal shape, and from oblique orientation and laterally facing to the perpendicular orientation and more ventrally facing. The changes have resulted in functional differences (Jenkins 1974, 1993): the half-saddle joint permits the protraction-retraction swing of the humerus relative to the scapulocoracoid but limits the ranges for elevation and depression, and for long-axis circumduction of the humerus. The more massive coracoid, as seen in monotremes, strengthens the glenohumeral joint for more forceful movement of the humerus yet is also more restrictive on some movement. Also the glenoid joint is stabilized by the supracoracoideus muscle known from other amniotes (Jenkins and Goslow 1983; Walter 1988), and by the infraspinous muscle—a plesiomorphy, as evidenced by the infraspinous fossae of cynodonts through mammaliaforms.

A supraspinous muscle of the therian proportion is absent in monotremes (Howell 1937; Jouffroy 1971; Walter 1988), and is inferred to be absent, or poorly developed in cynodonts and mammaliaforms (Jenkins 1971b; Jenkins and Weijs 1979; Sun and Li 1985; Martin 2005; Sues and Jenkins 2006). Theriimorph mammals have a sphero-ovoidal glenoid capable of a greater mobility, which is simultaneously less stable and more vulnerable to dislocation, to be stabilized, compensatorily, by a apomorphic supraspinous muscle (Jenkins and Goslow 1983; Williams et al. 1989; Fischer 2001).

Supraspinous fossa and scapular spine—Stem mammaliaforms and monotremes have a single-muscle (infraspinous) stabilization of the glenohumeral joint from the scapula, while therians and Mesozoic theriimorphs have a double-muscle (both infraspinous and supraspinous) stabilization with the diagnostic skeletal features of the supraspinous fossa and the spine (fig. 10.1). Historically, the therian spine was hypothesized to be homologous to the cranial margin of the scapula in non-mammalian amniotes and monotremes (Romer 1956; Jenkins 1971b; Jenkins and Weijs 1979). Presumably the large supraspinous muscle of extant therians was transformed from an ancestral condition in which an incipient supraspinous muscle was attached to a small area on the anterior surface of the medially flared acromion and spine (Walter 1988, although see the different opinion of Jouffroy and Lessertisseur 1971). In therian evolution, the incipient supraspinous muscle rotated and concomitantly expanded much larger on to the lateral surface of the scapula, an evolutionary process that was recapitulated in the ontogeny of therians, at least in part (Cheng 1955; Jouffroy 1971).

Recent studies demonstrated that the scapular spine of marsupials and placentals is a compound structure (Sánchez-Villagra and Maier 2002; Großmann et al. 2002). The septospine and the coraco-gleno-acromial region have distinctive embryonic development: the former is formed by appositional bone following the intermuscular septum dividing the common anlagen of the presumptive supraspinous and infraspinous muscles (Sánchez-Villagra and Maier 2003) while the latter is preformed in cartilage.

The anlage of the ventral part of the supraspinous muscle appears initially on the cranial margin of the acromial region, and subsequently rotates to the lateral surface of the scapula in early embryogenesis (Cheng 1955; Sánchez-Villagra and Maier 2003). However, anlage of the main part of the supraspinous muscle is on the lateral surface of the preformed, cartilaginous scapular blade upon its earliest mesenchymal condensation; the adult supraspinous muscle mass is developed through a vicariant separation from the infraspinous counterpart by the neomorphic septospine, not by rotation (Sánchez-Villagra and Maier 2003).

This new embryological observation suggests that not all of the therian (acromial + septal) spine is homologous to the cranial scapular margin of monotremes and sauropsids as previously believed (Romer 1956). Rather, only the acromial part of the therian spine is homologous to the cranial margin of scapula. Also, this suggests that the septospine is an evolutionary apomorphy, but the lateral surface of the supraspinous fossa, by itself, is not (Großmann et al. 2002; Sánchez-Villagra and Maier 2002).

Evolution of a therian-like septospine is homoplastic among the Mesozoic mammals: present in triconodonts and spalacotheroids, but absent in multituberculates that are phylogenetically intermediate between triconodonts and spalacotheroids. Either the septospine evolved convergently in triconodonts, and in spalacotheroids through crown therians, or the septospine had evolved in theriimorph ancestors but was secondarily lost in multituberculates (fig. 10.3). Either way it is clear that the septospine has a relatively independent evolutionary history from the acromion and scapular blade, consistent with its morphogenesis in living mammals.

Development and Evolution

Overall, therians and their more inclusive theriimorph clade have fewer shoulder girdle elements, which are also more gracile, than their mammaliaform homologues. The simpler scapulocoracoid of theriimorphs evolved by incorporation of the reduced metacoracoid into the adult scapula, and the loss of adult procoracoid. Concomitantly, the sternal structure underwent an evolutionary reduction, diminishing the manubrium and losing the interclavicle. The simpler shoulder girdle is more mobile, and the forelimb movement more versatile. By comparison, evolution of the shoulder and sternal structure in monotremes shows the opposite: the shoulder girdle and sternum, ancestrally strong in mammaliaforms, have become even more hypertrophied for highly specialized fossorial adaptations. These phylogenetic patterns, however, are not linear trends toward extant therians and monotremes, respectively. Rather, there are convergent acquisitions of derived features and prominent cases of parallel reduction of ancestral features.

Homoplasies of the shoulder girdle on a broad evolutionary scale are consistent with the labile development well documented by comparative embryology (Klima 1973, 1985, 1987) and the studies of cellular mechanism in skeletogenesis (Hall and Miyake 1992; Vickaryous and Hall 2006). A discrete element of a mature skeleton represents coalescence of clusters of embryonic progenitors, discernible by cell condensation. For example, adult interclavicle of monotremes is composed of embryonic components of the paired pars interclaviculae desmalis (presumptive lateral process) and the unpaired median pars chondralis interclaviculae (presumptive median plate) (Klima 1973). Adult manubrium of therians integrates components of the embryonic "procoracoid," the pars chondralis interclaviculae, and the paired sternal bands (Klima 1987). Coalescences of these cell condensations can change by mitosis (cell proliferation), by identity (differentiation), by the size of aggregation, and by localized heterochrony (Vickaryous and Hall 2006). The gain, loss, or change in shape and size in evolution are attributable to these labile cellular processes in skeletogenesis (Vickaryous and Hall 2006).

Several components of the shoulder girdle can be recognized as evolutionary and developmental modules in which the transformation of fossil mammals can be hypothesized as consequence of variation in skeletogenesis for which the genetic controls are already known to some extent:

Coracoids—The procoracoid of monotremes is developed from the medial part of the embryonic coracoid-scapular plate of fetal stage (Klima 1973). The three-element primary girdle (of the procoracoid, the metacoracoid, and the scapula) is a basic feature of basal synapsids, reptiles, and birds (Vickaryous and Hall 2006). Birds are now known to have conserved the deep homology of gene patterning for these coracoid elements (Huang et al. 2000, 2006). Absence of one of the adult coracoids in some amniote lineages can occur by differential coalescence of coracoids with the scapula into a single adult morphology (Vickaryous and Hall 2006). Thus the procoracoid and metacoracoid are ancestral features for mammaliaforms. Discrete adult procoracoid can be interpreted as a retention of an embryonic feature for monotremes, relative to the mammaliaform *Haldanodon*, and to *Fruitafossor* and theriimorphs, which have lost the procoracoid (fig. 10.3, node 1). Alternatively, the absence of the procoracoid and the greatly reduced coracoid process (metacoracoid) in *Haldanodon* (Martin 2005), and separately in *Fruitafossor* and theriimorphs, can result from an early coalescence of the procoracoid (*sensu* Vickaryous and Hall 2006) or as arrested growth of the coracoids, or both. Arrested growth of the coracoids in extant therians is documented by Hübler et al. (2010, 2013) who show that in the marsupial *Monodelphis*, growth rate of the coracoid has slowed down in late fetal development relative to the rest of scapula. Homoplasy of the procoracoid and the size change of the coracoid process in mammaliaform evolution can be hypothesized as manifestation of heterochrony of the embryonic coracoscapular plate.

Clavicle and sternum (fig. 10.5)—Development of the clavicle-sternum connection is influenced by *Hox5* and *Hox6*. Mutant mice with *Hox5* triple knockout (*Hox5abc*[-/-]) are missing the sternal manubrium and the first rib associated with the manubrium (McIntyre et al. 2007). The clavicle is disconnected from the sternum, except by connective tissue. In mutant mice with *Hox6* triple knockout (*Hox6abc*[-/-]), the manubrium and the first sternebra are miniscule and poorly developed, and their associated first rib fails to develop. This suggests that normal function of these genes is required for morphogenesis of the sternum (McIntyre et al. 2007). Formation of the sternum and the clavicle also requires the complementary function of *Pax1* (paired box 1) gene and *Hoxa5* (Timmons et al. 1994; Dietrich and Gruss 1995; Aubin et al. 2002), illustrated by the fact that mutant mice with deficient *Pax1* have a defective sternum (Timmons et al. 1994).

So far, influence of *Hox* genes and *Pax1* on the clavicle and sternum is demonstrated only in the mouse, a highly derived placental with a simple sternal manubrium. For marsupials, Keyte and Smith (2010) observed that widening of expression zones of *Hoxb5* and *Hoxc6* along the antero-posterior axis is related to the earlier and faster development of the shoulder girdle and forelimb in *Monodelphis*. However, monotremes have yet to be studied for the *Hox* patterning of the interclavicle. It is still unknown if the *Hox* and *Pax1* patterning in mouse, which has no interclavicle, can be extrapolated to monotremes with an interclavicle formed from composite embryonic progenitors, let alone to a wide diversity of Mesozoic nontherian clades that show great variation in the interclavicle.

Acromion-glenoid of scapula—With the exception of the septospine formed by appositional bone (Sánchez-Villagra and Maier 2003), the entire scapulocoracoid develops endochondrally. Within the endochondral part of the scapula, morphogenesis of the acromio-glenoid region is relatively independent from that of the blade and the septospine. The blade with its spine appears to be under different genetic control (fig. 10.5), from the fact that the scapular blade and septospine are missing in mutant mice with knockout *Emx2* and *Pax1* (*undulate* compound mutant), while the acromio-glenoid region can develop normally in the mutants (Timmons et al. 1994, fig. 5; Dietrich and Gruss 1995; Pellegrini et al. 2001).

The endochondral development of the entire scapula and the glenohumeral joint requires *Pbx1* (Selleri et al. 2001, fig. 8). It was further demonstrated that different combinations of the *Pbx* genes can influence the glenohumeral joint and different parts of the scapula (Capellini et al. 2011). *Pbx1* and *Emx2* cooperatively activate *Alx1* and *Alx4* and are parts of the gene network of the morphogenesis of these scapular features (Capellini et al. 2010; Hübler et al. 2013). Specific influence of *Pbx1* on the acromion and the glenoid is demonstrated by the hypoplasia of the acromion and the fusion of the glenohumeral joint in the *Pbx1* null mutant (*Pbx1$^{-/-}$*). Moreover, formation of the acromion and the acromion-clavicle joint requires normal function of *Pax1* and *Hox5*, demonstrated by the hypoplasia or total absence of acromion in the mutant mouse of *Pax1-/-* and *Hox5abc-/-* (Timmons et al. 1994; Dietrich and Gruss 1995; Aubin et al. 1998, 2002; Selleri et al. 2001; Kuijper et al. 2005; McIntyre et al. 2007; Capellini et al. 2010). The devel-

opment of the acromion, the coracoid process, and the glenohumeral joint is only modestly influenced by the genetic pathway that patterns the limb (Selleri et al. 2001; Capellini et al. 2010; Hübler et al. 2013), and not influenced by the genetic pathway for the scapula blade and the septospine.

Supraspinous fossa and septospine—The septospine defines the supraspinous fossa, and these two features have great functional significance, and also prominent homoplasy, in theriimorph evolution (fig. 10.3). Sánchez-Villagra and coworkers demonstrated that in most marsupials and some placentals, the septospine is formed by appositional bone in embryonic septum in the supraspinous and infraspinous muscle anlagen, whereas the acromial spine is developed endochondrally (Sánchez-Villagra and Maier 2002, 2003; Großmann et al. 2002). This distinction in embryogenesis is consistent with differences in gene patterning of development.

Pellegrini et al. (2001) showed that the scapular blade and the septospine are influenced by *Emx2*, a gene in the pathway for cell condensation and chondrogenesis and provides a positional signal for cell fates (Kuijper et al. 2005; Capellini et al. 2011), relatively independent of the acromio-glenoid region of the scapula. The scapular blade and the septospine fail to develop in the homozygous mutants of *Emx2* (*Emx2-/-*). Subsequent studies of mutants of deficient *Tbx15*, *Gli3*, *Alx4*, and related genes demonstrated that the anterior ("supraspinous") part corresponds to the functional area of *Alx1-4*. The septospine and middle part of the blade corresponds to the functional area of T-box transcription factors (*Tbx6*, *Tbx15*, and *Tbx18*) (Kuijper et al. 2005; Wehn and Chapman 2010), and *Pax3* (Farin et al. 2008). The posterior ("infraspinous") part is the functional area of *Gli3* (fig. 10.5) (Kuijper et al. 2005; Farin et al. 2008; Capellini et al. 2010). Mutant genotypes of these genes can result in hypoplasia of the scapular blade. In compound mutants of *Tbx* genes, an abnormal gap occurs where the septospine is expected in the scapular blade (Kuijper et al. 2005; Capellini et al. 2010).

The labile morphogenesis of the scapular blade and septospine under relatively independent genetic influence is consistent with the evolutionary homoplasy of the septospine among triconodonts, multituberculates, and spalacotheroids. For the evolution of the mammalian shoulder girdle, this is perhaps the best evolutionary and development module where fossil patterns can be well informed by developmental mechanisms.

Paleo Evo Devo

There is much new insight to be gained by integrating embryogenesis and developmental genetics into hypotheses for the phylogenetic evolution of fossils. The development of living mammals can provide mechanistic hypotheses for prominent homoplasies in early mammals. Reciprocally, the deep-time fossil record can help to demonstrate that phenotypes that are teratological, and demonstrable only by knockout genetic experiments on extant mammals, can actually occur in evolution. Morphological disparity of the sternum and shoulder girdle is especially prominent in Mesozoic mammal clades that have no living descendants (Kielan-Jaworowska et al. 2004; Luo 2007). The scope of this disparity has far exceeded what is possible for extant mammals (figs. 10.3–10.4). For example, the offset acromion (fig. 10.2D and fig. 10.3) common in Mesozoic theriimorphs is unknown for extant mammals. The unique scapular character combination of multituberculates of a perpendicular glenoid (derived) and the absence of the supraspinous fossa (primitive) can neither fit the scapular morphotype of living therians nor that of monotremes. The greater lineage diversity of Mesozoic mammals is a greater arena for repeated evolutionary experimentation of characters that are singular and unique among extant mammals. Fossils are milestones on what evolutionary transition was impacted by development at what time in geological history.

Caveats—There are limits on extrapolating development into fossils to account for macroevolutionary patterns. Interpretation of homoplasies in mammal phylogeny as repetitive developmental experimentations is ultimately dependent on a well-resolved phylogeny. But phylogenetic topology can differ by alternative analyses (e.g., Rowe et al. 2008 vs. Philips et al. 2009). The phylogenetic framework in which to interpret shoulder girdle evolution can change with new discoveries of fossils and new characteristics assimilated into analyses (e.g., see Luo et al. 2002; Luo and Wible 2005; Luo, Ji, and Yuan 2007; Wible et al. 2009; Rougier et al. 2011). Thus the development hypothesis for homoplasies in macroevolution has inherent uncertainty.

It is important to note that the genetic control of shoulder girdle development has not been studied in monotremes. Without genetic underpinning of their embryological observations (Klima 1973), it will be difficult to account for the evolutionary pattern of Mesozoic mammals that are intermediate between therians and monotremes. Despite much progress, the current understanding of genetic control of phenotype features is still rudimentary for the shoulder girdle of mouse, as widely acknowledged by many studies (e.g., Capellini et al. 2011), although gene patterning of some shoulder girdle features in marsupials are just beginning to be deciphered (Hübler et al. 2013). Many genes and pathways that can influence the development of the sternum and shoulder girdle can also have pleiotropic effects elsewhere. Phenotype areas of gene expression often overlap; thus a genotype may lack sufficiently specific correlation to a discrete feature seen in fossils. Perhaps more importantly the network of the genes influencing the same structure is not completely known. All are hurdles to overcome for using the gene patterning of morphogenesis of living mammals to account for evolutionary patterns.

Evolutionary development is not the only approach toward mechanistic understanding of phylogenetic evolution. Developmental interpretation of the homoplasies of the shoulder and sternal features is not mutually exclusive with the understanding of their functional evolution. Functional adaptation is also fundamental to convergent evolution. In the case of the supraspinous muscle of the scapula, functional adaptation to strengthen the glenohumeral joint of theriimorph mammals by the double muscle stabilization (Jenkins and Goslow 1983; Fischer 2001) can be a driving mechanism for convergent evolution, as is the known mechanism of labile morphogenesis of the related scapular features (fig. 10.5).

Summary

1. Transformation of the shoulder girdle from premammalian cynodonts through mammaliaforms to extant therians is characterized by the evolutionary reduction of the girdle elements, both by number and in size, ultimately simplifying the shoulder girdle in therians and their theriimorph relatives. But monotremes have retained the ancestral mammaliaform shoulder and

sternal structures and further developed these for fossorial adaptation.

2. Major transformation in theriimorph mammals includes the evolution of the mobile clavicle as a spoke for the arcuate movement of the shoulder girdle, and a reorientation of the glenohumeral joint stabilized by the derived supraspinous muscle, for greater mobility of the shoulder and forelimb.

3. Evolution of the shoulder girdle and sternal structure shows some prominent homoplasies, especially in the coracoid bones related to the pectoral and coracoid muscles in mammaliaforms and monotremes, and in the supraspinous fossa and septospine for muscle function to stabilize the glenohumeral joint in theriimorph mammals.

4. Genetic studies of the mammalian shoulder girdle can now attribute some phenotype features in increasingly specific ways to gene patterning of morphogenesis. The relatively independent and labile development of some features of the shoulder girdle can provide new insight on the mechanism of their evolution.

* * *

Acknowledgments

Professor Farish A. Jenkins Jr. is the inspiration for this review of the shoulder girdle evolution of mammals. His seminal studies on the evolution and function of the mammalian appendicular skeleton laid a foundation for our understanding of early mammal evolution. I want to thank Drs. Qiang Ji and Chongxi Yuan (Chinese Academy of Geological Sciences, Beijing), Peiji Chen and Gang Li (Nanjing Institute of Geology and Palaeontology), Qingjin Meng (Beijing Museum of Natural History), Ailin Sun and the late Yaoming Hu (Institute of Vertebrate Paleontology and Paleoanthropology, Beijing) for the opportunities to study the original fossils of Mesozoic mammals from China. During the course of this study, I benefited from many discussions with my colleagues K. Christopher Beard, Meng Chen, Thomas Martin, Guillermo Rougier, Timothy B. Rowe, Marcello Sánchez-Villagra, Karen Sears, Paul Sereno, Hans-Dieter Sues, Matthew Vickaryous, and John Wible. Paul Sereno and John Wible shared with me many observations from their own studies of the mammalian shoulder girdles. Susan Evans generously provided the specimen of the interclavicle of *Morganucodon*. Mary Dawson, Hans Sues, John Wible, and the editors of the volume helped to improve the manuscript. Lastly, I would like to thank Lauren Conroy and April Isch for assistance with illustrations. Research was supported by National Science Foundation (US), the Humboldt Foundation of Germany, and the University of Chicago.

References

Argot, C. 2001. Functional-adaptive anatomy of the forelimb in the Didelphidae, and the paleobiology of the Paleocene marsupials *Mayulestes ferox* and *Pucadelphys andinus*. Journal of Morphology 247:51–79.

Asher, R. J., I. Horovitz, T. Martin, and M. R. Sánchez-Villagra. 2007. Neither a rodent nor a platypus: a reexamination of *Necrolestes patagonensis* Ameghino. American Museum Novitates 3546:1–40.

Aubin, J., M. Lemieux, J. Moreau, J. Lapointe, and L. Jeannotte. 2002. Cooperation of *Hoxa5* and *Pax1* genes during formation of the pectoral girdle. Developmental Biology 244:96–113, doi: 10.1006/dbio.2002.0596.

Aubin, J., M. Lemieux, M. Tremblay, R. Behringer, and L. Jeannotte. 1998. Transcriptional interferences at the Hoxa4/Hoxa5 locus: importance of correct Hoxa5 expression for the proper specification of the axial skeleton. Developmental Dynamics 212:141–156.

Augee, M. L., B. A. Gooden, and A. M. Musser. 2006. Echidna—Extraordinary Egg-Laying Mammal. Collingwood, Victoria: CSIRO Publishing. Pp. 1–136.

Capellini, T. D., G. Vaccari, E. Ferretti, S. Fantini, M. He, M. Pellegrini, L. Quintana, G. Di Giacomo, J. Sharpe, L. Selleri, and V. Zappavigna. 2010. Scapula development is governed by genetic interactions of *Pbx1* with its family members and with *Emx2* via their cooperative control of *Alx1*. Development 137:2559–2569, doi:10.1242/dev.048819.

Capellini, T. D., V. Zappavigna, and L. Selleri. 2011. Pbx homeodomain proteins: *TALEnted* regulators of limb patterning and outgrowth. Developmental Dynamics 240:1063–1086.

Cave, A. J. E. 1970. Observation on the monotreme interclavicle. Journal of Zoology 160:297–312.

Chen, M., and Z.-X. Luo. 2013. Postcranial skeleton of the Cretaceous mammal *Akidolestes cifellii* and its locomotor adaptations. Journal of Mammalian Evolution 20:159–189, doi: 10.1007/s10914-012-9199-9.

Cheng, C. 1955. The development of the shoulder region of the opossum, *Didelphis virginiana*, with special reference to the musculature. Journal of Morphology 97:415–471.

Coues, E. 1871. On the myology of the *Ornithorhynchus*. Communications Essex Institute 6:127–137. (Digitized and redistributed by Google.)

Dietrich, S., and P. Gruss. 1995. *Undulated* phenotypes suggest a role of *Pax-1* for the development of vertebral and extravertebral structures. Developmental Biology 167:529–548.

Diogo, R., and V. Abdala. 2010. Muscles of Vertebrates: Comparative Anatomy, Evolution, Homologies and Development. Oxford: Taylor & Francis. Pp. 1–476.

Evans, H. E. 1993. Miller's Anatomy of the Dog. 3rd ed. Philadelphia: W. B. Saunders Company. Pp. 1–1113.

Evans, S. E. 1981. The postcranial skeleton of the Lower Jurassic eosuchian *Gephyrosaurus bridensis*. Zoological Journal of the Linnean Society 73:81–116.

Farin, H. F., A. Mansouri, M. Petry, and A. Kispert. 2008. *T-box* Protein *Tbx18* interacts with the Paired Box Protein *Pax3* in the development of the paraxial mesoderm. Journal of Biological Chemistry 283:25372–25380.

Fischer, M. S. 2001. Locomotory organs of mammals: new mechanics and feedback pathways but conservative central control. Zoology 103:230–239.

Gambaryan, P. P., and Z. Kielan-Jaworowska. 1997. Sprawling versus parasagittal stance in multituberculate mammals. Acta Paleontologica Polonica 42:13–44.

Gemmell, R. T., C. Veitch, and J. Nelson. 2002. Birth in marsupials. Comparative Biochemistry and Physiolology (B) 131:621–630.

Goswami, A., G. V. R. Prasad, P. Upchurch, D. M. Boyer, E. R. Seiffert, O. Verma, E. Gheerbrant, and J. J. Flynn. 2011. A radiation of arboreal basal eutherian mammals beginning in the Late Cretaceous of India. Proceedings of National Academy of Sciences, USA 108:16333–16338, doi: 10.1073/pnas.1108723108.

Gow, C. E. 2001. A partial skeleton of the trithelodontid *Pachygenelus* (Therapsida, Cynodontia). Palaeontologica Africana 37:93–97.

Großmann, M., M. R. Sánchez-Villagra, and W. Maier. 2002. On the development of the shoulder girdle in *Crocidura russula* (Soricidae) and other placental mammals: evolutionary and functional aspects. Journal of Anatomy 201:371–381.

Hall, B. K., and T. Miyake. 1992. The membranous skeleton: the role of cell condensation in vertebrate skeletogenesis. Anatomy and Embryology 186:107–124.

Hildebrand, M. 1985. Digging of quadrupeds. Pp. 89–109 in M. Hildebrand, D. M. Bramble, K. F. Liem and D. B. Wake, eds., Functional Morphology of Vertebrates. Cambridge: Belknap Press of Harvard University Press.

Hildebrand, M., and G. E. Goslow Jr. 2001. Analysis of Vertebrate Structure. 5th ed. New York: John Wiley. Pp. 1–635.

Horovitz, I. 2003. Postcranial skeleton of *Ukhaatherium nessovi* (Eutheria, Mammalia) from the Late Cretaceous of Mongolia. Journal of Vertebrate Paleontology 23:857–868.

Howell, A. B. 1937. Morphogenesis of the shoulder architecture Part V. Monotremata. Quarterly Review of Biology 12:191–205.

Hu, Y.-M. 2006. Postcranial morphology of *Repenomamus* (Eutriconodonta, Mammalia): implications for the higher-level phylogeny of mammals. PhD dissertation, the City University of New York. Pp. 1–405.

Hu, Y.-M., and Y.-Q. Wang. 2002. *Sinobataar* gen. nov.: first multituberculate from Jehol Biota of Liaoning, Northern China. Chinese Science Bulletin 47:933–938.

Hu, Y.-M., Y.-Q. Wang, Z.-X. Luo, and C.-K. Li. 1997. A new symmetrodont mammal from China and its implications for mammalian evolution. Nature 390:137–142.

Huang, R., B. Chris, and K. Patel. 2006. Regulation of scapula development. Anatomy and Embryology 211(Suppl. 1):S65–S71, doi: 10.1007/s00429-006-0126-9.

Huang, R., Q. Zhi, K. Patel, J. Wilting, and B. Chris. 2000. Dual origin and segmental organisation of the avian scapula. Development 127:3789–3794.

Hübler, M., A. C. Molineaux, A. Kayte, T. Schecker, and K. E. Sears. 2013. Development of the marsupial shoulder girdle complex: a case study in *Monodelphis domestica*. Evolution & Development 15:18–27, doi: 10.1111/ede.12011.

Hübler, M., L. A. Niswander, J. Peters, and K. E. Sears. 2010. The developmental reduction of the marsupial coracoid: a case study in *Monodelphis domestica*. Journal of Morphology 271:769–776.

Hurum, J. H., and Z. Kielan-Jaworowska. 2008. Postcranial skeleton of a Cretaceous multituberculate mammal *Catopsbaatar*. Acta Palaeontologica Polonica 53:545–566.

Jenkins, F. A., Jr. 1970a. The Chañares (Argentina) Triassic reptile fauna VII. The postcranial skeleton of the traversodontid *Massetognathus pascuali* (Therapsida, Cynodontia). Breviora 352:1–28.

Jenkins, F. A., Jr. 1970b. Limb movements in a monotreme (*Tachyglossus aculeatus*): a cineradiographic analysis. Science 168:1473–1475.

Jenkins, F. A., Jr. 1971a. Limb posture and locomotion in the Virginia opossum (*Didelphis marsupialis*) and in other non-cursorial mammals. Journal of Zoology (London) 165:303–315.

Jenkins, F. A., Jr. 1971b. The postcranial skeleton of African cynodonts. Peabody Museum of Natural History Bulletin (Yale University) 36:1–216

Jenkins, F. A., Jr. 1974. The movement of the shoulder in claviculate and aclaviculate mammals. Journal of Morphology 144:71–84.

Jenkins, F. A., Jr. 1993. The evolution of the avian shoulder joint. American Journal of Science 293A:253–267.

Jenkins, F. A., Jr., and G. E. Goslow Jr. 1983. The functional anatomy of the shoulder of the Savannah monitor lizard (*Varanus exanthematicus*). Journal of Morphology 175:195–216.

Jenkins, F. A., Jr., and D. W. Krause. 1983. Adaptations for climbing in North American multituberculates (Mammalia). Science 220:712–715.

Jenkins, F. A., Jr., and F. R. Parrington. 1976. The postcranial skeleton of the Triassic mammals *Eozostrodon*, *Megazostrodon* and *Erythrotherium*. Philosophical Transactions of Royal Society London B: Biolological Sciences 273:387–431.

Jenkins, F. A., Jr., and C. R. Schaff. 1988. The Early Cretaceous mammal *Gobiconodon* (Mammalia, Triconodonta) from the Cloverly Formation in Montana. Journal of Vertebrate Paleontology 8:1–24.

Jenkins, F. A., Jr., and W. A. Weijs. 1979. The functional anatomy of the shoulder in the Virginia opossum (*Didelphis virginiana*). Journal of Zoolology (London) 188:379–410.

Ji, Q., Z.-X. Luo, and S.-A. Ji. 1999. A Chinese triconodont mammal and mosaic evolution of the mammalian skeleton. Nature 398:326–330.

Ji, Q., Z.-X. Luo, C.-X. Yuan, and A. R. Tabrum. 2006. A swimming mammaliaform from the Middle Jurassic and ecomorphological diversification of early mammals. Science 311:1123–1127.

Ji, Q., Z.-X. Luo, C-X. Yuan, J. R. Wible, J.-P. Zhang, and J. A. Georgi. 2002. The earliest known eutherian mammal. Nature 416:816–822.

Jouffroy, F. K. 1971. Musclulature des membres. Pp. 1–475 in P. P. Grassé, ed., Traité de Zoologie, Tome XVI (Faciscle III). Paris: Masson.

Jouffroy, F. K., and J. Lessertisseur. 1971. Musculature post-cranienne. Pp. 679–836 in P. P. Grassé, ed., Traité de Zoologie, Tome XVI (Faciscle III). Paris: Masson.

Kardong, K. V. 1998. Vertebrates: Comparative Anatomy, Function, Evolution. 2nd edition. Dubuque: W. C. Brown Publishers. Pp. 1–777.

Keyte, A. L., and K. K. Smith. 2010. Developmental origins of precocial forelimbs in marsupial neonates. Development 137:4283–4294, doi: 10.1242/dev.049445.

Kielan-Jaworowska, Z., R. L. Cifelli, and Z.-X. Luo Z-X. 2004. Mammals from the Age of Dinosaurs. New York: Columbia University Press. Pp. 1–630.

Kielan-Jaworowska, Z., and P. P. Gambaryan. 1994. Postcranial anatomy and habits of Asian multituberculate mammals. Fossils and Strata 36:1–92.

Kielan-Jaworowska, Z., and J. H. Hurum. 2006. Limb posture in early mammals: sprawling or parasagittal. Acta Palaeontologica Polonica 51:393–406.

Kielan-Jaworowska, Z., and T. Qi. 1990. Fossorial adaptations of a taeniolabidoid multituberculate mammal from the Eocene of China. Vertebrata PalAsiatica 28:81–94.

Klima, M. 1973. Die Frühentwicklung des Schültergürtels und des Brustbeins bei den Monotremen (Mammalia: Prototheria). Advances in Anatomy, Embryology & Cell Biology 47:1–80.

Klima, M. 1985. Development of shoulder girdle and sternum in mammals. Forschritter der Zoologie 30:81–83.

Klima, M. 1987. Early development of the shoulder girdle and sternum in marsupials (Mammalia: Metatheria). Advances in Anatomy, Embryology & Cell Biology 109:1–91.

Krause, D. W., and F. A. Jenkins Jr. 1983. The postcranial skeleton of North American multituberculates. Bulletin of Museum of Comparative Zoology (Harvard University) 150:199–246.

Krebs, B. 1991. Das skelett von Henkelotherium guimarotae gen. et sp. nov. (Eupantotheria, Mammalia) aus dem Oberen Jura von Protugal. Berlin geowissenschaftliche, Abhandlungen A 133:1–110.

Kuijper, S., A. Beverdam, C. Kroon, A. Brouwer, S. Candille, G. Barsh, and F. Meijlink. 2005. Genetics of shoulder girdle formation: roles of Tbx15 and aristaless-like genes. Development 132:1601–1610, doi: 10.1242/dev.01735.

Lessertisseur, J., and R. Saban. 1967a. Squelette appendiculaire. Pp. 709–1078 in P.-P. Grassé, ed., Traité de Zoologie. Tome XVI (Fascicle I). Mammiferes: Teguments et Skelettes; Paris: Masson.

Lessertisseur, J., and R. Saban. 1967b. Squelette axial. Pp. 587–765 in P.-P. Grassé, ed., Traité de Zoologie. Tome XVI (Fasicle I). Mammiferes: Teguments et Skelettes; Paris: Masson.

Li, G., and Z.-X. Luo. 2006. A Cretaceous symmetrodont therian with some monotreme-like postcranial features. Nature 439:195–199.

Lillegraven, J. A. 1975. Biological considerations of the marsupial-placental dichotomy. Evolution 29:707–722.

Luo, Z.-X. 2007. Transformation and diversification in the early mammalian evolution. Nature 450:1011–1019.

Luo, Z.-X., P.-J. Chen, G. Li, and M. Chen. 2007. A new eutriconodont mammal and evolutionary development of early mammals. Nature 446:288–293.

Luo, Z.-X., Q. Ji, J. R. Wible, and C.-X. Yuan. 2003. An Early Cretaceous tribosphenic mammal and metatherian evolution. Science 302:1934–1940.

Luo, Z.-X., Q. Ji, and C.-X. Yuan. 2007b. Convergent dental evolution in pseudotribosphenic and tribosphenic mammals. Nature 450:93–97.

Luo, Z.-X., Z. Kielan-Jaworowska, and R. L. Cifelli. 2002. In quest for a phylogeny of Mesozoic mammals. Acta Palaeontologica Polonica 47:1–78.

Luo, Z.-X., and J. R. Wible. 2005. A Late Jurassic digging mammal and early mammalian diversification. Science 308:103–107.

Luo, Z.-X., C.-X. Yuan, Q.-J. Meng, and Q. Ji. 2011. A Jurassic eutherian mammal and the divergence of marsupials and placentals. Nature 476:442–445, doi: 10.1038/nature10291.

Martin, T. 2005. Postcranial anatomy of Haldanodon exspectatus (Mammalia, Docodonta) from the Late Jurassic (Kimmeridgian) of Portugal and its bearing for mammalian evolution. Zoological Journal of Linnean Society 145:219–248.

Matsuoka, T., P. E. Ahlberg, N. Kessaris, P. Iannarelli, U. Dennehy, W. D. Richardson, A. P. McMahon, and G. Koentges. 2005. Neural crest origins of the neck and shoulder. Nature 436:347–355.

McIntyre, D. C., S. Rakshi, A. R. Yallowitz, L. Loken, L. Jeannotte, M. R. Capecchi, and D. M. Wellik. 2007. Hox patterning of the vertebrate rib cage. Development 134:2981–2989, doi: 10.1242/dev.007567.

Meng, J., Y.-M. Hu, Y.-Q. Wang, and C.-K. Li. 2006. A Mesozoic gliding mammal from northeastern China. Nature 444:889–893.

Pellegrini, M., S. Pantano, M. P. Fumi, F. Lucchini, and A. U. Forabosco. 2001. Agenesis of the scapula in Emx2 homozygous mutants. Developmental Biology 232:149–156.

Phillips, M. J., T. H. Bennett, and M. S. Y. Lee. 2009. Molecules, morphology, and ecology indicate a recent, amphibious ancestry for echidnas. Proceedings of National Academy of Sciences, USA 106:17089–17094, doi: 10.1073/pnas.0904649106.

Polly, P. D. 2007. Limbs in mammalian evolution. Chapter 15, pp. 245–268, in B. K. Hall, ed., Fins into Limbs: Evolution, Development, and Transformation. Chicago: University of Chicago Press. Pp. 1–772.

Romer, A. S. 1956. Osteology of Reptiles. Chicago: University of Chicago Press. Pp. 1–772.

Romer, A. S., and A. D. Lewis. 1973. The Chañares (Argentina) Triassic reptile fauna. XIX. Postcranial materials of the cynodonts Probelesodon and Probainognathus. Breviora 407:1–26.

Rougier, G. W. 1993. Vincelestes neuquenianus Bonaparte (Mammalia, Theria), un primitivo mammifero del Cretacico Inferior de la Cuenca Neuqina. PhD thesis, Universidad Nacional de Buenos Aires, Facultad de Ciencias Exactas y Naturales, Buenos Aires. Pp. 1–720.

Rougier, G. W., S. Apesteguía, and L. C. Gaetano. 2011. Highly specialized mammalian skulls from the Late Cretaceous of South America. Nature 479:98–102, doi: 10.1038/nature10591.

Rougier, G. W., Q. Ji, M. J. Novacek. 2003. A new symmetrodont mammal with fur impressions from the Mesozoic of China. Acta Geologica Sinica 77:7–14.

Rowe, T. 1988. Definition, diagnosis, and origin of Mammalia. Journal of Vertebrate Paleontology 8:241–264.

Rowe, T., T. H. Rich, P. Vickers Rich, M. Springer, and M. O. Woodburne. 2008. The oldest platypus and its bearing on divergence timing of the platypus and echidna clades. Proceedings of National Academy of Sciences, USA 105:1238–1242.

Sánchez-Villagra, M. R., and W. Maier. 2002. Ontogenetic data and the evolutionary origin of the mammalian scapula. Naturwissenschaften 89:459–461, doi: 10.1007/s00114-002-0362-7.

Sánchez-Villagra, M. R., and W. Maier. 2003. Ontogenesis of the scapula in marsupial mammals, with special emphasis on perinatal stages of *Didelphis* and remarks on the origin of the therian scapula. Journal of Morphology 258:115–129.

Sargis, E. J. 2002. Functional morphology of the forelimb of tupaiids (Mammalia, Scandentia) and its phylogenetic implications. Journal of Morphology 253:10–42.

Sears, K. E. 2004. Constraints on the morphological evolution of marsupial shoulder girdles. Evolution 58:2353–2370.

Sears, K. E. 2005. Role of development in the evolution of the scapula of the giant sthenurine kangaroos (Macropodidae: Sthenurinae). Journal of Morphology 265:226–236.

Sears, K. E., C. Bianchi, L. Powers, and A. L. Beck. 2013. Integration of the mammalian shoulder girdle within populations and over evolutionary time. Journal of Evolutionary Biology 26:1536–1548.

Selleri, L., M. J. Depew, Y. Jacobs, S. K. Chandra, K. Y. Tsang, K. S. E. Cheah, J. L. R. Rubenstein, S. O'Gorman, and M. L. Cleary. 2001. Requirement for *Pbx1* in skeletal patterning and programming chondrocyte proliferation and differentiation. Development 128:3543–3557.

Sereno, P. C. 2006. Shoulder girdle and forelimb in a Cretaceous multituberculate: form, functional evolution, and a proposal for basal mammalian taxonomy. Pp. 315–370 in M. T. Carrano, T. J. Gaudin, R. W. Blob, and J. R. Wible, eds., Amniote Paleobiology: Perspectives on the Evolution of Mammals, Birds, and Reptiles. Chicago: University of Chicago Press.

Sereno, P. C., and M. C. McKenna. 1995. Cretaceous multituberculate skeleton and the early evolution of the mammalian shoulder girdle. Nature 377:144–147.

Sues, H.-D., and F. A. Jenkins Jr. 2006. Postcranial skeleton of *Kayentatherium wellesi* from the Lower Jurassic Kayenta Formation of Arizona and the phylogenetic significance of postcranial features in tritylodontid cynodonts. Pp. 114–152 in M. T. Carrano, T. J. Gaudin, R. W. Blob, and J. R. Wible, eds., Amniote Paleobiology: Perspectives on the Evolution of Mammals, Birds, and Reptiles. Chicago: University of Chicago Press.

Sun, A.-L., and Y.-H. Li. 1985. The postcranial skeleton of Jurassic tritylodonts from Sichuan Province. Vertebrata PalAsiatica 23:135–151.

Szalay, F. S. 1994. Evolutionary History of the Marsupials and an Analysis of Osteological Characters. Cambridge: Cambridge University Press. Pp. 1–481.

Taylor, M. E. 1974. The functional anatomy of the forelimb of some African Viverridae (Carnivora). Journal of Morphology 143:307–336.

Timmons, P. M., J. Wallin, P. W. J. Rigby, and R. Balling. 1994. Expression and function of *Pax1* during development of the pectoral girdle. Development 120:2773–2785.

Vickaryous, M. K., and B. K. Hall. 2006. Homology of the reptilian coracoid and a reappraisal of the evolution and development of the amniote pectoral apparatus. Journal of Anatomy 208:263–285.

Wehn, A. K., and D. L. Chapman. 2010. *Tbx18* and *Tbx15* null-like phenotypes in mouse embryos expressing *Tbx6* in somatic and lateral plate mesoderm. Developmental Biology 347:404–413.

Wible, J. R., G. W. Rougier, M. J. Novacek, and R. J. Asher. 2009. The eutherian mammal *Maelestes gobiensis* from the Late Cretaceous of Mongolia and the phylogeny of Cretaceous Eutheria. Bulletin of the American Museum of Natural History 327:1–123.

Walter, L. R. 1988. Appendicular musculature in the echidna *Tachyglossus aculeatus* (Monotremata: Tachyglossidae). Australian Journal of Zoology 36:65–81.

Williams, P. L., R. Williams, M. Dyson, and L. H. Bannister. 1989. Gray's Anatomy, 37th ed. New York: Churchill Livingstone. Pp. 1–1598.

Wilson, G. P., A. R. Evans, I. J. Corfe, P. D. Smits, M. Fortelius, and J. Jernvall. 2012. Adaptive radiation of multituberculate mammals before the extinction of dinosaurs. Nature 483:457–460.

11

Evolution of the Mammalian Nose

A. W. Crompton,* Catherine Musinsky,† and Tomasz Owerkowicz‡

Introduction

Endothermy requires an integrated set of morphological and physiological features (Kemp 2006) that evolved independently in birds and mammals (Ruben 1996). Morphologic evidence for the evolution of endothermy has partially been based upon the presence of the intranarial respiratory turbinals or conchae that act as temporal countercurrent exchange sites (Bennett and Ruben 1986). Since endotherms have a higher respiratory rate than ectotherms, evolution of the temporal countercurrent exchange mechanism allowed them to control the temperature of expired air when at rest and during mild exercise, and afforded them significant savings of heat and water (Owerkowicz et al., this volume). Birds achieved this by evolving cartilaginous nasal conchae (Bang 1961; Geist 2000), while mammals developed ossified maxillary turbinals (Macrini 2012; Rowe et al. 2005). Both mechanisms warm and humidify inhaled air, and cool exhaled air (Schmidt-Nielsen 1981), and both require a firm temporary seal between the nasal cavity and the trachea, so that inspired and expired air passes through the nose and bypasses the oral

* Museum of Comparative Zoology, and Department of Organismic and Evolutionary Biology, Harvard University
† Department of Organismic and Evolutionary Biology, Harvard University
‡ Department of Biology, California State University, San Bernardino

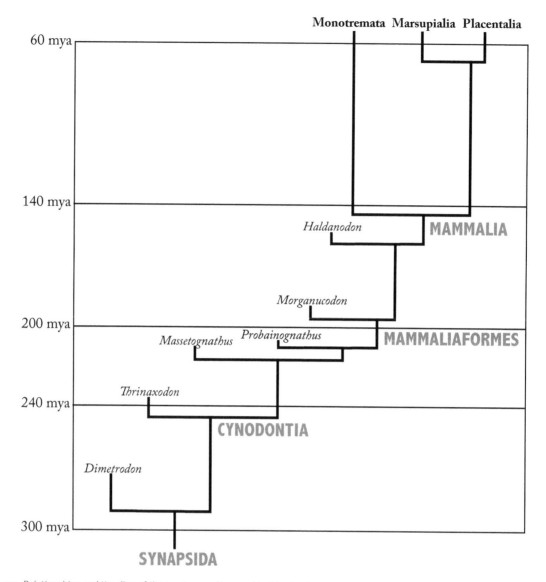

FIG. 11.1 Relationships and timeline of the specimens discussed in this chapter.

cavity. Conchae and maxilloturbinals also provide sites for evaporative cooling during exercise and high ambient temperatures (Taylor 1977).

Evolution of these countercurrent exchange sites was likely gradual and probably involved considerable reorganization of the skull skeleton. In this chapter, we focus on the evolutionary changes to the synapsid skull, which may have housed the incipient respiratory turbinates.

Within the nonmammalian therapsids (mammal-like reptiles) spanning more than 100 million years from the mid-Permian to the Early Jurassic (fig. 11.1) (Kemp 1982; Rowe and Gauthier 1992), some of the progressive changes leading to the mammalian skull—for example,

the three-boned middle ear, precise dental occlusion, determinate growth, and an enlarged braincase—are well documented. The evolution of the mammalian nose, however, remains relatively uncertain. Hillenius (1992, 1994), Hillenius and Ruben (2004), Ruben (1996), and Ruben et al. (2012) have pointed to ridges found on the inner surface of the nasal and maxillary bones to suggest that ossified turbinates were present in nonmammalian therapsids. On the basis of these characters—the purported maxillary ridges, the presence of a bony secondary palate, and an enlarged nasal respiratory chamber—the authors concluded that several groups of nonmammalian therapsids were either partially or fully endothermic.

In extant mammals the embryonic cartilaginous nasal capsule ossifies to form the mesethmoid and ethmoid bones. The latter include ossified maxilloturbinals. But such ossifications have never been found in the nasal region of nonmammalian synapsids, suggesting that the nasal capsule remained cartilaginous throughout life.

To understand the structure of the cartilaginous nasal capsule of nonmammalian synapsids, it is necessary to determine the shape of the space enclosed by the membrane bones that would have surrounded it. MicroCT scans contribute new insights to the findings of Hillenius (1994), Ruben (1996), and Ruben et al. (2012). Based on such scans, we offer suggestions as to the structure and function of some cartilaginous nasal capsules of Triassic nonmammalian cynodonts and

Jurassic mammaliaforms, and review important steps in the development of the mammalian nose.

Reptiles: Squamata

Olfaction is the primary function of the reptilian nose: olfactory nerves enter the posterodorsal region that lies above the flow of air through the nasal capsule. In typical living reptiles (Bellairs and Kamal 1981; Malan 1946; Parsons 1959; Pratt 1948), a cartilaginous nasal cavity opens directly into the oral cavity through primary choanae, or internal nares (fig. 11.2A, B); thus all air breathed through the nose must also pass through the oral cavity. The route of air through the nasal capsule between the nares and the primary choana is short (as indicated by the arrow in fig. 11.2C), limiting the space

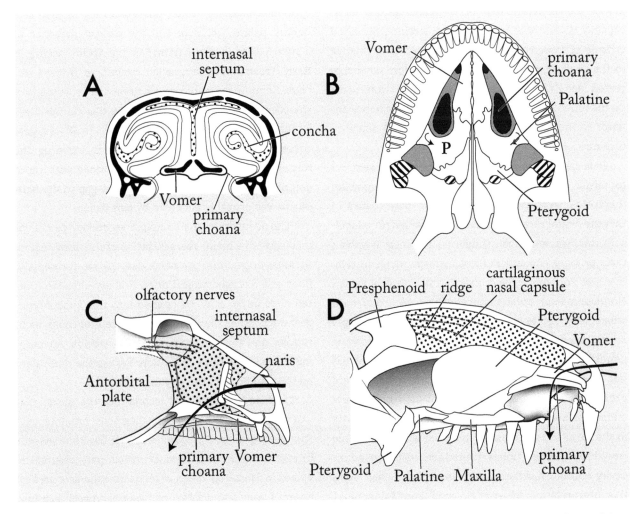

FIG. 11.2 Nasal region of an extant *Amblyrhynchus cristatus* (A–C) and an extinct *Dimetrodon limbatus* (D). A, Transverse section; B, ventral view of the palate; C, sagittal section; D, sagittal section through the nasal region.

for structures that might support respiratory epithelium. A tall, cartilaginous internasal septum between the vomer and the nasal roof separates the two sides of the nasal cavity (fig. 11.2A).

Early Nonmammalian Synapsids

In primitive nonmammalian synapsids, such as the pelycosaur *Dimetrodon* (fig. 11.2D), the nasal capsule remained cartilaginous. As in reptiles, the primary choanae open directly into the oral cavity (Hillenius 1992; Hillenius 1994; Romer and Price 1940), which limits the space available to house structures supporting respiratory epithelium. The posterior extent of a cartilaginous nasal capsule is constrained by the presphenoid bone. The presphenoid is Y-shaped in cross-section, its ventral keel forms an interorbital septum and its expanded dorsal area (Kemp 1980) encloses the olfactory nerves. The dorsal edge of the tall pterygoid bone probably marks the ventral extent of the cartilaginous nasal capsule. Given these volumetric constraints, the nasal capsule of *Dimetrodon* was long, narrow, and, relative to the skull height, shallow. Several authors have suggested (see Hillenius 1992; Ruben et al. 2012) that ridges on the inner surface of the intramembranous bones anterior to the presphenoid supported ossified olfactory turbinals.

It is generally accepted that pelycosaurs were ectothermic (Hopson 2012; Kemp 2006). Therocephalian therapsids were more advanced than pelycosaurs in that the dental row differentiated into well-defined incisors, canines, and small simple postcanines; however, their primary choanae did not increase much in relative size and still opened directly into the oral cavity. Nonmammalian basal therocephalians lacked a hard palate, although Maier et al. (1996) claimed that ridges on the maxilla and palatine parallel to the postcanine dentition supported choanal soft tissue folds that did not meet in the midline. When their free edges were pressed against the ventral surface of the vomer, a secondary palate was created that continued posteriorly in the form of a mammalian muscular soft palate. This would have allowed uninterrupted activity in the mouth cavity without interfering with breathing. Based upon this interpretation, Maier et al. (1996) and Maier (1999) claimed that nonmammalian therocephalians and more advanced nonmammalian therapsids suckled their young, possessed hair, and were endothermic. However, this conclusion invites reexamination because nonmammalian therocephalians had multiple generations of tooth replacement throughout life, rather than the diphyodont dentition of mammals, which correlates with rapid growth and early dependence on maternal milk for nourishment (Hopson 1973).

Nonmammalian Cynodonts

The description of the nasal capsule of nonmammalian cynodonts given here is based upon MicroCT scans of *Massetognathus* and *Probainognathus* the description of the skull of *Thrinaxodon* upon a reconstruction from serial grinding (Fourie 1974); and that of several cynodonts upon manually prepared cynodont skulls (Brink 1955; Estes 1961; Hillenius 1992; Kemp 1980; Sues 1986).

Advanced nonmammalian synapsids, such as the cynodont *Massetognathus,* possess a fully ossified secondary palate (fig. 11.3B–E). The earliest presence of a hard palate in nonmammalian therapsids occurs in Early Triassic nonmammalian cynodonts (*Procynosuchus*; Kemp 1979), and consists of medial projections of the premaxilla, maxilla, and palatine that do not meet in the midline. Thomason and Russell (1986) suggest these played a mechanical function in resisting the forces of mastication that probably arose with more complex postcanine teeth. The median gap in the hard palate was probably occupied by soft tissue.

The acquisition of a complete secondary palate coincided with a major reorganization of the nasal region. In *Massetognathus*, the secondary palate formed the floor of a nasopharyngeal duct. Primary choanae (dotted lines in fig. 11.3B–D) opened into this nasopharyngeal duct rather than straight into the oral cavity as in reptiles and early nonmammalian synapsids. Air passing through the nose could now bypass the oral cavity en route to or from the larynx.

It is now possible to reconstruct the space that would have contained the cartilaginous nasal capsule (fig. 11.4; fig. 11.5A, B). Unlike the long flat bone present in extant reptiles and nonmammalian pelycosaurs, the vomer in *Massetognathus* extends dorsally as a vertical sheet of bone and divides the nasopharyngeal duct into left and right regions (figs. 11.3B, C, 11.5A). The increase in

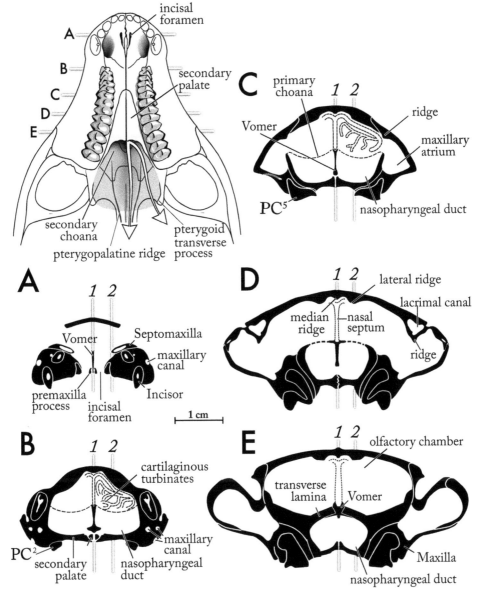

FIG. 11.3 Nasal region of a nonmammalian cynodont (*Massetognathus*). Ventral view of the palate (left) and transverse sections of the nasal cavity (A–E). Capital letters to the left of the palate indicate the position of the transverse sections shown in A–E. The two vertical lines labeled 1 and 2 indicate the planes shown in figure 11.5A and B.

height of this bone appears to have occurred as the premaxilla, maxilla, and palatine extended below the floor of the primary palate to form the hard secondary palate.

Compared to those of primitive nonmammalian therapsids, *Massetognathus* and *Probainognathus* have longer and wider primary choanae that extend back to the anterior edge of the primary palate (heretofore referred to as the "transverse lamina") (figs. 11.3E, 11.5B). The primary choanae are bordered laterally by the maxilla and palatine, medially by the vomer, and posteriorly by the palatine and vomer (fig. 11.3B–D). The posterior portion of the vomer forms a strut that, together with the medially directed plates of the palatines, forms the anterior part of the transverse lamina (fig. 11.3E) and the floor to the posterior (olfactory) chamber of the nasal cavity (fig. 11.5A, B). The grooved dorsal border of the vomer and the medial ridge on the ventral side of the nasal bone in *Massetognathus* suggest the presence of a tall cartilaginous internasal septum (figs. 11.3B–E and 11.6D). The posterior extension of the primary choanae

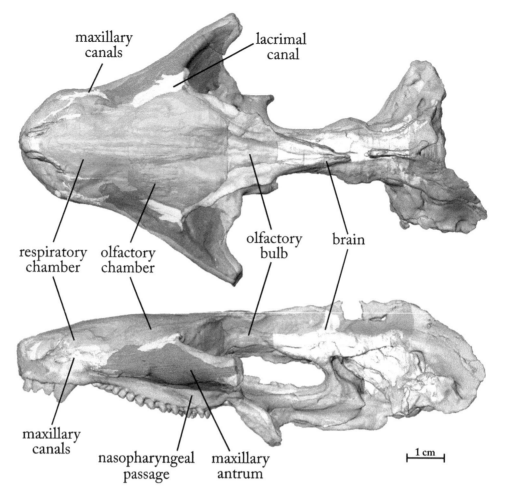

maxillary canals

lacrimal canal

respiratory chamber

olfactory chamber

olfactory bulb

brain

maxillary canals

nasopharyngeal passage

maxillary antrum

1 cm

FIG. 11.4 Division of the nasal cavity as seen in dorsal and lateral views (see also supplementary video, Musinsky 2012, http://vimeo.com/43930404).

greatly increases the volume of a respiratory chamber through which inhaled and exhaled air pass. Arrows in figure 11.5B indicate airflow through the primary choana into the nasopharyngeal duct.

The internal surfaces of the membrane bones surrounding the nasal cavity of *Massetognathus* have a series of ridges: a median ridge, two lateral ridges—one on either side (fig. 11.3C–E)—and one ridge on each inner surface of the lacrimals and maxillae (figs. 11.3C, D and 11.5A, B). These and some additional ridges have been described in several nonmammalian synapsids (Brink 1955, 1957; Fourie 1974; Hillenius 1994; Kermack et al. 1981). All these authors described these ridges as the remnants of the bases of either ossified olfactory or respiratory turbinals. In modern mammals when the maxilloturbinals are lost due to damage in dried skulls, a ridge terminated by a fracture line remains, indicating the presence of maxilloturbinals. In nonmammalian

cynodonts the ridges have a smooth surface, and no such fracture line is present. It is likely that the ridge on the ventral edge of the nasal bone that Hillenius (1992) claimed supported a respiratory turbinate in *Massetognathus* is an artifact of dorsoventral compression before fossilization (Hopson 2012 contra Hillenius 1994). The medial ridge on the inner surface of the nasal bone in *Massetognathus* almost certainly lay above a cartilaginous internasal septum (fig. 11.3C, D). This supports the view that the ridges on the inner surfaces of bones in the nasal region may, in fact, indicate the presence of cartilaginous structures. Similarly, the lateral ridges parallel to the median ridge may point to the presence of cartilaginous nasal turbinals. However, we could find no ridges in the CT scans of *Massetognathus* that could indicate the number or orientation of cartilaginous ethmoturbinals; so no attempt to reconstruct them has been made in figure 11.3D or E. The long ridge that forms the base for the maxilloturbinals in mammals lies close to the opening of the nasolacrimal duct (fig. 11.6E). The posterior edge of a similar ridge in *Massetognathus* (fig. 11.5B) lies above the internal opening of the nasolacrimal duct, suggesting that this ridge lies proximal to the base of a cartilaginous maxilloturbinal (figs. 11.3C, D

and 11.6D). In reptiles, the nasal capsule does not extend below the level of the dorsal border of the vomer (fig. 11.2A) (Malan 1946); likewise, in *Massetognathus* and other cynodonts the cartilaginous capsule probably lay above the dorsal border of the tall vomer. Respiratory turbinals may have filled the respiratory chamber

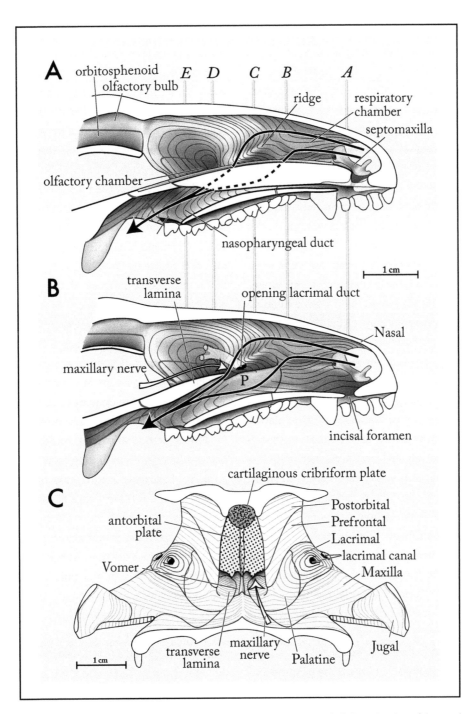

FIG. 11.5 A, Sagittal and B, parasagittal sections through the snout of *Massetognathus pascuali*. C, Posterior view of the nasal region of the same specimen. Vertical lines labeled A through E indicate the position of the sections included in figure 11.3. The two vertical lines in figure 11.3 indicate the planes shown here in A and B.

above the large primary internal choana (fig. 11.6D), but they may have extended ventrally into the nasopharyngeal duct. We cannot calculate the surface area of hypothetical respiratory turbinals, if they were in fact present, but since cartilaginous laminae tend to be thicker than ossified turbinals, it is unlikely that their surface area could have approached that of the highly branched respiratory turbinals in many terrestrial and some aquatic mammals (Van Valkenburgh et al. 2011).

Respiratory turbinates in mammals are effective at controlling heat and water loss only if air to and from the lungs can bypass the oral cavity (Biewener et al. 1985; Schmidt-Nielsen 1981; Taylor 1977). In mammals, a palatopharyngeal muscle extending from the hard and soft palates (Wood Jones 1940) can surround the larynx and tip of the epiglottis and seal off from the oral cavity the passage of air from the nasal capsule to the trachea (Crompton et al. 2008) while the animal is either mildly exercising or at rest. In most mammals the larynx remains in an intranarial position except during swallowing, panting, or vocalization. Nothing is known about the structure, position, or even the very presence of an intranarial larynx in nonmammalian synapsids. Barghusen (1986) suggested that ridges extending backward from the posterior edge of the hard palate to the outer edge of the transverse process of the pterygoid bone in nonmammalian cynodonts supported a non-muscular soft palate (shaded area in fig. 11.3). Barghusen (1986) further suggested that this would form the floor of a nasopharynx. Parallel pterygo-palatine ridges on the roof of the nasopharynx indicate two medial and two lateral channels (arrows, fig. 11.3). The former, he proposed, led to the opening of the larynx and the latter to the medial surface of the angular bone of the lower jaw and a possible homologue of the mammalian auditory (Eustachian) tube. In all nonmammalian cynodonts the transverse processes of the pterygoid abut against the lower jaw, restricting jaw movement to the sagittal plane. Barghusen (1986) claimed that a "reptilian" pterygoideus muscle originated on the posterior surface of the transverse processes and inserted on the postdentary bones as it does in lizards.

The shape of the ventral surface of the palatine and pterygoid changed fundamentally between nonmammalian cynodonts and mammals. The transverse process of the pterygoid reduced in size and lost contact with the lower jaw, thus permitting more complex jaw movements (Crompton 1995). The pterygoid hamulus could be either a remnant of the large transverse process of the pterygoid of nonmammalian therapsids or a modification of the ectopterygoid (Barghusen 1986). In mammals, the palatine and the pterygoid hamulus form the roof and sidewall of a dome-shaped nasopharynx with a muscular soft palate as its floor. The palatopharyngeal muscle originates on the posterior edge of the hard palate (palatine), hamulus, and soft palate and extends backward around the larynx to insert in the dorsal wall of the pharynx (Crompton et al. 2008; Wood Jones 1940). The tensor veli palatini is positioned lateral to the pterygoid hamulus and extends medially around the hamulus's ventral surface to insert in the anterior portion of the soft palate. The levator palatini runs ventrally to insert on the soft palate behind the insertion of the tensor veli palatini. Barghusen (1986) has shown that the tensor veli palatini and tensor tympani muscles derived from the "reptilian" pterygoideus muscle. The former could only have inserted on a soft palate by wrapping around the ventral margin of the transverse process of the pterygoid, and could only have achieved this once the transverse process no longer acted as a guide for lower jaw movement. Given these marked differences it is unlikely that the mammalian arrangement of pharyngeal muscles or an intranarial larynx was present in nonmammalian cynodonts. The origin and development of mammalian pharyngeal musculature requires further study.

Finally, temporal countercurrent exchange on turbinals works only if inspiration is quickly followed by expiration (Owerkowicz et al., this volume). This rhythm requires that a relaxed phase of breathing follows expiration (breathing out), when the lungs are in a deflated state and the animal is momentarily apnoeic (not breathing). Such is the case in extant mammals, in which the relatively stiff ribcage and muscular diaphragm allow maintenance of subatmospheric pleural pressure and prevent collapse of alveolar lungs. Hence, the ability to conserve heat and water in the nasal cavity required the presence of turbinals, but also depended on postcranial morphology. Klein and Owerkowicz (2006) argued that the costal plates and thoracolumbar differentiation in *Thrinaxodon* point to the presence of a muscular diaphragm before the origin of mammals, as originally suggested by Brink (1955). We now find this scenario unlikely, because other cynodonts do not show such postcranial morphology (Jenkins 1971; Kemp 1982), and the earliest record of a complex metameric sternum (with sternebrae) is in the Early Jurassic tritylodontids *Oligokyphus* (Kühne 1956) and *Kayentatherium* (Sues and Jenkins 2006). The costal plates of *Thrinaxodon* have previously been interpreted to allow the animal to brace the trunk against external forces (Jenkins 1971), similar to extant edentates (Jenkins 1970). This agrees with a current interpretation of *Thrinaxodon* as an adept burrower, adapted to fossorial habitats (Damiani et al. 2003).

The olfactory region of *Massetognathus* opens posteriorly into the orbital region through a wide foramen (fig. 11.5C). The frontal and postorbital form the roof of the foramen; the ventrally directed wings of the postorbital, the prefrontal, and the ascending lamina of the palatine form its lateral borders; and the transverse lamina, its floor. All blood vessels and nerves entering the nasal region passed through it. The maxillary nerve probably passed through this opening (fig. 11.5C), forward through the maxillary canal and then through the maxilla to exit onto the external surface through numerous small foramina (fig. 11.3A, B), as it does in *Procynosuchus* (Kemp 1979), *Thrinaxodon*, and several therocephalians (Estes 1961; Fourie 1974), rather than from a large infraorbital foramen as in most extant mammals (Miller et al. 1968; Wible 2003).

The *Massetognathus* orbitosphenoid is a long trough-shaped bone, which joins the frontal to form a space for the olfactory bulb (fig. 11.5A). The olfactory bulb opens into the posterodorsal aspect of the olfactory chamber. In reptiles, embryonic mammals, and presumably cynodonts as well (fig. 11.6A), the cartilaginous antorbital plate forms the part of the posterior wall of the nasal capsule (Bellairs et al. 1981; De Beer 1937; Kuhn 1971; Zeller 1987) that lies below the entrance of the olfactory nerves. In *Massetognathus*, the gap between the orbitosphenoid and the nasal cavity may have remained open as it does in reptiles and the monotreme *Ornithorhynchus* (De Beer 1937); or it may have contained a cartilaginous cribriform plate as in embryonic *Tachyglossus* (Kuhn 1971). An interorbital septum below the orbitosphenoid, ossified in some other nonmammalian synapsids (Kemp 1972, 1969), appears to have

remained cartilaginous in *Massetognathus* and other cynodonts.

The climate of the Middle to Late Triassic included arid to semiarid environments, as well as wet temperate conditions with cool winters and warm summers (Preto et al. 2010). Fauna included numerous diurnal nonmammalian cynodonts and early archosaurs. Hopson (2012) has characterized most nonmammalian cynodonts as "widely foraging" (WF), enabling carnivorous forms to search and hunt prey, and herbivorous forms to search a broad area for suitable vegetation. He concludes that WF nonmammalian cynodonts' low basal metabolic rate exceeded the basal metabolic rate of WF reptiles (e.g., varanids), and that they possessed a high maximum aerobic metabolism necessary for sustained activity over long distances. It is doubtful that free-ranging nonmammalian cynodonts could have maintained a constant body temperature if their basal metabolic rate was low. Diurnal mammals possess a high basal metabolic and breathing rate, and maintain their body temperature slightly above the average ambient temperature. This maintains a temperature gradient for the loss of body heat to the environment and minimizes the need for excessive evaporative cooling when at rest or during mild exercise. On the other hand, maintaining a constant body temperature and breathing rate when the ambient temperature drops below the preferred body temperature requires insulation, control of blood flow to the skin, and a mechanism to control the loss of water and heat in the expired air.

The structure of the palatine and pterygoid bones of nonmammalian cynodonts suggests that they did not possess the pharyngeal musculature to support an intranarial larynx, necessary to restrict the passage of respiratory airflow through the nasal cavity when at rest. It is, therefore, unlikely that cartilaginous maxilloturbinals could have acted as temporal countercurrent exchange sites. Instead, turbinates could have provided a surface for evaporative cooling as they do during panting in mammals (air in through the nose and out through the mouth) when high levels of heat are generated during exercise or when the ambient temperature exceeds the preferred body temperature. Evaporative cooling on the surface of the maxilloturbinals of mammals have been shown to reduce body and brain temperatures (Taylor and Lyman 1972). We conclude that cartilaginous respiratory turbinals arose initially to aid in cooling and only later were co-opted to act as countercurrent exchangers when an intranarial larynx arose. Although adult humans have lost the intranarial larynx, lips can seal the entrance to the oral cavity and cause both inhaled and exhaled air to pass instead through the nasal cavity.

Mammaliaformes

Mammaliaforms such as the Late Triassic/Early Jurassic *Morganucodon* show several more mammalian nasal features. We MicroCT scanned the *Morganucodon* specimen (Institute of Paleontology and Paleoanthropology, Beijing, #V8682) referred to in Rowe et al. (2011). The individual bones of this specimen are displaced and a reconstruction is currently being undertaken, but the scans show a vomer dramatically reduced in height, almost to mammalian levels, compared to nonmammalian cynodonts. The much-flattened vomer no longer divides the nasopharyngeal duct; consequently the respiratory chamber of the nasal capsule is much larger than in nonmammalian cynodonts. Kermack et al. (1981) reconstructed the internal structure of the nasal capsule of *Morganucodon* to include maxillary turbinals, a cribriform plate and a mesethmoid bone. No evidence of any of these ossified structures appear in our MicroCT scans of *Morganucodon*. Perhaps the turbinals were too fragile to have remained intact before fossilization, but if the bulkier parts of the nasal capsule such as the cribriform plate or the mesethmoid had ossified in *Morganucodon*, surely some part of these bones would have been fossilized, especially as the remaining bones of the skull are so well preserved. It appears, therefore, that in mammaliaforms the nasal capsule remained cartilaginous.

The orbitosphenoid and the olfactory bulb were, relative to skull size, considerably larger in *Morganucodon* than in *Massetognathus* (Rowe et al. 2011). It is generally agreed that some of the smaller mammaliaforms such as *Morganucodon* and *Kuehneotherium* had a diphyodont dentition and definitive growth (Hopson 1973; Luo et al. 2004). Van Nievelt and Smith (2005) suggested that this indicated that these animals suckled their young because "the period of lactation which requires no teeth eliminates the need for functional replacement early during growth or at a small size." These animals were considerably smaller than their nonmammalian ancestors. It has been suggested that they were nocturnal (Kemp 1982; Kielan-Jaworowska et al. 2004). If this is true and they retained a combination of a low basal metabolic rate and high aerobic metabolism, they may have been capable of maintaining a low constant body temperature, as several nocturnal mammals such as tenrecs and hedgehogs do, by combining a low basal metabolic rate and resting body temperatures (Crompton et al. 1978). We agree with Kemp (2006) that these extant species secondarily modified their metabolic rates to enter a nocturnal environment, but suggest that low basal metabolic and high maximum metabolic rates could have also been a strategy adopted by small nocturnal mammaliaforms and early mammals.

The vacant space in the nasal cavity of *Massetognathus* (fig. 11.6A, D) resembles the space that contains an ossified nasal capsule in the marsupial *Didelphis* (fig. 11.6B–C, 11.6E–F). The embryonic nasal septum in mammals ossifies to form the mesethmoid (fig. 11.6B, E) (Moore 1981). The compression of the vomer from a tall bone in nonmammalian cynodonts to a thin splint of bone lying below the mesethmoid increases the relative size of the respiratory chamber, possibly allowing for increased space for ossified maxilloturbinals (fig. 11.6).

The posterior region of the embryonic cartilaginous nasal capsule in mammals ossifies to form the ethmoid bone that includes the ossified ethmoturbinals, nasoturbinals, and cribriform and lateral plates (Macrini 2012; Miller et al. 1968; Rowe et al. 2005).

The ethmoturbinals occupy the chamber that forms a blind alley above the transverse lamina, above and beyond direct airflow from the nares to the nasopharyngeal passage (fig. 11.6C). This space in cynodonts undoubtedly also included olfactory turbinals that formed in the same way as they do in mammals—that is, as medial extensions of the nasal capsule's cartilaginous side wall—except that they remained cartilaginous. In embryonic mammals the cartilaginous anterior region of the nasal capsule appears close to the inner surface of the maxilla. During development, the ventral edges of the nasal capsule extend into the nasal cavity. These and membranous extensions later ossify to form maxilloturbinals, which in turn fuse directly to the maxillary bones (fig. 11.6E).

Maxilloturbinals branch repeatedly to create a large surface area, and are organized so as to reduce resistance to the airflow to and from the nasopharyngeal

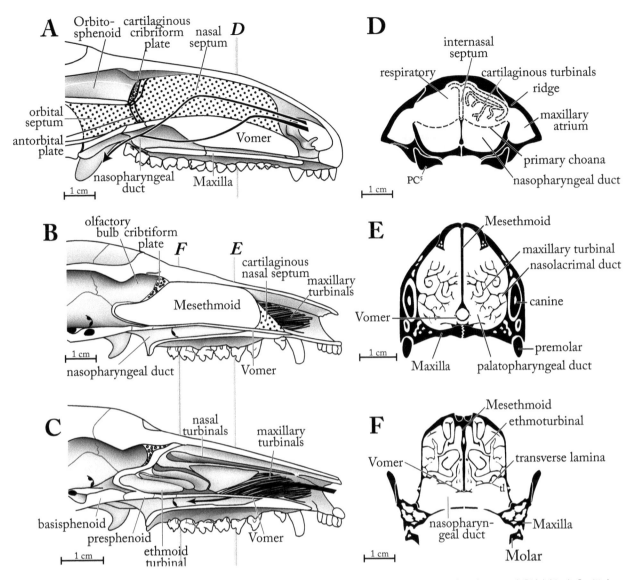

FIG. 11.6 Comparison of the nasal region of a nonmammalian cynodont, *Massetognathus*, and an extant basal mammal, *Didelphis*. A, Sagittal section through the nasal region of *Massetognathus*. Vertical line shows the position of the section shown in D; B, sagittal section through the nasal region of *Didelphis*; C, parasagittal section through the nasal region of *Didelphis*, slightly lateral of B. Vertical lines indicate the position of the sections shown in E and F.

duct. The long ridge that forms the base for the maxilloturbinals in mammals lies close to the opening of the naso-lacrimal duct (fig. 11.6E).

The first suggestion of ossified turbinals and a mesethmoid bone appear in the late Jurassic mammaliaforms docodont *Haldanodon*, as described by Lilligraven and Krusat (1991). Although their paper does not distinguish between olfactory and respiratory turbinals, its authors claim that turbinals filled the nasal cavity and that the foundations of some of these turbinal scrolls are preserved as ridges on the maxillae.

The pterygoid bones of multituberculates are, as in all mammals, widely separated from the lower jaw. The structure of the palatal region behind the hard palate in the multituberculate, *Paulchoffatia delgao* (Hahn 1987) is essentially mammalian, and suggests the presence in mid-Jurassic mammals of pharyngeal muscles and an intranarial larynx. Fragments of bone have been described in the nasal cavities of the Late Cretaceous multituberculates, *Nemegtbataar* and *Chulsanbaatar* (Hurum 1994). Hurum claims that these are the remnants of ossified turbinates. All three groups of living mammals—

Theria

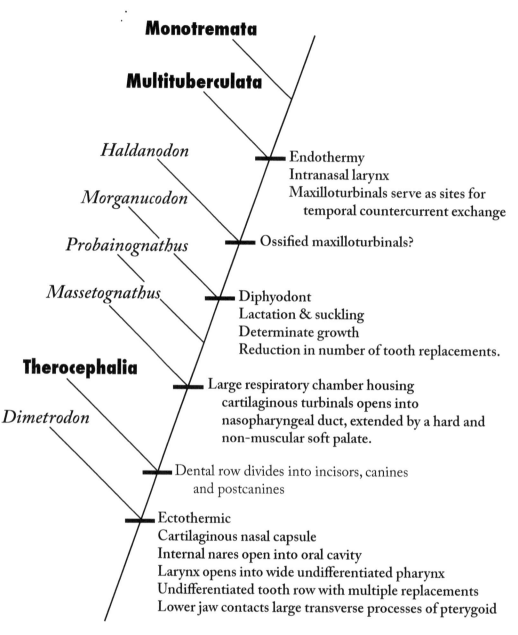

Monotremata

Multituberculata

Haldanodon

Morganucodon

Probainognathus

Massetognathus

Therocephalia

Dimetrodon

Endothermy
Intranasal larynx
Maxilloturbinals serve as sites for
    temporal countercurrent exchange

Ossified maxilloturbinals?

Diphyodont
Lactation & suckling
Determinate growth
Reduction in number of tooth replacements.

Large respiratory chamber housing
    cartilaginous turbinals opens into
    nasopharyngeal duct, extended by a hard and
    non-muscular soft palate.

Dental row divides into incisors, canines
    and postcanines

Ectothermic
Cartilaginous nasal capsule
Internal nares open into oral cavity
Larynx opens into wide undifferentiated pharynx
Undifferentiated tooth row with multiple replacements
Lower jaw contacts large transverse processes of pterygoid

FIG. 11.7 Progressive changes in the nasal, oral, and pharyngeal regions in nonmammalian synapsids.

monotremes, marsupials, and placentals (fig. 11.1)—possess ossified maxilloturbinals, so it is probable that their common ancestor, which dates back to the Middle Jurassic (Luo 2007), and probably existed alongside some mammaliaforms (e.g., *Haldanodon*), also possessed ossified maxilloturbinals and an intranarial larynx.

Diurnal mammals have constant body temperature, high basal and maximum aerobic metabolic rates, de-creased compliance of the chest matching an increased volume of the nasal cavity, an intranarial larynx, and ossified maxilloturbinals that can function either for cooling or as temporal countercurrent exchange sites. Conservation of water and heat represent only one feature necessary to support endothermy, but, we argue, maintaining constant body temperature in an environment with significant shifts in daily and seasonal ambient temperatures would only have been possible when nasal temporal countercurrent exchange sites were present together with other features necessary to support high basal and maximum metabolic rates (Hopson 2012; Kemp 2006).

Summary and Conclusions

Many questions remain about the successive changes involved in the evolution of the mammalian nose. The following steps are proposed (fig. 11.7):

1. Early Nonmammalian Synapsids

Permian nonmammalian pelycosaurs and therocephalians possessed a cartilaginous nasal capsule that opened directly into the oral cavity. This limited the space for respiratory turbinals within the airflow from the external nares to the primary choanae. They were ectothermic, and the nose was designed primarily for olfaction.

2. Nonmammalian Cynodonts

Nonmammalian cynodonts added a hard palate that may have been extended posteriorly by a non-muscular soft palate. Their respiratory chamber increased dramatically in size. They probably lacked an intranarial larynx, and, if this is so, the path of respiratory airflow during rest could not have been temporarily restricted to the nasal cavity. Consequently, they lacked temporal countercurrent exchange sites for the conservation of water and heat, necessary to maintain a constant body temperature and rhythmic breathing rate. However, their cartilaginous maxilloturbinals and oral cavity provided surfaces for evaporative cooling during panting.

3. Mammaliaformes

The structure of the nose in mammaliaforms appears to have been the same as in cynodonts, but the olfactory bulb was larger. A reduction in the height of the vomer increased the relative volume of the respiratory chamber. Their diphyodont dentition suggests determinate growth and suckling of their young.

4. Mammalia

The common ancestor of Mammalia probably had a muscular soft palate and a palatopharyngeal muscle that could hold the larynx in an intranarial position. The embryonic cartilaginous nasal capsule, including respiratory turbinates, completely ossifies in the adult. This animal had a higher basal metabolic and breathing rate than in typical reptiles and its maxilloturbinals, operating as temporal countercurrent exchange sites, controlled the loss of water and heat. Mammalian endothermy arose when these changes in nasal structure combined with a host of other specializations necessary to support a high resting metabolic rate and constant body temperature.

Future Directions

The conclusions reached in this paper are of necessity speculative due to the spotty fossil record upon which they are based, an incomplete analysis of the internal and external structures of critical regions in known fossils, and the difficulty of relying on bony structures alone for the reconstruction of soft tissues. Such shortcomings can be ameliorated in the future by using advanced imaging techniques to reexamine existing material, and continuing field work to fill in gaps in the fossil record, especially from the Early to Middle Jurassic, when the final steps in the evolution of mammals occurred. Understanding the morphology of new specimens will still require classic comparative anatomical and physiological approaches aided by the study of the function and development of comparable regions in extant animals.

* * *

References

Bang, B. G. 1961. The surface pattern of the nasal mucosa and its relation to mucous flow—a study of chicken and herring gull nasal mucosae. J. Morphol. 109:57.

Barghusen, H. R. 1986. On the evolutionary origin of the therian tensor veli palatini and tensor tympani muscles. P. 253 in The Ecology and Biology of Mammal-Like Reptiles. Washington, DC, and London: Smithsonian Institution Press.

Bellairs, A. d. A., and A. M. Kamal. 1981. The chondrocranium and the development of the skull in recent reptiles. P. 1 in C. Gans and T. S. Parsons, eds., Biology of the Reptilia: Morphology. Toronto: Academic Press.

Bennett, A. F., and J. A. Ruben. 1986. The metabolic and thermoregulatory status of therapsids. P. 207 in The Ecology and Biology of Mammal-Like Reptiles. Washington, DC, and London: Smithsonian Institution Press.

Biewener, A. A., G. W. Soghikian, and A. W. Crompton. 1985. Regulation of respiratory airflow during panting and feeding in the dog. Respir. Physiol. 61:185.

Brink, A. S. 1955. A study on the skeleton of *Diademodon*. Palaeontol. Afr. 3:3.

Brink, A. S. 1957. Speculations on some advanced mammalian characteristics in the higher mammal-like reptiles. Palaeontol. Afr. 4:77.

Crompton, A. W. 1995. Masticatory function in nonmammalian cynodonts and early mammals. P. 55 in J. Thomason, ed., Functional Morphology in Vertebrate Paleontology. Cambridge: Cambridge University Press.

Crompton, A. W., R. Z. German, and A. J. Thexton. 2008. Development of the movement of the epiglottis in infant and juvenile pigs. Zoology (Jena), 111:339.

Crompton, A. W., C. R. Taylor, and J. A. Jagger. 1978. Evolution of homeothermy in mammals. Nature (London) 272:333.

Damiani, R., S. Modesto, A. Yates, and J. Neveling. 2003. Earliest evidence of cynodont burrowing. Proc. R. Soc. Lond. Biol. 270:1747.

De Beer, G. 1937. The Development of the Vertebrate Skull. Oxford: Clarendon Press.

Estes, R. 1961. Cranial anatomy of the cynodont reptile *Thrinaxodon liorhinus*. Bull. Mus. Comp. Zool. Harvard College 125–26.

Fourie, S. 1974. The cranial morphology of *Thrinaxodon liorhinus*. Ann. South Afr. Mus. 65:337.

Geist, N. R. 2000. Nasal respiratory turbinate function in birds. Physiol. Biochem. Zool. 73:581.

Hahn, G. 1987. Neue Beobachtungen zum Schadel- und Gebiss-Bau der Paulchoffatiidae (Multituberculata, Ober-Jura). Palaeovertebrata (Montpellier) 17:155.

Hillenius, W. J. 1992. The evolution of nasal turbinates and mammalian endothermy. Paleobiology 18:17.

Hillenius, W. J. 1994. Turbinates in therapsids: evidence for Late Permian origins of mammalian endothermy. Evolution 48:207.

Hillenius, W. J., and J. A. Ruben. 2004. The evolution of endothermy in terrestrial vertebrates: who? when? why? Physiol. Biochem. Zool. 77:1019.

Hopson, J. A. 1973. Endothermy, small size, and the origin of mammalian reproduction. Amer. Nat. 107:446.

Hopson, J. A. 2012. The role of foraging mode in the origin of therapsids: implications for the origin of mammalian endothermy. Fieldiana Life Earth Sci. 5:126.

Hurum, J. H. 1994. Snout and orbit of Cretaceous Asian multituberculates studied by serial sections. Acta Palaeontol. Pol. 39:181.

Jenkins, F. A., Jr. 1970. Anatomy and function of expanded ribs in certain edentates and primates. J. Mammal. 51:288.

Jenkins, F. A., Jr. 1971. The postcranial skeleton of African cynodonts. Bull. Peabody Mus. Nat. Hist. 36:i.

Kemp, T. S. 1969. On the functional morphology of the gorgonopsid skull. Philos. Trans. R. Soc. Biol. 256:1.

Kemp, T. S. 1972. Whaitsiid Therocephalia and the origin of cynodonts. Philos. Trans. R. Soc. London Biol. 264:1.

Kemp, T. S. 1979. The primitive cynodont *Procynosuchus*: functional anatomy of the skull and relationships. Philos. Trans. R. Soc. London Biol. 285:73.

Kemp, T. S. 1980. Aspects of the structure and functional anatomy of the Middle Triassic cynodont *Luangwa*. J. Zool. (Lond.) 191:193.

Kemp, T. S. 1982. Mammal Like Reptiles and the Origin of Mammals. London and New York: Academic Press.

Kemp, T. S. 2006. The origin of mammalian endothermy: a paradigm for the evolution of complex biological structure. Zool. J. Linn. Soc. 147:473.

Kermack, K. A., F. Musset, and H. W. Rigney. 1981. The skull of *Morganucodon*. Zool. J. Linn. Soc. 71:1.

Kielan-Jaworowska, Z., R. Cifelli, and Z.-X. Luo. 2004. Mammals from the Age of Dinosaurs: Origins, Evolution, and Structure. New York: Columbia University Press.

Klein, W., and T. Owerkowicz. 2006. Function of intracoelomic septa in lung ventilation of amniotes: lessons from lizards. Physiol. Biochem. Zool. 79:1019.

Kuhn, H. J. 1971. Die Entwicklung und Morphologie des Schädels von *Tachyglossus aculeatus*. Abh Senckenb. Naturf. Ges. 528:1.

Kühne, W. G. 1956. The Liassic Therapsid *Oligokyphus*. London: Order of the Trustees of the British Museum.

Lillegraven, J. A., and G. Krusat. 1991. Cranio-mandibular anatomy of *Haldanodon exspectatus* (Docodonta; Mammalia) from the Late Jurassic of Portugal and its implication to the evolution of mammalian characters. Contrib. Geol. Univ. Wyoming, 28:39.

Luo, Z. -X. 2007. Transformation and diversification in early mammal evolution. Nature 450:1011.

Luo, Z. -X., Z. Kielan-Jaworowska, and R. L. Cifelli. 2004. Evolution of dental replacement in mammals. Bull. Carn. Mus. Nat. Hist. 36:159.

Macrini, T. E. 2012. Comparative morphology of the internal nasal skeleton of adult marsupials based on x-ray computed tomography. Bull. Am. Mus. Nat. Hist. 365:1.

Maier, W. 1999. On the evolutionary biology of early mammals—with methodological remarks on the interaction between ontogenetic adaptation and phylogenetic transformation. Zool. Anzeiger 238:55.

Maier, W., J. van den Heever, and F. Durand. 1996. New therapsid specimens and the origin of the secondary hard and soft palate of mammals. J. Zool. Sys. Evol. Res. 34:9.

Malan, M. E. 1946. Contributions to the comparative anatomy of the nasal capsule and the organ of Jacobson of the Lacertilia. Ann. Univ. Stellenbosch 24A:69.

Miller, M. E., G. C. Christensen, and H. E. Evans. 1968. Anatomy of the Dog. Philadelphia: W. B. Saunders.

Moore, W. J. 1981. The Mammalian Skull. Cambridge: Cambridge University Press.

Musinsky, C. 2012. Evolution of the Mammalian Nose. Vimeo .com/43930404.

Parsons, T. S. 1959. Nasal anatomy and the phylogeny of reptiles. Evolution 13:175.

Pratt, C. W. M. 1948. The morphology of the ethmoidal region of *Sphenodon* and lizards. Proc. Zool. Soc. London 118: 171.

Preto, N., E. Kustatscher, and P. B. Wignall. 2010. Triassic climates—state of the art and perspectives. Palaeogeog. Palaeoclim. Palaeoecol. 290:1.

Romer, A. S., and L. I. Price. 1940. Review of the Pelycosauria. Geol. Soc. Am. Washington Spec. Pap. 28:18.

Rowe, T. B., T. P. Eiting, T. E. Macrini, and R. A. Ketcham. 2005. Organization of the olfactory and respiratory skeleton in the nose of the gray short-tailed opossum *Monodelphis domestica*. J. Mamm. Evol. 12:303.

Rowe, T., and J. Gauthier. 1992. Ancestry, palaeontology, and definition of the name Mammalia. Syst. Biol. 41:372.

Rowe, T. B., T. E. Macrini, and Z.-X. Luo. 2011. Fossil evidence on origin of the mammalian brain. Science 332:958.

Ruben, J., 1996. Evolution of endothermy in mammals, birds and their ancestors. P. 349 in I. A. Johnston and A. F. Bennett, eds., Animals and Temperature: Phenotypic and Evolutionary Adaptation. Cambridge: Cambridge University Press.

Ruben, J. A., W. J. Hillenius, N. R. Geist, A. Leitch, T. D. Jones, P. J. Currie, J. R. Horner, and G.E. Espe III. 1996. The metabolic status of some Late Cretaceous dinosaurs. Science 273:1204.

Ruben, J. A., W. J. Hillenius, T. S. Kemp, and D. E. Quick. 2012. The evolution of mammalian endothermy. P. 272 in A. Chinsamy-Turan, ed., Forerunners of Mammals: Radiation—Histology—Biology. Bloomington: Indiana University Press.

Schmidt-Nielsen, K. 1981. Countercurrent systems in animals. Sci. Am. 244:100.

Sues, H.-D. 1986. The skull and dentition of two tritylodontid synapsids from the Lower Jurassic of western North America. Bull. Mus. Comp. Zool. 151:217.

Sues, H.-D., and F. A. Jenkins Jr. 2006. The postcranial skeleton of *Kayentatherium wellesi* from the Lower Jurassic Kayenta Formation of Arizona and the phylogenetic significance of postcranial features in tritylodontid cynodonts. P. 114 in M. T. Carrano, T. J. Gaudin, R. W. Blob, and J. R. Wible, eds., Amniote Paleobiology: Perspectives on the Evolution of Mammals, Birds, and Reptiles. Chicago: University of Chicago Press.

Taylor, C. R. 1977. Exercise and environmental heat loads: different mechanisms for solving different problems? Int. Rev. Physiol. 15:119.

Taylor, C. R., and C. Lyman 1972. Heat storage in running antelopes: independence of brain and body temperatures. Am. J. Physiol.—Legacy Content, 222:114.

Thomason, J. J., and A. P. Russell 1986. Mechanical factors in the evolution of the mammalian secondary palate: a theoretical analysis. J. Morphol. 189:199.

van Nievelt, A. F. H., and K. K. Smith 2005. To replace or not to replace: the significance of reduced functional tooth replacement in marsupial and placental mammals. Paleobiology 31:324.

Van Valkenburgh, B., A. Curtis, J. X. Samuels, D. Bird, B. Fulkerson, J. Meachen-Samuels, and G. J. Slater 2011. Aquatic adaptations in the nose of carnivorans: evidence from the turbinates. J. Anat. 218:298.

Wible, J. R. 2003. On the cranial osteology of the short-tailed opossum *Monodelphis brevicaudata* (Didelphidae, Marsupialia). Ann. Carnegie Mus. 72:137.

Wood, Jones, F. 1940. The nature of the soft palate. J. Anat. 74: 147.

Zeller, U. 1987. Morphogenesis of the mammalian skull with special reference to *Tupaia*. P. 17 in H. J. Kuhn, D. A. N. Hoyte, and U. Zeller, eds., Morphogenesis of the Mammalian Skull. Göttingen: Paul Parey.

12

Placental Evolution in Therian Mammals

Kathleen K. Smith*

Introduction

One of the great transformations in the history of vertebrates involves the origin and early diversification of the mammals. Mammals are distinguished from other vertebrates by almost countless aspects of their anatomy, physiology, behavior, reproduction, and life history. Many of these features appear to have arisen as an integrated complex during the early evolution of mammals. An exceedingly detailed fossil record has aided our understanding of this transition, and decades of work have led to an understanding of many of the unique adaptations of this group. In particular, the intimate interdependence of reproduction, anatomy, and physiology in mammals and their roles during the transition to the mammalian condition has presented a fascinating puzzle to vertebrate biologists (e.g., Crompton 1980; Crompton and Jenkins 1973; Crompton et al. 1978; Farmer 2000; Guillette and Hotton 1986; Hopson 1973; Jenkins 1984; Kemp 2006, 2007; Koteja 2012; McNab 1978; Pond 1977; Ruben 1995).

While the origin of mammals has provided one set of puzzles, the diversification of mammals is also ripe with questions. Three extant major groups of mammals exist. Monotremes retain many primitive features and are generally considered to have diverged from the therians in the Triassic to Jurassic periods (e.g., O'Leary

* Department of Biology, Duke University

et al. 2013, Luo 2007). The therians comprise the eutherians[1] or placentals and the metatherians or marsupials; these groups diverged sometime in the Jurassic to early Cretaceous periods (e.g., O'Leary et al. 2013, Luo 2007). Among the living mammals differences in reproductive strategies are perhaps the most intriguing. By definition all mammals nourish young postnatally by maternal provision of milk; however, other aspects of reproduction differ. Monotremes are **oviparous**[2] and the hatchling is **altricial**. Both therian groups are **viviparous**, but differ in other aspects of reproductive strategy. Marsupials give birth to highly altricial young after a short period of maternal nutrition via a **placenta**. Eutherian newborns are all born after a relatively long or intensive period of intrauterine development with a well-developed placenta and may be altricial or **precocial**. These differences in reproductive strategy are among the most consistent defining characteristic of these clades. Although these differences had been recognized for well over a century, in the late 1970s and early 1980s a particularly lively debate on the potential reasons for and consequence of these differences arose.

Much of this debate was stimulated by Lillegraven (Lillegraven 1969, 1975, 1984, 2004; Lillegraven et al. 1987) who sought the causes and consequences of the "marsupial-placental dichotomy." Lillegraven proposed that the critical elements differentiating the two clades arose from the differences in the capacities of the placenta. He maintained that the eutherian placenta differed from all other vertebrates in allowing "intimate apposition of fetal and maternal tissues and circulatory systems" with "sustained, active morphogenesis" accompanied by immunological protection of the fetus from the mother. Lillegraven proposed that this capacity was due to the **trophoblast**, which he believed was a unique "invention" of eutherians. He argued that because marsupials lacked a trophoblast they were unable to maintain a prolonged, active intrauterine gestation and therefore were constrained to a primitive state of reproductive capacity. Lillegraven further hypothesized

that this reproductive limitation imposed constraints on the subsequent development and evolution of marsupials. Finally he argued that the eutherian placenta was a key innovation and the ultimate source of many of the characteristics that have led to the great diversification and relative dominance of eutherians today (Lillegraven et al. 1987).

This view was countered by several authors (e.g., Hayssen et al. 1985; Kirsch 1977a, 1977b; Low 1978; Parker 1977) who believed that it was unlikely that marsupial biology was the result of being stuck at an intermediate stage of evolution or was fundamentally constrained. Instead, they proposed, the marsupial mode of reproduction should be thought of as an alternative strategy that provided greater maternal control of reproductive resources. Additionally, these authors argued that although marsupials exhibit less overall diversity than eutherians, they exhibited a large range of anatomical, ecological, and physiological adaptations, casting doubt on the hypothesis that their evolution was fundamentally constrained. Tyndale-Biscoe and Renfree (e.g., Renfree 1983, 1993, 1995, 2010; Tyndale-Biscoe 2005; Tyndale-Biscoe and Renfree 1987) portrayed reproduction in monotremes, marsupials, and eutherians as a continuum, with underlying similarity but differing emphases on maternal nutrition via the placenta or lactation. They emphasized that marsupials, by placing emphasis on lactation, lacked some of the adaptations of eutherians for intrauterine gestation, but also possessed a number of distinct adaptations of their own, particularly for specialized lactation. Examination of the consequences of the reproductive differences has continued to the present time with a large number of recent studies exploring comparative development in marsupial and eutherian mammals and its evolutionary consequences (e.g., Bennett and Goswami 2011; Cooper and Steppen 2010; Goswami et al. 2009, 2012; Kelly and Sears 2011; Keyte and Smith 2010, 2012; Sánchez-Villagra 2013; Sánchez-Villagra et al. 2008; Sánchez-Villagra and Maier 2003; Sears 2004, 2009; Smith 1997, 2001, 2006; Vaglia and Smith 2003; van Nievelt and Smith 2005; Weisbecker 2011; Weisbecker and Goswami 2010; Weisbecker et al. 2008; Weisbecker and Sánchez-Villagra 2007).

While there have been a number of studies that have addressed hypotheses on the development, adaptations, and diversity of marsupials, there has not been as much focus on the hypothesis that the trophoblast and placenta are unique innovations of eutherians. In

---

1. The terms marsupial and eutherian will be used in an informal sense throughout this paper. Properly the crown group of extant eutherians should be termed the Placentalia (Rougier et al. 1998; Asher and Helgen 2010), equivalent to the crown group Marsupialia. However, as this paper is on the evolution of the placenta, the term eutherian will be used for the crown group to avoid awkward constructions such as "the placenta in placentals".

2. Terms in bold are defined in table 12.1.

recent years a great deal of new data on the basic biology of reproduction in mammals has emerged. These data provide new insight into the anatomy, physiology, and phylogeny of the placenta within mammals and in nonmammalian amniotes and allow evaluation of this fundamental hypothesis. In addition, in the last several decades, entirely new research areas arising from the discovery of genomic imprinting and the proposal of the theory of maternal-fetal conflict have emerged. Therefore the time is ripe for a fresh look at the issues surrounding the evolutionary patterns and significance of reproductive strategies in mammals.

In this review I will examine data on the function and evolution of the placenta in therian mammals. My intent is not to revisit the original debate or to provide a detailed critique of past discussions, but instead I will focus on the evolution of the placenta. I will summarize new data on the variation in placental morphology in both marsupials and eutherians, the relation between placental form and development, and new phylogenetic analyses of placental form. I will then discuss comparative aspects of reproduction in marsupials and eutherians with particular focus on maternal-fetal conflict and maternal control of reproductive effort. My hope is to shed new light on the significance of variation in reproductive patterns in therian mammals and ultimately on the role that reproduction may have played in the origin and diversification of mammals.

The Placenta: A Brief Overview

In vertebrates **viviparity** has evolved many times. In some taxa this includes the evolution of a specialized interface, the placenta, between fetal and maternal tissues for the transfer of nutrients. In others it consists of retention of the egg or embryo with little or no nutrition provided by the mother beyond the initial yolk. In a few taxa viviparity involves adaptations for embryonic consumption of maternal tissues, specialized eggs, or littermates (reviewed in Blackburn 2006; Shine 1995; Wake 1993). In amniotes viviparity is common in squamates, where it has evolved independently at least 100 times, and characteristic of therian mammals. In most viviparous squamates nutrition is still largely provided by the yolk, but in many, fetal membranes are specialized for gas, water, and mineral exchange (reviews in Blackburn 2006 and Blackburn and Flemming 2009;

Flemming and Blackburn 2003; Stewart and Thompson 2003; Thompson and Speake 2006). In all mammals and many squamates the mother provides nutrition beyond the yolk. This nutrition can be provided in two major ways: **histotrophically**, where specialized uterine secretions are absorbed by extraembryonic or placental tissues, and **hematrophically** where nutrients, waste products, and gas are exchanged between maternal and fetal blood supplies.

In therian mammals and many squamates the **placenta** facilitates the transfer of maternal provision to the fetus. Placentae are by most definitions structures consisting of the apposition of specialized maternal and fetal tissues for physiological exchange (e.g., Mossman 1937). In mammals and other amniotes the fetal side of the placenta generally develops from the specialized extraembryonic membranes of amniotes (the primitive **yolk sac** along with the **chorion**, **amnion**, and **allantois**) while the maternal side involves specializations of the endometrium. It is important to note that a specialized placenta is not a eutherian or even a mammalian invention, but instead is seen in many vertebrates. In squamates the convergence is "astonishing" (Blackburn and Flemming 2009) and can exist on the morphological, physiological, and genetic level (e.g., Brandley et al. 2012). The anatomy and function of the placenta in mammals have been well studied; these studies reveal extraordinary morphological and functional variability. Comprehensive reviews of placental morphology may be found in many sources (e.g., Carter 2001; Carter and Mess 2007, 2008; Carter et al. 2004; Enders and Carter 2004, 2012a, 2012b; Freyer et al. 2001, 2003, 2007; Freyer and Renfree 2009; Luckett 1977; Mess and Carter 2006, 2007; Mossman 1937; Wooding and Burton 2008 and references therein). Here I briefly introduce terminology important for the remaining discussion.

As mentioned above the placenta is a structure that depends on the interaction of fetal and maternal elements. The fetal elements arise from the extraembryonic membranes of amniotes, the **chorion**, the **allantois**, and the **amnion.** Together with the yolk sac these membranes combine to form the membranes of the egg in egg-laying species, or the fetal contribution to the placenta in animals with a placenta. In both egg-laying and placental amniotes the amnion surrounds the embryo, while the chorion may fuse with either the yolk sac (vitelline membrane) or allantois to form the choriovitelline or chorioallantoic membrane, respectively. In animals with placentae, these membranes are closely apposed to maternal epithelium to form chorioallantoic or choriovitelline placentae (table 12.1; fig. 12.1). In many egg-laying amniotes, including most squamates as well as monotremes, the chorioallantoic and choriovitelline membranes are both well developed; the chorioallantois thought to be largely responsible for gas exchange and the choriovitelline membrane for nutrition.

A yolk sac or **choriovitelline placenta** is present in early development in both marsupials and eutherians (Freyer and Renfree 2009). In addition to serving as an early site of nutrient exchange, it is important in production of growth factors, binding proteins, and receptors, as well as hematopoiesis, cholesterol production, and transfer of nutrients in all mammals. It forms the final functional placenta in most marsupials. In most eutherians, the yolk sac is reduced, and the **chorioallantoic placenta** is the final functional placenta.

Chorioallantoic placentae are found in several marsupials as well, at least transitorily, and in peramelids the chorioallantois contributes to the final functional placenta.

In therians there is also significant variation in the degree that fetal tissues invade the uterine lining. This trait is independent of the specific fetal membranes contributing to the placenta (table 12.1; fig. 12.2). The least invasive type, the **epitheliochorial placenta**, involves contact between the epithelia of the chorion and the uterus. There is no erosion or invasion of the uterine lining by fetal tissues, and nutrients diffuse through several cell layers. In **endotheliochorial** placentae embryonic tissues invade the uterine lining and are in contact with the endothelium of the maternal blood vessels. Finally, in the most invasive condition, the **hemochorial placenta**, fetal tissues have eroded maternal blood vessels, so that maternal blood accumulates in spaces within the **chorioallantois**. These categories are often further subdivided. Tremendous variation and also

TABLE 12.1 A glossary of important terminology

General characteristic	Important terminology	Functional notes
Reproductive type	Viviparous	Female gives birth to live young. This term includes a wide range of conditions from ovoviviparity, where eggs are retained within oviduct or uterus with no provision of nutrients beyond initial yolk, to placental viviparity where elaborate adaptations exist for maternal nutrient provision. Many intermediate forms exist.
	Oviparous	An egg is laid and no further nutrition provided by mother before hatching.
Form of newborn young	Altricial	Young is small and poorly developed at birth. In general eyes are closed, and minimal thermoregulation and locomotion capacity exist. Requires a long period of postnatal care by parents.
	Precocial	Young is relatively independent at birth: eyes open, generally capable of locomotion. May be provisioned for variable period by parent.
Placenta		A structure composed of fetal and maternal tissue adapted for physiological exchange. May have many specific forms and may be responsible for many kinds of exchange.
Type of intrauterine provision of nutrients	Histotrophic	Secretions from the uterine lining are absorbed by vasculature in fetal membranes.
	Hematrophic	Placental adaptations exist for fetal blood supply to interact with maternal blood supply and absorb nutrients and gasses directly from it.
Precursor to extraembryonic membrane	Trophectoderm or trophoblast	Portion of blastocyst that gives rise to ectodermal portions of the chorion. Two terms generally considered to be homologous (for further information see table note).

TABLE 12.1 (continued)

General characteristic	Important terminology	Functional notes
Extraembryonic membrane (primitive)	Yolk sac membrane	Initially formed by enclosure of yolk (or hollow blastocyst) by mesodermal cells from embryonic disc; migration of endoderm forms bilaminar yolk sac. Primitive function nutritive.
Extraembryonic membrane (derived in amniotes; their presence defines the cleidoic egg)	Chorion	Develops from trophectoderm and extraembryonic mesoderm. Generally nonvascular. Primitive function gas exchange.
	Allantois	Forms from extraembryonic endoderm and mesoderm. Readily vascularizes. Primitive function waste removal.
	Amnion	Forms either from the chorion or extraembryonic ectoderm. Forms protective cavity around developing embryo.
Membrane/placental type	Choriovitelline	Membrane is formed from the yolk sac membrane and chorion. In some species/some regions a second invasion of mesoderm vascularizes the yolk sac to form trilaminar yolk sac. When this membrane forms the embryonic contribution to the placenta, it is called a choriovitelline placenta.
	Chorioallantoic	Membrane is formed from chorion and allantois. Vascularization is via allantois. When this membrane forms the embryonic contribution to the placenta, it is called a chorioallantoic placenta.
Fetal-maternal interface	Epitheliochorial	No layers removed; epithelium of uterus in contact with epithelium of chorion.
	Synepitheliochorial (syndesmochorial)	Syncytium of maternal-fetal cells at interface (a subdivision of epitheliochorial placentation).
	Endotheliochorial	Erosion of uterine lining by trophectoderm so that chorion is in contact with the endothelium of maternal blood vessels.
	Hemochorial	Maternal blood in direct contact with chorion and placental tissues derived from fetus.
Nature of interdigitation	Villous	The geometry of the contact between maternal and fetal tissues in the placenta. The categories listed represent increasing surface area of contact between maternal and fetal tissues. Many intermediate conditions.
	Folded trabecular	
	Lamellar trabecular	
	Labyrinthine	
Shape of placenta	Diffuse	Maternal-fetal interface scattered on surface of chorion.
	Cotyledonary	Areas of contact limited to distinct cotyledons.
	Zonary	Fetal portion of placenta contacts uterus in a central band.
	Discoid	Distinct disc for contact between fetus and uterus.
Cell fusion	Syncytium	Fusion of cells into a multinucleate tissue. May be fusion of fetal (chorional) cells only, or fusion of fetal cells with maternal cells. Appears to require capture of retroviral gene syncytium by host genome.

Note: Major terms and features of reproductive biology and placental morphology used in this paper. See Wooding and Burton (2008) and references in text for more information on placental types. The terminology for the trophectoderm/trophoblast deserves particular note. Although Lillegraven (2004) has continued to argue that the trophoblast of eutherians is unique, virtually all workers on early development consider the trophoblast and the trophectoderm, a more general term, to be homologous (e.g., Johnson 1996; Johnson and Selwood 1996; Moffett et al. 2006; Renfree 2010; Selwood and Johnson 2006; Taylor and Padykula 1978; Wooding and Burton 2008). Indeed, Hubrecht, who first coined the term trophoblast, explicitly includes the extraembryonic membrane of marsupials in his definition of the trophoblast (Pijnenborg and Vercruysse 2013).

homoplasy exists within this simple scheme (e.g., Enders and Carter 2012a, 2012b; Wooding and Burton 2008). It is important to note that hematrophic nutrition does not require a hemochorial placenta.

In addition to the degree of tissue invasion there is variation in the way maternal and fetal tissues interdigitate as well as in the overall shape of the placenta (table 12.1). For example, the specific geometry of contact between maternal and fetal tissues may be **villous**, **trabecular**, or **labyrinthine** (table 12.1). These different arrangements of interdigitation change the relative surface area in contact between maternal and fetal tissues and may be of major significance in placental

function (Capellini 2012). The three features of placental morphology—degree of invasiveness, type of interdigitation, and general shape—show some phylogenetic correlation, but also independent evolution (e.g., Capellini 2012; Elliot and Crespi 2009; Wildman et al. 2006).

A final feature of interest is the formation of **syncytia** within placental tissues. A syncytium is the fusion of cells to form a multinucleate tissue. In many species syncytia form in placental tissues, and in a few (e.g., eutherian ruminants, marsupial peramelids) a syncytium of maternal and fetal cells forms (Wooding and Burton 2008). It is thought that this acellular tissue allows more efficient transfer of nutrients. In recent years it has been

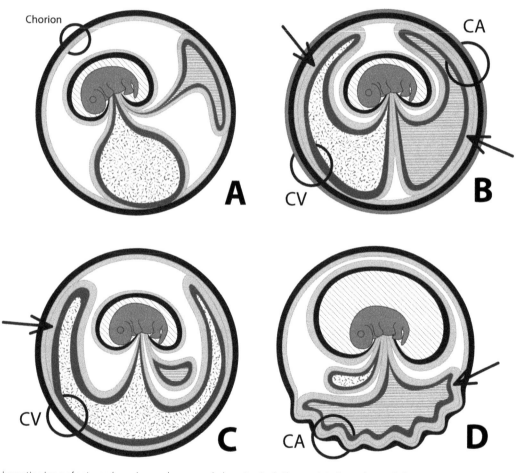

FIG. 12.1 Schematic views of extraembryonic membranes and placentae in A, the amniote "morphotype"; B, monotremes (a similar arrangement is seen in squamates); C, most marsupials; and D, eutherians. The primitive amniote condition with both chorioallantoic (CA) and choriovitelline (CV) membranes are also seen in most squamates and also monotremes. In these animals the chorioallantoic membranes are largely respiratory while the choriovitelline membranes are important in nutrition. In marsupials and eutherians (and some squamates) these extraembryonic membranes join with the uterine lining to form placentae. In many marsupials the chorioallantoic membrane is present for part of gestation and is replaced by the choriovitelline membrane as the final placenta. In all illustrations the darkest gray represents membranes derived from ectoderm, medium gray from endoderm, and light gray from mesoderm. The cavity with stippled lines is the amnion, irregular stippling is the yolk sac, and solid lines indicate the allantois. The chorion forms the outermost layer in all forms. The arrows point to regions where a second layer of mesoderm has invaded, providing vascularization and contributing to a trilaminar placenta. Redrawn from Ferner and Mess (2011).

maternal tissues

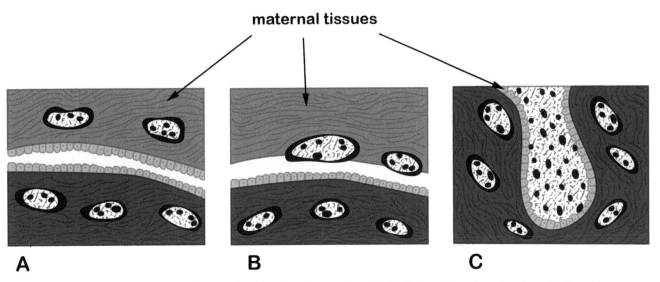

FIG. 12.2 Schematic view of the layers of maternal and fetal tissue in placentae. A, Epitheliochorial placentae, where the epithelium of the maternal (uterine lining/decidua) and fetal tissues (chorion) are in contact; B, endotheliochorial placentae, where the fetal tissue (chorion) is in contact with the endothelium of the maternal blood vessels; C, hemochorial placentae where maternal blood circulates in spaces within the fetal tissue. In all illustrations, small cells represent the epithelial layer; light gray, the tissue the maternal placenta/decidua; and dark gray, the fetal/placental tissue. The circular forms are red blood vessels or blood cells. Redrawn from Ferner and Mess (2011).

shown that in several eutherians the formation of syncytia depends on the presence of the retroviral gene syncytin. Interestingly, retroviral syncytin genes seem to have been independently captured by primates, rodents (mice), lagomorphs, and carnivores, and it is presumed that they will be present in all mammals with placental syncytia although they have not yet been identified in marsupials (e.g., Blaise et al. 2003; Cornelis et al. 2012; Harris 1998; Heidmann et al. 2009; Mi et al. 2000). The capture of retroviral genes has been proposed to have a general role in the evolution of the placenta (e.g., Chuong 2013; Haig 2012).

Evolutionary Patterns: Marsupials

The marsupial constraint hypothesis claims that marsupials are limited to a short intrauterine period, an inefficient yolk sac placenta, and altricial young because they are unable to develop the critical adaptations for intimate and prolonged gestation seen in eutherians. It is almost impossible to confirm or refute a claim of evolutionary constraint; however, one may look at the corollaries of this hypothesis to see if they are supported by variation in existing organisms. In particular, an implication of this hypothesis is that more "eutherian-like" conditions are derived, while the primitive condition in marsupials would reflect the constraints of

short gestation, highly altricial young, and simple placentae. More specifically, we can make the following predictions:

1) in marsupials, highly altricial young would be the primitive condition and more precocial young the derived condition.

2) close maternal-fetal interaction (i.e., invasive, syncytial placentae) would be a derived condition within marsupials.

3) increased invasiveness of the placenta would correlate with a more precocial young, as would elaboration of mechanisms for hematrophic, as opposed to histotrophic, nutrition.

The Altricial Precocial Spectrum in Marsupials

When compared to eutherians, the marsupial neonate appears highly altricial and relatively invariant. Nonetheless, variation exists. Hughes and Hall (1988) define three grades of marsupial newborns that they termed G1, G2, and G3 (table 12.2; fig. 12.3). These three grades correlate with the time between primitive streak formation and birth and also neonatal size. They represent distinct levels of development, with external features, musculoskeletal elements, and internal organs (e.g., lungs, digestive

TABLE 12.2 Major characteristics of grades 1, 2, and 3 in marsupial neonates

	Grade 1	Grade 2	Grade 3
Examples	Dasyuridae *Tarsipes* (?)	Peramelids Didelphids *Trichosurus*	*Macropus* species
Neonatal weight	3–20 mg	100–300 mg	300–900 mg
Eye	No retinal pigmentation, no visible eye primordial or lids	Pigmentation, visible primordial	Prominent primordial, pigmented ring
Manus	Metacarpals just differentiating	Intermediate	Phalanges and metacarpals separated by joint capsules
Hind limb primordial	Undifferentiated paddle	Some digit differentiation	Early digital separation
Lung	Few partitions, superficial capillaries	Intermediate	Enlarged lung, highly subdivided, rich vascularization
Tongue muscles	Multinucleate tubes; immature striations	Striations	Mature striations
Metanephric kidney	Ureteric bud	Primitive ureter, secondary branching	Terminal branching, collecting ducts

Sources: Characters from Hughes and Hall (1988); body weight from Tyndale Biscoe and Renfree (1987).

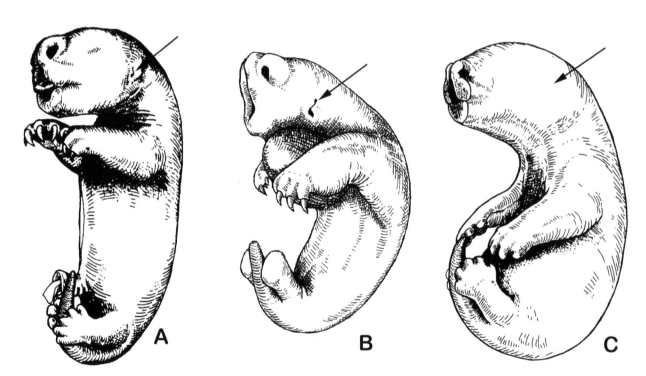

FIG. 12.3 Neonates representing the altricial precocial spectrum in marsupials. A, Precocial marsupial, *Macropus eugenii*; B, ultra-altricial marsupial neonate, *Dasyurus viverrinus*; C, "intermediate" marsupial, *Monodelphis domestica*. These are not drawn to scale, but illustrate the relative differentiation of cranial and limb structures at birth in each taxon. The arrows represent the ear opening; note the lack of features between the ear and the nasal opening in the dasyurid. Also note the extreme difference in development of the fore- and hind limb in the ultra-altricial dasyurid.

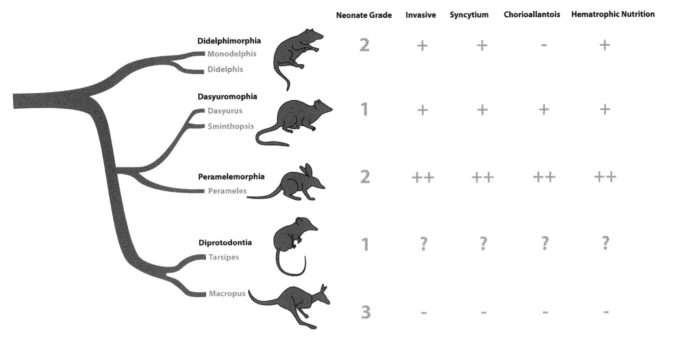

	Neonate Grade	Invasive	Syncytium	Chorioallantois	Hematrophic Nutrition
Didelphimorphia Monodelphis Didelphis	2	+	+	−	+
Dasyuromophia Dasyurus Sminthopsis	1	+	+	+	+
Peramelemorphia Perameles	2	++	++	++	++
Diprotodontia Tarsipes	1	?	?	?	?
Macropus	3	−	−	−	−

FIG. 12.4 Major characteristics of marsupial development and placenta form plotted on a phylogeny (redrawn from Nilsson et al. 2010). The cladogram of marsupials is highly reduced, with many clades missing. Neonatal grade represents the altricial precocial spectrum as shown in figure 12.3, with 1 representing the most altricial neonates, and 3 the best developed. The other columns represent invasive placentae (i.e., evidence of an endotheliochorial placenta), the presence of syncytium, the presence (at least transitorily) of a chorioallantoic membrane. A plus sign (+) represents a character that is closer to the eutherian condition (with ++ representing much closer), while a *dash* (-) represents a character that is further from the condition seen in eutherians. The final category represents whether or not evidence exists for hematrophic nutrition, again a condition thought to be primitive for eutherians. This illustration demonstrates that the most precocious neonates develop with placentae least like the primitive eutherian condition.

tract, and UG systems) at differing levels of functionality at birth. At one end of the spectrum are ultra-altricial species such as *Sarcophilus* (the Tasmanian devil) and other dasyurids. These animals often weigh significantly less than 20 mg at birth and are born after an extremely short gestation. Anatomical systems are minimally developed. At the other end are many macropodids, weighing between 300 and 900 mg at birth, with relatively long periods of development and relatively well-developed anatomy and physiology. The intermediate grade is most common and includes peramelids, most phalangers, didelphids, and representatives of many other taxa. These neonates are generally intermediate in developmental time, morphological development, and weight (table 12.2).

These three grades of altriciality do not represent an evolutionary gradation within marsupials. The "intermediate" condition appears to be primitive (fig. 12.4). It is found consistently in the didelphids, the basal group of marsupials, and is also widespread across the marsupials. Both the relatively precocial grade and also the ultra-altricial state are derived. The ultra-altricial

condition appears to have evolved twice as it is seen in dasyurids and also in the honey possum, *Tarsipes rostratus*. This latter species is embedded within the clade that includes the macropodids (Kavanagh et al. 2004). Newborn *T. rostratus* weigh 3–6 mg (about the weight of a single, dry grain of rice), with an unknown, but "very brief" period of active gestation (Renfree et al. 1984). The fact that an extremely short gestation period and an ultra-altricial young is derived within marsupials, more than once, strongly suggests that ultra-altriciality is of evolutionary advantage, the result of selection, and not simply the result of primitive constraints due to an inefficient placenta.

Placentation in Marsupials

Although elements of the anatomy of the placenta in marsupials have been studied since Hill in the early 1900s, recent work has added considerably to our understanding and allows a deeper and more comprehensive view of placental evolution in marsupials. This work has shown that in marsupials although the placenta is functional for a relatively brief time, it shows a wide range of variation in morphology and function and is well developed in many species.

A **choriovitelline** or yolk sac placenta is present at least as a transitory structure in both marsupials and eutherians (Freyer and Renfree 2009). In most marsupials the choriovitelline placenta forms the functional placenta throughout gestation. In bandicoots, dasyurids, and wombats a **chorioallantois** is also present at least transitionally (Freyer et al. 2003; Freyer and Renfree 2009; Hughes 1974; Tyndale-Biscoe and Renfree 1987). The broad phylogenetic distribution of a chorioallantois as well as its importance in monotremes and viviparous squamates has led Freyer et al. (2003) to speculate that a chorioallantoic contribution to the placenta may have been primitively present in marsupials and subsequently lost in many.

In many marsupials the placenta is **epitheliochorial** as there is no loss of tissue layers between the chorion and the maternal circulation. However, in many marsupials there is erosion of the epithelium of the uterus by the fetal tissues, to produce at least a partial **endotheliochorial** placenta. Hill (1900) first described the invasive placenta of dasyurids (see also Hughes 1974 and Roberts and Breed 1994). In *Dasyurus viverrinus* placental cell processes grow to enclose the capillaries underlying the maternal epithelium. Fetal tissue is fused into a **syncytium**, and maternal blood is in close approximation to the yolk sac and chorion. The same condition of eroded maternal epithelium by placental tissues is also seen in the dasyurid, *Sminthopsis crassicaudata* (Hughes 1974; Roberts and Breed 1994). In *Monodelphis* the choriovitelline placenta is also invasive (Zeller and Freyer 2001); it penetrates the maternal epithelium at regular intervals, leading to close association of fetal tissues and maternal capillaries. In *Monodelphis* as in dasyurids, syncytia form in fetal tissues. Recent studies (i.e., McGowen et al. 2013) have shown that the placenta in marsupials such as *Monodelphis* expresses a wide range of genes and that it is metabolically active with many adaptations for the transfer of nutrients between maternal and fetal tissues.

Perameles has the most invasive placenta of any marsupial with a highly vascularized invasive **chorioallantoic** membrane. This **endotheliochorial chorioallantoic** placenta forms the functional placenta through to the final stages of gestation (Freyer et al.

2003; Hughes 1974; Padykula and Taylor 1976, 1982). As in some other taxa a syncytium forms, but, in *Perameles* the syncytium results from a fusion of maternal and fetal cells. This forms a single tissue layer, which provides intimate contact between fetal cells and maternal circulation.

In contrast to the above species, *Macropus eugenii* and other macropodids lack most of these adaptations for close maternal-fetal interaction (Freyer et al. 2007). The placenta is solely formed from a bilaminar choriovitelline membrane, is not invasive, and does not form syncytia.

Histotrophic versus Hematrophic Nutrition

In monotremes the egg receives **histotrophic** nutrition before being laid, and it is often assumed that histotrophic nutrition is primitive in therians. Indeed, Lillegraven (1969) asserted that with the possible exception of *Perameles*, marsupials were limited to histotrophic nutrition while in utero. It appears that initially there is significant histotrophic nutrition in all marsupial taxa. However, in *Monodelphis*, for example, histotrophic secretions are replaced by **hematrophic** nutrition as the trilaminar choriovitelline membrane invades the uterine lining and establishes close association between fetal tissues and maternal circulation (Zeller and Freyer 2001). It is likely that all marsupials with invasive, vascularized placentae such as peramelids, dasyurids, and some didelphids utilize significant hematrophic nutrition. In contrast in *M. eugenii* and other macropodids, hematrophic nutrition never develops, but instead, *M. eugenii* develops extensive and enhanced adaptations for increased histotrophic nutrition (Freyer et al. 2007). These include well-developed uterine glands for secretion and high levels of nutrients in the fluids of the uterus.

The phylogenetic distribution of these traits allows us to trace evolutionary patterns in marsupials and arrive at tentative conclusions about potential constraints. First, both chorioallantoic and invasive (endotheliochorial) placentae appear broadly in marsupials. This type of placenta has been thought to be characteristic of, and indeed limited to, eutherians, but it is possible that both a functional chorioallantois and an endotheliochorial placenta are primitive in marsupials (Freyer et al. 2003). Second, peramelids are uniformly recognized as having the most "advanced" or eutherian-like placentae. However, they do not use their advanced placentae to produce precocial young, but instead the invasive placenta allows them to reproduce particularly fast. As pointed out by Russell (1982) the rapidity of reproduction in bandicoots is remarkable. Their gestation length is short (12–13 days); weaning time is early (50–60 days), and the delay between weaning and the birth of the next litter inconsequential (peramelids mate while lactating and give birth immediately after a litter is weaned). The possession of highly invasive placentae does not correlate with the supposed advantageous long gestation or precocial young but instead rapid reproduction. Third, invasive, syncytial placentae are also seen in dasyurids, whose young are ultra-altricial, again arguing against a relation between precociality and placenta type.

Finally, the most precocial marsupial young are produced by the placentae that least resembles the "eutherian condition" as envisioned by Lillegraven. Macropodids produce relatively large, relatively precocial, single young. The placenta is bilaminar, noninvasive, and does not form syncytia. Importantly, hematrophic nutrition never develops, and a sophisticated method of nutrient delivery via histotrophic means is present (Freyer et al. 2007). In addition, macropodids also appear to have the most advanced level of hormonal interplay between the placenta and the mother so that the "feto-placental unit is capable as in all eutherian mammals of redirecting maternal physiology" (Renfree 2010; see also Renfree 2000).

These patterns of placental function provide very little to support the view that there is an overriding or basic constraint on marsupial placental function. Marsupials exhibit a wide range of placental adaptations that appear related to variation in reproduction patterns that match complex life history requirements (Fisher et al. 2001). There is no correlation between neonate grade and any particular characteristic of placental function. Further, marsupials are clearly capable of developing sophisticated maternal-fetal contact via placental structure, significant histotrophic and/or hematrophic nutrition, and significant hormonal interplay between the fetus, placenta, and mother (reviewed in Bradshaw and Bradshaw 2011; Renfree 2010; Tyndale-Biscoe and Renfree 1987). All evidence points to a strategy of maternal-fetal interaction under selection rather than limited by fundamental constraints.

One more point often thought to be unique to eutherians is the immune protection of the fetus. Lillegraven (1975, 713) proposed that "the major barrier between the fetal antigens and maternal antibodies is an anatomical one, consisting of the placental trophoblast and its noncellular components." Recent work suggests that such protection is not a unique feature of the trophoblast, but involves wide-scale modulation of the immune system and maternal physiology. It is impossible to directly test whether marsupials have the evolutionary capacity to develop mechanisms for immunological protection. Attempts to induce an immune reaction by multiple mating of a female to the same male produce no heightened immunological response, suggesting immune protection is not a current constraint (e.g., Rodger et al. 1985; Walker and Tyndale-Biscoe 1978).

There is some question about whether animals with epitheliochorial placentae (as in the majority of marsupials) face a large immunological challenge as contact is limited to an interface between epithelia. Embryos in such cases have been likened to "commensal bacteria in the gut," provoking little or no immune response (Moffett and Loke 2006). It is a critical issue in mammals with hemochorial placentae; however, even in these cases, recent work demonstrates that such protection is due not only to characteristics of the trophoblast (chorion) but instead complex regulation and modulation of gene expression in the decidua of the uterus, the maternal immune system as well as placental tissues (e.g., Arck et al. 2007; Kumpel and Manoussaka 2012; Moffett and Loke 2004, 2006; Mor and Cardenas 2010; Nancy et al. 2012; Yoshinaga 2012). This multisystem response makes it unlikely that the eutherian chorion is an invention that is responsible for immune protection during gestation.

Placenta Evolution in Eutherians

The placenta in eutherian mammals is well developed and also highly variable (table 12.1). For many years it was thought that the epitheliochorial condition was primitive and that the more intimate connections of the endotheliochorial and hemochorial were derived in eutherians (reviewed in Luckett 1993). There were a number of reasons this hypothesis was appealing. It was thought that marsupials were limited to

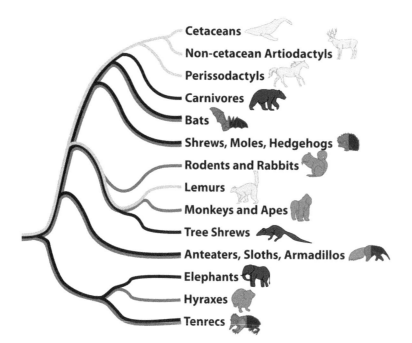

Cetaceans
Non-cetacean Artiodactyls
Perissodactyls
Carnivores
Bats
Shrews, Moles, Hedgehogs
Rodents and Rabbits
Lemurs
Monkeys and Apes
Tree Shrews
Anteaters, Sloths, Armadillos
Elephants
Hyraxes
Tenrecs

FIG. 12.5 Placental variation in the Eutheria. Light gray represents taxa with epitheliochorial placentae; medium gray, taxa with hemochorial placentae; and dark gray, taxa with endotheliochorial placentae. Taxa with two colors represent clades with significant variation within the clade. Far more variation exists than is represented here. For example, within the rodents there are taxa with endotheliochorial placentae (Heteromyidae); individual taxa with epitheliochorial placentae are found in the shrews and the clade Afrotheria; hyenas, within the carnivores, have hemochorial placentae, etc. The colors represent character states of terminal taxa; no attempt has been made to represent ancestral character states. Compiled from the more complete representations of several sources (Elliot and Crespi 2009; Mess and Carter 2007; Vogel 2005; Wildman et al. 2006).

epitheliochorial placentae and thus this type characterized the outgroup to eutherians. Further, the evolutionary transition from less elaborate maternal-fetal contact to more elaborate and intimate interchange appeared to be parsimonious as a functional scenario. Finally, and perhaps not incidentally, humans and other anthropoid primates exhibit an intimate hemochorial placenta; the condition in humans was assumed to be the advanced and not the primitive condition. Recently our understanding of the evolution of the placenta in eutherians has been reversed, largely because of new understanding of mammalian phylogeny. When modern phylogenies have been used to reconstruct patterns of placenta evolution, the most parsimonious reconstruction places the epitheliochorial (least invasive) condition as derived (fig. 12.5). Some models have reconstructed the hemochorial condition as primitive while others cannot resolve whether the hemochorial or endotheliochorial placenta is primitive (e.g., Elliot and Crespi 2009; Martin 2008; Mess and Carter 2007; Vogel 2005; Wildman et al. 2006).[3] Thus, it seems almost certain that invasive placentae are primitive in eutherians, and the less invasive epitheliochorial condition is derived.

Character state reconstructions of other aspects of placental morphology including the shape and nature of contact between maternal and fetal tissues also have reached the conclusion that more intimate maternal-fetal contact is primitive in eutherians and the derived condition represents evolution away from intimate contact. Most phylogenies reconstruct the **labyrinthine**, **discoid** placenta (see table 12.1) as primitive (e.g., Capellini 2012; Elliot and Crespi 2009; Wildman et al. 2006).

This somewhat unexpected result has stimulated a number of studies on the evolutionary and functional significance of differing placental morphologies. Lewitus and Soligo (2011) find two predominant "constellations" of placental morphology and life history characteristics in eutherians (table 12.3). Hemochorial, labyrinthine placentae are most often found in eutherians with small, altricial young, born after short gestations. These animals tend to have large litters, short times to weaning, and high reproductive turnover. In contrast, epitheliochorial (noninvasive), villous, or trabecular (less interdigitated) placentae are most often found in animals with long life spans, large precocial young, and long periods of gestation and lactation.

3. There is tremendous homoplasy in all characters of the placenta, and it is likely that the simple categories of epitheliochorial, endotheliochorial, and hemochorial are not adequate to fully describe character state evolution. For example, Enders and Carter (2012a, 2012b) present multiple examples of clear homoplasy and possible independent evolu-

tion in both endotheliochorial and hemochorial placentae. Further, many taxa exhibit more than one condition through the course of pregnancy, so that the primitive condition for that taxon is difficult to reconstruct. Therefore it is likely that placenta evolution is more complex than appears in most phylogenetic reconstructions.

TABLE 12.3 Examples of the constellations of characters of placental morphology and life history traits in eutherian mammals identified by Lewitus and Soligo (2011)

		Type I	Type II
Placenta		Epitheliochorial	Hemochorial
		Villous to trabecular	Labyrinthine
		Diffuse	Discoid
Life history		Long lifespan	Short lifespan
		Precocial young	Altricial young
		Small litter	Large litter
		Long gestation	Short gestation
		Large neonate	Small neonate
		Late weaning and maturity	Early weaning and maturity
		High interbirth interval	Low interbirth interval

Capellini et al. (2011), who studied all therians including marsupials, reached a similar conclusion. They found that degree of placental invasiveness and interdigitation type correlated with gestation length and fetal growth rates. More interdigitated and more invasive placentae were associated with shorter gestation times, holding fetal brain and body mass constant. They found no correlation with placental type and either neonatal size or maternal size. Capellini et al. (2011) confirmed Elliot and Crespi's (2008) result that there is a positive relation between more invasive placentae and neonatal brain size; however, they state that this relation is "minimal." They concluded that the form of interdigitation was the critical factor in determining gestation time and not invasiveness as hemochorial and endotheliochorial placentae were similar (see also Capellini 2012). Several overall conclusions are important. Fast gestation is highly correlated with invasive placentae, and particularly placentae with a highly interdigitated interface between maternal and fetal tissues. Further, the production of large, fast developing litters, generally associated with altriciality is highly associated with labyrinthine (highly interdigitated) placentae.

Epitheliochorial placentae, with less interdigitation and less intimate contact between maternal and fetal placental layers, have independently evolved in several taxa (fig. 12.5). They are most often found in large-bodied species such as cetaceans or perissodactyls with long gestation times and large, precocial young. Such taxa also appear to depend on increased histotrophic nutrition. This observation has led to the conclusion that less intensive placental types are advantageous in species with large young and long gestations because they protect the mother from excessive energy demands from the fetus (e.g., Capellini 2012; Capellini et al. 2011; Carter and Mess 2007; Klisch and Mess 2007; Wildman et al. 2006). Further, the switch to histotrophic from hematrophic nutrition may allow the female greater control of her reproductive resources, while allowing efficient and prolonged energy transfer between mother and fetus.[4] There is no evidence that histotrophic nutrition is less efficient than hematrophic nutrition (Vogel 2005).

Thus, in eutherians there is tremendous variation in placental form and function. Further it appears that as in marsupials, in eutherians less invasive placentae are derived and more precocial young rely more on histotrophic nutrition. Highly invasive placentae

4. There are obvious exceptions to this broad overview. In particular the pattern seen in primates challenges this simple model. Hemochorial, highly invasive placentae are found in the anthropoid primates; the condition in the hominoids, and in particular humans, is among the most intimate of maternal-fetal interactions seen in mammals. On the other hand, the lemuriformes are the only major group of mammals outside the perissodactyl/cetartiodactyl clade to have epitheliochorial placentae. They do not appear to fit the general correlation of large highly precocial young. It has been proposed that the possession of hemochorial placentae in anthropoids is related to large neonatal brain size (Martin 1996, 2008), which requires high rates of nutrient transfer during gestation.

are generally correlated with the most altricial young and in particular animals with the fastest reproductive turnover. Why have we seen an evolution away from the highly invasive placentation types?

The Dark Side of the Placenta

In addition to new insight arising from our changing view of eutherian phylogeny and therefore the evolution of the placenta, our view of placental function has changed in the last 20 or so years with the development of new theories regarding maternal-fetal conflict and genomic imprinting.

These issues were first put forth clearly in a series of papers by David Haig (e.g., 1993). In short, the theory of maternal-fetal conflict states that a fetus is a hybrid of maternal and paternal genes. In any case where the fetus has direct access to maternal nutrient stores (as in placentation), a potential evolutionary conflict exists because the fetus (which is half paternal genes) and mother do not necessarily share the same long-term reproductive strategy. It is in the evolutionary interest of the fetal genes to consume the maximum maternal resources necessary for its own growth during its own developmental cycle. This is particularly true in cases where the mother mates with more than one male (in a breeding or over a lifetime). In contrast, it is in the evolutionary interest of the mother to allocate her energetic investment across the sum of all reproductive efforts to increase her lifetime fitness. This is particularly important in animals that produce multiple litters. This conflict of interest may be easily controlled by the mother in cases where she directly provisions the young through active feeding, or lactation (as long as she can control access to milk). However, when the placenta—whose tissues are largely produced by the embryo—is responsible for the nutritional interchange, the mother's ability to control her resources may be reduced.

Haig's hypothesis took the maternal-fetal conflict theory a step further when he proposed that it was an evolutionary explanation for the then newly discovered phenomenon of genomic imprinting (e.g., Haig 1996, 1998, 2000, 2004). Genomic imprinting was discovered about 25 years ago, and arose from the discovery that for some genes only one copy is expressed in the fetus. Depending on the gene, either the copy inherited from

the mother or father is expressed. Two observations bolster the hypothesis that gene imprinting is related to maternal-fetal conflict. First, it has only been seen in clades in which the embryo has access to maternal resources during development: some insects, angiosperms, and therian mammals. It has been documented in both marsupials and eutherians but not in monotremes or any other vertebrate (e.g., Renfree et al. 2012).

The second feature is that the specific genes exhibiting imprinting are often found in the placenta or are important in fetal growth (e.g., Coan et al. 2005; Cross et al. 2003; Ferguson-Smith 2011; Ferguson-Smith et al. 2006; Fowden et al. 2006; Frost and Moore 2010; Gootwine 2004; Haig 1996, 2008; Hemberger 2007; Jirtle and Weidman 2007; Reik et al. 2003; Renfree et al. 2008, 2009; Wang et al. 2013). Frequently when the paternal copy is the one that is expressed, fetal or placental growth is enhanced. The "canonical" example (e.g., Frost and Moore 2010) is the well-known *Igf2* (insulin growth factor) system. *Igf2* is a potent enhancer of fetal growth, and disruptions in its function have been shown to disrupt normal growth in fetal mice. In humans and mice (as well as many other mammals) the paternal copy of the *Igf2* gene is expressed, and fetal growth is enhanced substantially with higher levels of this product. The "tug of war" is made more complex in mice because the maternal copy of another gene is expressed, which functions to reduce the bioavailability of this protein. In humans there may be a parallel negative regulation of *Igf2*, but it is not as well characterized. *Igf2* has also been shown to be imprinted in marsupials as well as the other eutherians in which it has been studied (Ager et al. 2007; Killian et al. 2000).

We are still far from a complete understanding of the extent, function, and significance of imprinting for the vast majority of genes. The relation of imprinted genes to maternal-fetal conflict is still somewhat controversial (e.g., Ashbrook and Hager 2013; Moore 2012), and not all imprinted genes discovered so far fit easily into the model of maternal-fetal conflict. For example, Keverne (2013) portrays some imprinted genes as being involved in a complex process of coadaptation between the mother and fetus. But both theoretical and empirical work suggests that in animals with close maternal-fetal exchange, genetic conflict should, and in fact does, exist. Many workers believe that both the tremendous variability in placental type seen in eutherians and the overall evolutionary trend toward reduced intimacy of interaction in many mammals is in part maternal responses to this conflict (e.g., Capellini et al. 2011; Elliot and Crespi 2009; Enders and Carter 2012a; Garratt et al. 2013; Vogel 2005; Wildman et al. 2006).

These hypotheses and data then allow us to examine the evolution of reproductive strategies of therian mammals in a new light. Previous workers (Hayssen et al. 1985; Kirsch 1977a, 1977b; Low 1978; Parker 1977; Tyndale-Biscoe and Renfree 1987) emphasized that the advantage of the marsupial strategy is that it allows the female more control of her reproductive efforts. This hypothesis was largely couched in the context of ecology and overall energetics. Specifically it was argued that in the face of highly unpredictable resources the marsupial reproductive strategy would be strongly advantageous because a female could adjust her litter size or terminate her investment more easily during lactation than during internal gestation. The evolutionary significance of this hypothesis was criticized (e.g., Morton et al. 1982; Lee and Cockburn 1985), but there has been no real refutation of the statement that a female marsupial does retain greater control of litter size and her ability to allocate resources during a reproductive bout than do eutherians. The emergence of the maternal-fetal conflict theory in the past 20 years suggests that the control of resources by the mother during gestation is a general issue in therians and not dependent on specific reproductive conditions. The maternal-fetal battle is a potential cost of prolonged active, intrauterine gestation, which may have had significant impact on the evolution of both marsupial and eutherian reproductive strategy.

Discussion and Summary

I started this review by revisiting the debate on the marsupial-placental dichotomy of the 1970s. However, our understanding of many aspects of mammalian reproduction has changed since that time. Marsupials were characterized as having minimal development of the placenta and, with the possible exception of peramelids, were thought to be largely dependent on histotrophic nutrition. Histotrophic nutrition was thought to be a primitive means of nourishing the young and generally less efficient. Marsupials were assumed to produce uniformly highly altricial young and, because of limitations arising from the lack of a trophoblast, to be

unable to increase the complexity, length, or intensity of the maternal-fetal interface. Eutherians, on the other hand, were seen as initially having effective epitheliochorial and endotheliochorial placentae and eventually evolving an intense hemochorial placentation, which provided efficient nutrition for rapid development of precocial, large-brained young. It was thought that eutherians were able to protect the fetus from allograft rejection and also to establish hormonal communication from the placenta to the mother while marsupials could not.

Many years of work by a large number of workers have turned this simple model on its head. In both marsupials and eutherians the structure and function of the placenta is complex and variable. In both, variation is influenced by life history strategy (as the rate of reproduction and relative altriciality of the neonate is determined by ecological and evolutionary contexts in both groups). There is little evidence that marsupials are fundamentally constrained in their placental structure and function.

Instead as emphasized by many, it appears that short intrauterine gestation is a specific evolutionary strategy that has been paired with exceptional adaptations for lactation. Likewise, in eutherians, although there is a much more intense intrauterine interaction, a great deal of the variation in placental form and function is also best interpreted in the light of the need for maternal control of her reproductive resources and reduction of direct fetal access to maternal circulatory supply. In many eutherians there has been evolution away from a strategy that emphasizes a highly intensive maternal-fetal interface and returns control of resources to the mother (e.g., Klisch and Mess 2007). In short, this new perspective allows new appreciation for the issues surrounding maternal control of reproductive resources in all therians, and lends support to hypotheses that it may be of fundamental importance in the evolutionary choices made by marsupials.

The degree of difference in the two strategies should not be minimized. They do not represent a continuum but are quite distinct strategies. Some of the best evidence for the distinctness of the strategies comes from the study of early development, where it can be clearly shown that marsupial development is modified from the earliest stages of differentiation to meet the needs of the highly altricial young (e.g., Keyte and Smith 2010,

2012; Smith 1997, 2001, 2006; Vaglia and Smith 2003). These studies further show that the marsupial neonate is not just embryonic at birth, but exhibits a very elaborate complex of adaptations that allows it to survive after an exceedingly short gestation. The adaptations of the marsupial neonate are derived, and unlike that seen in any other amniote, including monotremes (Smith 1997, 2001; Weisbecker 2011).

It is undeniable that the placenta in eutherians consistently develops far beyond any stage seen in marsupials. Internal gestation is prolonged, and we see many unique adaptations of the uterus, decidua, and hormonal and immune system (e.g., Carter 2012). As just one example, Lynch et al. (2008) have demonstrated that the regulatory gene *HoxA-11* has taken on a unique role in eutherian mammals as a regulator of prolactin expression in the uterus in eutherians. One function of this prolactin expression is the silencing of expression of several genes involved in the inflammatory response. This suppression of the inflammatory response in eutherians is hypothesized to help maintain pregnancy. Prolactin is not expressed in the uterus of *Monodelphis*, and presumably other marsupials. On the other hand, just as it is clear eutherians have developed a number of adaptations that allow prolonged intrauterine gestation, marsupials exhibit adaptations for lactation not seen in eutherians, as has long been argued by Renfree (1983, 2010). For example, recent comparative genomic studies demonstrate more milk-specific proteins in the milk of marsupials than either monotremes or eutherians. Further, these studies reveal complex changes over time in the expression of specific milk proteins uniquely in marsupials, to meet the changing nutritional needs of the young through the very long gestation (Lefèvre et al. 2010).

There is little data to support the hypothesis that reproductive strategies in marsupials are simply the retention of a primitive pattern. But, if the differences we observe do not grow out of constraints, why do we see such differences in the two taxa? Ultimately, the question of why a clade does or does not take a certain evolutionary path is virtually impossible to answer, especially when dealing with a single case or comparison and with events long in the past that leave little or no trace in the fossil record.

One of the most important points summarized in this paper is that internal gestation via a placenta places significant potential costs on a mother. Evolutionary theory predicts that selection should favor those embryos that develop greater capacity to extract disproportional resources from the mother for the current pregnancy even at the cost of her lifetime fitness (Haig 2000). Early mammals were small, homeothermic, and nocturnal and likely to be under considerable energetic stresses. Viviparity evolved in this context, and it is likely that it placed significant additional energetic stresses on the mother in early marsupials and eutherians. But it appears that the response to these energetic stresses led to two different paths. In marsupials control of reproduction was strongly biased toward the mother. Gestation was kept short, and intense maternal-fetal intrauterine contact limited. Reproductive effort was largely devoted to lactation, during which the mother maintained greater control of duration of a given reproductive effort and the number of young raised to weaning. In addition this mode extended maternal energetic output over a longer time period lessening the energy demands at any given time. While this basic point has been made by Renfree, Parker, Low, and Hayssen, it is generally in the context of highly fluctuating resources such as those faced by many extant Australian marsupials. Here I argue that internal gestation via a placenta in and of itself may be a significant energy constraint. Supporting the idea that marsupials evolved away from intense maternal-fetal contact is the evidence that more intense maternal-fetal contact was primitive in marsupials. Additionally, it is clear that ultra-altriciality is a derived state in the group, and not a primitive condition for marsupials, again suggesting selection away from maternal investment in utero in some taxa.

In eutherians, if current reconstructions are correct, a very intense form of maternal-fetal contact—hemochorial, labyrinthine placentae—evolved early. Today we see such placentae for the most part in animals with very fast reproductive turnover: rapid production of large litters with short gestations. Although these young are often altricial by eutherian standards, they are relatively well developed, and can have a very rapid time to weaning. Subsequently, within eutherians we see evolution away from this intense form of placentation, in particular as mammals entered environments in which it was necessary for the young to be precocial, and highly independent at birth.

These scenarios leave the question of "why" unanswered. Did early marsupials and placentals face slightly different ecological conditions so that rapid turnover was favored in eutherians, and longer, less intense investment favored in marsupials? Did chance events early in the evolution of the lineages lead to different paths? Were there particular adaptations toward improved lactation early in the marsupial line, or did particular innovations appear in early eutherians for internal gestation? Understanding of the full complexity of the patterns of form and function, of phylogeny and variation allows us to ask, and perhaps someday answer, questions of intricacy beyond a simple hypothesis of constraint.

Finally, the evolutionary consequences of the "marsupial-placental dichotomy" in reproductive choice continue to puzzle biologists. Eutherians are more diverse phylogenetically and evolutionarily, having produced radiations in the air and the sea never utilized by marsupials. It has tempted biologists for many years to attribute this difference in evolutionary diversity to the differences in reproductive strategy. However, as pointed out recently by Sánchez-Villagra (2013), the factors determining the evolution of major lineages are complex, and include many aspects of ecology, geology, biogeography, and perhaps chance. It might be argued, for example, that "eutherian" diversity is really nested within the Boreoeutheria (Asher et al. 2009), and that all other mammalian clades, eutherian and marsupial, are far less diverse taxonomically and morphologically. It is interesting that these other clades (Afrotheria, Xenarthra, Marsupialia, as well as early radiations of South American mammals) all evolved on smaller southern continents rather than the large Laurasian mass, which gave rise to the bulk of eutherian diversity. The fact remains that after long years of study we still have not resolved many of the most interesting issues concerning evolutionary patterns in therian mammals. The origin and early diversification of the Mammalia remains one of the great stories of transformations in vertebrate evolution, ripe for future study.

✳ ✳ ✳

Acknowledgments

I am deeply grateful to the late Professor Farish A. Jenkins Jr. for recruiting me into the wonderful world of vertebrate evolution and for having been a mentor and friend for over 35 years; I dedicate this paper to his memory. I thank Drs. Elizabeth Brainerd, Ken Dial, and Neil Shubin for the invitation to contribute to this volume, Drs. Anna Keyte, V. Louise Roth, Marcelo Sánchez-Villagra, Christine Wall, and Anne Yoder for comments on earlier versions of this manuscript, and Anya Broverman-Wray for assistance with the illustrations.

References

Ager, E., S. Suzuki, A. Pask, G. Shaw, F. Ishino, and M. B. Renfree. 2007. Insulin is imprinted in the placenta of the marsupial, *Macropus eugenii*. Developmental Biology 309:317–328.

Arck, P., P. J. Hansen, B. M. Jericevic, M. P. Piccinni, and J. Szekeres-Bartho. 2007. Progesterone during pregnancy: endocrine-immune cross talk in mammalian species and the role of stress. American Journal of Reproductive Immunology 58:268–279.

Ashbrook, D. G., and R. Hager. 2013. Empirical testing of hypotheses about the evolution of genomic imprinting in mammals. Frontiers in Neuroanatomy 7:1–6.

Asher, R. J., N. Bennett, and T. Lehmann. 2009. The new framework for understanding placental mammal evolution. BioEssays 31:853–864.

Asher, R. J., and K. M. Helgen. 2010. Nomenclature and placental mammal phylogeny. BMC Evolutionary Biology 10:1–9.

Bennett, C. V., and A. Goswami. 2011. Does developmental strategy drive limb integration in marsupials and monotremes? Mammalian Biology 76:79–83.

Blackburn, D. G. 2006. Squamate reptiles as model organisms for the evolution of viviparity. Herpetological Monographs 20:131–146.

Blackburn, D. G., and A. F. Flemming. 2009. Morphology, development, and evolution of fetal membranes and placentation in squamate reptiles. Journal of Experimental Zoology Part B–Molecular and Developmental Evolution 312B:579–589.

Blaise, S., N. de Parseval, L. Benit, and T. Heidmann. 2003. Genome wide screening for fusogenic human endogenous retrovirus envelopes identifies syncytin 2, a gene conserved on primate evolution. Proceedings of the National Academy of Sciences USA 100:13013–13018.

Bradshaw, F. J., and D. Bradshaw. 2011. Progesterone and reproduction in marsupials: a review. General and Comparative Endocrinology 170:18–40.

Brandley, M. C., R. L. Young, D. L. Warren, M. B. Thompson, and G. P. Wagner. 2012. Uterine gene expression in the live-bearing lizard, *Chalcides ocellatus*, reveals convergence of squamate reptile and mammalian pregnancy mechanisms. Genome Biology and Evolution 4:394–411.

Capellini, I. 2012. The evolutionary significance of placental interdigitation in mammalian reproduction: contributions from comparative studies. Placenta 33:763–768.

Capellini, I., C. Venditti, and R. A. Barton. 2011. Placentation and maternal investment in mammals. American Naturalist 177:86–98.

Carter, A. M. 2001. Evolution of the placenta and fetal membranes seen in the light of molecular phylogenetics. Placenta 22:800–807.

Carter, A. M. 2012. Evolution of placental function in mammals: the molecular basis of gas and nutrient transfer, hormone secretion and immune responses. Physiological Reviews 92:1543–1576.

Carter, A. M., A. C. Enders, H. Kunzle, D. Oduor-Okelo, and P. Vogel. 2004. Placentation in species of phylogenetic importance: the Afrotheria. Animal Reproduction Science 82–83:35–48.

Carter, A. M., and A. Mess. 2007. Evolution of the placenta in eutherian mammals. Placenta 28:259–262.

Carter, A. M., and A. Mess. 2008. Evolution of the placenta and associated reproductive characters in bats. Journal of Experimental Zoology Part B-Molecular and Developmental Evolution 310B:428–449.

Chuong, E. B. 2013. Retroviruses facilitate the rapid evolution of the mammalian placenta. BioEssays 35:853–861.

Coan, P. M., G. J. Burton, and A. C. Ferguson-Smith. 2005. Imprinted genes in the placenta—a review. Placenta 26:S10–S20.

Cooper, W. J., and S. J. Steppan. 2010. Developmental constraint on the evolution of marsupial forelimb morphology. Australian Journal of Zoology 58:1–15.

Cornelis, G., O. Heidmann, S. Bernard-Stoecklin, K. Reynaud, G. Veron, B. Mulot, A. Dupressoir, and T. Heidmann. 2012. Ancestral capture of syncytin-Car1, a fusogenic endogenous retroviral envelope gene involved in placentation and conserved in Carnivora. Proceedings of the National Academy of Sciences USA 109:E432–E441.

Crompton, A. W. 1980. Biology of the earliest mammals. Pp. 1–12 in K. Schmidt-Nielsen, L. Bolis, and C. R. Taylor, eds., Comparative Physiology: Primitive Mammals. Cambridge: Cambridge University Press.

Crompton, A. W., and F. A. Jenkins. 1973. Mammals from reptiles: a review of mammalian origins. Annual Review of Earth and Planetary Sciences 1:131–153.

Crompton, A. W., C. R. Taylor, and J. A. Jagger. 1978. Evolution of homeothermy in mammals. Nature 272:333–337.

Cross, J. C., D. Baczyk, N. Dobric, M. Hemberger, M. Hughes, D. G. Simmons, H. Yamamoto, and J. C. P. Kingdom. 2003. Genes, development and evolution of the placenta. Placenta 24:123–130.

Elliot, M. G., and B. J. Crespi. 2008. Placental invasiveness and brain-body allometry in eutherian mammals. Journal of Evolutionary Biology 21:1763–1778.

Elliot, M. G., and B. J. Crespi. 2009. Phylogenetic evidence for early hemochorial placentation in Eutheria. Placenta 30:949–967.

Enders, A. C., and A. M. Carter. 2004. What can comparative studies of placental structure tell us? a review. Placenta 25:S3–S9.

Enders, A. C., and A. M. Carter. 2012a. The evolving placenta: convergent evolution of variations in the endotheliochorial relationship. Placenta 33:319–326.

Enders, A. C., and A. M. Carter. 2012b. Review: the evolving placenta; different developmental paths to a hemochorial relationship. Placenta 33:S92–S98.

Farmer, C. G. 2000. Parental care: the key to understanding endothermy and other convergent features in birds and mammals. American Naturalist 155:326–334.

Ferguson-Smith, A. C. 2011. Genomic imprinting: the emergence of an epigenetic paradigm. Nature Reviews Genetics 12:565–575.

Ferguson-Smith, A. C., T. Moore, J. Detmar, A. Lewis, M. Hemberger, H. Jammes, G. Kelsey, C. T. Roberts, H. Jones, and M. Constancia. 2006. Epigenetics and imprinting of the trophoblast—a workshop report. Placenta 27:S122–S126.

Ferner, K., and A. Mess. 2011. Evolution and development of fetal membranes and placentation in amniote vertebrates. Respiratory Physiology and Neurobiology 178:39–50.

Fisher, D. O., I. P. F. Owens, and C. N. Johnson. 2001. The ecological basis of life history variation in marsupials. Ecology 82:3531–3540.

Flemming, A. F., and D. G. Blackburn. 2003. Evolution of placental specializations in viviparous African and South American lizards. Journal of Experimental Zoology Part A-Comparative Experimental Biology 299A:33–47.

Fowden, A. L., C. Sibley, W. Reik, and M. Constancia. 2006. Imprinted genes, placental development and fetal growth. Hormone Research 65:50–58.

Freyer, C., and M. B. Renfree. 2009. The mammalian yolk sac placenta. Journal of Experimental Zoology Part B-Molecular and Developmental Evolution 312B:545–554.

Freyer, C., U. Zeller, and M. B. Renfree. 2001. Placentation in marsupials: contribution to the marsupial stem species pattern. Journal of Morphology 248:231.

Freyer, C., U. Zeller, and M. B. Renfree. 2002. Ultrastructure of the placenta of the tammar wallaby, *Macropus eugenii*: comparison with the grey short-tailed opossum, *Monodelphis domestica*. Journal of Anatomy 201:101–119.

Freyer, C., U. Zeller, and M. B. Renfree. 2003. The marsupial placenta: a phylogenetic analysis. Journal of Experimental Zoology Part A-Comparative Experimental Biology 299A:59–77.

Freyer, C., U. Zeller, and M. B. Renfree. 2007. Placental function in two distantly related marsupials. Placenta 28:249–257.

Frost, J. M., and G. E. Moore. 2010. The importance of imprinting in the human placenta. Plos Genetics 6.

Garratt, M., J.-M. Gaillard, R. C. Brooks, and J.-F. Lemaître. 2013. Diversification of the eutherian placenta is associated with changes in the pace of life. Proceedings of the National Academy of Sciences 110: 7760–7765.

Gootwine, E. 2004. Placental hormones and fetal-placental development. Animal Reproduction Science 82–83:551–566.

Goswami, A., P. D. Polly, O. B. Mock, and M. R. Sánchez-Villagra. 2012. Shape, variance and integration during craniogenesis: contrasting marsupial and placental mammals. Journal of Evolutionary Biology 25:862–872.

Goswami, A., V. Weisbecker, and M. R. Sánchez-Villagra. 2009. Developmental modularity and the marsupial-placental dichotomy. Journal of Experimental Zoology Part B-Molecular and Developmental Evolution 312B:186–195.

Guillette, L. J., and N. Hotton. 1986. The evolution of mammalian reproductive characteristics in therapsid reptiles. Pp. 239–262 in N. Hotton, P. D. MacLean, J. J. Roth, and E. C. Roth, eds., The Ecology and Biology of Mammal-Like Reptiles. Washington, DC: Smithsonian Institution Press.

Haig, D. 1993. Genetic conflicts in human pregnancy. Quarterly Review of Biology 68:495–532.

Haig, D. 1996. Placental hormones, genomic imprinting, and maternal-fetal communication. Journal of Evolutionary Biology 9:357–380.

Haig, D. 1998. Genomic imprinting. American Journal of Human Biology 10:679–680.

Haig, D. 2000. The kinship theory of genomic imprinting. Annual Review of Ecology and Systematics 31:9–32.

Haig, D. 2004. Genomic imprinting and kinship: how good is the evidence? Annual Review of Genetics 38:553–585.

Haig, D. 2008. Placental growth hormone-related proteins and prolactin-related proteins. Placenta 29:S36–S41.

Haig, D. 2012. Retroviruses and the placenta. Current Biology 22:R609–R613.

Harris, J. R. 1998. Placental endogenous retrovirus (ERV): structural, functional, and evolutionary significance. Bioessays 20:307–316.

Hayssen, V., R. C. Lacy, and P. J. Parker. 1985. Metatherian reproduction: transitional or transcending? American Naturalist 126:617–632.

Heidmann, O., C. Vernochet, A. Dupressoir, and T. Heidmann. 2009. Identification of an endogenous retroviral envelope gene with fusogenic activity and placenta-specific expression in the rabbit: a new "syncytin" in a third order of mammals. Retrovirology 6.

Hemberger, M. 2007. Epigenetic landscape required for placental development. Cellular and Molecular Life Sciences 64:2422–2436.

Hill, J. P. 1900. On the foetal membranes, placentation and parturition of the native cat (Dasyurus viverrinus). Anatomischer Anzeiger 18:364–373.

Hopson, J. A. 1973. Endothermy, small size and the origin of mammalian reproduction. American Naturalist 107:446–451.

Hughes, R. L. 1974. Morphological studies on implantation in marsupials. Journal of Reproduction and Fertility 39:173–186.

Hughes, R. L., and L. S. Hall. 1988. Structural adaptations of the newborn marsupial. In H. Tyndale-Biscoe and P. A. Janssens, eds., The Developing Marsupial. New York: Springer-Verlag.

Jenkins, F. A. 1984. A survey of mammalian origins. Pp. 32–47 in P. D. Gingerich and C. E. Badgley, eds., Mammals: Notes for a Short Course. Knoxville: University of Tennessee.

Jirtle, R. L., and J. R. Weidman. 2007. Imprinted and more equal. American Scientist 95:143–149.

Johnson, M. H. 1996. Origins of pluriblast and trophoblast in the eutherian conceptus. Reproduction, Fertility and Development 8:699–709.

Johnson, M. H., and L. Selwood. 1996. Nomenclature of early development in mammals. Reproduction, Fertility and Development 8:759–764.

Kavanagh, J. R., A. Burk-Herrick, M. Westerman, and M. S. Springer. 2004. Relationships among families of Diprotodontia (Marsupialia) and the phylogenetic position of the Autapomorphic honey possum (Tarsipes rostratus). Journal of Mammalian Evolution 11:207–222.

Kelly, E. M., and K. E. Sears. 2011. Limb specialization in living marsupial and eutherian mammals: constraints on mammalian limb evolution. Journal of Mammalogy 92:1038–1049.

Kemp, T. S. 2006. The origin of mammalian endothermy: a paradigm for the evolution of complex biological structure. Zoological Journal of the Linnean Society 147:473–488.

Kemp, T. S. 2007. The origin of higher taxa: macroevolutionary processes, and the case of the mammals. Acta Zoologica 88:3–22.

Keyte, A. L., and K. K. Smith. 2010. Developmental origins of precocial forelimbs in marsupial neonates. Development 137:4283–4294.

Keyte, A. L., and K. K. Smith. 2012. Heterochrony in somitogenesis rate in a model marsupial, Monodelphis domestica. Evolution & Development 14:93–103.

Keverne, E. B. 2013. Importance of the matriline for genomic imprinting, brain development and behaviour. Philosophical Transactions of the Royal Society B 368:1–10.

Killian, J. E., J. C. Byrd, V. Jirtle, B. L. Munday, M. K. Stoskopf, and R. L. Jirtle. 2000. M6P/IGF2R imprinting evolution in mammals. Molecular Cell 5:707–716.

Kirsch, J. A. W. 1977a. Biological aspects of the marsupial-placental dichotomy: a reply to Lillegraven. Evolution 31:898–900.

Kirsch, J. A. W. 1977b. The six-percent solution: second thoughts on the adaptedness of the Marsupialia. American Scientist 65:276–288.

Klisch, K., and A. Mess. 2007. Evolutionary differentiation of cetartiodactyl placentae in the light of the viviparity-driven conflict hypothesis. Placenta 28:353–360.

Koteja, P. 2012. Energy assimilation, parental care and the evolution of endothermy. Proceedings of the Royal Society of London B 267:479–484.

Kumpel, B. M., and M. S. Manoussaka. 2012. Placental immunology and maternal alloimmune responses. Vox Sanguinis 102:2–12.

Lee, A. K., and A. Cockburn. 1985. Evolutionary Ecology of Marsupials. Cambridge: Cambridge University Press.

Lefèvre, C. M., J. A. Sharp, and K. R. Nicholas. 2010. Evolution of lactation: ancient origin and extreme adaptations of the lactation system. Annual Review of Genomics and Human Genetics 11:219–238.

Lewitus, E., and C. Soligo. 2011. Life-history correlates of placental structure in eutherian evolution. Evolutionary Biology 38:287–305.

Lillegraven, J. A. 1969. Latest Cretaceous mammals of the upper part of the Edmonton formation of Alberta, Canada, and a review of the Marsupial-placental dichotomy in mammalian evolution. University of Kansas Paleontological Contributions 50:1–1222.

Lillegraven, J. A. 1975. Biological considerations of the marsupial-placental dichotomy. Evolution 29:707–722.

Lillegraven, J. A. 1984. Why was there a "marsupial-placental dichotomy"? Pp. 72–86 in P. D. Gingerich and C. E. Badgley, eds., Mammals: Notes for a Short Course. Knoxville: University of Tennessee.

Lillegraven, J. A. 2004. Polarities in mammalian evolution seen through the homologs of the inner cell mass. Journal of Mammalian Evolution 10:277–330.

Lillegraven, J. A., S. D. Thompson, B. K. McNab, and J. L. Patton. 1987. The origin of eutherian mammals. Biological Journal of the Linnaean Society 32:281–336.

Low, B. S. 1978. Environmental uncertainty and the parental strategies of marsupials and placentals. American Naturalist 112:197–213.

Luckett, W. P. 1977. Ontogeny of amniote fetal membranes and their application to phylogeny. Pp. 439–516 in M. K. Hecht, P. C. Goody, and B. M. Hecht, eds., Major Patterns in Vertebrate Evolution. New York: Plenum Press.

Luckett, W. P. 1993. Uses and limitations of mammalian fetal membranes and placenta for phylogenetic reconstruction. Journal of Experimental Zoology 266:514–527.

Luo, Zhe-Xi. 2007. Transformation and diversification in early mammal evolution. Nature 450:1011–1019.

Lynch, V. J., A. Tanzer, Y. J. Wang, F. C. Leung, B. Gellersen, D. Emera, and G. P. Wagner. 2008. Adaptive changes in the transcription factor HoxA-11 are essential for the evolution of pregnancy in mammals. Proceedings of the National Academy of Sciences USA 105:14928–14933.

Martin, R. D. 1996. Scaling of the mammalian brain: the maternal energy hypothesis. News in Physiological Sciences 11:149–156.

Martin, R. D. 2008. Evolution of placentation in primates: implications of mammalian phylogeny. Evolutionary Biology 35:125–145.

McGowen, M. R. 2013. Gene expression in the term placenta of the opossum Monodelphis domestica and the evolution of the therian placenta. Placenta 34:A24–25.

McNab, B. K. 1978. Evolution of endothermy in phylogeny of mammals. American Naturalist 112:1–21.

Mess, A., and A. M. Carter. 2006. Evolutionary transformations of fetal membrane characters in Eutheria with special reference to Afrotheria. Journal of Experimental Zoology Part B-Molecular and Developmental Evolution 306B:140–163.

Mess, A., and A. M. Carter. 2007. Evolution of the placenta during the early radiation of placental mammals. Comparative Biochemistry and Physiology A–Molecular & Integrative Physiology 148:769–779.

Mi, S., X. Lee, X. P. Li, G. M. Veldman, H. Finnerty, L. Racie, E. LaVallie, X. Y. Tang, P. Edouard, S. Howes, J. C. Keith, and J. M. McCoy. 2000. Syncytin is a captive retroviral envelope protein involved in human placental morphogenesis. Nature 403:785–789.

Moffett, A., and C. Loke. 2006. Immunology of placentation in eutherian mammals. Nature Reviews Immunology 6:584–594.

Moffett, A., C. Loke, and A. McLaren. 2006. Biology and Pathology of Trophoblast. Cambridge: Cambridge University Press.

Moffett, A., and Y. W. Loke. 2004. The immunological paradox of pregnancy: a reappraisal. Placenta 25:1–8.

Moore, T. 2012. Review: parent-offspring conflict and the control of placental function. Placenta 33:S33–S36.

Mor, G., and I. Cardenas. 2010. The immune system in pregnancy: a unique complexity. American Journal of Reproductive Immunology 63:425–433.

Morton, S. R., H. F. Recher, S. D. Thompson, and R. W. Braithwaite. 1982. Comments on the relative advantages of marsupial and eutherian reproduction. American Naturalist 120:128–134.

Mossman, H. W. 1937. Comparative morphogenesis of the fetal membranes and accessory uterine structures. Carnegie Institute Contributions to Embryology 26:129–246.

Nancy, P., E. Tagliani, C.-S. Tay, P. Asp, D. Levy, and A. Erlebacher. 2012. Chemokine gene silencing in decidual stromal cells limits T cell access to the maternal-fetal interface. Science 336:1317–1321.

Nilsson, M. A., G. Churakov, M. Sommer, N. Van Tran, A. Zemann, J. Brosius, and J. Schmitz. 2010. Tracking marsupial evolution using archaic genomic retroposon insertions. Plos Biology 8.

O'Leary, M. A., J. I. Bloch, J. J. Flynn, T. J. Gaudin, A. Giallombardo, N. P. Giannini, S. L. Goldberg, B. P. Kraatz, Z. Luo, J. Meng, et al. 2013. The placental mammal ancestor and the post K-Pg radiation of placentals. Science 339:662–667.

Padykula, H. A., and J. M. Taylor. 1976. Ultrastructural evidence for loss of the trophoblastic layer in the chorioallantoic placenta of Australian bandicoots (Marsupialia: Peramelidae). Anatomical Record 186:357–386.

Padykula, H. A., and J. M. Taylor. 1982. Marsupial placentation and its evolutionary significance. Journal Reproduction and Fertility (Suppl.) 31:95–104.

Parker, P. J. 1977. An ecological comparison of marsupial and placental patterns of reproduction. Pp. 273–286 in B. Stonehouse and E. Gilmore, eds., The Biology of Marsupials. Baltimore: University Park Press.

Pijnenborg, R., and L. Vercruysse. 2013. A. A. W. Hubrecht and the naming of the trophoblast. Placenta 34: 314–319.

Pond, C. M. 1977. The significance of lactation in the evolution of mammals. Evolution 31:177–199.

Reik, W., M. Constancia, A. L. Fowden, N. Anderson, W. Dean, A. C. Ferguson-Smith, B. Tycko, and C. Sibley. 2003. Regulation of supply and demand for maternal nutrients in mammals by imprinted genes. Journal of Physiology 547:35–44.

Renfree, M. B. 1983. Marsupial reproduction: the choice between placentation and lactation. Pp. 1–29 in C. A. Finn, ed., Oxford Reviews of Reproductive Biology. Oxford: Oxford University Press.

Renfree, M. B. 1993. Ontogeny, genetic control, and phylogeny of female reproduction in monotreme and therian mammals in F. S. Szalay, ed., Mammal Phylogeny: Mesozoic Differentiation, Multituberculates, Monotremes, Early Therians, and Marsupials. New York: Springer-Verlag.

Renfree, M. B. 1995. Monotreme and marsupial reproduction. Reproduction Fertility and Development 7:1003–1020.

Renfree, M. B. 2000. Maternal recognition of pregnancy in marsupials. Reviews of Reproduction 5:6–11.

Renfree, M. B. 2010. Marsupials: placental mammals with a difference. Placenta 31:S21–S26.

Renfree, M. B., E. I. Ager, G. Shaw, and A. J. Pask. 2008. Genomic imprinting in marsupial placentation. Reproduction 136:523–531.

Renfree, M. B., T. A. Hore, G. Shaw, J. A. Graves, and A. J. Pask. 2009. Evolution of genomic imprinting: insights from marsupials and monotremes. Annual Review Genomics and Human Genetics 10:241–262.

Renfree, M. B., E. M. Russell, and R. D. Wooller. 1984. Reproduction and life history of the honey possum, Tarsipes rostratus. Pp. 427–437 in A. P. Smith and I. D. Hume, eds., Possums and Gliders. Sydney: Australian Mammal Society.

Renfree, M. B., S. Suzuki, and T. Kaneko-Ishino. 2012. The origin and evolution of genomic imprinting and viviparity in mammals. Philosophical Transactions of the Royal Society B 368:20120151

Roberts, C. T., and W. G. Breed. 1994. Placentation in the dasyurid marsupial, Sminthopsis crassicaudata, the fat-tailed dunnart, and notes on placentation of the didelphid, *Monodelphis domestica*. Journal of Reproduction and Fertility 100:105–113.

Rodger, J. C., T. P. Fletcher, and C. H. Tyndale-Biscoe. 1985. Active anti-paternal immunization does not affect the success of marsupial pregnancy. Journal of Reproductive Immunology 8:249–256.

Rougier, G. W., J. R. Wible, and M. J. Novacek. 1998. Implications of *Deltatheridium* specimens for early marsupial history. Nature 396:459–463.

Ruben, J. 1995. The evolution of endothermy in mammals and birds—from physiology to fossils. Annual Review of Physiology 57:69–95.

Russell, E. M. 1982. Patterns of parental care and parental investment in marsupials. Biological Reviews 57:423–485.

Sánchez-Villagra, M. R. 2013. Why are there fewer marsupials than placentals? on the relevance of geography and physiology to evolutionary patterns of mammalian diversity and disparity. Journal of Mammalian Evolution 20:279–290.

Sánchez-Villagra, M. R., A. Goswami, V. Weisbecker, O. Mock, and S. Kuratani. 2008. Conserved relative timing of cranial ossification patterns in early mammalian evolution. Evolution & Development 10:519–530.

Sánchez-Villagra, M. R., and W. Maier. 2003. Ontogenesis of the scapula in marsupial mammals, with special emphasis on perinatal stages of didelphids and remarks on the origin of the therian scapula. Journal of Morphology 258:115–129.

Sears, K. E. 2004. Constraints on the morphological evolution of marsupial shoulder girdles. Evolution 58:2353–2370.

Sears, K. E. 2009. Differences in the timing of prechondrogenic limb development in mammals: the marsupial-placental dichotomy resolved. Evolution 63:2193–2200.

Selwood, L., and M. H. Johnson. 2006. Trophoblast and hypoblast in the monotreme, marsupial and eutherian mammal: evolution and origins. BioEssays 28:128–145.

Shine, R. 1995. A new hypothesis for the evolution of viviparity in reptiles. American Naturalist 145:809–823.

Smith, K. K. 1997. Comparative patterns of craniofacial development in eutherian and metatherian mammals. Evolution 51:1663–1678.

Smith, K. K. 2001. The evolution of mammalian development. Bulletin of the Museum of Comparative Zoology 156:119–135.

Smith, K. K. 2006. Craniofacial development in marsupial mammals: developmental origins of evolutionary change. Developmental Dynamics 235:1181–1193.

Stewart, J. R., and M. B. Thompson. 2003. Evolutionary transformations of the fetal membranes of viviparous reptiles: a case study of two lineages. Journal of Experimental Zoology Part A-Comparative Experimental Biology 299A:13–32.

Taylor, J. M., and H. A. Padykula. 1978. Marsupial trophoblast and mammalian evolution. Nature 271:588.

Thompson, M. B., and B. K. Speake. 2006. A review of the evolution of viviparity in lizards: structure, function and physiology of the placenta. Journal of Comparative Physiology B-Biochemical Systemic and Environmental Physiology 176:179–189.

Tyndale-Biscoe, H. 2005. Life of Marsupials. Collingwood, Australia: CSIRO Publishing.

Tyndale-Biscoe, H., and M. Renfree. 1987. Reproductive physiology of marsupials. Cambridge: Cambridge University Press.

Vaglia, J. L., and K. K. Smith. 2003. Early differentiation and migration of cranial neural crest in the opossum, *Monodelphis domestica*. Evolution & Development 5:121–135.

van Nievelt, A. F. H., and K. K. Smith. 2005. To replace or not to replace: the significance of reduced functional tooth replacement in marsupial and placental mammals. Paleobiology 31:324–346.

Vogel, P. 2005. The current molecular phylogeny of eutherian mammals challenges previous interpretations of placental evolution. Placenta 26:591–596.

Wake, M. H. 1993. Evolution of oviductal gestation in amphibians. Journal of Experimental Zoology 266:394–413.

Walker, K. Z., and C. H. Tyndale-Biscoe. 1978. Immunological aspects of gestation in the Tammar wallaby, *Macropus eugenii*. Australian Journal of Biological Science 31:173–182.

Wang, X., D. C. Miller, R. Harman, D. F. Antczak, and A. G. Clark. 2013. Paternally expressed genes predominate in the placenta. Proceedings of the National Academy of Sciences. 110:10705–10710.

Weisbecker, V. 2011. Monotreme ossification sequences and the riddle of mammalian skeletal development. Evolution 65:1323–1335.

Weisbecker, V., and A. Goswami. 2010. Brain size, life history, and metabolism at the marsupial/placental dichotomy. Proceedings of the National Academy of Sciences USA 107:16216–16221.

Weisbecker, V., A. Goswami, S. Wroe, and M. R. Sánchez-Villagra. 2008. Ossification heterochrony in the therian postcranial skeleton and the marsupial-placental dichotomy. Evolution 62:2027–2041.

Weisbecker, V., and M. Sánchez-Villagra. 2007. Postcranial sequence heterochrony and the marsupial-placental dichotomy. Journal of Vertebrate Paleontology 27:164A..

Wildman, D. E., C. Y. Chen, O. Erez, L. I. Grossman, M. Goodman, and R. Romero. 2006. Evolution of the mammalian placenta revealed by phylogenetic analysis. Proceedings of the National Academy of Sciences USA 103:3203–3208.

Wooding, P., and G. Burton. 2008. Comparative Placentation. Berlin: Springer-Verlag.

Yoshinaga, K. 2012. Two concepts on the immunological aspect of blastocyst implantation. Journal of Reproduction and Development 58:196–203.

Zeller, U., and C. Freyer. 2001. Early ontogeny and placentation of the grey short-tailed opossum, *Monodelphis domestica* (Didelphidae : Marsupialia): contribution to the reconstruction of the marsupial morphotype. Journal of Zoological Systematics and Evolutionary Research 39:137–158.

13

Going from Small to Large:
Mechanical Implications of Body Size Diversity in Terrestrial Mammals

Andrew A. Biewener*

An animal's limb posture represents a characteristic feature of its body form, fundamentally influencing the manner in which it moves and the habitats it can exploit. These movements and the underlying biomechanical mechanisms have been studied since the time of Borelli and Marey (Borelli 1685; Marey 1901). Over the course of tetrapod evolution, changes in locomotor limb posture have followed a general trend from sprawling to more erect limb movement patterns (Gray 1968; Gregory 1912; Jenkins 1971a). This trend has been interpreted as reflecting, in part, selection for more economical and faster speeds of terrestrial locomotion. The lateral sprawling gaits of early tetrapods are commonly linked with lateral undulation of the body axis to enhance forward progression (e.g., Gray 1968). Although lateral sprawling gaits have been associated with an added cost of body support and constraints on ventilation for aerobic endurance (Carrier 1987), mechanical constraints due to heightened musculoskeletal loading have been less apparent (Blob and Biewener 1999). Nevertheless, faster moving terrestrial vertebrates have generally evolved more erect limb postures, emphasizing parasagittal limb movement and dorso-ventral undulation of the vertebral column, in order to minimize lateral body sway, facilitate ventilation, and enhance forward progression and endurance capability (Gray 1968; Gregory 1912; Carrier 1987).

* Concord Field Station and the Museum of Comparative Zoology, Department of Organismic and Evolutionary Biology, Harvard University

Great transformations within mammalian evolution therefore were facilitated by shifts in locomotor limb posture that allowed a broad diversification of body size across terrestrial mammalian taxa. The earliest late Triassic mammals (Jenkins and Parrington 1976) were small nocturnal animals, likely possessing a "semierect" though still crouched limb posture similar to their cynodont ancestors (Jenkins 1971b). These small early mammals subsequently gave rise to repeated diversifications of mammalian taxa that evolved larger body size. The broad size ranges of terrestrial mammals were enabled, in large part, by shifts from crouched to more erect locomotor limb posture. As a point of contrast, when freed from gravitational constraints linked to a terrestrial lifestyle, aquatic mammals evolved extremely large size, achieving sizes well beyond those of any other mammalian group. Similar patterns of an evolutionary increase in body size have notably occurred within terrestrial archosaurs and extant birds. However, whether or not these diversifications in body size evolution were facilitated by shifts in locomotor limb posture similar to those observed within extant mammals demands further study. Recent innovative musculoskeletal modeling analyses (Hutchinson 2004a, 2004b) suggest that such shifts in posture likely occurred in extinct bipedal theropods and extant birds, facilitating the diversification of size within these groups.

In general, the ability of terrestrial animals to move across a range of different speeds has been facilitated by the evolution of a characteristic series of gaits. Across their diversity, individual gaits are distinguished by timing patterns of limb movement and mechanical energy changes of the animal's center of mass (Cavagna et al. 1977; Gambaryan 1974; Hildebrand 1989). Despite the evolutionary conservation of terrestrial gait patterns across species, different-sized mammals adopt varying limb postures to support their weight while moving over ground. These changes in limb posture are broadly linked to scale effects on musculoskeletal design due to differences in body size (Biewener 1989). In general, smaller mammals utilize more crouched limb postures, whereas large mammals move with upright postures during limb support. Size-related changes in limb posture affect the magnitude of limb joint torques that are produced by the ground reaction force (G) exerted when an animal's limb is in contact with the ground. With a more upright posture, G operates closer to the joints of the limb (lower moment arm, R), reducing the joint torques developed during limb support (fig. 13.1). Smaller animals with more crouched postures, therefore, must withstand larger joint torques (for their size) than large animals.

The joint torques generated by locomotion must be countered by moments produced by the limb muscles that transmit force across a joint during limb support. The moment arms (r) of these muscles can also vary across species, such that shorter muscle moment arms require larger muscle forces to counter a given joint torque. The magnitude of muscle force (F) that must be developed during the support phase of gait, therefore, depends on the relative magnitude of the muscles' moment arms to the ground reaction force moment arm (r/R), as well as the magnitude of G. In past work (Biewener 1989; Biewener et al. 2004), the lever arm ratio r/R has been defined as the muscle's *effective mechanical advantage* (EMA) because it relates the magnitude of muscle force required to counter the torque produced by the ground reaction force at a particular joint. This can be integrated over the time period of limb support as the ratio of ground reaction force impulse to muscle impulse:

$$r/R = \int G / \int F \quad (1)$$

Eq(1) is most accurately evaluated for changes in muscle moment arm over time, along with changes in R, G, and F over time. However, for most muscles and joints, changes in r are small over the ranges of joint motion that occur during limb support relative to changes in R, G, and F. In instances where more than one muscle acts as an agonist to produce the joint torque, evaluation of r/R can be further refined by determining a weighted mean muscle moment arm based on the relative fiber cross-sectional areas of the individual muscles, normalized to total agonist muscle fiber cross-sectional area (Biewener 1989). An advantage of using the ratio of ground impulse to muscle impulse to quantify muscle EMA is that it encompasses the entirety of limb support (Biewener et al. 2004; Roberts et al. 1998). Because the moment arm ratio r/R approaches infinity when the ground reaction force passes through a joint center and $R \to 0$, in past studies muscle EMA has been determined for the period of support when muscle force exceeds 25% F_{max} for a stride (Biewener 1989; Biewener 1990).

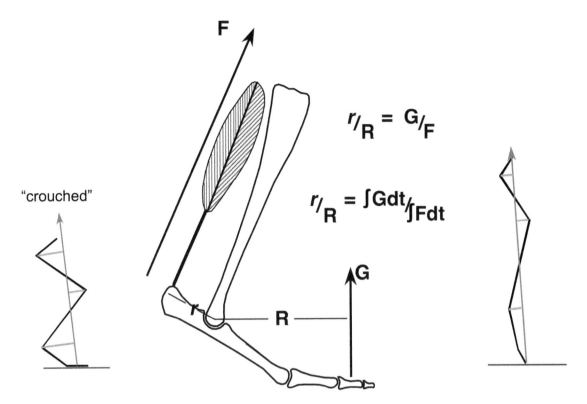

"crouched"

$$r/R = G/F$$

$$r/R = \int Gdt / \int Fdt$$

FIG. 13.1 (A) Limb muscle EMA is defined as the moment arm ratio r/R, which equals the impulse ratio $\int Gdt/\int Fdt$. (B) Limb muscle EMA increases for animals with more erect limb postures (shown at the hip, knee, and ankle joints) due to the reduction in R (green lines) relative to r over the stance phase of the stride.

Muscle EMA is the inverse of muscle gearing (R/r), which has been used to characterize the ratio of segmental angular motion produced by a given muscle shortening velocity (Carrier et al. 1998; Carrier et al. 1994). Past work shows that muscle gearing changes to offset force-velocity effects of ankle extensor muscles and enhances the ability of animals to accelerate (Carrier et al. 1994; Williams et al. 2009). Low muscle gearing corresponds to slower movements, but results in a high muscle EMA (i.e., low muscle force requirement for resisting external loads, such as the ground reaction force G). Changes in muscle EMA have also been shown to underlie the jumping ability of frogs. Roberts and Marsh (2003) showed that a bullfrog's plantaris (ankle extensor) begins to contract and develop force before the frog begins to move, with its muscle operating with a low EMA (or high gear ratio) due to its crouched prejump posture. This enables the plantaris to stretch its tendon, storing elastic energy that is released at a much higher rate than the muscle can provide directly, enhancing the frog's jumping performance.

In this chapter, I reappraise the scaling of muscle EMA across different mammalian groups with the goal of addressing how broadly muscle EMA scaling serves as a mechanism to maintain musculoskeletal stresses (force/area of bone or muscle tissue) within safe limits to ensure that large and small terrestrial mammals operate with similar safety factors (most commonly defined as the ratio of failure stress/peak functional stress) (Alexander 1981). I also compare how muscle EMA changes across gait and running speed in humans, as compared to large and small mammalian quadrupeds, and how it changes when animals accelerate versus when moving at a steady speed.

Scaling of Muscle EMA within Mammals

The scaling of muscle EMA depends on obtaining time-varying force platform recordings of ground reaction force G, in relation to limb kinematics and functional measures of muscle moment arms at different joints within the limb (fig. 13.2). This involves experimental

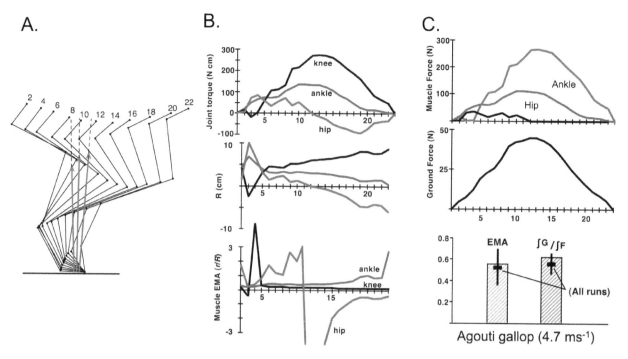

FIG. 13.2 (A) Shows variation in limb position over several frames of film (2 to 22) together with changes in the ground reaction force G (red vector arrows) for three successive frames from which the position of hind limb joints of an agouti galloping at 4.7 ms⁻¹ were obtained and synchronized to determine external joint torques shown in (B), along with changes in R and in muscle EMA through the stance phase (for all frames 1-24). Because G passes through the hip joint in frame 12, hip EMA approaches infinity (EMA at a joint is therefore only summed when muscle forces exceed 25% of Fmax). (C) Shows extensor muscle forces determined at each joint through time along with changes in G, as well as the computed hind limb EMA compared with impulse ratio ∫G/∫F (the mean ± SD determined by each method is also shown for all hind limb experimental trials; N = 22).

measurements of animals moving at a variety of speeds and different gaits over a force platform embedded in a runway, while the animal's joint centers are imaged in lateral view at high speed (100–250 fps). The analysis of external joint moments, muscle forces, and muscle EMA (fig. 13.2) has focused on parasagittal limb movements and fore-aft horizontal and vertical components of G, ignoring limb movement and forces in the mediolateral plane. Although mediolateral loading has been indicated in some crouched mammalian quadrupeds (Gosnell et al. 2011), for most terrestrial mammals with mainly parasagittal limb movements this is a reasonable simplifying approach, given the challenge of obtaining such recordings for a variety of different sized mammals. Synchronization of ground force components to film or video frames by an electronic shutter enables a frame-by-frame free-body analysis of external joint torques and muscle forces required to resist those torques. Overall limb muscle EMA is then determined

by averaging muscle EMA calculated for individual joints (forelimb: shoulder, elbow, wrist, and metacarpophalangeal joints; hind limb: hip, knee, ankle, and metatarsophalangeal joints) over the period of limb support. Figure 13.2C shows that muscle EMA, determined as r/R, is slightly less (7% for all agouti hind limb runs analyzed) than the impulse ratio $\int G/\int F$, due to the need to exclude periods of limb support when G passes through or close to a joint center, as it does in frame 12 of the example agouti gallop trial shown in figure 13.2.

For my original evaluation of muscle EMA scaling within terrestrial mammals (Biewener 1989), I examined four quadrupedal rodents (mouse, elephant shrew, chipmunk, and ground squirrel), one bipedal rodent (kangaroo rat), a carnivoran (dog), and two ungulates (goat and horse). Results for these species yielded a scaling of forelimb EMA = $0.233M^{0.246\ (\pm\ 0.087\ 95\%CI)}$ similar to hind limb EMA = $0.256M^{0.271\ (\pm\ 0.079\ 95\%CI)}$. Within four of the species (chipmunk, ground squirrel, dog, and horse), for which a range of speeds and gaits (walk, trot, and gallop) were recorded, forelimb and hind limb EMA did not vary significantly with speed or gait ($p > 0.05$ in all cases). A subsequent review of terrestrial biomechanics (Biewener 1990) included two additional species (prairie dog and white-tailed deer), which did not significantly alter the scaling patterns that emerged (forelimb EMA α $M^{0.251\ (\pm 0.065\ 95\%CI)}$ and hind limb EMA α $M^{0.273\ (\pm 0.072\ 95\%CI)}$, with an overall scaling of muscle EMA α $M^{0258\ \pm\ (0.043\ 95\%\ CI)}$).

Because of EMA scaling, muscle forces decrease relative to muscle mass and body mass (α $M/M^{0.258}$ α $M^{0.742}$) in larger terrestrial mammals. When compared with the scaling of muscle fiber cross-sectional area (α $M^{0.80}$) (Alexander et al. 1981) and bone cross-sectional area (α $M^{0.72}$) (Alexander et al. 1979; Biewener 1982), the reduction in muscle force relative to body weight enables terrestrial mammals to maintain similar peak muscle stresses (α $M^{-0.06}$) and bone stresses (α $M^{0.02}$) at comparable levels of performance (e.g., the quadruped trot-gallop transition speed). Thus, changes in locomotor limb posture compensate for the adverse scaling of force transmission by muscles and bones (Biewener 1990), which principally depends on cross-sectional area, relative to changes in loading that depend on body volume (or weight).

One limitation of this earlier work was the biased sample of rodents at the small end and ungulates on the large end of the mammalian size range. In part, this reflects the fact that taxonomic differences for terrestrial mammals are unavoidably size-biased, given that rodents occupy smaller body sizes than ungulate species, with limited overlap of their size ranges. Carnivorans occupy intermediate size ranges, as do primates, but both groups are either difficult to study across a broad taxonomic sample, or exhibit diverse locomotor adaptations that include arboreal and fossorial habits.

Does Limb Muscle EMA Scale Differently within More Closely Related Taxonomic Groups?

In order to test whether rodents, as a group, exhibit limb muscle EMA scaling similar to mammals more generally, two larger species (agoutis—mean body mass: 2.94 kg; and capybaras—mean body mass: 34.4 kg) were studied using methods similar to those described above. Measurements of muscle EMA for different forelimb

and hind limb joints within the two species, similar to other sampled quadrupeds, exhibited little change with speed and gait (fig. 13.3). With the inclusion of these two larger rodent species, rodent forelimb EMA scaled as $0.27M^{0.24\ (\pm 0.12\ 95\%\ CI)}$ and rodent hind limb EMA scaled as $0.27M^{0.25\ (\pm 0.14\ 95\%\ CI)}$—not significantly different for all mammals that were sampled (fig. 13.4) (Biewener 2005). However, this comparison is again limited by the fact that 9 of the 14 studied species are rodents.

In a study of felids based on 30 Hz video recordings of walking cats ranging from 0.4 to 200 kg, Day and Jayne (2007) found little kinematic evidence for a size-related change to more erect limb posture within this group of feline predators. Although Day and Jayne's study did not record ground reaction forces, muscle moment arms, or record faster speeds and gaits, and estimated joint centers from markers placed on the animal's fur, their results suggest the possibility that more closely related taxa may not show size-dependent changes in limb posture. Analysis of muscle moment arm scaling across different-sized terrestrial mammals (Biewener 1990) indicates that r scales α $M^{0.43}$, demonstrating that a substantial component of muscle EMA scaling derives from allometric changes in muscle moment arms with size, in addition to changes in limb posture that affect external joint moments. An analysis of how muscle moment arms (r) scale within felids that could test for this effect, however, has not yet been performed. With no change in postural effects on muscle EMA for the 500-fold change in felid body mass studied by Day and Jayne (2007), and no allometric change in muscle moment arms, musculoskeletal force requirements would be predicted to increase 7.8-fold to maintain comparable locomotor performance assuming geometric scaling (for which stress is predicted to increase α $M^{1/3}$). This is substantial and would indicate that larger cats cannot achieve accelerations comparable to smaller cats. Changes in r across this size range (similar to those reported for other quadrupedal mammals) (Biewener 1990) would reduce weight-related stresses, but at the expense of accelerating limb segments for higher speed movement due to reduced muscle gearing (Carrier et al. 1994).

Another group in which little change of size-related limb posture and its effect on muscle EMA has been observed is the Macropoididae (Bennett and Taylor 1995; McGowan et al. 2008) (fig. 13.4). Experimental measurements of ground reaction forces and limb

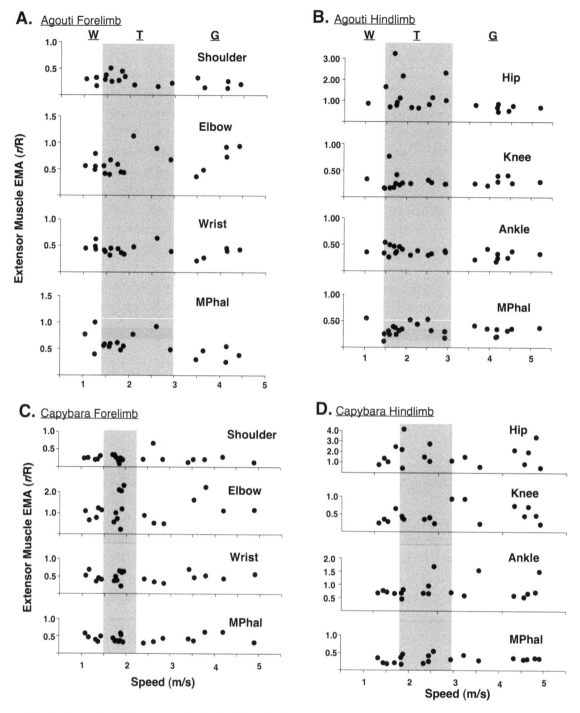

FIG. 13.3 Variation in muscle EMA versus speed and gait (W = walk, T = trot, G = gallop) at different joints for the forelimb (A) and hind limb (B) of agoutis and the forelimb (C) and hind limb (D) of capybaras. For all limb joints of both species, no significant change in muscle EMA (r/R) is observed versus speed and gait (p > 0.05 in all cases).

kinematics (McGowan et al. 2008) or estimates based on the assumption of a vertical G acting on the limb at mid-support (Bennett and Taylor 1995) indicate that macropodoids exhibit a slight positive scaling of limb EMA ($\alpha M^{0.09}$) or no significant scaling ($\alpha M^{0.0}$). In either case, the lack of a substantial increase in limb muscle EMA within this group appears to correlate with the reliance on elastic energy savings in distal hind limb tendons of larger macropodoids (Alexander and Vernon 1975; Biewener and Baudinette 1995). This energy

ANDREW A. BIEWENER

savings was likely linked to the evolution of grazing for dispersed food resources in arid regions, as well as the ability to carry pouch young at reduced energy cost (Baudinette and Biewener 1998).

Changes in Human Muscle EMA with Gait Are Linked to Differences in Transport Cost

In contrast to quadrupedal mammals that exhibit little or no change in limb EMA with speed and gait, humans exhibit a significant reduction in limb EMA when they change from a walking to a running gait (Biewener et al. 2004) (fig. 13.4). The high limb EMA of humans during walking may reflect selection for economical walking. The decrease in limb EMA during running results mainly from the change in limb posture that occurs at the knee joint. Walking humans employ an "inverted pendulum" mechanism to exchange potential and kinetic energy of the center of mass (Cavagna et al. 1977), in which the knee is maintained in an extended configuration (162° at peak torque). However, when humans run the knee is more flexed (135° at peak torque), resulting in a 4.9-fold increase in knee torque and a 68% reduction of knee extensor EMA. Although this enables humans to achieve

spring-mass bouncing mechanics to conserve center of mass potential and kinetic energy within elastic tissues of the limb during running (Blickhan 1989; Cavagna et al. 1977; McMahon 1985), it incurs an increased cost for muscle force generation. During running the volume of active muscle increases 2.3 fold for the limb as a whole, and increases 4.9 fold within knee extensors, compared to walking (Biewener et al. 2004). This increase in muscle force generating requirements at the knee correlates with the 50% increase in the energy cost of transport (Joules/m) of human running compared with walking (Farley and McMahon 1992; Margaria 1976). Interestingly, the decrease in muscle EMA that occurs from walking to running in humans (fig. 13.4) also corresponds to the higher mass-specific cost of human running compared to that predicted for a quadruped of similar size (Taylor et al. 1970).

Limb and Muscle EMA during Acceleration and Changes in Running Grade

In contrast to steady speed running, limb and muscle EMA is expected to change when animals decelerate or accelerate due to shifts in ground reaction force

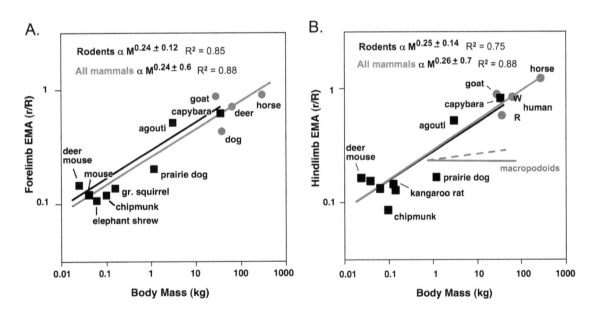

FIG. 13.4 Scaling of forelimb (A) and hind limb (B) EMA in rodents (black points and least-squares regression line) versus more general pattern across mammalian taxa (red points and least-squares regression line). The regression equations with scaling exponents (± the 95% confidence intervals) and correlation coefficients are indicated. In (B) the scaling of hind limb EMA in macropodoids is shown in blue solid (α M⁰; Bennett and Taylor 1995) and blue dashed (α M⁰·⁰⁹; McGowan et al. 2008) lines. Also shown in (B) is the mean hind limb muscle EMA scaling of humans during walking (W) versus running (R) (Biewener et al. 2004). Human running EMA falls significantly below the 95% CI for the overall mammal scaling slope (not shown).

orientation and corresponding changes in limb posture. In a recent study, Williams and colleagues (2009) studied changes in hind limb muscle EMA and gearing as greyhounds accelerated over a series of force platforms. Consistent with earlier results for other quadrupeds (Biewener 1989), Williams et al. found no significant change in limb muscle EMA when greyhounds moved at different speeds and gaits. However, limb EMA decreased substantially (33%) with increasing acceleration (from 0 to 5 ms^{-2}). A decrease in limb EMA likely enables the limb to remain in contact with the ground for a longer time period, increasing the duration of force generation for acceleration. More importantly, Williams et al. note that a more crouched posture during acceleration increases the limb's mechanical advantage for generating horizontal forces, which are critical to enhanced acceleration performance. The reduction in overall limb EMA, favoring enhanced acceleration in the horizontal direction, also corresponds to most of the required power for acceleration being generated by the greyhound's hip extensor muscles during accelerating strides (Williams et al. 2009).

Muscle EMA is also likely to change when animals move over different grades due to shifts in locomotor limb posture relative to the ground reaction forces produced. In one study, Roberts and Belliveau (2005) observed a reduction in hip extensor muscle EMA when humans ran up a 12° incline compared with level running. This reflected the increase in muscle and joint work required for incline running. Nearly all of the required increase in limb work was produced at the hip by means of a substantial increase in hip joint torque.

Summary and Suggestions for Future Study

Changes in overall limb and muscle EMA are critical for effective body weight support in terrestrial mammals that vary over an enormous range in size. Measurements of terrestrial quadrupeds that span four orders of magnitude in body mass (0.3 to 300 kg) show that a size-dependent increase in limb EMA achieved by moving with more erect locomotor postures and an allometric increase in extensor muscle moment arms enable a reduction in the magnitude of force that must be generated and transmitted by the musculoskeletal system (Biewener 1990). This allows terrestrial mammals to operate with similar peak muscle and bone stresses,

assuring an adequate margin of safety for their musculoskeletal system. Although this sample is biased by the number of rodent species studied (9 of 14) at the lower range of size and by the ungulate taxa sampled (4 of 14) at the upper range of size, limb EMA scaling within rodents is the same as that observed for the entire mammalian sample (Biewener 2005). Nevertheless, more closely related taxa may well exhibit reduced effects of size on locomotor posture, with attendant biomechanical consequences. There is some evidence for this within felids (Day and Jayne 2007), and clearer evidence for minimal change in limb EMA within macropodoids (Bennett and Taylor 1995; McGowan et al. 2008). Within larger macropodoids, this appears to reflect selection for increased elastic energy recovery during hopping.

Scaling comparisons of musculoskeletal mechanics and limb posture within more closely versus more broadly related groups represent an area that is ripe for future study. Despite the obvious challenges of working with felids, obtaining ground reaction force data and kinematics at faster speeds and gaits would help to answer the question of whether evolutionary shifts in body size within this clade have retained a characteristic limb mechanical advantage, distinct from other mammalian groups, and if so, why? Ground-running birds represent a second group of experimentally accessible animals that have evolved a large size range (0.02 to 60 kg: painted quail to ostrich) and would be expected to face similar size scaling demands for shifts in locomotor posture to maintain effective weight support. Innovative musculoskeletal modeling analyses carried out by Hutchinson (2004a) suggest this is the case. Finally, fossil archosaurs are another group of terrestrial vertebrates that display considerable diversity of size, post-cranial anatomy and bipedal versus quadrupedal gait patterns (Gatesy 1991; Carrano 2000). Reconstruction of archosaur limb anatomy, combined with computational modeling approaches, as have been applied to *T. rex* (Hutchinson and Garcia 2002) and other bipedal archosaurs (Hutchinson 2004b), would help to illuminate how reconstructed limb postures and musculoskeletal mechanics interact to predict patterns of locomotor evolution within this diverse group.

For quadrupedal mammals that have been studied, little change in limb muscle EMA is apparent with changes in speed and gait. However, as with certain

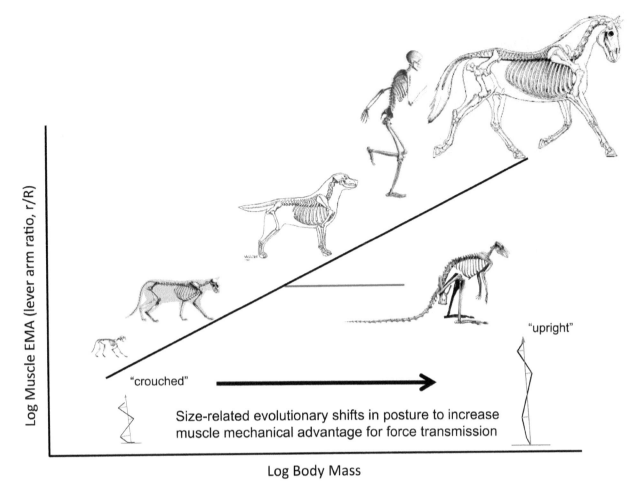

FIG. 13.5 Evolutionary changes in body size across terrestrial mammals are linked to shifts in limb posture that increase muscle mechanical advantage relative to ground reaction forces, providing similar safety factors (peak musculoskeletal stresses/failure stress limits) for large and small mammals. More erect locomotor postures in larger animals, however, are likely linked to reduced maneuverability and acceleration capacity.

other features (Bramble and Lieberman 2004), humans appear to be unique in regard to shifts in kinematics (Gatesy and Biewener 1991) and underlying biomechanics that occur with a change of gait from walking to running. The reduction in limb EMA when humans run (Biewener et al. 2004) corresponds to an increased energetic transport cost of running compared to walking (Farley and McMahon 1992; Margaria 1976) and to the higher cost of human running compared to trotting in quadrupeds of similar size (Taylor et al. 1970). Still, the reduction in human limb EMA during running appears to favor effective spring-mass mechanics to increase elastic energy recovery, reducing muscle work and overall metabolic cost.

Studies of limb and muscle EMA are challenging to carry out on diverse taxa of different size given the need for obtaining ground reaction forces and high-speed limb kinematics, together with musculoskeletal anatomical data. As a result, past work has mainly focused on steady speed locomotion. Studies of greyhounds (Williams et al. 2009) and humans (Roberts and Belliveau 2005), however, show that limb EMA decreases when greater joint work is required for acceleration or incline running. Reduced muscle EMA results from changes in limb posture that increase the magnitude of joint torque and work output compared with steady speed locomotion. Although not well studied, it seems likely that other terrestrial animals show similar shifts in limb EMA linked to changes in limb posture and ground reaction force orientation that occur during acceleration and deceleration. This represents another area needing further investigation. Many animals regularly change speed and direction in their natural environment. Consequently, this must influence how evolution has shaped the biomechanics of limb function.

As past studies of tetrapod locomotor evolution have shown (Gray 1968; Gregory 1912; Jenkins 1971a, 1971b), shifts in limb posture underlie key transitions in locomotor performance, affecting both the biomechanics and energetic cost of movement. Much of this work has focused on more cursorial taxa that move with upright postures and parasagittal limb excursion. As demonstrated in a seminal study (Jenkins 1971a), studies of non-cursorial mammals are critical and also serve to illuminate evolutionary patterns that underlie the diversification of terrestrial mammals from their tetrapod ancestors.

* * *

Acknowledgments

The author thanks David R. Carrier and Richard W. Blob for critical reviews of an earlier manuscript draft. The author is indebted to Professor Farish A. Jenkins Jr. for his mentorship, sage and entertaining advice, collegial support, and scientific exploration of all things vertebrate over many years of shared fascination of their biology; and for his dedication to students, teaching, and scholarly research. Farish, you are and will remain an inspiration to us all.

References

Alexander, R. M. 1981. Factors of safety in the structure of animals. Sci. Prog. 67:119–140.

Alexander, R. M., G. Goldspink, A. S. Jayes, G. M. O. Maloiy, and E. M. Wathuta. 1979. Allometry of the limb bones of mammals from shrew (Sorex) to elephant (Loxodonta). J. Zool., Lond. 189:305–314.

Alexander, R. M., A. S. Jayes, G. M. O. Maloiy, and E. M. Wathuta. 1981. Allometry of the leg muscles of mammals. J. Zool., Lond. 194:539–552.

Alexander, R. M., and A. Vernon. 1975. The mechanics of hopping by kangaroos (Macropodidae). J. Zool., Lond. 177:265–303.

Baudinette, R. V., and A. A. Biewener. 1998. Young wallabies get a free ride. Nature 395:653–654.

Bennett, M. B., and G. C. Taylor. 1995. Scaling elastic strain energy in kangaroos and the benefits of being big. Nature 378:56–59.

Biewener, A. A. 1982. Bone strength in small mammals and bipedal birds: do safety factors change with body size? J. Exp. Biol. 98:289–301.

Biewener, A. A. 1989. Scaling body support in mammals: limb posture and muscle mechanics. Science 245:45–48.

Biewener, A. A. 1990. Biomechanics of mammalian terrestrial locomotion. Science 250:1097–1103.

Biewener, A. A. 2005. Biomechanical consequences of scaling. J. Exp. Biol. 208:1665–1676.

Biewener, A. A., and R. V. Baudinette. 1995. In vivo muscle force and elastic energy storage during steady-speed hopping of tammar wallabies (Macropus eugenii). J. Exp. Biol. 198:1829–1841.

Biewener, A. A., C. T. Farley, T. J. Roberts, and M. Temaner. 2004. Muscle mechanical advantage of human walking and running: implications for energy cost. J. Appl. Physiol. 97:2266–2274.

Blickhan, R. 1989. The spring-mass model for running and hopping. J. Biomech. 22:1217–1227.

Blob, R. W., and A. A. Biewener. 1999. In vivo locomotor strain in the hindlimb bones of Alligator mississippiensis and Iguana iguana: implications for the evolution of limb bone safety factor and non-sprawling limb posture. J. Exp. Biol. 202:1023–1046.

Borelli, G. A. 1685. De Motu Animalium. Rome.

Bramble, D. M., and D. E. Lieberman. 2004. Endurance running and the evolution of Homo. Nature 432:345–352.

Carrano, M. T. 2000. Homoplasy and the evolution of dinosaur locomotion. Paleobiol. 26:489–512.

Carrier, D. R. 1987. The evolution of locomotor stamina in tetrapods: circumventing a mechanical constraint. Paleobiol. 13:326–341.

Carrier, D. R., C. S. Gregersen, and N. A. Silverton. 1998. Dynamic gearing in running dogs. J. Exp. Biol. 201:3185–3195.

Carrier, D. R., N. C. Heglund, and K. D. Earls. 1994. Variable gearing during locomotion in the human musculoskeletal system. Science 265:651–653.

Cavagna G. A., N. C. Heglund, and C. R. Taylor. 1977. Mechanical work in terrestrial locomotion: two basic mechanisms for minimizing energy expenditures. Am. J. Physiol. 233:R243–261.

Day, L. M., and B. C. Jayne. 2007. Interspecific scaling of the morphology and posture of the limbs during the locomotion of cats (Felidae). J. Exp. Biol. 210:642–654.

Farley, C. T., and T. A. McMahon. 1992. Energetics of walking and running: insights from simulated reduced-gravity experiments. J. Appl. Physiol. 73(6):2709–2712.

Gambaryan, P. 1974. How Mammals Run: Anatomical Adaptations. New York: Wiley.

Gatesy, S. M. 1991. Hind limb scaling in birds and other theropods: implications for terrestrial locomotion. J. Morph. 209: 83–96.

Gatesy, S. M., and A. A. Biewener. 1991. Bipedal locomotion: effects of speed, size and limb posture in birds and humans. J. Zool., Lond. 224:127–147.

Gosnell, W. V., M. T. Butcher, T. Maie T, and R. W. Blob. 2011. Femoral loading mechanics in the Virginia opossum, *Didelphis virginiana*: torsion and mediolateral bending in mammalian locomotion. J. Exp. Biol. 214:3455–3466.

Gray, J. 1968. Animal Locomotion. New York: Norton & Co.

Gregory, W. K. 1912. Notes on the principles of quadrupedal locomotion and on the mechanism of the limbs in hoofed animals. Ann. N.Y. Acad. Sci. 22:267–294.

Hildebrand, M. 1989. The quadrupedal gaits of vertebrates. Bioscience 39:766–775.

Hutchinson, J. R. 2004a. Biomechanical modeling and sensitivity analysis of bipedal running ability. I. Extant taxa. J. Morph. 262:421–440.

Hutchinson, J. R. 2004b. Biomechanical modeling and sensitivity analysis of bipedal running ability. II. Extinct taxa. J. Morph. 262:441–460.

Hutchinson, J. R., and M. Garcia. 2002. Tyrannosaurus was not a fast runner. Nature 415:1018–1021.

Jenkins, F. A., Jr. 1971a. Limb posture and locomotion in the Virginia opossum (*Didelphis marsupialis*) and in other non-cursorial mammals. J. Zool., Lond. 165:303–315.

Jenkins, F. A., Jr. 1971b. The postcranial skeleton of African cynodonts. Bull. Peabody Mus. Nat. Hist. 36:1–216.

Jenkins, F. A., Jr., and Parrington, F. R. 1976. The postcranial skeletons of Triassic mammals *Eozostrodon, Megazostrodon* and *Erythrotherium*. Phil. Trans. Roy. Soc. Series B 173:387–431.

Marey, E. J. 1901. Animal Mechanism: A Treatise on Terrestrial and Aerial Locomotion. New York: D. Appleton & Co. 283 pp.

Margaria, R. 1976. Biomechanics and Energetics of Muscular Exercise. Oxford: Oxford University Press. 146 pp.

McGowan, C. P., J. Skinner, and A. A. Biewener. 2008. Hind limb scaling of kangaroos and wallabies (superfamily Macropodoidea): implications for hopping performance, safety factor and elastic savings. J. Anat. 212:153–163.

McMahon, T. A. 1985. The role of compliance in mammalian running gaits. J. Exp. Biol. 115:263–282.

Roberts, T. J., and R. A. Belliveau. 2005. Sources of mechanical power for uphill running in humans. J. Exp. Biol. 208:1963–1970.

Roberts, T. J., M. S. Chen, and C. R. Taylor. 1998. Energetics of bipedal running. II. Limb design and running mechanics. J. Exp. Biol. 210:2753–2762.

Roberts, T. J., and R. L. Marsh. 2003. Probing the limits to muscle-powered accelerations: lessons from jumping bullfrogs. J. Exp. Biol. 206:2567–2580.

Taylor, C. R., K. Schmidt-Nielsen, and J. L. Raab. 1970. Scaling of energetic cost of running to body size in mammals. Am. J. Physiol. 219(4):1104–1107.

Williams, S. B., J. R. Usherwood, K. Jespers, A. J. Channon, and A. M. Wilson. 2009. Exploring the mechanical basis for acceleration: pelvic limb locomotor function during accelerations in racing greyhounds (*Canis familiaris*). J. Exp. Biol. 212(4):550–565.

14

Evolution of Whales from Land to Sea

Philip D. Gingerich*

Whales have always been mysterious. In the biblical story of creation, Genesis 1:1–31, whales were brought forth in the water one full creation-day before cattle and other beasts on land. Aristotle called whales "the strangest of animals." He grouped baleen and toothed whales together, and referred to them collectively as Cete or Cetacea. Aristotle separated whales from fish and recognized that they had many mammalian characteristics. In *De Partibus Animalium*, Aristotle distinguished whales by their spouting respiration and by their possession of lungs, mammary glands, and testes. Further, Aristotle wrote that if we define water-animals as those that take in water and define land-animals as those that take in air, then Cetacea are not easily classified as one or the other because they "do partly one and partly the other" (Ogle 1882). John Ray (1671) regarded the porpoise he dissected as a fish with characteristics of quadrupedal mammals. Later, in his *Synopsis Methodica Animalium Quadrupedum et Serpentini Generis*, Ray (1693) distinguished air-breathing cetaceans from the other gill-breathing fishes, and distinguished aquatic cetaceans from terrestrial fur-bearing "Quadrupedia." Comparisons were made, but the unity of cetaceans and land-living mammals was not yet clear.

* Museum of Paleontology, and Department of Earth and Environmental Sciences, University of Michigan

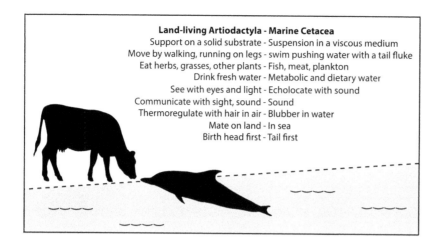

FIG. 14.1 A summary of anatomical, physiological, and behavioral contrasts between land-living Artiodactyla and marine Cetacea. Cows and dolphins are among the most divergently specialized of terrestrial artiodactyls and sea-living cetaceans, but there is an evolutionary connection. The transition from land to sea, when Artiodactyla became Cetacea, was made by early representatives of both groups (see text).

Study of fossil whales, featured here, goes back to the very beginnings of paleontology in *La vana speculazione disingannata dal senso*, or "Vain speculation undeceived by sense" of Agostino Scilla (1670). Scilla argued that petrified fossils, like the dentary of *Squalodon* he illustrated from a tuff in Malta, were not ruses of God to test our faith but rather objects worthy of contemplation. The dentary didn't help Scilla distinguish whales from fish, but finding a fossil cetacean was nevertheless, as we shall see, an early step toward understanding the systematic position of whales.

The origin of Cetacea, long shrouded in mystery, has required the careful attention and comparative observation of investigators in many disciplines. The reason is clear: whales truly are among "the strangest of animals." Cetaceans have mammalian characteristics but live like fish. Whales meet their land-mammal cousins at the water's edge today, changed in so many anatomical, physiological, and behavioral characteristics that a connection is hard to imagine (fig. 14.1). The evolution of whales from land to sea is one of the great macroevolutionary transformations, and a transformation for which there is now much fossil evidence.

Phylogeny and Classification of Cete or Cetacea

Three competing ideas emerge as we review the history of ideas on the phylogeny and classification of Cete or Cetacea. Early investigators thought: (1) whales are very different morphologically from other animals, and possibly had a long history of independence since the Mesozoic; (2) whales share some features and appear most closely related to Cenozoic hoofed mammals, specifically Artiodactyla; or, alternatively, (3) whales share some features and appear most closely related to Cenozoic clawed mammals, specifically Creodonta or Carnivora. One form of synthesis that appears from time to time is the inference that hoofed and clawed mammals are themselves closely related phylogenetically. These ideas are intertwined in development to a degree that defies simple exposition.

The tenth edition of Carolus Linnaeus's *Systemae Naturae* (1758) is taken as the starting point of formal biological nomenclature. Here Linnaeus classified mammals in eight orders, from primal to primeval, starting at the top with Primates and working down to Cete. He cited a spiracula or blow hole on top of the head, pectoral fins, a horizontal tail, and the absence of claws or hooves, to distinguish whales from other mammals. Linnaeus used the conformation of the teeth (or lack of teeth), and the position of the "fistula" or spiracula to distinguish members within the order. Reading Linnaeus today, we are given no sense of phylogenetic relationships unless we consider his ranking of "Bellua" comprising *Hippopotamus* ("river horses") and *Equus* (true horses) a step above Cete to imply somehow their derivation from Cete, and his ranking of "Pecora" (Artiodactyla) a step above Bellua to imply somehow their derivation from Bellua and derivation in turn from Cete. One of Linnaeus's greatest contributions to taxonomy and biological classification was recognition that milk-producing mammae are important, which led to formal inclusion of Cete or Cetacea within the class Mammalia. Both are innovations of the tenth edition of Linnaeus

(1758). In previous editions, starting with the first, Linnaeus (1735) classified whales (and manatees) as fish.

Georges Cuvier (1817) outlined a scheme similar to that of Linnaeus, with primates at the top, and cetaceans (again grouped with sirenians) in a basal group. Henri de Blainville (in Gervais 1836) took a different tack, and grouped cetaceans with armadillos, aardvarks, anteaters, and pangolins as "Edentés." Ernst Haeckel (1866) grouped Cetacea (including sirenians) with Ungulata (Perissodactyla and Artiodactyla) as orders of "indeciduate" placental mammals. Richard Owen (1868) included Cetacea and Sirenia in "Mutilata" and separated these from Unguiculata (clawed mammals) and Ungulata (hoofed mammals). Owen grouped all three in "Gyrencephala," by which he meant mammals with a convoluted cerebrum. Theodore Gill (1870) classified Cete (zeuglodonts, odontocetes, and mysticetes) and Ferae (fissiped carnivores and pinnipeds), as orders, in what he called a "Feral Series." Edward Cope (1891) followed Owen, setting Cetacea and Sirenia aside in Mutilata.

In the 20th century, Max Weber (1904) placed Cetacea as a distinct order between Carnivora on one hand and all of the hoofed orders on the other. Frederick True (1908) regarded zeuglodonts or archaeocetes as unrelated to Cetacea, and Cetacea to be an extremely ancient branch unconnected to any known land mammals. William Gregory (1910) included zeuglodonts in Cetacea and regarded Cetacea to be closer to creodonts than to ungulates. His phylogenetic diagram represented Carnivora and Artiodactyla as sister taxa, with Cetacea equidistant from both (Gregory 1910, fig. 31).

Jumping back in time, John Hunter (1787) was seemingly the first to associate whales with mammalian Artiodactyla when he argued that the digestive system and genitalia of modern whales are most similar to those of ruminants. These were interpreted, by implication, as shared-derived specializations. William Flower (1883a, 1883b) repeated the idea of a Cetacea-Artiodactyla relationship a century later, citing the elongated larynx, complex stomach, simple liver, respiratory organs, male and female reproductive organs, and fetal membranes. Flower, following earlier authors, attributed meaning to the common name "sea-pig," *meereswine*, or *porcpoisson*, which he thought the etymological origin of "porpoise." But Flower also wrote, realistically: "in the present state of our knowledge, the Cetacea are absolutely isolated, and

little satisfactory reason has ever been given for deriving them from any one of the existing divisions of the class [Mammalia] rather than from any other" (Flower 1883a).

Fossil whales gave a different perspective, as they became better known, with nothing to suggest relationship to Artiodactyla. Fossils came into play when Richard Owen, focused on dinosaurs, recognized Eocene *Basilosaurus* or "*Zeuglodon*" to be an archaic cetacean-like mammal (Owen 1841). Ernst Stromer (1903) collected new material and considered Owen's "*Zeuglodon*" to have triconodont teeth reminiscent of Jurassic mammals. *Protocetus*, described by Eberhard Fraas (1904), was even more interesting. It had upper molars with remnant protocones, which led Fraas to propose:

> Considered systematically I remove the Archaeoceti entirely from the Cetacea and regard them as a subset of Creodonta. Remaining whales can be retained as an independent group for as long as desired until we know their evolutionary history . . . Archaeoceti are closer to Pinnipedia, as they share a common terrestrial ancestry, namely again Creodonta. (Fraas 1904, 220; my translation)

There was a general consensus in the 1920s and 1930s, based on living and fossil whales, that the ancestry of cetaceans was to be sought in an ancient and primitive insectivore-carnivore stock (Winge 1921; Miller 1923; Kellogg 1928; Howell 1930). In 1936, Remington Kellogg published *A Review of the Archaeoceti* and considered the possible relationships of Archaeoceti to marine reptiles, to "Promammalia," and then to triconodont mammals, Marsupialia, Insectivora, Creodonta, Pinnipedia, Edentata, Perissodactyla, Sirenia, and finally mysticete and odontocete Cetacea. Kellogg concluded:

> In summation, it would appear that the evidence seems to point toward the concept that the archaeocetes are related to if not descended from some primitive insectivore-creodont stock, but they branched off from that stock before the several orders of mammals that reached the flood tide of their evolutionary advance during the Cenozoic era were sufficiently differentiated to be recognized as such. Morphologically the archaeocetes seem to stand relatively near to the typical Mysticeti and Odontoceti,

although all three suborders were separated from each other during a long interval of geologic time. It is not necessary to assume that any known archaeocete is ancestral to some particular kind of whale, for the archaeocete skull in its general structure appears to be divergent from rather than antecedent to the line of development that led to the telescoped condition of the braincase seen in skulls of typical cetaceans. On the contrary it is more probable that the archaeocetes are collateral derivatives of the same blood-related stock from which the Mysticeti and the Odontoceti sprang. (Kellogg 1936, 343)

Everhard Slijper (1936) had a different view in *Die Cetaceen*. Slijper regarded modern whales to be diphyletic, with Mysticeti and Odontoceti being derived, independently, from a Cretaceous insectivoran. Slijper's phylogeny is reproduced in figure 14.2. Slijper viewed Mysticeti and Archaeoceti as being more closely related to each other than either was to Odontoceti. The ideas of Kellogg and Slijper were combined more recently by Yablokov (1964), who wrote that "the similarity between the living suborders [Mysticeti and Odontoceti] is the result of convergence, and Archaeoceti are not ancestral to the other two."

The next study of importance, following those of Kellogg and Slijper, was George Simpson's classic *Principles of Classification, with a Classification of Mammals* (Simpson 1945). Here he listed Cetacea as an order in its own cohort Mutica, inserted between two other cohorts: Glires (Lagomorpha and Rodentia) on one hand and Ferungulata (Creodonta, Carnivora, Condylarthra, Proboscidea, Perissodactyla, Artiodactyla, etc.) on the other. Simpson explained this by writing:

Because of their perfected adaptation to a completely aquatic life, with all its attendant conditions of respiration, circulation, dentition, locomotion, etc., the cetaceans are on the whole the most peculiar and aberrant of mammals. Their place in the sequence of cohorts and orders is open to question and is indeed quite impossible to determine in any purely objective way. There is no proper place for them in a scala naturae or in the necessarily one-dimensional sequence of a written classification. Because of their strong specialization, they might be placed at the end, but this would remove them far from any possible ancestral or

related forms and might be taken to imply that they are the culmination of the Mammalia or the highest mammals instead of merely being the most atypical. A position at the beginning of the eutherian series would be even more misleading. They are, therefore, inserted into this series in a more or less parenthetical sense. They may be imagined as extending into a different dimension from any of the surrounding orders or cohorts.

It is clear that the Cetacea are extremely ancient as such and that none of the various proposals of exact source, such as that deriving them from certain creodonts, is very probable. They probably arose very early and from a relatively undifferentiated eutherian ancestral stock. On this basis they deserve to rank as a separate cohort, for which a Linnaean name is available and deserves resurrection. (Simpson 1945, 213–214)

Simpson's clear statement of the problem stimulated Alan Boyden and Douglas Gemeroy to seek resolution by taking a new and different approach. Boyden and Gemeroy made careful comparisons of the immunological reactions of cetacean blood serum proteins with those of other mammals, hoping to find evidence from a new source bearing on the relationships of Cetacea. Boyden and Gemeroy (1950) showed that serum proteins, especially albumins, of Cetacea and Artiodactyla are more similar than the corresponding proteins of the other orders tested. As a result, Cetacea were interpreted to have a closer blood and genetic relationship to Artiodactyla than to other orders of mammals. This evidence, and Boyden and Gemeroy's interpretation of it, were influential and became one of the earliest successful demonstrations of evolutionary relationships using something other than morphology.

Everhard Slijper (1966) revised his earlier interpretation of cetacean relationships (fig. 14.2) and, following Hunter and Flower, considered a great many reproductive characteristics of Cetacea to indicate relationship with ungulates, especially Artiodactyla. These characteristics include the position and the shape of the penis, the presence of a *musculus retractor penis*, the absence of an *os penis*, the microscopic structure of the penis (fibroelastic) associated with a very quick erection and rapid coitus, the size and shape of the spermatozoa, the structure of the uterine cervix, the fact

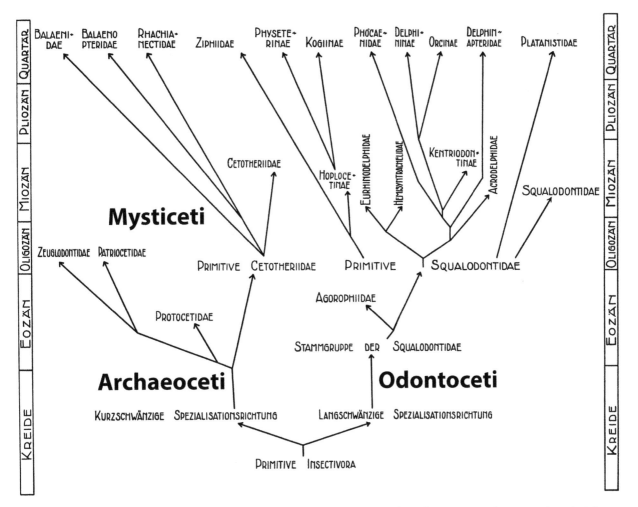

FIG. 14.2 Everhard Slijper's (1936) schematic representation of the phylogenetic relationships of cetaceans to other mammals, and relationships within Cetacea. Slijper hypothesized that Mysticeti and Odontoceti had independent origins from Cretaceous insectivores specialized with shorter and longer tails, respectively. Note Slijper's grouping of Archaeoceti with Mysticeti. In contrast, Kellogg (1936) regarded Archaeoceti, Mysticeti, and Odontoceti as having been separate through a long interval of geological time, and interpreted Mysticeti and Odontoceti as more closely related to each other than either was to Archaeoceti. Evidence today constrains cetacean history to Eocene and later times: Artiodactyla ancestral to whales appeared during the thermal maximum at the Paleocene-Eocene boundary (PETM; 55.8 Ma); Archaeoceti derived from Artiodactyla are known by the late early Eocene (ca. 53 Ma); and Mysticeti and Odontoceti derived from Archaeoceti appeared by the early Oligocene (ca. 33 Ma).

that cetaceans and many artiodactyls are uniparous and give birth to very well-developed precocial young, the shape of the chorionic vesicle, the development of the embryonic membranes, the diffuse epitheliochorial placenta, the presence of amniotic pearls, the presence of fructose in the fetal blood (and probably in the amniotic fluid), and the short umbilical cord.

Van Valen (1966) linked Cetacea to ungulates, including Artiodactyla, through Paleocene Mesonychidae and Arctocyonidae (later Mesonychia and Arctocyonia; Van Valen 1969). *Protocetus* was a key taxon, with

dental resemblances to mesonychid condylarths or "Acreodi," as other authors noted previously. Van Valen's evidence was not extensive, but he wrote in a concluding summary that "whales were probably derived from mesonychids in the Paleocene." The argument was bolstered by citation of Boyden and Gemeroy's (1950) serological evidence linking Cetacea and Artiodactyla, and Van Valen's hypothesized Cetacea-Mesonychidae-Arctocyonidae-Artiodactyla connection provided plausible reconciliation with the known fossil record. McKenna (1975) grouped Acreodi with Cetacea as orders in a "Mirorder" Cete. McKenna and Bell (1997) made the same grouping but as suborders in an order Cete.

Some more recent morphological studies have found Cetacea or Cete (including mesonychians) to be more closely related to Perissodactyla than to Artiodactyla (Prothero et al. 1988; Thewissen 1994; Heyning 1997), but other studies have found cetaceans to be closer to Artiodactyla than to Perissodactyla (Novacek 1992; O'Leary 1999; Luo and Gingerich 1999; Geisler and Uhen 2005; Spaulding et al. 2009).

Boyden and Gemeroy's (1950) immunological study of blood serum proteins inspired a series of later investigations of the phylogeny and classification of Cete or Cetacea that involved first immunology and immunological distance, cladistic analysis of amino acid sequence replacements, measurement of nucleic acid hybridization, and cladistic analysis of nucleotide sequence substitutions, and finally cladistic analysis of single-nucleotide polymorphisms or SNPs. Sarich (1993) pursued the immunology of serum albumins and transferrins, and found that cetaceans are close to non-suid artiodactyls, with hippos as a possible sister group. Irwin and Arnason (1994) compared mitochondrial cytochrome *b* gene sequences and found too that hippos are most closely related to cetaceans. Gatesy et al. (1996) sequenced casein milk protein genes, found whales closest to hippos, and predicted that new fossil discoveries should include basal cetaceans that have typical artiodactyl skeletal characters. Montgelard et al. (1997) compared complete mitochondrial cytochrome *b* and 12s rDNA sequences for 17 representatives of Cetacea and Artiodactyla, found the cetacean-hippo link, and coined the name Cetartiodactyla for the combined order. These findings were supported by other molecular studies in the 1990s (e.g., Shimamura et al. 1997, 1999; Ursing and Arnason 1998; Gatesy et al. 1999; Kleineidam et al. 1999; Nikaido et al. 1999).

It is safe to say that by the end of the 1990s the first of our three competing ideas was dead: no one any longer believed that whales are so different from other mammals that they had a long and independent history. This may be because the morphology and anatomical systems of mammals were now better studied, or it may be because they were supplemented with evidence from molecular genetics. Certainly the fossil record was known better in the late 1990s than it was when Simpson wrote in 1945—in terms of geographic coverage,

paleoenvironmental representation, and temporal resolution. The controversy at the end of the 1990s was whether the earliest whales with piscivorous teeth evolved from carnivorous ancestors as paleontologists believed, embodied in Van Valen's mesonychian hypothesis, or whether they evolved from herbivorous ancestors as molecular systematicists believed, embodied in the derivation from Artiodactyla and possibly from a hippo-related branch within Artiodactyla.

Fossil Whales and Evolution from Land to Sea

It is surprising in hindsight to see that in 1936 so notable a cetacean expert as Everhard Slijper hypothesized that Mysticeti and Odontoceti had independent origins from Cretaceous insectivores (fig. 14.2). It is surprising to see that Remington Kellogg could write, again in 1936, that archaeocetes are collateral derivatives of a blood-related stock from which the Mysticeti and the Odontoceti sprang. And it is surprising too to read that George Simpson thought, in 1945, that the place of whales within Mammalia was open to question and indeed impossible to determine in any purely objective way. A lot has changed in 70 years in our understanding of the genetic relationships of living animals and in our documentation of intermediate fossils linking living groups.

Our understanding of the evolution of whales from land to sea started with the description of *Protocetus atavus* by Fraas (1904). The cranium and primitive dentition of this genus received the most attention, but the specimen included postcranial remains as well. Van Valen wrote of Fraas's middle Eocene *Protocetus*:

> I see nothing whatever in the known anatomy of *Protocetus* . . . to preclude this part of the Protocetidae from having given rise to both suborders of recent whales. It is beautifully intermediate between primitive mesonychids and recent whales, although in most respects more similar to the latter. . . . It may be relevant that in *Protocetus* the sacrum is relatively well developed, indicating a relatively large pelvis and perhaps more or less functional hind legs. Adaptation to water was clearly imperfect at this stage from other evidence also (position of external nares, tooth shape, jaw form and musculature, relatively mesonychid-like vertebrae [notably the uncompressed cervicals], retention of a large tensor tympani, and presumably

flipper flexibility and lack of polyphalangy). (Van Valen 1968, 40).

Stromer (1908) recognized the distinct primitiveness of *Protocetus* relative to other archaic whales and placed this in its own family Protocetidae. Additional genera and species based on fragmentary specimens were added to Protocetidae over the years, and the divergently specialized families Remingtonocetidae and Ambulocetidae were recognized as well (fig. 14.3). The core of archaeocete evolution in the middle Eocene, Protocetidae, includes a broad range of feeding types. There are generalized forms like *Protocetus*, *Artiocetus*, and *Rodhocetus*, narrow-snouted forms like *Gaviacetus* and *Makaracetus*, and broad-snouted forms such as *Takracetus*. Several are represented by skeletons as well as skulls.

I became interested in fossil whales in 1978 when fieldwork in Pakistan yielded the posterior part of an enigmatic cranium representing a mammal a little larger than a wolf. The specimen was found encased in matrix, but even in the field it was clear that the size of the braincase was small, the sagittal crest was high and narrow, and the temporal fossae were large. The specimen was found by Jean-Louis Hartenberger on an expedition that I led with Donald Russell. Graduate student Neil Wells was working on a dissertation at the time on the Kuldana Formation yielding this and other fossils. Wells interpreted the Kuldana Formation as a wedge of fluvial clastic sediment intercalated between marine strata below and marine strata above, during a brief low-stand of the sea. Proximity to marine strata led us to think the enigmatic cranium might be a primitive whale of some kind. When the cranium was cleaned and prepared for study, I showed it to Ewan Fordyce and Lawrence Barnes during a visit to the Smithsonian Institution. Fordyce and Barnes confirmed the identification by pointing to the sigmoid process on the tympanic bulla, which is a distinctive feature of fossil and extant whales.

Russell and I named the Pakistan cranium and the species it represents *Pakicetus inachus* (Gingerich and Russell 1981). The low sea stand is generally regarded as being very early in the middle Eocene, which is the age of *Pakicetus* as well (ca. 48 Ma; Gingerich 2003b). *Pakicetus* was the oldest fossil whale known at the time, it came from a fluvial red-bed formation, and it was

associated with continental land mammals. Comparative study showed that there was no development of pterygoid sinuses in front of the petrosal or periotic ear bones, meaning that the bones were not isolated from each other as would be required to hear directionally in water (Gingerich et al. 1983). Postcranial elements were rare, and some from the *Pakicetus* site were initially misidentified as mammalian Artiodactyla (e.g., Thewissen et al. 1987), a mistake that wouldn't be corrected until skeletons of *Artiocetus* and *Rodhocetus* were found with associated skulls and postcranial skeletons.

All remains of *Pakicetus* found to date come from beds of calcarenite representing reworked calcareous soil nodules, with vertebrate bones swept from a weathered soil surface and deposited together by transient streams (Wells 1983). This makes taxonomic identification of the bones difficult because postcranial elements are not associated with identifiable skulls and teeth, nor are they associated with each other. Thewissen et al. (2001) attempted reconstruction of a skeleton of *Pakicetus*, but very few bones are known that preserve enough morphology to enable reliable identification. Thewissen et al. (2001) concluded that "pakicetids were terrestrial mammals, no more amphibious than a tapir." However, the pelvis of *Pakicetus* identified by Thewissen et al. has a very short ilium like that of semiaquatic protocetids (e.g., *Rodhocetus*; Gingerich 2003a), and unlike that of land mammals. Similarly, study of bone microstructure in the elements described by Thewissen et al. demonstrated that, contrary to the terrestrial interpretation, "pakicetids were highly derived semiaquatic mammals" (Madar 2007). *Indohyus* is an Indo-Pakistan contemporary of *Pakicetus* that Thewissen et al. (2007) identified as (1) an aquatic artiodactyl and (2) the sister group of *Pakicetus* and other Cetacea. This can be interpreted to mean that aquatic adaptations evolved in Cetacea independently before and after *Pakicetus*; that *Pakicetus* reacquired its terrestrial adaptations convergently with some artiodactyls; or, more parsimoniously, that *Pakicetus* was semiaquatic rather than terrestrial. The evidence from Gingerich (2003a), from Madar (2007), and from Thewissen et al. (2007) is consistent with interpretation of *Pakicetus* as a semiaquatic foot-powered swimmer rather than a terrestrial cursor (fig. 14.4).

Pakicetus and other pakicetids are archaeocetes because they have a sigmoid process on the tympanic bulla, and they have premolar and molar teeth like

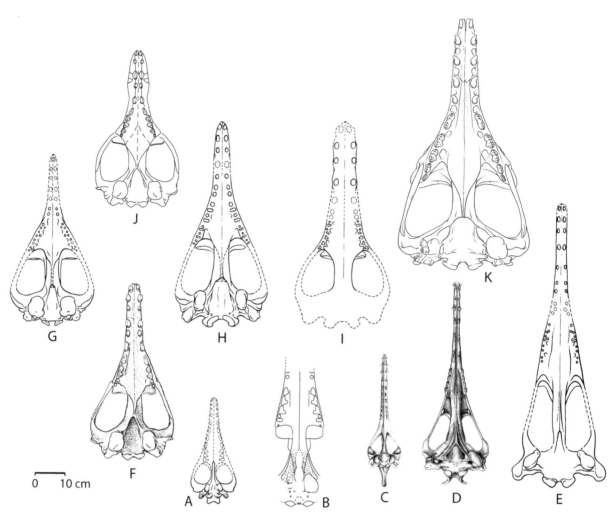

FIG. 14.3 Skulls of Eocene Archaeoceti illustrating the diversity of feeding types. All are shown in palatal view, drawn at the same scale. A, Pakicetid *Pakicetus inachus* from the beginning of the middle Eocene (ca. 49 Ma; Gingerich et al. 1983). B, Ambulocetid *Ambulocetus natans*, early middle Eocene (48 Ma; Thewissen et al. 1996). C, Remingtonocetid *Andrewsiphius sloani* (42–41 Ma; Thewissen and Bajpai 2009). D, Remingtonocetid *Remingtonocetus harudiensis* (42–41 Ma; Bajpai et al. 2011). E, Remingtonocetid *Dalanistes ahmedi* (46 Ma; Gingerich et al. 1995). F, Protocetid *Artiocetus clavis* (47 Ma; Gingerich et al. 2009). G, Protocetid *Gaviacetus razai* (46 Ma, Gingerich et al. 1995). H, Protocetid *Rodhocetus kasranii* (46 Ma, Gingerich et al. 1994). I, Protocetid *Takracetus simus* (46 Ma, Gingerich et al. 1995). J, Protocetid *Makaracetus bidens* (Gingerich et al. 2005). K, Basilosaurid *Dorudon atrox* (37 Ma; Uhen 2004). Middle Eocene protocetids include narrow- and broad-snouted forms, but most had rostra of intermediate width like those of late Eocene basilosaurids. Middle Eocene remingtonocetids had unusually long and narrow rostra suggesting a distinctive mode of feeding.

later protocetids. *Pakicetus* was initially regarded as a protocetid, but the position of the orbits on the dorsal surface of the skull (Nummela et al. 2006), absence of pterygoid sinuses in front of the petrosals in the middle ear, small size of the tympanic bullae, and absence of enlarged mandibular canals (Gingerich et al. 1983; Bajpai and Gingerich 1998) indicate a difference in sensory structure and function that is sufficient to warrant placement in a separate family Pakicetidae.

The first good skeletons of Protocetidae were those of *Rodhocetus kasranii* (Gingerich et al. 1994) and *Georgiacetus vogtlensis* (Hulbert et al. 1998); however, both

lacked forelimbs and most of the hind limbs. A breakthrough in understanding the origin and systematic relationships of whales came in 2000 and 2001 with discovery and description of *Rodhocetus balochistanensis* and *Artiocetus clavis* (Gingerich et al. 2001). Both were found as articulated or partially articulated skulls and skeletons with well-preserved ankle bones. *Artiocetus* was found on October 22, 2000, the first day of our first exploration of the Kunvit area west of Rakhni in Balochistan. Iyad Zalmout found the body of an astragalus weathering out on a dry grassy slope, and later Munir ul Haq found the head of the astragalus. When the body and head of the bone snapped together in my hands, it was clear that *Artiocetus* had an astragalus ankle bone with the double trochlea or "double-pulley" shape characteristic and diagnostic of one group of mammals: Artiodactyla. Excavation yielded more of the skeleton including an exceptionally well-preserved protocetid cranium.

As a paleontologist, I was thoroughly familiar with both the theory and evidence behind Van Valen's idea that archaic whales were linked to mesonychid condylarths, mesonychids to arctocyonid condylarths, and arctocyonids to Artiodactyla. I expected the astragalus of protocetids, when found, to look like that of a mesonychid. Here, in the form of a double-pulley astragalus associated with an early protocetid skeleton, was paleontological evidence that whales were not closely related to mesonychids but rather to artiodactyls. This corroborated the observations of comparative anatomists going back to Hunter (1787) and molecular systematists going back to Boyden and Gemeroy (1950), Sarich (1993), and Irwin and Arnason (1994). This confirmed Gatesy et al.'s (1996) prediction that new fossil discoveries would include basal cetaceans with typical artiodactyl skeletal characters. Here was evidence that whales in some sense *are* artiodactyls (at least in a cladistic sense). As fieldwork continued that season, it was interesting to see how I struggled to believe the evidence that had snapped together in my own fingers: Initially I simply could not accept it. However, a few days later I realized that we had found the cuboid of the *Artiocetus* ankle too, and it was notched for articulation with the calcaneum just like the cuboid of artiodactyls. So, Van Valen and many of the rest of us in paleontology were wrong when we argued that whales originated from mesonychid condylarths. No one likes to be wrong, but in this case there was a silver lining because the discoveries put an end to disagreement and uncertainty about the origin of whales. A synthesis was achieved, with evidence from the fossil record.

Further discoveries important for understanding protocetid locomotion and life history came from Pakistan in 2000 and 2004 in the form of additional skeletons discovered by Iyad Zalmout and Munir ul Haq. The first find included the skull and partial skeleton of a near-term fetus in utero within an adult skeleton. This showed that (1) the adult skeleton was female and (2) the near-term fetus was about to be born headfirst as is characteristic of mammalian birth on land. The second discovery was a complete skeleton with bones some 12% larger in linear dimensions compared to the skeleton known to be female, with a more robust upper canine tooth. Thus the second skeleton was identified as male. Both were named *Maiacetus inuus* and described by Gingerich et al. (2009).

The male skeleton of *Maiacetus inuus* (fig. 14.5) was found fully articulated in the field, and it is extraordinary in being complete from the slightly weathered cranium to the tips of the toes and the tip of the tail. The articulated skeleton had 7 cervical vertebrae, 13 thoracics, 6 lumbars, 4 sacrals, and 21 caudals, for a total of 51 vertebrae. The formula 7:13:6:4 for precaudal vertebrae is characteristic of all early protocetids, but *Maiacetus* was the first to yield a complete caudal series. *Maiacetus* has a large skull compared to the rest of its body, and it was undoubtedly a piscivorus predator. It has large mandibular canals and large tympanic bullae. Contemporary protocetids had asymmetrical skulls (Fahlke et al. 2011), and there is little doubt that protocetids had an auditory system for hearing in water advanced over that seen in pakicetids.

FIG. 14.4 Reconstruction of the 48-million-year-old (48 Ma) pakicetid archaeocete *Pakicetus attocki* based on the isolated postcranial bones described by Thewissen et al. (2001) and Madar (2007). This is a composite of fragmentary specimens, but enough is known of individual elements to show that it was semiaquatic (Madar 2007), and not terrestrial (Thewissen et al. 2001). These indicate too that *Pakicetus* was a foot-powered swimmer like *Maiacetus*. Illustration is by John Klausmeyer, University of Michigan Natural History Museum.

The sacrum of *Maiacetus* has three vertebrae fully co-ossified and a fourth with pleurapophyses fused to the third. Robust auricular processes are present on the first sacral for articulation with ilia of the pelvis. Hind limbs are connected to the backbone, and the fused sacrum provided a platform for powerful swimming. Principal components analysis showed *Maiacetus* to have the trunk and limb skeletal proportions of a semiaquatic mammal, and we interpreted *Maiacetus*, like other protocetids, to have been a foot-powered swimmer. Thrust was provided by powerful muscular extension of the hind limbs, pushing water with webbed toes extended. Hind-limb extension was alternating, left and right, to minimize pitch. The heavy tail, narrow and deep, served as an inertial stabilizer to minimize yaw.

The final transitional stage in the evolution of whales from land to sea is represented by fully aquatic middle and late Eocene Basilosauridae. *Basilosaurus* was the first archaic whale known, and it is the typical form on which first *Zeuglodon* (Owen 1839) and then Archaeoceti were originally based (Flower 1883b). *Basilosaurus*, the senior objective synonym of *Zeuglodon*, is clearly the same grade as other genera referred to the family, but *Basilosaurus* itself is highly specialized and divergent, with a large 15- to 18-meter-long body, serpentine or anguilliform (eel-like) in form. *Dorudon atrox* is a more generalized basilosaurid, similar in almost every feature of morphology, but only 5-meters long with a more compact body. Here *Dorudon* (fig. 14.6) is taken to represent late Eocene basilosaurid whales. *Zygorhiza*, *Saghacetus*, *Cynthiacetus*, *Stromerius*, and *Masracetus* are similar but less well known (see Kellogg 1936, on

Zygorhiza; and Martínez-Cáceres and Muizon 2011, on *Cynthiacetus*).

The skull of *Dorudon* is smaller than that of *Maiacetus* in relation to the rest of its body, but the narial opening is higher on the skull. The dentition is similar but the shearing premolars are larger, and the molars are reduced in number and simpler in form. The postcranial skeleton had 7 cervical vertebrae, 17 thoracics, 20 lumbars, 0 sacrals distinct as such, and 21 caudals, for a total of 65 vertebrae (Uhen 2004). The posteriormost lumbar vertebrae can be considered homologous with sacrals, and in this case the vertebral formula 7:17:(16):(4):21 compares well with 7:13:6:4:21 for early protocetids. The difference is evolutionary addition of 14 vertebrae in the posterior thoracic and lumbar series. Basilosaurid vertebrae are more uniform in size and more uniform in development of neural spines and transverse processes than vertebrae of protocetids, and in basilosaurids there is no sacrum as such interrupting the flexibility of the vertebral column. Hind limbs were retained, but these were no longer connected to the vertebral column in any known basilosaurid. Finally, the centra of distal caudal vertebrae are flattened dorsoventrally indicating the presence of a tail fluke like that of modern cetaceans. Late Eocene *Dorudon* was clearly fully aquatic, with no ability to come out of the water or move on land, and it was undoubtedly a fluke-driven, tail-powered swimmer like modern whales.

Much progress has been made in understanding the relationships of Eocene Archaeoceti to more recent Mysticeti and Odontoceti. This has come through reinterpretation of long-known intermediates in the fossil

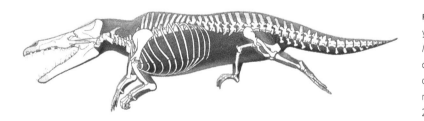

FIG. 14.5 Reconstruction of the 47-million-year-old (47 Ma) protocetid archaeocete *Maiacetus inuus* based on a complete skeleton described by Gingerich et al. (2009; University of Michigan–Geological Survey of Pakistan specimen 3551). The male skeleton shown here is 2.6 meters long. *Maiacetus inuus* is estimated to have weighed 280 kg (female) to 390 kg (male). *Maiacetus* was a semiaquatic foot-powered swimmer, swimming with alternating power strokes, and the deep narrow tail is interpreted to have been an inertial stabilizer. Terminal caudal vertebrae are not flattened dorsoventrally, and *Maiacetus* clearly lacked the tail fluke of modern cetaceans. Illustration is by John Klausmeyer, University of Michigan Natural History Museum.

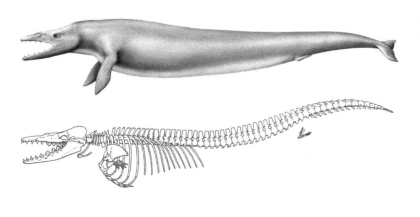

FIG. 14.6 Reconstruction of the 37-million-year-old (37 Ma) late Eocene basilosaurid archaeocete *Dorudon atrox* from Wadi Al Hitan in Egypt. Several virtually complete skeletons are known, described by Uhen (2004; skeleton here is based on University of Michigan specimens 101222 and 101215). These were the first to preserve all vertebrae, complete forelimbs with hands, and *Basilosaurus*-like hind limbs. Terminal caudal vertebrae are flattened dorsoventrally like those of modern cetaceans, indicating the presence of a tail fluke. *Dorudon* was a fully aquatic tail-powered swimmer. Illustration at the top is by John Klausmeyer, University of Michigan Natural History Museum.

record like *Squalodon*, *Agorophius*, and *Mammalodon*, and through recovery of new intermediate fossils. One of the first of the new fossils was late Oligocene *Aetiocetus*, named by Emlong (1966). Emlong described *Aetiocetus* as an archaic whale because it retained teeth, but Van Valen (1968) correctly interpreted *Aetiocetus* as a primitive modern whale with distinctively mysticete characteristics. Van Valen noted that "the ancestors of mysticetes must have had teeth, and *Aetiocetus* is in other respects similar to mysticetes." Another new mysticete of note is Oligocene *Janjucetus hunderi* with closely spaced teeth in a short broad rostrum (Fitzgerald 2006, 2010). New odontocetes include Oligocene *Simocetus* (Fordyce 2002) and *Albertocetus* (Uhen 2008), which retain primitive dental and cranial characteristics. I will not summarize the extensive literature of recent years on the transition from Archaeoceti to Mysticeti and Odontoceti, but refer the reader to reviews by Whitmore and Sanders (1976), Barnes et al. (1985), Fordyce and Barnes (1994), Barnes et al. (1995), Fordyce and Muizon (2001), Geisler and Sanders (2003), Gingerich (2005), and Uhen (2008, 2010). Morphological and temporal intermediates link mysticetes and odontocetes to archaeocetes, demonstrating that all are part of one monophyletic diversification, and it is easy to envision modern whales evolving from a generalized *Dorudon*-like basilosaurid (Fordyce and Barnes 1994; Fordyce and Muizon 2001; Uhen 2010).

Fossils and a Synthesis

Whales were simply mysterious in the early annals of our science. This was so because whales are so different from other animals, and because whales are so different from even other Mammalia where they are now recognized to belong. Whales have gone from being "fish" because they live in the sea, to inclusion with mammals because they breathe air, bear fur, and, most importantly, nurse their young with specialized mammary glands. Once whales were finally recognized as mammals, by Linnaeus (1758), then controversy shifted to whether they were most closely related to herbivorous Artiodactyla on one hand or related to carnivorous Creodonta and early Carnivora on the other. Comparative anatomy and eventually molecular genetics of living whales favored Artiodactyla, while early fossils like *Protocetus* with its carnivorous premolars and molars favored meat-eating mammals of some kind. Whales today are planktivorous, piscivorous, and carnivorous, and the modern whale diet too influenced our sense of their probable ancestry.

It may be simplistic to point to October 22, 2000, as a moment of synthesis. This is when the body and head of the *Artiocetus* astragalus snapped together and I couldn't believe the evidence in my own hands because it was so surprising. Whales were related to artiodactyls, but I didn't expect them to be real artiodactyls in the sense of having an ankle with an artiodactyl double-pulley astragalus. I didn't expect to find a basal cetacean with typical artiodactyl skeletal characteristics, as Gatesy et al. (1996) predicted. Munir ul Haq found the skeleton of *Rodhocetus balochistanensis*, with its articulated hands and articulated feet with double-pulley astragali, six days later on October 28, 2000. By this time I knew that advocates deriving whales from herbivorous Artiodactyla had won the intellectual tug of war, and

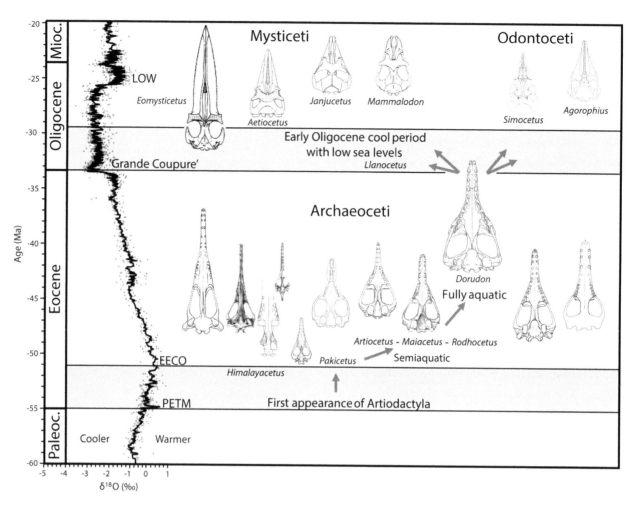

FIG. 14.7 Origin and early evolution of whales in geological time and environmental context. The times of major transition during the early Eocene and the early Oligocene are shown with shaded bars. The origin of whales happened between about 55 and 51 million years ago. The first appearance of the ancestral group Artiodactyla in the Paleocene-Eocene thermal maximum (PETM) global greenhouse warming event provides a lower early bound on the time of origin of Archaeoceti (Gingerich 2006), and the first appearance of *Himalayacetus* provides an upper late bound on its time of origin (Bajpai and Gingerich 1998). The great diversification of Archaeoceti took place in warm shallow seas of the middle and late Eocene, fostered by global warming during the early Eocene climatic optimum (EECO). Viewed in hindsight, the main line of archaeocete evolution was through or near *Pakicetus*, *Maiacetus*, and *Dorudon*. The first appearance of modern Mysticeti (baleen whales) and Odontoceti (toothed whales) was in the very latest Eocene or early Oligocene "Grande Coupure" when global temperatures declined, polar ice caps expanded, sea level fell, and ocean circulation changed. One group, Mysticeti, responded by moving down the trophic pyramid to feed on zooplankton, while the other, Odontoceti, retained its position feeding on fish or continued to move up to feed more on other predators. Low sea levels in the early Oligocene may explain a paucity of early mysticete and odontocete fossils, but much new diversity is evident in the late Oligocene. The oxygen isotope temperature curve for the Eocene is from Zachos et al. (2001). From the late Eocene onward the curve is influenced by ice volume as well as temperature.

advocates for deriving whales from carnivorous Mesonychia, myself included, had lost. The origin of whales was solved.

There are, however, several questions more. When, where, why, and how did an herbivorous artiodactyl become a carnivorous cetacean? I focused on *Pakicetus*, *Maiacetus*, and *Dorudon* above, spanning a 20 million year interval of early Eocene through late Eocene time,

from about 53 Ma to 33 Ma. The diversity of semiaquatic archaeocetes early in the transition (Pakicetidae, Ambulocetidae, and Protocetidae) is based on a disparity of cranial forms that appear related to differences in feeding (fig. 14.7), while the postcranial skeletons were more narrowly similar. The diversity of fully aquatic archaeocetes later in the transition (Basilosauridae) is based on a disparity of vertebral forms, with the crania now being the narrowly similar elements. Experimentation was seemingly associated with feeding in a new environment early in the transition, in the middle Eocene. This variation was then winnowed to a narrower *Dorudon*-like range of cranial proportions later in the transition, in the late Eocene, when experimentation shifted to the postcranial skeleton and to modes of swimming associated with pursuit predation.

The transition from land to sea did not happen in an environmental vacuum. There are three climatic events of importance in the Eocene: (1) the Paleocene-Eocene thermal maximum or PETM; (2) the early Eocene climatic optimum or EECO; and (3) cooling in the terminal Eocene event, TEE, or here Grande Coupure (fig. 14.7). The PETM is the rapid global greenhouse warming event responsible for the origin of most of the modern orders of mammals, including Artiodactyla (and subsequently Cetacea; Gingerich 2006). The EECO is the longest period of sustained warmth in the Eocene (fig. 14.7; Zachos et al. 2001), and it is probably no accident that both Cetacea and Sirenia, the two fully aquatic mammalian orders living today, made their initial transition into the water during the EECO interval of global warming. The rate of convective heat loss for an endotherm in water is 90 times the rate of heat loss in air (Downhower and Blumer 1988), and a transition from land to sea would have been easier during the EECO when the sea was warm (rapid heat loss in neonates may explain why protocetids remained semiaquatic, giving birth on land, for so long during the transition). The fossil record we have, starting with *Himalayacetus* in India and *Pakicetus* in Pakistan, suggests that the transition was made in the eastern Tethys Sea (Neotethys) as the Indian subcontinent closed against the rest of Asia (Gingerich et al. 1983; Bajpai and Gingerich 1998). Finally, the Grande Coupure at the end of the Eocene was a time of global cooling, polar ice formation, and lowered sea levels (fig. 14.7): this is the time when archaeocetes were replaced by mysticetes and odontocetes.

Why Did a Lineage of Artiodactyls Make the Transition to Become Cetaceans in the Sea?

Here the observations of wildlife biologists are helpful. David Case and Dale McCullough (1987) published an interesting study showing that white-tailed deer (*Odocoileus*) on North Manitou Island in Lake Michigan consume large quantities of dead alewives (*Alosa*). Alewives are a river herring invasive in Lake Michigan. Case and McCullough (1987) concluded that deer on North Manitou Island learned to exploit and purposefully seek alewives as a readily available, easily digestible, and highly nutritious food source. Pietz and Granfors (2000) reported that white-tailed deer feed opportunistically on songbird nestlings, predation they interpreted as deliberate and more common than generally believed. Nack and Ribic (2005) reported domestic cattle doing the same. African hippos (*Hippopotamus*) are known to prey on impala (*Aepyceros*), wildebeast (*Connochaetes*), and even other hippos (Dudley 1996, 1998). Pigs too are notoriously omnivorous. Artiodactyls are more predatory and carnivorous than generally believed, and knowing this it is easy to envision early artiodactyls stepping their way evolutionarily from scavenging dead fish, to preying on dying fish, to active pursuit of healthy fish. The general incentive was opportunity, and the opportunity was access to a new source of food. The sea would be a hostile refuge for land mammals if competition or predation pushed them in this direction, and it is more likely that opportunity drew them instead. This is corroborated to some extent by the slow pace of the transition, and the long interval when archaeocetes were semiaquatic before they became fully aquatic.

Finally we can consider how the transition took place through time. The problems to be solved were listed in figure 14.1, and the timings of smaller changes are listed in figure 14.8. Note that two or more changes were generally involved for each characteristic of interest. Land-living artiodactyls in the early Eocene first became semiaquatic archaeocetes in the early and middle Eocene, and then fully aquatic archaeocetes in the middle and late Eocene. Artiodactyls that walked and ran on legs first became foot-powered archaeocetes, and then tail-powered archaeocetes. Artiodactyls that ate plants first became fish-eating archaeocetes, and then added other vertebrates in later archaeocetes and odontocetes. Mysticetes became even more specialized, removing fish as economic "middle men" and processing marine plankton in bulk themselves. Most land-dwelling artiodactyls are precocial, giving birth to a single large and well-developed fetus that emerges headfirst on land ready to move and forage within hours. This was evidently the pattern in middle Eocene protocetids too (Gingerich et al. 2009). There is no evidence of birth or precociality

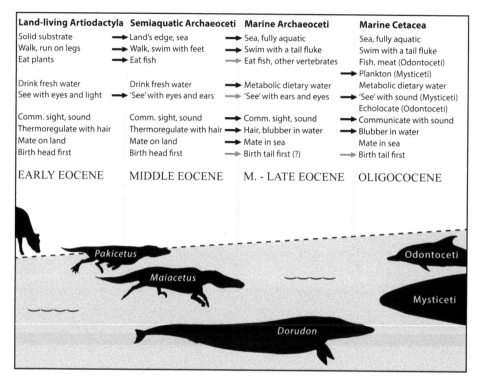

Land-living Artiodactyla	Semiaquatic Archaeoceti	Marine Archaeoceti	Marine Cetacea
Solid substrate →	Land's edge, sea →	Sea, fully aquatic →	Sea, fully aquatic
Walk, run on legs →	Walk, swim with feet →	Swim with a tail fluke →	Swim with a tail fluke
Eat plants →	Eat fish →	Eat fish, other vertebrates →	Fish, meat (Odontoceti)
			→ Plankton (Mysticeti)
Drink fresh water	Drink fresh water →	Metabolic dietary water →	Metabolic dietary water
See with eyes and light →	'See' with eyes and ears →	'See' with ears and eyes →	'See' with sound (Mysticeti)
			Echolocate (Odontoceti)
Comm. sight, sound	Comm. sight, sound →	Comm. sight, sound →	Communicate with sound
Thermoregulate with hair	Thermoregulate with hair →	Hair, blubber in water →	Blubber in water
Mate on land	Mate on land →	Mate in sea →	Mate in sea
Birth head first	Birth head first →	Birth tail first (?) →	Birth tail first
EARLY EOCENE	MIDDLE EOCENE	M. - LATE EOCENE	OLIGOCOCENE

FIG. 14.8 Mosaic pattern of change in anatomical, physiological, and behavioral characteristics during the transition from early land-living Artiodactyla to modern fully marine Cetacea. Arrows here show when and how characteristics contrasted in figure 14.1 changed through Eocene time. Note that two or more changes were generally involved for each characteristic of interest. For example, land-living artiodactyls in the early Eocene first became semiaquatic archaeocetes in the early and middle Eocene, and then became fully aquatic archaeocetes in the middle and late Eocene. Change took place through the whole transition. Some characteristics changed earlier, and some changed later. Some characteristics changed in coordination with others, and some changed independently.

in basilosaurids, but the fact that they were fully aquatic and had a heavily muscled torso means that they probably gave birth to a single large and well-developed fetus that was born tail first like modern whales (reflecting the simple geometry of fitting a large torpedo-shaped fetus into an already torpedo-shaped muscular mother). The arrows in figure 14.8 show that individual characteristics changed step by step during the transition from land to sea. The overall pattern is mosaic, with some characteristics changing earlier in the transition and some changing later, some changing in coordination and others changing independently.

The evolution of whales is macroevolutionary by any definition, with an origin in one adaptive zone, on land, and subsequent diversification in a different adaptive zone, the sea. Fossil skeletons allow us to trace the stages by which this transition happened, stepping through time—through the Eocene; stepping through space—from land, to river, to sea; and stepping through diet and nutrition—from plants, to fish, and eventually,

in some odontocetes, to meat, or, in mysticetes, to zooplankton. The transition is backward in the sense that it takes a vertebrate group back to the sea from whence we came (Shubin et al. 2006). The transition is backward too, in part, because while odontocetes continued the climb up the trophic pyramid initiated by archaeocetes, mysticetes moved down a step. Both reversals support our understanding that evolution as a process is opportunistic as well as competitive, and nothing about rate or direction is preordained.

Whales are as interesting an example of evolution, evolutionary transition, and evolutionary transformation through time as anyone could hope to find in the fossil record. Fossils document important intermediate stages in this history, but there will always be room for more intermediates in the future. New archaic whales are being found in Africa, Antarctica, Asia, greater Australia, Europe, North America, and South America, by investigators young and old. Thus a fine fossil record is becoming even better. There is opportunity too beyond field exploration and discovery. Interpretation of form in terms of function is still in its infancy for early cetaceans. If form and time are keys to understanding whale evolution in terms of history, then form, function, and time together are keys to understanding whale evolution in terms of adaptation.

* * *

Acknowledgments

I wrote this review with Farish Jenkins in mind. Farish was a professional role model for me since we first met 45 years ago. I always admired and sought to emulate the clear prose of his writing and the simple elegance of his illustrations. In addition, I frequently benefitted from his encouragement, even when working on subjects like whale evolution far from his own.

I haven't gone into detail, but this review is a story of fieldwork and fossils. Most who helped have been coauthors at one time or another. I thank Bonnie Miljour and John Klausmeyer for bringing the fossils to life in illustrations. Colleagues Zhe-Xi Luo and Mark Uhen reviewed the manuscript, and I am grateful for their suggestions for improvement. Support for many years of fieldwork came from the National Geographic Society and from the US National Science Foundation.

References

Bajpai, S., and P. D. Gingerich. 1998. A new Eocene archaeocete (Mammalia, Cetacea) from India and the time of origin of whales. Proceedings of the National Academy of Sciences USA 95:15464–15468.

Bajpai, S., J. G. M. Thewissen, and R. W. Conley. 2011. Cranial anatomy of Middle Eocene Remingtonocetus (Cetacea, Mammalia) from Kutch, India. Journal of Paleontology 85:703–718.

Barnes, L. G., D. P. Domning, and C. E. Ray. 1985. Status of studies on fossil marine mammals. Marine Mammal Science 1:15–53.

Barnes, L. G., M. Kimura, H. Furusawa, and H. Sawamura. 1995. Classification and distribution of Oligocene Aetiocetidae (Mammalia; Cetacea; Mysticeti) from western North America and Japan. Island Arc 3:392–431.

Boyden, A. A., and D. G. Gemeroy. 1950. The relative position of the Cetacea among the orders of Mammalia as indicated by precipitin tests. Zoologica, New York Zoological Society 35:145–151.

Case, D. J., and D. R. McCullough. 1987. White-tailed deer forage on alewives. Journal of Mammalogy 68:195–197.

Cope, E. D. 1891. Syllabus of Lectures on Geology and Paleontology. Part III. Paleontology of the Vertebrata. Philadelphia: Ferris Brothers. 90 pp.

Cuvier, G. 1817. Le Règne Animal, Distribué d'Aprés son Organisation, pour Servir de Base à l'Histoire Naturelle des Animaux et d'Introduction à l'Anatomie Comparée, Tome I. Paris: Deterville. 540 pp.

Downhower, J. F., and L. S. Blumer. 1988. Calculating just how small a whale can be. Nature 335:675.

Dudley, J. P. 1996. Record of carnivory, scavenging and predation for Hippopotamus amphibius in Hwange National Park, Zimbabwe. Mammalia 60:486–488.

Dudley, J. P. 1998. Reports of carnivory by the common hippo Hippopotamus amphibius. South African Journal of Wildlife Research 28:58–59.

Emlong, D. 1966. A new archaic cetacean from the Oligocene of north-west Oregon. Bulletin of the Museum of Natural History, University of Oregon 3:1–51.

Fahlke, J. M., P. D. Gingerich, R. C. Welsh, and A. R. Wood. 2011. Cranial asymmetry in Eocene archaeocete whales and the evolution of directional hearing in water. Proceedings of the National Academy of Sciences USA 108:14545–14548.

Fitzgerald, E. M. G. 2006. A bizarre new toothed mysticete (Cetacea) from Australia and the early evolution of baleen whales. Proceedings of the Royal Society of London, Series B 273:2955–2963.

Fitzgerald, E. M. G. 2010. The morphology and systematics of Mammalodon colliveri (Cetacea: Mysticeti), a toothed mysticete from the Oligocene of Australia. Zoological Journal of the Linnean Society 158:367–476.

Flower, W. H. 1883a. On whales, past and present, and their probable origin. Notices of the Proceedings of the Royal Institution of Great Britain, London 10:360–376.

Flower, W. H. 1883b. On the arrangement of the orders and families of existing Mammalia. Proceedings of the Zoological Society of London 1883:178–186.

Fordyce, R. E. 2002. Simocetus rayi (Odontoceti: Simocetidae, new family): a bizarre new archaic Oligocene dolphin from the eastern North Pacific. In R. J. Emry, ed., Cenozoic Mammals of Land and Sea: Tributes to the Career of Clayton E. Ray. Smithsonian Contributions to Paleobiology 93:185–222.

Fordyce, R. E., and L. G. Barnes. 1994. The evolutionary history of whales and dolphins. Annual Review of Earth and Planetary Sciences 22:419–455.

Fordyce, R. E., and C. d. Muizon. 2001. Evolutionary history of cetaceans: a review. Pp. 169–233 in J.-M. Mazin and V. d. Buffrénil, eds., Secondary Adaptation of Tetrapods to Life in Water. Munich: Verlag Friedrich Pfeil.

Fraas, E. 1904. Neue Zeuglodonten aus dem unteren Mitteleocän vom Mokattam bei Cairo. Geologische und Paläontologische Abhandlungen, Jena 6:197–220.

Gatesy, J. E., C. Hayashi, M. A. Cronin, and P. Arctander. 1996. Evidence from milk casein genes that cetaceans are close relatives of hippopotamid artiodactyls. Molecular Biology and Evolution 13:954–963.

Gatesy, J. E., M. C. Milinkovitch, V. G. Waddell, and M. J. Stanhope. 1999. Stability of cladistic relationships between Cetacea and higher-level artiodactyl taxa. Systematic Biology 48:6–20.

Geisler, J. H., and A. E. Sanders. 2003. Morphological evidence for the phylogeny of Cetacea. Journal of Mammalian Evolution 10:23–129.

Geisler, J. H., and M. D. Uhen. 2005. Phylogenetic relationships of extinct cetartiodactyls: results of simultaneous analyses of molecular, morphological, and stratigraphic data. Journal of Mammalian Evolution 12:145–160.

Gervais, P. 1836. Dictionnaire pittoresque d'histoire naturelle et des phénomènes de la nature, Tome IV. Paris: Cosson. 619 pp.

Gill, T. 1870. On the relations of the orders of mammals (abstract). Proceedings of the American Association for the Advancement of Science 18:268–270.

Gingerich, P. D. 2003a. Land-to-sea transition of early whales: evolution of Eocene Archaeoceti (Cetacea) in relation to skeletal proportions and locomotion of living semiaquatic mammals. Paleobiology 29:429–454.

Gingerich, P. D. 2003b. Stratigraphic and micropaleontologic constraints on the middle Eocene age of the mammal-bearing Kuldana Formation of Pakistan. Journal of Vertebrate Paleontology 23:643–651.

Gingerich, P. D. 2005. Cetacea. Pp. 234–252 in K. D. Rose and J. D. Archibald, eds., Placental Mammals: Origin, Timing, and Relationships of the Major Extant Clades. Baltimore: Johns Hopkins University Press.

Gingerich, P. D. 2006. Environment and evolution through the Paleocene-Eocene thermal maximum. Trends in Ecology and Evolution 21:246–253.

Gingerich, P. D., M. Arif, and W. C. Clyde. 1995. New archaeocetes (Mammalia, Cetacea) from the middle Eocene Domanda Formation of the Sulaiman Range, Punjab (Pakistan). Contributions from the Museum of Paleontology, University of Michigan 29:291–330.

Gingerich, P. D., M. Haq, W. v. Koenigswald, W. J. Sanders, B. H. Smith, and I. S. Zalmout. 2009. New protocetid whale from the middle Eocene of Pakistan: birth on land, precocial development, and sexual dimorphism. PLoS One 4 (e4366):1–20.

Gingerich, P. D., M. Haq, I. S. Zalmout, I. H. Khan, and M. S. Malkani. 2001. Origin of whales from early artiodactyls: hands and feet of Eocene Protocetidae from Pakistan. Science 293:2239–2242.

Gingerich, P. D., S. M. Raza, M. Arif, M. Anwar, and X. Zhou. 1994. New whale from the Eocene of Pakistan and the origin of cetacean swimming. Nature 368:844–847.

Gingerich, P. D., and D. E. Russell. 1981. Pakicetus inachus, a new archaeocete (Mammalia, Cetacea) from the early-middle Eocene Kuldana Formation of Kohat (Pakistan). Contributions from the Museum of Paleontology, University of Michigan 25:235–246.

Gingerich, P. D., N. A. Wells, D. E. Russell, and S. M. I. Shah. 1983. Origin of whales in epicontinental remnant seas: new evidence from the early Eocene of Pakistan. Science 220:403–406.

Gingerich, P. D., I. S. Zalmout, M. Haq, and M. A. Bhatti. 2005. Makaracetus bidens, a new protocetid archaeocete (Mammalia, Cetacea) from the early middle Eocene of Balochistan (Pakistan). Contributions from the Museum of Paleontology, University of Michigan 31(9):197–210.

Gregory, W. K. 1910. The orders of mammals. Bulletin of the American Museum of Natural History 27:1–524.

Haeckel, E. H. P. A. 1866. Generelle Morphologie der Organismen. Zweiter Band. Allgemeine Entwicklungsgeschichte der Organismen. Berlin: Georg Reimer. clx + 462 pp.

Heyning, J. E. 1997. Sperm whale phylogeny revisited: analysis of the morphological evidence. Marine Mammal Science 13:596–613.

Howell, A. B. 1930. Aquatic mammals. Springfield, IL: Charles C. Thomas. 1–338 pp.

Hulbert, R. C., R. M. Petkewich, G. A. Bishop, D. Bukry, and D. P. Aleshire. 1998. A new middle Eocene protocetid whale (Mammalia: Cetacea: Archaeoceti) and associated biota from Georgia. Journal of Paleontology 72:907–927.

Hunter, J. 1787. Observations on the structure and oeconomy of whales. Philosophical Transactions of the Royal Society of London 77:371–450.

Irwin, D. M., and U. Arnason. 1994. Cytochrome b gene of marine mammals: phylogeny and evolution. Journal of Mammalian Evolution 2:37–55.

Kellogg, R. 1928. The history of whales—their adaptation to life in the water. Quarterly Review of Biology 3:29–76, 174–208.

Kellogg, R. 1936. A review of the Archaeoceti. Carnegie Institution of Washington Publications 482:1–366.

Kleineidam, R. G., G. Pesole, H. J. Breukelman, J. J. Beintema, and R. A. Kastelein. 1999. Inclusion of Cetaceans within the order Artiodactyla based on phylogenetic analysis of pancreatic ribonuclease genes. Journal of Molecular Evolution 48:360–368.

Linnaeus, C. 1735. Systema Naturae, sive regna tria naturae systematice proposita per classes, ordines, genera, species. Leiden: Lugduni Batavorum. 12 pp.

Linnaeus, C. 1758. Systema Naturae, 10th edition (Systema naturae per regna tria naturae, secundum classes, ordines, genera, species, cum characteribus, differentiis, synonymis, locis. Tomus I. Editio decima, reformata). Holmiae (Stockholm): Laurentii Salvii. 824 pp.

Luo, Z., and P. D. Gingerich. 1999. Terrestrial Mesonychia to aquatic Cetacea: transformation of the basicranium and evolution of hearing in whales. University of Michigan Papers on Paleontology 31:1–98.

Madar, S. I. 2007. The postcranial skeleton of early Eocene pakicetid cetaceans. Journal of Paleontology 81:176–200.

Martínez-Cáceres, M., and C. de Muizon. 2011. A new basilosaurid (Cetacea, Pelagiceti) from the late Eocene to early Oligocene Otuma Formation of Peru. Comptes Rendus Palevol 10:517–526.

McKenna, M. C. 1975. Toward a phylogenetic classification of the Mammalia. Pp. 21–46 in W. P. Luckett and F. S. Szalay, eds., Phylogeny of the Primates. New York: Plenum Publishing Corporation.

McKenna, M. C., and S. K. Bell. 1997. Classification of Mammals above the Species Level. New York: Columbia University Press. 631 pp.

Miller, G. S. 1923. The telescoping of the cetacean skull. Smithsonian Miscellaneous Collections 76:1–70.

Montgelard, C., F. M. Catzeflis, and E. Douzery. 1997. Phylogenetic relationships of artiodactyls and cetaceans as deduced from the comparison of cytochrome *b* and 12S rRNA mitochondrial sequences. Molecular Biology and Evolution 14:550–559.

Nack, J. L., and C. A. Ribic. 2005. Apparent predation by cattle at grassland bird nests. Wilson Bulletin 117:56–62.

Nikaido, M., A. P. Rooney, and N. Okada. 1999. Phylogenetic relationships among cetartiodactyls based on evidence from insertions of SINEs and LINEs: hippopotamuses are the closest extant relatives of whales. Proceedings of the National Academy of Sciences USA 96:10261–10266.

Novacek, M. J. 1992. Mammalian phylogeny: shaking the tree. Nature 356:121–125.

Nummela, S., S. T. Hussain, and J. G. M. Thewissen. 2006. Cranial anatomy of Pakicetidae (Mammalia, Cetacea). Journal of Vertebrate Paleontology 26:746–759.

Ogle, W. 1882. Aristotle on the Parts of Animals, Translated with Introduction and Notes. London: Kegan Paul, French. 263 pp.

O'Leary, M. A. 1999. Parsimony analysis of total evidence from extinct and extant taxa and the cetacean-artiodactyl question (Mammalia, Ungulata). Cladistics 15:315–330.

Owen, R. 1839. Observations on the teeth of the *Zeuglodon, Basilosaurus* of Dr. Harlan. Proceedings of the Geological Society of London 1839:24–28.

Owen, R. 1841. Observations on the *Basilosaurus* of Dr. Harlan (*Zeuglodon cetoides* Owen). Transactions of the Geological Society of London 2:69–79.

Owen, R. 1868. Anatomy of Vertebrates. Volume III. Mammals. London: Longmans, Green, and Co. 915 pp.

Pietz, P. J., and D. A. Granfors. 2000. White-tailed deer (*Odocoileus virginianus*) predation on grassland songbird nestlings. American Midland Naturalist 144:419–422.

Prothero, D. R., E. M. Manning, and M. S. Fischer. 1988. The phylogeny of the ungulates. Pp. 201–234 in M. J. Benton, ed., The Phylogeny and Classification of the Tetrapods, vol. 2. Oxford: Clarendon Press.

Ray, J. 1671. An account of the dissection of a porpess, having therein observed some things omitted by Rondeletius. Philosophical Transactions of the Royal Society of London 6:2274–2279.

Ray, J. 1693. Synopsis Methodica Animalium Quadrupedum et Serpentini Generis. London: Robert Southwell. 336 pp.

Sarich, V. M. 1993. Mammalian systematics: twenty-five years among their albumins and transferrins. Pp. 103–114 in F. S. Szalay, M. J. Novacek, and M. C. McKenna, eds., Mammal Phylogeny: Placentals. New York: Springer.

Scilla, A. 1670. La vana speculazione disingannata dal senso: lettera risponsiva circa i Corpi Marini, che Petrificati si trovano in varij luoghi terrestri. Naples: Andrea Colicchia. 168 pp.

Shimamura, M., H. Abe, M. Nikaido, K. Ohshima, and N. Okada. 1999. Genealogy of families of SINEs in cetaceans and artiodactyls: the presence of a huge superfamily of tRNA^Glu-derived families of SINEs. Molecular Biology and Evolution 16:1046–1060.

Shimamura, M., H. Yasue, K. Ohshima, H. Abe, H. Kato, T. Kishiro, M. Goto, I. Munechika, and N. Okada. 1997. Molecular evidence from retroposons that whales form a clade within even-toed ungulates. Nature 388:666–670.

Shubin, N. H., E. B. Daeschler, and F. A. Jenkins. 2006. The pectoral fin of *Tiktaalik roseae* and the origin of the tetrapod limb. Nature 440:764–771.

Simpson, G. G. 1945. The principles of classification and a classification of mammals. Bulletin of the American Museum of Natural History 85:1–350.

Slijper, E. J. 1936. Die Cetaceen, Vergleichend-Anatomisch und Systematisch. Capita Zoologica 6–7:1–590.

Slijper, E. J. 1966. Functional morphology of the reproductive system in Cetacea. Pp. 278–319 in K. S. Norris, ed., Whales, Dolphins, and Porpoises. Berkeley: University of California Press.

Spaulding, M., M. A. O'Leary, and J. E. Gatesy. 2009. Relationships of Cetacea (Artiodactyla) among mammals: increased taxon sampling alters interpretations of key fossils and character evolution. PLoS One 4:e7062.

Stromer von Reichenbach, E. 1908. Die Archaeoceti des ägyptischen Eozäns. Beiträge zur Paläontologie und Geologie Österreich-Ungarns und des Orients, Vienna 21:106–178.

Stromer von Reichenbach, E. 1903. *Zeuglodon*-Reste aus dem oberen Mitteleocän des Fajum. Beiträge zur Paläontologie und Geologie Österreich-Ungarns und des Orients, Vienna, 15:65–100.

Thewissen, J. G. M. 1994. Phylogenetic aspects of cetacean origins: a morphological perspective. Journal of Mammalian Evolution 2:157–184.

Thewissen, J. G. M., and S. Bajpai. 2009. New skeletal material of *Andrewsiphius* and *Kutchicetus*, two Eocene cetaceans from India. Journal of Paleontology 83:635–663.

Thewissen, J. G. M., L. N. Cooper, M. T. Clementz, S. Bajpai, and B. N. Tiwari. 2007. Whales originated from aquatic artiodactyls in the Eocene epoch of India. Nature 450:1190–1195.

Thewissen, J. G. M., P. D. Gingerich, and D. E. Russell. 1987. Artiodactyla and Perissodactyla (Mammalia) from the early-middle Eocene Kuldana Formation of Kohat (Pakistan). Contributions from the Museum of Paleontology, University of Michigan 27:247–274.

Thewissen, J. G. M., S. I. Madar, and S. T. Hussain. 1996. *Ambulocetus natans*, an Eocene cetacean (Mammalia) from Pakistan. Courier Forschungsinstitut Senckenberg, Frankfurt am Main 191:1–86.

Thewissen, J. G. M., E. M. Williams, L. J. Roe, and S. T. Hussain. 2001. Skeletons of terrestrial cetaceans and the relationship of whales to artiodactyls. Nature 413:277–281.

True, F. W. 1908. On the classification of the Cetacea. Proceedings of the American Philosophical Society 47:385–391.

Uhen, M. D. 2004. Form, function, and anatomy of *Dorudon atrox* (Mammalia, Cetacea): an archaeocete from the middle to late Eocene of Egypt. University of Michigan Papers on Paleontology 34:1–222.

Uhen, M. D. 2008. A new *Xenorophus*-like odontocete cetacean from the Oligocene of North Carolina and a discussion of the basal odontocete radiation. Journal of Systematic Palaeontology 6:433–452.

Uhen, M. D. 2010. The origin(s) of whales. Annual Review of Earth and Planetary Sciences 38:189–219.

Ursing, B., and U. Arnason. 1998. Analyses of mitochondrial genomes strongly support a hippopotamus-whale clade. Proceedings of the Royal Society of London, Series B 265:2251–2255.

Van Valen, L. M. 1966. Deltatheridia, a new order of mammals. Bulletin of the American Museum of Natural History 132:1–126.

Van Valen, L. M. 1968. Monophyly or diphyly in the origin of whales. Evolution 22:37–41.

Van Valen, L. M. 1969. The multiple origins of the placental carnivores. Evolution 23:118–130.

Weber, M. 1904. Die Säugetiere. Einführung in die Anatomie und Systematik der Recenten und Fossillen Mammalia. Jena: Gustav Fischer. 866 pp.

Wells, N. A. 1983. Transient streams in sand-poor redbeds: early-middle Eocene Kuldana Formation in northern Pakistan. Special Publication of the International Association of Sedimentologists 6:393–403.

Whitmore, F. C., and A. E. Sanders. 1976. Review of the Oligocene Cetacea. Systematic Zoology 25:304–320.

Winge, H. 1921. A review of the interrelationships of the Cetacea (translated by Gerrit S. Miller). Smithsonian Miscellaneous Collections 72 (8):1–97.

Yablokov, A. V. 1964. Convergence or parallelism in the evolution of cetaceans. International Geological Review 7:1461–1468.

Zachos, J. C., M. Pagani, L. C. Sloan, E. Thomas, and K. Billups. 2001. Trends, rhythms, and aberrations in global climate 65 Ma to present. Science 292:686–693.

15

Major Transformations in the Evolution of Primate Locomotion

John G. Fleagle[*] and Daniel E. Lieberman[†]

Introduction

Compared to other mammalian orders, Primates use an extraordinary diversity of locomotor behaviors, which are made possible by a complementary diversity of musculoskeletal adaptations. Primate locomotor repertoires include various kinds of suspension, bipedalism, leaping, and quadrupedalism using multiple pronograde and orthograde postures and employing numerous gaits such as walking, trotting, galloping, and brachiation. In addition to using different locomotor modes, primates regularly climb, leap, run, swing, and more in extremely diverse ways. As one might expect, the expansion of the field of primatology in the 1960s stimulated efforts to make sense of this diversity by classifying the locomotor behavior of living primates and identifying major evolutionary trends in primate locomotion. The most notable and enduring of these efforts were by the British physician and comparative anatomist John Napier (e.g., Napier 1963, 1967b; Napier and Napier 1967; Napier and Walker 1967). Napier's seminal 1967 paper, "Evolutionary Aspects of Primate Locomotion," drew on the work of earlier comparative anatomists such as LeGros Clark, Wood Jones, Straus, and Washburn. By synthesizing the anatomy and behavior of extant primates with the primate fossil record, Napier argued that

* Department of Anatomical Sciences, Health Sciences Center, Stony Brook University
† Department of Human Evolutionary Biology, Harvard University

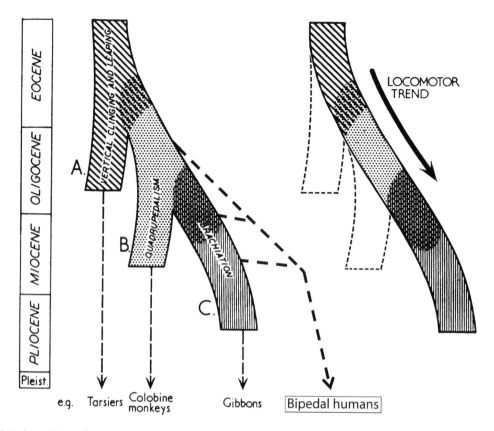

FIG. 15.1 Trends in the evolution of primate locomotion. Napier portrayed the evolution of primate locomotion as a progression or trend. Vertical clinging and leaping was the initial locomotor behavior for Primates in the Eocene, and gave rise to quadrupedalism in the Oligocene, which, in turn gave rise to brachiation in the Miocene. Each of these stages is still present among living taxa. It is unclear which form of locomotion gave rise to human bipedalism. Modified from Napier (1967b).

the evolution of primate locomotion could be characterized by a series of distinct locomotor stages (fig. 15.1). According to this scheme, the earliest primates were *vertical clingers and leapers*, similar to living galagos, tarsiers, and indriid lemuroids; vertical clingers and leapers subsequently gave rise to *arboreal quadrupeds* similar to living Old World monkeys; and arboreal quadrupeds gave rise to brachiators or *suspensory* primates, like the living apes. Last but not least, *bipedal* humans evolved from these suspensory apes (Napier 1967a).

Napier's scheme has influenced, directly and indirectly, much thinking about the evolution of primate locomotion, but it is based on the assumption that the behaviors and morphologies of extant primates are sufficient to characterize the kinds of locomotor behavior, hence morphological adaptations, found in fossil primates. But is this assumption warranted? Can present day diversity usefully predict the ancient diversity that gave rise to the patterns we observe today? How

can the fossil record help reevaluate Napier's scheme? In this contribution, we examine major transitions in primate evolution to consider what the primate fossil record tells us about the evolution of primate locomotor diversity, drawing especially upon insights derived from Farish A. Jenkins Jr.'s research. Jenkins repeatedly showed that the locomotor behavior of primates and other vertebrates is closely related to details of the joint surfaces involved in specific movements. Jenkins's meticulous research on the mechanics and kinematics of virtually all joints, including the elbow (Jenkins 1973), wrist (Jenkins and Fleagle 1975; Jenkins 1981), hip (Jenkins 1972; Jenkins and Camazine 1977), and ankle (Jenkins and McLearn 1984) provide a fundamental basis for inferring the locomotor behavior of fossil primates and for thinking about the evolution of extant primates. Using this perspective, we focus on four major transitions: (1) the origins of vertical clinging and leaping; (2) the evolution of terrestrial quadrupedalism; (3) the origin and evolution of suspensory locomotion; and (4) the origin and evolution of bipedalism in humans. Our unsurprising, but often unappreciated, conclusion is that locomotor diversity among primates was much greater in the past than it is today, and that many morphological features used to infer locomotor adaptations appear in past species in different combinations than are found today (e.g., Jungers et al. 2002; Fleagle 2013). Put differently, many locomotor adaptations evolved independently, in different combinations, and in different ways. One could not possibly reconstruct the evolution of primate locomotion correctly without an integrated approach that includes studying the fossil record, laboratory studies of the locomotor biomechanics, and careful comparative analyses of morphology with a special focus on joint function.

Vertical Clinging and Leaping, and Primate Origins

In 1967, Napier and Walker named and described a distinctive type of locomotion among living primates, "vertical clinging and leaping." Their designation of this behavior notes, "The animals concerned are arboreal, have a vertical clinging posture at rest and are well adapted to a leaping mode of progression during which the hind limbs, used together, provide the propulsive force" (Napier and Walker 1967, 204) for leaping between vertical supports. As examples, they identified *Tarsius*; the indriids *Indri, Propithecus*, and *Avahi*; lemurids such as *Lepilemur* and *Hapalemur*; and the lorisoids *Galago* and *Euoticus*. In addition to describing and drawing locomotor behaviors, they identified a collection of bony features in the skeleton that were found in various living leapers and seemed associated with the behavior. In addition, they suggested, "The special interest of this locomotor group of Vertical Clinging and Leaping is that it appears in a preliminary study to constitute the only known locomotor adaptation of Eocene primates; possibly it is to be regarded as the earliest locomotor specialization of primates and therefore preadaptive to some or possibly all of the later patterns of primate locomotion" (Napier and Walker 1967, 204).

The vertical clinging and leaping (VCL) hypothesis has received much criticism (e.g., Cartmill 1972; Martin 1972; Szalay 1972; Stern and Oxnard 1973; Anemone 1990; Martin 1990). Many critics argued that this "locomotor group" actually contains animals with very different morphological features; others argued that it wasn't sufficiently inclusive because there are vertical clingers among other groups of primates, especially New World monkeys (e.g., Kinzey et al. 1975). Perhaps most damning was Szalay's (1972, 33) claim that "the osteological characters that might be *invariably* associated with vertical clinging and leaping have not been deduced as yet, if such clear-cut features exist at all." In a later review of the topic, Anemone (1990) found that many musculoskeletal features shared by vertical clingers and leapers are also found in related primates that do not engage in VCL. Most critics of the VCL hypothesis also noted that leaping among primates is not a single type of locomotion with a single set of anatomical correlates. Primate leapers have evolved in many different clades, each with different initial morphologies. In addition, leaping may be combined with many other types of movements, leading to morphological compromises. Finally, primates may show adaptations for leaping to and from substrates with very different orientations and sizes.

Despite these criticisms, VCL behaviors are still widely attributed to fossil primates, especially early primates. Following Jenkins's approach, it is useful to evaluate the skeletal indicators of VCL by focusing on the appropriate joints and the specific behaviors that take place at those joints: clinging, which involves the forelimb as well as hands and feet; and leaping from vertical supports, which involves the hind limb, especially the hip. In 1980, Szalay and Dagosto published an important review of the evolution of the elbow in early primates. Among their observations they noted the presence in extant indriids (*Propithecus*) of a "secondary" articulation, termed a "clinging facet," above the humeral trochlea that seems to be the result of habitual use of clinging postures. Importantly, this "clinging facet" is commonly found in other primates that have been identified as "vertical clingers," including tarsiers, some galagos, pygmy marmosets, and white-faced sakis (Fleagle and Meldrum 1988), and it is not found in habitually quadrupedal primates.

Almost all primates can leap to some degree, but habitual leapers are usually characterized by a set of musculo-skeletal adaptations that enable them to rapidly extend their hind limbs for propulsion (e.g., Stern 1971; Fleagle 1977a, 1977b; Anemone 1990). As a rule, leapers have a long ischium for increasing the moment

FIG. 15.2 Leaping from a vertical support, as illustrated by the sifaka on the left, involves a different excursion of the femur relative to the axis of the trunk and pelvis, and a different orientation of the ischium, than either quadrupedal walking or leaping from a horizontal support, as illustrated by the lemur on the right. Modified from Fleagle and Anapol (1992.)

arm of hip extensors (Smith and Savage 1956). They also tend to have a deep knee joint to increase leverage of the quadriceps via the patella. In addition, leapers usually have a very proximal attachment for the hamstrings on the tibia so that hip extension is not accompanied by knee flexion, as it is in quadrupeds (e.g., Haxton 1947; Stern 1971; Fleagle 1977a, 1977b). However, as shown in figure 15.2, the mechanics of leaping from vertical supports are considerably different from the mechanics of leaping from a horizontal support (Fleagle and Anapol 1992). In quadrupedal walking or leaping from a horizontal support, the excursion of the femur is approximately perpendicular to the axis of the trunk, and the hamstrings achieve maximum leverage by extension of the ischium distally along the axis of the trunk or the blade of the ilium. However, during leaping from a vertical support with a vertical (or orthograde) body position, maximum extension of the hind limb brings the femur in line with the axis of the trunk or the blade of the ilium. For this kind of movement, distal extension of the ischium is not only unhelpful

but also makes contraction of the hamstrings impossible because it greatly shortens the distance between their origin and insertion. Thus, primates that leap from a vertical posture have a dorsally rather than distally elongated ischium.

Oddly, Napier and Walker (1967) and other authors (Zuckerman et al. 1973; Walker 1974; McArdle 1981) noted that VCL primates are unusual among mammalian leapers in having a distally short ischium, but the mechanical significance of this feature was not appreciated. Dorsal rather than distal elongation of the ischium in primates leaping from a vertical or orthograde posture evolved for the same mechanical reason that upright, bipedal humans also have an ischium that extends dorsally rather than distally (fig. 15.3; e.g., Robinson 1972). The distally extended ischium of chimpanzees explains why, even when you force a chimpanzee to walk bipedally with its trunk in an orthograde posture, its femur remains nearly perpendicular to the axis of its trunk, as Jenkins (1972) elegantly documented using cineradiography. A broad comparison of primate pelves (fig. 15.4; also other figures in Fleagle and Anapol 1992) shows that habitual vertical clingers from many different families are characterized by an ischium that extends dorsally rather than distally, whereas quadrupeds and primates that normally leap from horizontal supports have a distally extended ischium (Fleagle and Anapol 1992). Confirmation for this observation comes from primates (*Lepilemur*, *Hapalemur*, and *Eulemur rubriventor*) that have a more mixed locomotor repertoire—sometimes running and leaping from horizontal supports, sometimes clinging and leaping from vertical supports. In these species, the ischium tends to be extended both distally and dorsally.

We can therefore conclude that, contra Szalay's (1972) pessimistic conclusion 40 years ago, there are at least two anatomical features that distinguish habitual vertical clingers and leapers from primates that rely more extensively on other modes of locomotion: clinging facets and dorsally extended ischia. If we apply these features to the available skeletal material of Eocene primates from North America and Europe, we find no early primates with the distinctive short, dorsally extended ischium of habitual vertical clingers and leapers (fig. 15.4) and none show a clinging facet on the humerus. However, there are several early Eocene primates (*Notharctus*, *Smilodectes*, and *Omomys*) in which

the ischium is extended both distally as well as dorsally, suggesting that these species were probably capable of leaping from either horizontal or vertical supports (Fleagle and Anapol 1992). In addition, evidence of the hamstring attachments on the tibia suggest that none of the notharctines (*Notharctus*, *Smilodectes*, or *Cantius*)—regularly described as vertical clingers and leapers on the basis of limb proportions—had a proximal attachment for the hamstrings in the tibia. Instead, they display a distal insertion more characteristic of arboreal quadrupeds. Other early Eocene taxa, notably *Shoshonius* (see Dagosto et al. 1999), also seem to have been primarily arboreal quadrupeds. It should be noted that many early primates are thought to have been leapers of some kind based on their ankle morphology. However, as numerous workers have noted (Gebo et al. 2012; Ni et al. 2013), there is no indication from ankle morphology that these taxa were as committed to VCL as are extant indriids, tarsiers, and some galagos. Moreover, the presence of elongated tarsals in cheirogaleids indicates that tarsal elongation does not necessarily imply leaping behavior.

Almost 50 years after its initial proposal, we can draw several conclusions about the VCL hypothesis. First, vertical clinging and leaping is just one of many different types of leaping locomotion used by living primates, many of which have a wide range of musculoskeletal

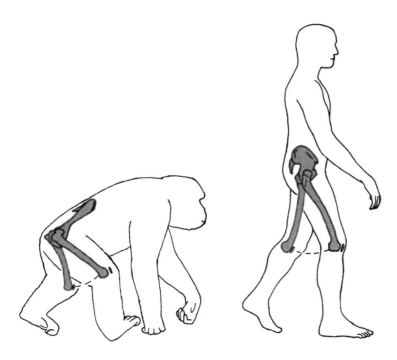

FIG. 15.3 Human bipedalism, shown on the right, involves a different excursion of the femur than quadrupedal walking in a chimpanzee, shown on the left. The femoral excursion in bipedal walking is similar to that used in leaping from a vertical support (fig. 15.2) and is made possible by a dorsal, rather than a distal, extension of the ischium. Modified from Fleagle and Anapol (1992).

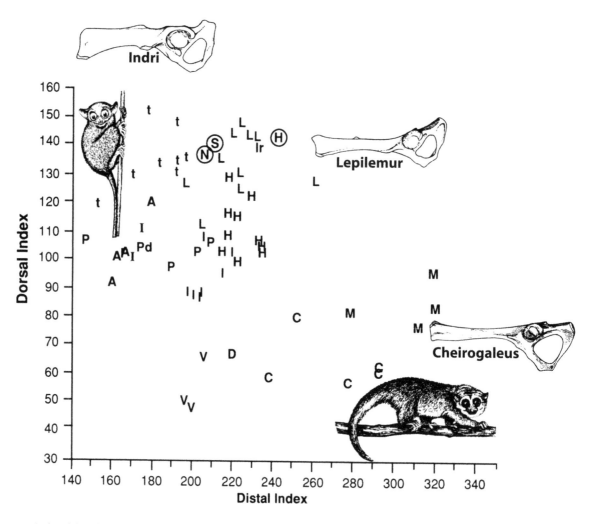

FIG. 15.4 A plot of the relative distal projection of the ischium and the relative dorsal projection of the ischium separates taxa that habitually leap from vertical supports such as tarsiers (t, upper left) from more quadrupedal taxa such as dwarf lemurs (C, lower right). Some taxa, such as *Lepilemur* (L) or *Hapalemur* (H), have an ischium that is elongated in both directions, and both taxa have been reported to be adept at leaping from both vertical and horizontal supports. The pelves of Eocene primates (circled letters) all plot near these "all purpose leapers" rather than with habitual vertical clingers and leapers. Pelves of Eocene fossils: N = *Notharctus*; H = *Hemiacodon*, S = *Smilodectes*. For other taxa see Fleagle and Anapol (1992). Redrawn from Fleagle and Anapol (1992).

features. Second, VCL behaviors evolved independently several times in the Order Primates, each time slightly differently. Third, there is no indication that habitual VCL as practiced by extant indriids, tarsiers, or galagos was the primitive locomotor behavior for the earliest primates or was practiced by any known Eocene primates, including the basal haplorhine *Archicebus*, known from a nearly complete skeleton (Ni et al. 2013). The mosaic combinations of features evident in Eocene primates are not found in any single living primate taxon. This suggests that early primates tended to have a highly generalized locomotor anatomy capable of a wide range of locomotor behaviors including some vertical leaping, but they did not show the morphology

or behaviors seen in the specialized extant taxa as envisioned by Napier and Walker (1967).

The Evolution of Terrestrial Quadrupedal Locomotion

It is paradoxical to discuss quadrupedal locomotion in primates in the context of transformations because, as discussed above, it seems most likely that the earliest primates were some type of arboreal quadruped or quadrupedal leaper (Jenkins 1974; Kirk et al. 2008). Similarly, the earliest platyrrhines, the earliest catarrhines, the earliest hominoids, and perhaps even the earliest hominins were arboreal quadrupeds (see below). Put differently, arboreal quadrupedalism is probably the baseline from which other locomotor modes evolved in primates.

That said, one key locomotor transformation in primate evolution is from arboreal to terrestrial quadrupedalism. As Jenkins noted in 1974, the difference between being an arboreal and terrestrial quadruped is less clear for smaller animals than for larger ones for two reasons. For a small animal, the ground in a forested environment is not a uniform substrate because small irregularities in terrain and vegetation pose greater hurdles for a small animal than a larger one. Similarly, in an arboreal environment, small animals would more likely encounter branches and boughs much larger than their body that could be traversed without grasping. For larger primates, however, the main functional challenges posed by an arboreal environment are balance and grasping on relatively smaller supports (e.g., Napier 1967a; Fleagle and Mittermeier 1980). Consequently, primates who are much larger than the supports upon which they are walking or running maintain their balance by lowering their center of gravity with flexed and abducted limbs (Grand 1968; Schmitt 2003b), by using their tail as a counterbalance, and by maintaining a firm grasp on several supports throughout each step cycle, which likely involves some degree of hyperpronation at the wrist (Jenkins and Fleagle 1975; Jenkins 1981).

These size-related challenges change on the ground. For terrestrial primates, the problem of balance and the associated need for digital grasping are relaxed, making possible selection for other performance tasks. This shift is reflected in many aspects of the anatomy of terrestrial quadrupeds such as the frequent reduction or loss of the tail, and the evolution of longer, subequal fore- and hind limbs. In addition, several well-documented features of the forelimb are likely adaptations for terrestrial quadrupedalism (Jenkins 1973; Jolly 1967; Bown et al. 1982; Fleagle 1999). In these primates, the greater tubercle of the humerus extends well above the relatively narrow humeral head, and the olecranon process tends to extend dorsally rather than proximally because the action of the triceps normally takes place with the elbow extended rather than flexed as in arboreal quadrupeds. Adaptations for terrestrial quadrupedalism in the elbow are particularly well documented and understood (Jenkins 1973). The muscles involved in wrist and digit flexion used during locomotion arise largely from the medial epicondyle of the humerus. During terrestrial quadrupedalism, the forearm is normally pronated, and contraction of the wrist and digit flexors generates a supinating torque around the humero-ulnar joint (Jenkins 1973). Because the ulna, unlike the radius, is not free to rotate, many terrestrial quadrupeds counteract these torques with two osseous adaptations. First, the medial epicondyle tends to project dorsally rather than medially, directing the action of the muscles more in line with the pronated forearm. Second, there is a prominent medial lip on the trochlea of the humerus (and also on the lateral side of the olecranon fossa) that helps keep the ulna in the joint. Because weight bearing is likely more important than mobility in the midcarpal joint of terrestrial quadrupeds, the proximal surface of the hamate usually faces more proximally (Jenkins and Fleagle 1975).

As one might expect, the suite of forelimb adaptations for terrestrial quadrupedalism outlined above apparently evolved independently at least three times in primates. Among catarrhines, these adaptations are most prominent in terrestrial Old World monkeys such as baboons, macaques, mandrills, and geladas, plus numerous extinct monkeys, including *Victoriapithecus* (Harrison 1989) and *Dolichopithecus* (Szalay and Delson 1979). These adaptations are also present in some of the stem hominoids in the Miocene of Africa such as *Proconsul nyanzae* (Fleagle 1999). Among the Malagasy lemurs, the archeolemurine indriids, *Archaeolemur* and *Hadropithecus*, are very similar to baboons and macaques in their elbow anatomy (Jouffroy 1963; Walker 1974; Jungers et al. 2002). Finally, the most terrestrial of the extant apes, gorillas and chimpanzees, show a mosaic of features related to a combination of arboreal and terrestrial quadrupedal locomotion. Like other terrestrial primates, they have fore- and hind limbs that are more similar in length than those of other apes, and the midcarpal joints of their wrists are more suitable for bearing weight than those of gibbons or orangutans (Jenkins and Fleagle 1975; Richmond et al. 2001). However, all apes have a distinctive elbow joint that provides both stability and mobility in all types of locomotion (Jenkins 1973). Thus, as with VCL, terrestrial quadrupedalism has evolved in many different lineages of primates, all of which share common features related to the uniform mechanics of the behavior, but the similarities are often mixed with clade-specific differences.

The Evolution of Suspensory Locomotion

Brachiation, hanging, or more broadly, suspensory locomotion, in which animals regularly suspend themselves by their arms (and usually other appendages as well) has long been a prominent topic in discussions of primate evolution as a whole, but especially for the evolution of apes and humans. The transition from quadrupedalism to brachiation was the second event in Napier's scheme, and followed on Sir Arthur Keith's observations of the behavior and anatomy of gibbons in Southeast Asia, which led him to suggest that many of the anatomical specializations shared by living apes and humans in the trunk, viscera, and other parts of the musculoskeletal system were adaptations to forelimb suspension and orthograde postures (e.g., Keith 1891, 1923; Washburn 1968). As a group, the living apes (Hominoidea) including the gibbons (Hylobatidae) and the great apes (orangutans, gorillas, chimpanzees, and humans) share many anatomical features of their limbs and trunk that have been related to suspensory behavior. These features include loss of the tail; derived lumbar vertebrae with transverse processes coming off the arch rather than the vertebral body (in siamangs but not small gibbons); long forelimbs relative to hind limb and trunk length; a spherical humeral head; a reduced ulnar styloid process, which prevents the ulna from articulating with the carpus; relatively long, curved, manual and pedal phalanges (except in humans); a femur whose head extends above the greater trochanter; broad, shallow femoral condyles (except in humans); a talus with a low shallow trochlea; and a calcaneus with a broad sustentaculum. Finally, the elbow joint of all living apes is also distinctive. Unlike other primates, the distal articulation of the ape humerus has a relatively deep and narrow trochlea with prominent borders on both the medial and lateral side, clearly demarcating the trochlea from the capitulum, which is normally rounded or spherical with a distinct *zona conoidea* medially. The olecranon fossa on the joint's posterior aspect tends to be relatively deep. The complementary features on the antebrachium include an ulna sigmoid cavity that is often keeled to match the deep trochlea on the humerus along with a very short olecranon process that fits into the olecranon fossa to permit full extension at the elbow. The ape radius has a rounded head that permits extensive rotation with the rounded capitulum (Rose

1988). As suggested by Jenkins (1973; also Washburn 1968), this morphology provides apes with a stable elbow with extensive rotational capability at all positions of flexion and extension, an important adaptation for an animal that spends most of its day supporting and moving its body using its forelimb.

Although it is generally assumed that the above described anatomical features related to suspension in hominoids are shared derived features of the superfamily Hominoidea, there are numerous variations in the distribution of these features and their expression among extant apes, leading some authors to suggest that many may be parallelisms (Larson 1998; but see Young 2003). Moreover, the fossil record of ape evolution samples much more locomotor diversity than we can observe among extant taxa. These differences among extant apes, together with a consideration of the fossil record, call for a reinterpretation of the evolution of suspensory locomotion in hominoids, a group that first appeared in Africa and initially diverged from Old World monkeys in the late Oligocene, approximately 25 million years ago. In fact, evidence of suspensory behavior in ape evolution is remarkably rare in the Miocene of Africa despite an abundance of fossil anthropoids that are certainly not Old World monkeys and which most authorities believe are apes, perhaps even great apes. Only three or four of the ape features commonly related to suspensory behavior have been reported in the fossil apes from the African Miocene. Loss of the tail is known for one late Miocene taxon (*Nacholapithecus*) and is likely for several early Miocene species of *Proconsul*, although clear evidence in the form of a sacrum is missing. The much-discussed functional significance of tail loss in suspensory hominoids most likely results from the lack of any positive selection for a long tail as a balancing organ combined with the re-use of tail musculature (e.g., the levator ani) into the pelvic diaphragm of species that were perhaps more orthograde and less strictly quadrupedal.

Other early indications of some suspensory behavior in African fossil apes are found in *Morotopithecus* from the early Miocene of Uganda (e.g. Gebo et al. 1997; MacLatchy et al. 2000; MacLatchy 2004). This taxon is unique among fossil apes from the Miocene of Africa in having lumbar vertebrae that resemble those of extant apes in many features, including transverse processes that emerge from the pedicle rather than from the vertebral body (Sanders and Bodenbender 1994). This dorsal repositioning of the transverse process presumably facilitates stability of the lower back in suspensory behavior. It is also found in the extinct sloth lemurs of Madagascar, another lineage of suspensory primates (Shapiro et al. 2005). *Morotopithecus* has also been argued to be unique among large African Miocene apes in having an ape-like glenoid process of the scapula that is rounded, rather than pear-shaped as in most monkeys. A rounded glenoid is associated with a mobile glenohumeral joint. Although the larger apes from the Miocene of Africa appear to have been mostly arboreal and terrestrial quadrupeds, the smaller proconsuloids such as *Dendropithecus* and *Simiolus* have features including long slender limbs, an *Ateles*-like wrist morphology, and long manual phalanges that suggest some suspensory behavior (e.g., Fleagle 1983; Rossie et al. 2012). However, beyond these hints, the Miocene record of ape evolution in Africa, especially among the larger taxa, seems to document predominantly quadrupedal animals throughout the epoch. The middle and late Miocene fossil record of hominoids in Europe is strikingly different and indicates a wider, more mosaic range of suspensory abilities. *Pierolapithecus*, a late Miocene hominoid from Spain, has many ape-like features missing in *Proconsul* including a broad shallow thorax, lumbar vertebrae with the transverse processes coming off the base of the pedicle, and absence of an articulation between the ulna and the wrist (Moyà-Solà et al. 2004). However, the hands of *Pierolapithecus* may be less elongated and less curved than those of extant Asian apes (Alba 2012; but see Deane and Begun 2010). *Hispanopithecus* and *Rudapithecus* also show similarities to suspensory extant apes in aspects of their digital length and curvature, but unlike extant apes, have short metacarpals. Moreover, in none of these taxa is the olecranon process of the ulna as reduced as in extant hominoids. *Oreopithecus* from the latest Miocene of Italy shows the most extensive skeletal similarities to extant apes in features indicative of suspensory behavior, but is widely considered the most distant in its likely phylogenetic relationships (e.g., Begun et al. 2012). Thus while the middle and late Miocene hominoids of Europe show more evidence of suspensory behavior than those from the early Miocene of Africa, they are nevertheless different from any modern apes in many features. Rather, they show a mosaic of features not found in any extant apes or monkeys.

The hominoid fossil record in Asia is even less similar to what one observes among extant apes. In fact, while gibbons and siamangs (Hylobatidae) and orangutans (Ponginae) of Southeast Asia are the most suspensory of the apes, and also of all living primates, there is little evidence for suspensory behavior in any Asian fossil primates. Although *Sivapithecus*, from the late Miocene of Indo-Pakistan, appears to have an elbow region similar to that of extant great apes in the shape of the distal humerus, most aspects of the postcranial anatomy of *Sivapithecus*, and also *Ankarapithecus* from the late Miocene of Turkey (Kappelman et al. 2003), suggest arboreal quadrupedal behavior. While the overall morphology of the postcranial remains of *Sivapithecus* is clearly similar to that of extant apes, *Sivapithecus* shows a mosaic of hominoid features but no specific similarities to any one extant taxon and no indications of suspension as seen in extant orangutans (e.g., Madar et al. 2002). The contrast between the striking similarities in facial anatomy shared by *Sivapithecus* and *Pongo* (e.g., Pilbeam 1982), and the apparent differences in their postcranial anatomy (e.g., Pilbeam et al. 1990; Mader et al. 2002), has been called the "*Sivapithecus* Dilemma" (Pilbeam and Young 2001), and there are alternate views on how *Sivapithecus* and other Asian fossil hominoids are related to living apes.

Regardless of how one reconstructs phylogenetic relationships among fossil and living apes, there has been considerable homoplasy in the evolution of suspensory behavior within Hominoidea, and postcranial evidence for habitual suspensory locomotion, which is characteristic of all living apes, is not found in most fossil representatives of the clade. Only in the middle-late Miocene of Europe are there indications of significant suspensory behavior. Living apes are thus poor referential models for interpreting much of the locomotor history of hominoids as they are known from the fossil record.

Although much attention is paid to the evolution of suspensory locomotion and orthograde posture and locomotion in hominoids, largely because of its relevance to the origins of human bipedalism (see below), suspensory behavior has evolved several times in other primate radiations. Pliopithecids, a widespread radiation of Old World anthropoids from the Miocene of Asia and Europe (and maybe Africa), were once thought to be related to gibbons, but are now usually considered stem catarrhines. However, in many aspects of their postcranial anatomy, including a relatively short olecranon, a high brachial index, and a very mobile hip joint, they are very similar to the New World atelides such as *Alouatta*, *Lagothrix*, and *Ateles*, suggesting some suspensory behavior (e.g., Fleagle 1984a). The New World platyrrhine monkeys have been separated from the Old World catarrhines for over 30 million years. Among the platyrrhines, the largest taxa, the atelids, are all characterized by suspensory behavior. As initially documented by Erikson (1963) half a century ago and documented in much more detail since, the New World monkeys *Alouatta*, *Lagothrix*, *Brachyteles*, and *Ateles* show a cline in both suspensory behavior and morphology, with limb proportions and aspects of forelimb anatomy that often resemble the Old World apes. In addition, two related fossil taxa, *Protopithecus* (= *Cartelles*) and *Caipora*, are twice as large as any living New World monkeys and also show similar adaptations for suspensory behavior (Cartelle and Hartwig 1996; Hartwig and Cartelle 1996; Halenar 2011).

Suspensory behavior also evolved independently in the lemurs of Madagascar, another radiation of geographically isolated primates. Although the extant fauna of Malagasy primates are all less than 10 kg and are almost all primarily arboreal quadrupeds or leapers, many much larger extinct taxa are known from the fossil record from the last 26,000 years, and some of these species probably went extinct only in the last few hundred years. Among the most unusual of these extinct giant lemurs are the sloth lemurs of the subfamily Palaeopropithecinae (Godfrey and Jungers 2003). These lemurs ranged in size from about 10 kg for *Mesopropithecus*, the smallest, to about 200 kg for *Archaeoindris*, the largest. *Babakotia* (17 kg) and *Palaeopropithecus* (~40 kg) are intermediate in size. Sloth lemurs are the sister taxa of the living indriids. Although they are best known for their VCL behaviors in trees and bipedal hopping on the ground, the larger living indriids, *Propithecus* and *Indri*, frequently engage in suspensory behavior when they hang upside down by all four limbs or suspend themselves using only their feet. On the basis of limb proportions and many aspects of their skeletal anatomy, it seems that *Mesopropithecus* was a loris-like quadruped that also had some suspensory abilities, while *Babakotia* and *Palaeopropithecus* were dedicated arboreal suspensory folivores like living sloths (e.g., Jungers et al.

1997). The postcranial anatomy of the giant *Archaeoindris* is known from a single femur that resembles that of a ground sloth (Vuillaume-Randriamanantena 1988); at 200 kg it was probably largely terrestrial. *Babakotia* and *Palaeopropithecus* share many anatomical features with extant apes, including reduction or loss of a tail; a high intermembral index; an extremely short ulnar olecranon process; lumbar vertebrae with transverse processes arising from the arch; very long, curved phalanges; reduction or loss of the thumb; and a femur with the head literally on top of the shaft.

Thus, suspensory locomotion evolved many times in primates, often in anatomically similar ways involving long forelimbs, extreme mobility of joints, and enhanced grasping abilities. However, there are also unique differences in the nature of the suspensory behavior and associated anatomical structures in each primate radiation. In apes, the evolution of suspensory behavior has been associated with development of a mobile, but extremely stable elbow morphology that enables apes to support their weight in many different positions or in some cases to propel themselves between arboreal supports using only their arms. In addition, suspensory behavior in hominoids has involved major changes in the structure of the vertebrae, the loss of the tail, as well as rearrangements of the shape of the thorax and the suspension of the abdominal viscera. In contrast, suspensory platyrrhines all have a prehensile tail, which is an extra grasping organ. Like the platyrrhines, the sloth lemurs evolved a range of suspensory adaptations from more quadrupedal taxa to others resembling extant tree sloths that were probably incapable of any type of locomotion other than suspension. Thus the evolution of suspensory adaptations in primates has been very clade specific. As discussed below, it is difficult to determine the extent to which these different adaptations reflect phylogenetic contingencies, environmental differences under which the suspensory adaptations evolved, or compromises with selection for other behaviors.

Evolution of Human Bipedalism

Perhaps no locomotor transition in primate evolution is more contentious than the origin of bipedalism. And, as with the other transformations, major theories about hominin locomotor evolution have mostly used extant primate species—especially brachiating gibbons, knuckle-walking African apes, and, of course, striding bipedal humans—as models for reconstructing ancestral locomotor behaviors. To some extent, these analogies have been necessitated by the deficiencies of the fossil record, but they have also been influenced by two strong preconceptions. First, bipedalism has long been posited as a driving force for many aspects of hominin evolutionary change. In Darwin's words: "If it be an advantage to man to stand firmly on his feet and to have his arms free . . . then I can see no reason why it should not have been advantageous to the progenitors of man to have become more erect or bipedal" (Darwin 1871, 141). Since then, efforts to explain human uniqueness often project modern human-like bipedalism back into the fossil record. Fossils that are neither ape-like nor human-like have tended to be interpreted, sometimes wrongly, as simply intermediate. Second, ever since Darwin, researchers who study human evolution have held strong views about the nature of the last common ancestor (LCA) of humans and apes, often based on assumptions about how and why hominin bipedalism evolved. Not surprisingly, preconceived notions about the LCA have then influenced interpretations of the fossil evidence.

The actual fossil evidence for the evolution of hominin bipedalism, however, challenges this way of thinking. In particular, analyses of early hominin functional morphology suggest that locomotor diversity in hominin evolution was considerably greater than has often been hypothesized. In the last two decades, new discoveries have led to the recognition of more than 19 putative hominin species (fig. 15.5), eight of which were unknown prior to 1994. These species are diverse not only cranially but also postcranially, indicating a variety of locomotor behaviors, many of which were unique, not necessarily ancestral to *H. sapiens*, and difficult to classify as intermediate between the LCA and modern humans. We explore this diversity in the context of two issues: the locomotor adaptations of the LCA of humans and chimps, and locomotor diversity within the hominin lineage.

The Origin of Hominin Bipedalism

Hypotheses about the origin of hominin bipedalism have largely been based on speculation necessitated by the absence of direct fossil evidence. Molecular analyses indicate that the chimpanzee and human lineages probably diverged between 5 and 8 Ma (Kumar et al. 2005; Patterson et al. 2006), but prior to 1994 there were almost no hominin fossils older than 4 Ma. In the absence of fossils from the first few million years of hominin evolution, species from the genus *Australopithecus* were typically considered representative of all early hominins, even though the genus mostly dates to between 4 and 2 Ma. The absence of fossils from the first several million years of hominin evolution was, and still is, compounded by a nearly complete absence of any fossils attributed to *Gorilla* and *Pan* (but see McBrearty and Jablonski 2005; Suwa et al. 2007). As a result, reconstructions of the locomotor behavior of the LCA relied very heavily on models and inferences from extant apes. For a long time, the dominant model was that the LCA was gibbon-like: orthograde and suspensory (e.g., Keith 1923; Napier 1964). This idea is still favored by some (e.g., Crompton et al. 2010). An alternative view, first strongly advocated by Washburn (1968, 1973) and his students, is that the LCA was like an African great ape: a terrestrial quadruped as well as a suspensory climber. This reconstruction was bolstered by molecular evidence that chimpanzees and humans are monophyletic (Ruvolo 1997), and by scaling studies that have found that chimp and gorilla skulls are largely scaled versions of each other (Shea 1985; Berge and Penin 2004; Guy et al. 2005). Such scaling is less conserved in the postcranium (Jungers and Hartman 1988), but unless the many similarities between chimps and gorillas evolved independently, then the LCA of humans and chimpanzees must have been somewhat chimp- or gorilla-like. If so, the LCA was probably a knuckle-walker that also engaged in suspensory, orthograde climbing.

In the last two decades, three new genera of hominins from the late Miocene and early Pliocene have been discovered. *Sahelanthropus tchadensis*, which is from Chad and dated to between 6 and 7.2 Ma, is so far known only from cranial material (Brunet et al. 2002, 2005), although undescribed postcranial remains exist (Beauvillian and Watté 2009). The one *Sahelanthropus* cranium, however, is sufficiently complete to indicate that when the species was locomoting with its orbital plane perpendicular to earth horizontal (as primates usually do), its foramen magnum was oriented inferiorly (Zollikofer et al. 2005). Because the long axis of the upper cervical vertebral column in great apes and

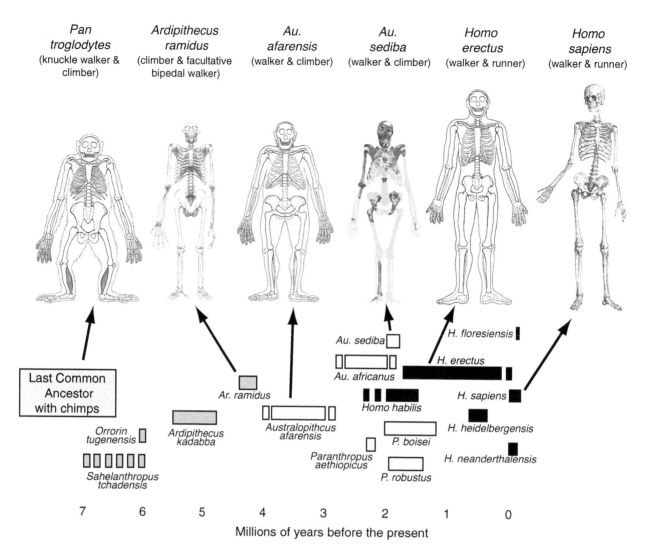

FIG. 15.5 Major hominin species and approximate ages (bottom) divided into three general grades: early hominins (gray), australopiths (white), and the genus *Homo* (black). Shown above are reconstructions of the skeleton for four fossil hominin species: *Ardipithecus ramidus*, a facultative biped and climber; *Australopithecus afarensis*, a habitual biped and climber; *Australopithecus sediba*, also a habitual biped and climber; *Homo erectus*, a walker and runner. Also shown are *Pan troglodytes*, a knuckle-walker climber; and *Homo sapiens*, a walker and runner. Reconstructions of *P. troglodytes*, *Au. afarensis*, and *H. erectus* are from Bramble and Lieberman (2004); reconstruction of *Au. sediba* courtesy of Lee Berger; reconstruction of *Ar. ramidus* courtesy of Tim White.

humans is always within 10° of perpendicular to the foramen magnum, this is indirect evidence for some form of bipedalism (Lieberman 2011). *Orrorin tugenensis*, which is from Kenya and is dated to approximately 6 Ma (Pickford and Senut 2001), includes a femur that has a number of features (a relatively large femoral head, a long femoral neck, and a transversely broad upper femoral shaft) that align it with later hominins rather than any ape, suggesting some form of bipedal locomotion (Richmond and Jungers 2008; but see Almécija et al. 2013). Finally, *Ardipithecus* is known from two Ethiopian

species: *Ar. kadabba*, dated to 5.2–5.8 Ma, and *Ar. ramidus*, dated to 4.3–4.4 Ma. *Ar. ramidus* is the best known of these species, and includes a partial skeleton, ARA-VP-6/500, which has a suite of pelvic and hind limb features that suggest bipedalism, the most compelling being a superoinferiorly short ilium (see below).

Although *Sahelanthropus*, *Orrorin*, and *Ardipithecus* likely support the hypothesis that bipedalism is a key derived early feature of the hominin lineage, these three genera have sparked considerable debate regarding the LCA and the nature of early hominin bipedalism. Most arguments center on interpretations of the *Ardipithecus* skeleton. White, Lovejoy, and colleagues have proposed that *Ardipithecus* was a "facultative" biped that lacks features typical of *Pan* and *Gorilla* (White et al. 2009; Lovejoy, Suwa, Simpson, et al. 2009; Lovejoy, Latimer, et al. 2009; Lovejoy, Suwa, Spurlock, et al., 2009; Lovejoy, Simpson, et al., 2009). In particular, the skeleton is purported to have equally long fore- and hind limbs (based on the radius-tibia ratio), relatively short metacarpals, proximal metacarpal joints capable of substantial dorsiflexion at the metacarpophalangeal joints, and no adaptations for knuckle-walking. These observations along with other inferences led to the interpretation of the LCA as a non-suspensory, pronograde, above-branch quadruped—more like reconstructions of *Proconsul* than any extant great ape. This interpretation, if correct, requires considerable levels of homoplasy in recent hominoid evolution, including most of the features shared between chimps and gorillas, as well as others shared between chimps, gorillas, and orangutans. Moreover, Lovejoy and colleagues' reconstruction of *Ardipithecus* is based on many inferences from anatomical features that are not preserved (e.g., a long, flexible lumbar column) or difficult to estimate reliably from the available evidence (e.g., body mass and limb proportions), and it disregards those features that *Ardipithecus* shares with *Pan* and *Gorilla* in the cranium, foot, and upper limb. Ongoing analyses may lead to substantial reinterpretations of the species' anatomy and locomotor adaptations. One possibility is that *Ardipithecus* retains many similarities to chimpanzees and gorillas with adaptations for both orthograde suspensory and bipedal locomotion. An alternative hypothesis is that *Ardipithecus*, *Sahelanthropus*, and *Orrorin* are not hominins at all (Wood and Harrison 2011). Homoplasy is extremely common in all phylogenetic reconstructions (e.g., Lockwood and Fleagle 1999), and just as it is difficult to sort out the first fossil mammals (Cartmill 2012; Luo, this volume), it may also prove to be a challenge to distinguish reliably the earliest hominins from closely related late Miocene hominoid taxa. Further, if these proposed early hominins are more closely related to chimps or gorillas or members of extinct clades, then bipedalism evolved more than once in the Hominoidea.

More fossils and more analyses will undoubtedly shed needed light on *Ardipithecus* as well as *Sahelanthropus* and *Orrorin*. Regardless of what we will eventually learn about these species, it seems likely that the LCA's postural repertoire included more than orthogrady, and that its locomotor repertoire included suspensory hanging and climbing as well as quadrupedal and bipedal gaits—all of which one can observe in chimpanzees and gorillas. Even though there is no morphological evidence for knuckle-walking in early hominins, and despite arguments to the contrary (e.g., Kivell and Schmitt 2009; White et al. 2009), parsimony suggests that it is more likely than not that the LCA sometimes knuckle-walked (Richmond et al 2001), perhaps as an adaptation to allow a suspensory primate with relatively long forelimbs to engage in terrestrial quadrupedalism.

Locomotor Diversity among Hominins

For many decades, alternative theories about locomotor evolution within the hominin clade have focused on where the australopiths fall on a continuum between two endpoints. One endpoint is the hypothetical LCA, which must have been a partly arboreal ape with some adaptations for orthograde and pronograde locomotion (see above). The other, uncontested endpoint is *H. sapiens*. A common view is that australopith locomotion was intermediate between these two endpoints, and included habitual bipedal walking as well as tree climbing. How australopiths walked, however, remains unclear. Some scholars have proposed that australopiths did not walk with a completely modern gait with relatively extended hips and extended knees (EHEK), but instead with a bent-hip bent-knee (BHBK) gait, not unlike modern chimps when they walk bipedally, because they retained ancestral adaptations for arboreality that compromised bipedal performance (e.g., Susman et al. 1984; Schmitt 2003a). Others have argued that australopiths had a very modern human-like EHEK gait that might even have been more efficient than modern humans, whose performance has been compromised by adaptations for giving birth to large-brained infants (e.g., Lovejoy 1988; Ward 2002). Few question the inference that since *H. erectus*, the genus *Homo* has had an essentially, modern human-like style of locomotion (Wood and Collard 1999).

The fossil record requires us to recognize more diversity in hominin locomotion than these schemes hypothesize (fig. 15.5). A major cause of this reevaluation has been evidence for facultative bipedalism in

Ardipithecus. Although fossils ascribed to this genus have many adaptations for arboreality, such as a highly abducted and internally rotated, short hallux, that would have been effective at grasping, they also preserve several features that are likely adaptations for bipedal posture and locomotion. The *Ardipithecus* pelvis, albeit distorted, has a superoinferiorly short ilium that appears to have laterally oriented blades that would have allowed the gluteus medius and minimus to function as hip abductors during stance (Lovejoy, Suwa, Spurlock, et al., 2009). In addition, *Ardipithecus* feet have dorsally canted metatarsophalangeal joints capable of hyperextension at toe-off, and an apparently semirigid midtarsal region that would have helped permit powered toe-off, although more along the lateral column of the foot than in humans (Lovejoy, Latimer, et al. 2009; Haile-Selassie et al. 2012). If *Ardipithecus* is a hominin, it was probably not an obligate terrestrial biped but instead was often arboreal, and probably incorporated bipedalism only as part of a broader locomotor repertoire.

There is also new evidence for considerable locomotor diversity within *Australopithecus. Au. afarensis,* long the focus of attention, has some adaptations for arboreality as well as habitual bipedalism (see Aiello and Dean 1990; Stern 2000; Ward 2002), but the extent to which it or other australopiths had a BHBK or EHEK gait is unresolved. The need to counter the high hip and knee moments required by a BHBK explains most of the fourfold greater cost of transport in bipedal chimpanzees versus modern humans (Sockol et al. 2007), suggesting there would have been strong selection against BHBK gaits, which provide little performance benefit for an habitual biped (Sellers et al. 2005). Analyses of trabecular orientation of the distal tibia in *Au. africanus* strongly point to human-like ankle hence knee and hip orientations during peak loading (Barak et al. 2013), and analyses of the Laetoli footprints combined with estimates of effective leg length (Raichlen et al. 2008), limited ankle dorsiflexion (DeSilva 2009), evidence for a heel-strike during walking (Latimer and Lovejoy 1989), and a dorsally projecting ischium (Pontzer et al. 2009) are commensurate with an EHEK in *Au. afarensis.* Although some researchers have hypothesized that the relatively long toes of *Au. afarensis* would have compromised its ability to walk with a human-like gait (Susman et al. 1984), inverse dynamic analyses indicate that the high moments generated around the metatarsophalangeal joints in *Au. afarensis* would have been a challenge during only running not walking (Rolian et al. 2009).

Au. afarensis, moreover, is not the sole representative of australopith locomotion as previously assumed. A 3.4 Ma foot from Burtele, Ethiopia has a short, abducted hallux and very long, curved phalanges, which indicate that hominins with *Ardipithecus*-like feet were contemporary with *Au. afarensis* (Haile-Selassie et al. 2012). Another interesting contrast is provided by two partial skeletons of *Au. sediba* from South Africa dated to approximately 2 Ma (Berger et al. 2010). This species has many adaptations for habitual bipedalism, including a *Homo*-like pelvis (Kibii et al. 2011), but it has an inverted subtalar joint and its tuber calcaneus is relatively small and chimp-like without a weight-bearing lateral plantar process (Zipfel et al. 2011). These and other pedal features, which are less derived than in *Au. afarensis,* suggest that *Au. sediba* walked with an inverted foot without a distinct heel strike and with less medial weight transfer during stance (DeSilva et al. 2013). *Au. sediba*'s upper limb is also primitive with a relatively high brachial index; long, curved manual phalanges; and a short clavicle with a cranially oriented glenoid joint indicative of a narrow ape-like shoulder adapted for suspensory climbing (Berger et al. 2010; Kivell et al. 2011; Churchill et al. 2013). Although there is much research to be done, and postcranial anatomy is poorly known for many species, it is reasonable to hypothesize that there was considerable locomotor diversity within the genus *Australopithecus,* with more than one kind of striding gait still sometimes associated with substantial arboreality.

Finally, it has become evident that locomotor diversity within the genus *Homo* is also greater than previously credited. Many scholars of the human fossil record have interpreted the postcranial differences between *Australopithecus* and *Homo* as a shift to more committed, effective walking (e.g., Aiello and Dean 1990), but an alternative hypothesis is that some of these shifts reflect selection for endurance running—a gait that has been largely ignored in theories about hominin locomotor evolution. Like other primates, apes gallop infrequently and for only short distances (Hunt 1992), but humans are unusual among mammals and unique among primates in being exceptional long distance runners (Carrier 1984; Bramble and Lieberman 2004). Bramble and Lieberman proposed that selection for endurance running capabilities in the genus *Homo* explains many novel features (e.g., enlarged anterior and posterior semicircular canals, an expanded gluteus maximus proprius, shortened toes) that improve running performance, but have little effect on walking. Endurance running was probably crucial for early hunting before the comparatively recent (<500 Ka) invention of lithic projectile technology (Lieberman et al. 2008). These authors also proposed that trade-offs between adaptations for climbing and running, not climbing and walking, have resulted in humans being the only extant primate that is rarely arboreal. Adding to diversity, moreover, is evidence for variation within the genus *Homo*. For example, an early pelvis attributed to *H. erectus* is relatively wide (Simpson et al. 2008), and archaic *Homo* species such as *H. heidelbergensis* and *H. neanderthalensis* have relatively wider interacetabular distances and more laterally flared ilia than *H. sapiens* (Arsuaga et al. 1999; Bonmatí et al. 2010). The biomechanical consequences of these differences remain speculative.

Discussion and Conclusions

Napier's (1967) "locomotor trend" in primate evolution was just that: a sequence of proposed evolutionarily successive developments, with primates from the Eocene and Oligocene being primarily vertical clingers and leapers, then quadrupedal monkeys appearing in the Oligocene and Miocene, followed by the evolution of brachiators in the Miocene and Pliocene, and topped off by bipedal humans. Each of these locomotor types was based on living representatives, and his scheme made sense given the very limited record of fossil primates from Europe and Africa known at the time. Today, it is more accurate to describe the evolution of these different patterns of locomotion as major transformations that occurred at least three times independently in isolated radiations of primates: the haplorhines of Eurasia and Africa, the Malagasy strepsirrhines, and the platyrrhines of Central and South America (fig. 15.6). In each of these radiations we find leapers, arboreal quadrupeds, and suspensory species, and in one, we also find bipeds. Napier (and others such as Straus [1962]) also saw primate evolution as a progression toward forelimb suspension and increasing length of the forelimb relative to the hind limb. That unidirectional view is no longer tenable, especially when one considers locomo-

tor evolution within clades. As described above, within Old World haplorhines the earliest members were probably leapers, from which quadrupeds evolved, and then later gave rise to both suspensory and bipedal forms. Among Malagasy strepsirrhines, it is unclear whether the ancestral taxa were leapers or quadrupeds, and it is not totally clear whether the suspensory sloth lemurs evolved from more quadrupedal ancestors or from leapers. Among the New World platyrrhines, it is almost certain that both the specialized leapers and specialized suspensory taxa evolved from quadrupeds.

Just as locomotor specializations evolved independently many times in different primate clades, there is also considerable diversity in the details of locomotor behavior and morphology both among and between the leapers, quadrupeds, and suspensors in different clades. It is not clear whether these clade-specific differences are the result of contingency caused by ancestry, effects of body size, selection for different behaviors and morphological compromises due to competition or habitat, or, most likely, some combination of these factors. For example, differences in the limb proportions and locomotor behaviors between leapers, quadrupeds, and suspensors are less extreme among platyrrhines than among catarrhines or strepsirrhines (Fleagle and Mittermeier 1980). Similarly, within the strepsirrhines, both quadrupeds and leapers tend to have relatively long hind limbs compared with their counterparts in other radiations.

The high degree of convergence and diversity among primates is phylogenetic "noise" to taxonomists but a boon to functional anatomists because these "natural experiments" allow us to use the comparative method to identify potential morphological correlates of specific behaviors and hopefully assess the relative contributions of phylogeny and function. As Jenkins repeatedly showed, however, such inferences still need to be tested experimentally in the lab by evaluating the biomechanical models that underlie hypotheses about how variations in anatomy affect performance. Such experiments are especially necessary to test functional interpretations of rare or unique combinations of features evident in fossils that are not represented by modern taxa. This problem is of special relevance to obligate bipedalism, which is present in just one extant taxon, *Homo sapiens*, limiting our abilities to interpret the behavior of fossil hominins that are generally similar to modern humans, but differ in details (Cartmill 1992). One cannot hope to explain the evolution

FIG. 15.6 Independent transformations have yielded similar locomotor diversity among major primate radiations. The primates of Eurasia and Africa (above) include leapers (*Tarsius* and *Galago*), both arboreal (*Cercopithecus*) and terrestrial (*Papio*) quadrupeds, and hangers or suspensory taxa (*Hylobates* and *Pongo*). Similarly, the endemic fauna of living and fossil lemurifroms from Madagascar includes a variety of leapers (illustrated by *Indri* and *Lepilemur*), arboreal (*Varecia*) and terrestrial (*Lemur* and *Hadropithecus*) quadrupeds, and suspensory "sloth lemurs" (*Palaeopropithecus* and *Babakotia*). Likewise, the radiation of platyrrhine monkeys in Central and South America includes leapers (*Cebuella* and *Pithecia*), arboreal quadrupeds (*Chiropotes*, *Saimiri*, and *Cebus*), and suspensory taxa (*Ateles* and *Brachyteles*). Drawings are not to scale. A cross indicates that the genus is extinct. Illustration by Stephen Nash.

of bipedalism simply by using modern species as simple referential models, and a better understanding of bipedal evolution will require new kinds of integration between laboratory and paleontological analyses.

Evidence that primates in many different clades independently evolved diverse locomotor specializations including different types of quadrupedalism, leaping, brachiation, and (in at least one case) bipedalism raises interesting questions about the selective forces and mechanisms that led to the evolution of this diversity. To address this question, we return to the three factors Napier identified in 1967: body size, habitat, and diet. Each of these has been addressed in subsequent research, and they are not totally independent.

Body size is unquestionably a major factor, but cannot be evaluated without also considering habitat. As outlined by Napier (1967; see also Fleagle and Mittermeier 1980; Fleagle 1984b, 1985) the relationship between body size and locomotion for arboreal creatures often concerns problems of balance, access to available supports, and gaps in the forest. For a quadrupedal and arboreal primate walking through a network on supports of variable size and flexibility, balance is major concern. In any given environment, small animals will find more supports that are relatively stable, but with increasing size, animals are less likely to find themselves in an environment that they can easily traverse quadrupedally. Suspensory postures, in which an animal hangs below supports, solve the problem of maintaining balance on relatively smaller mobile supports, which explains why larger primates tend to be suspensory. However, with increasing size, an individual needs to locate numerous strong supports, thus we find that larger species who use suspensory locomotion and postures are characterized by long forelimbs as well as mobile joints and grasping hands and feet enabling them to reach far and wide to locate sufficient supports. For small arboreal primates, the critical issue is not balance, but gaps in the canopy. Small primates are therefore more likely to be leapers. In addition, should they fall to the ground during a leap, a smaller animal will suffer less from the momentum of the fall (Cartmill and Milton 1977). In addition, leapers should be favored in habitats with more vertical supports, for example, those that frequent bamboo such a *Callimico* (Pook and Pook 1981) or *Hapalemur* (Tan 1999).

The effects of size on locomotion differ for more terrestrial primates, and probably for no clade more so than hominins, which includes the only primate that rarely uses trees, and does so clumsily: *H. sapiens*. Although large body mass can be a problem in trees because of balance and the need for larger supports, larger terrestrial quadrupeds are both faster and more economical (Alexander 1991), which may explain why there was a trend toward increasing body size, especially leg length, within hominins.

Finally, there is little evidence for any simple relationship between diet and locomotion. Napier (1967) argued that the locomotor differences between largely quadrupedal Old World monkeys and more suspensory apes related to a specialization for a diet of leaves in the former and fruits in the latter. However, both Old World monkeys and apes contain species that are predominately frugivorous and predominately folivorous with no clear relationship to locomotion. The same is true for Malagasy lemurs and platyrrhines. Indeed, Fleagle and Mittermeier (1980; also Fleagle 1984b) suggested that within specific faunas, species with similar diets tend use different locomotor and postural behaviors to access those foods in different parts of the habitat. Similarly, species with similar locomotor and postural abilities tend to differ in their diet. Thus, among frugivores, there are leaping, quadrupedal, and suspensory taxa. Conversely, among suspensory species there are frugivores and folivores. Nevertheless, there are certainly cases in which specific diets entail specialized locomotor or postural abilities. For example, gum-eating species need the ability to cling to tree trunks. Also animals feeding on terrestrial foods, need to have locomotor abilities for traveling on the ground.

The Tyranny of the Present

Regardless of whether one stands, hangs, or leaps back from the details, it is evident why our understanding of primate locomotor evolution, especially comparative and functional anatomy, is strongly rooted in our understanding of extant primates. These are the animals we can observe in the wild, study their muscles, and bring into the lab for experimental studies of their locomotor biomechanics and physiology. Fossil primates are, alas, mostly sampled from teeth and jaws; limb bones are rare and often not associated with other body parts. Complete skeletons are exceptionally rare, and even then, just bones. Nevertheless, despite its limitations, the fossil record offers necessary insights into the history of primate locomotion that are not available from the study of extant taxa alone, and which suggest that, in many respects, extant primates offer a distorted view of the history of primate locomotion. Most significantly, the fossil record provides evidence of many locomotor morphologies and combinations of features that are not present in any extant species. This disparity between the present and the past is perhaps most extreme in Madagascar, where all extant taxa are small (<8 kg) leapers and arboreal quadrupeds. However, the fossil record from just the past few thousand years documents terrestrial quadrupeds, a whole radiation of large suspensory sloth lemurs, and some taxa such as the koala-like *Megaladapis* or *Archaeoindris*, a giant relative of the sloth lemurs with a femur that suggests it may have been most comparable to a ground sloth. Another source of extreme disparity are the hominoids, which used to include a wide variety of non-suspensory apes that are known only from the fossil record.

If we were to wander through a forest in the Eocene, Miocene, or Pliocene of Eurasia or Africa, we would see primates very different from those we observe today. Many of them would be quadrupedal, and we would see fewer leapers and suspensory primates, especially among the apes. We would see, however, a much greater range of intermediate morphologies and mosaic combinations. If any one clade might surprise us the most, it would probably be our own. We live in an unusual era in which just one species of hominin is present (and few of us have to locomote much anymore), but our unusual form of walking and running should not blind us to the greater diversity of bipedal and climbing behaviors that must have characterized our ancestors and cousins. In hindsight, the diversity of locomotor morphologies evident from the fossil record of primate evolution is a wonderful fortune. Paleontology would be very boring if the past were just like the present. However, ancient diversity poses substantial challenges, especially when trying to understand great transformations of the past, of which the extant species we observe today are just mere samples. The only effective approach to understanding the present is through an integrated research program that combines fieldwork, experimental studies of biomechanics, and detailed studies of morphology, especially the joint surfaces of bones that actually permit animals to move as they do.

* * *

Acknowledgments

We thank, commemorate, and dedicate this to Farish A Jenkins Jr. for many wonderful years of mentorship, friendship, and inspiration. We are also grateful to W. Jungers and D. Pilbeam for comments, Luci Betti-Nash and Stephen Nash for assistance with figures, Ian Wallace for editorial help, and to E. Brainerd, K. Dial, and N. Shubin for organizing this volume.

References

Aiello, L., and C. Dean. 1990. An Introduction to Human Evolutionary Anatomy. London: Academic Press.

Alba, D. M., S. Almécija, I. Casanovas-Vilar, et al. 2012. Fossil apes from the Vallès-Penedès Basin. Evolutionary Anthropology 147:135–140.

Alexander, R. M. 1991. Energy-saving mechanisms in walking and running. Journal of Experimental Biology 160:55–69.

Alexander, R. M. 2006. Principles of Animal Locomotion. Princeton, NJ: Princeton University Press.

Almécija, S., M. Tallman, D. M. Alba, M. Pina, S. Mòya-Sòla, and W. L. Jungers. 2013. The femur of *Orrorin tugenensis* exhibits morphometric affinities with both Miocene apes and later hominins. Nature Comm. doi:10.1038/ncomms3888

Anemone, R. L. 1990. The VCL hypothesis revisited: patterns of femoral morphology among quadrupedal and saltatorial prosimian primates. American Journal of Physical Anthropology 83:373–393.

Arsuaga, J. L., C. Lorenzo, J. M. Carretero, et al. 1999. A complete human pelvis from the Middle Pleistocene of Spain. Nature 399:255–258.

Barak, M. M., D. E. Lieberman, D. Raichlen, H. Pontzer, A. G. Warrener, and J. J. Hublin. 2013. Trabecular evidence for a human-like gait in *Australopithecus africanus*. PLoS One 8:e77687.

Beauvilain, A., and J.-P. Watté. 2009. Toumaï (Sahelanthropus tchadensis) a-t-il été inhumé? Bulletin de la Société Géologique de Normandie et des Amis du Muséum du Havre 96:19–26.

Begun, D. R., M. Nargolwalla, and L. Kordos. 2012. European Miocene hominoids and the origin of the African ape and human clade. Evolutionary Anthropology 21:10–23.

Berge, C., and X. Penin. 2004. Ontogenetic allometry, heterochrony, and interspecific differences in the skull of African apes, using tridimensional Procrustes analysis. American Journal of Physical Anthropology 124:124–138.

Berger, L. R., D. J. de Ruiter, S. E. Churchill, et al. 2010. Australopithecus sediba: a new species of Homo-like australopith from South Africa. Science 328:195–204.

Bonmatí, A., A. Gómez-Olivencia, J. L. Arsuaga, et al. 2010. Middle Pleistocene lower back and pelvis from an aged human individual from the Sima de los Huesos site, Spain. Proceedings of the National Academy of Sciences USA 107:18386–18391.

Bown, T. M., M. J. Kraus, S. L. Wing, et al. 1982. The Fayum primate forest revisited. Journal of Human Evolution 11:603–632.

Bramble, D. M., and D. E. Lieberman. 2004. Endurance running and the evolution of Homo. Nature 432:345–352.

Brunet, M., F. Guy, D. Pilbeam, et al. 2002. A new hominid from the upper Miocene of Chad, central Africa. Nature 418:145–151.

Brunet, M., F. Guy, D. Pilbeam, et al. 2005. New material of the earliest hominid from the Upper Miocene of Chad. Nature 434:752–755.

Carrier, D. R. 1984. The energetic paradox of human running and hominid evolution. Current Anthropology 25:483–495.

Cartelle, C., and W. C. Hartwig. 1996. A new extinct primate among the Pleistocene megafauna of Bahia, Brazil. Proceedings of the National Academy of Sciences USA 93:6405–6409.

Cartmill, M. 1972. Arboreal adaptations and the origin of the Order Primates. Pp. 97–122 in R. Tuttle, ed., The Functional and Evolutionary Biology of Primates. Chicago: Aldine.

Cartmill, M. 1992. Human uniqueness and theoretical content in paleoanthropology. International Journal of Primatology 11:173–192.

Cartmill, M. 2012. Primate origins, human origins and the end of higher taxa. Evolutionary Anthropology 21:208–220.

Cartmill, M., and K. Milton. 1977. The lorisiform write and the evolution of "brachiating" adaptations in the Hominoidea. American Journal of Physical Anthropology 47:249–271.

Churchill, S. E., T. W. Holliday, K. J. Carlson, et al. 2013. The upper limb of Australopithecus sediba. Science 340:1233477.

Crompton, R. H., W. I. Sellers, and S. K. Thorpe. 2010. Arboreality, terrestriality and bipedalism. Philosophical Transactions of the Royal Society B: Biological Sciences 365:3301–3314.

Dagosto, M., D. L. Gebo, and C. Beard. 1999. Revision of the Wind River faunas, early Eocene of central Wyoming. Part 14. Postcranium of Shoshonius cooperi (Mammalia: Primates). Annals of Carnegie Museum 68:175–211.

Darwin, C. 1871. The Descent of Man and Selection in Relation to Sex. London: J. Murray.

Deane, A. S., and D. R. Begun. 2010. Pierolapithecus locomotor adaptations: a reply to Alba et al.'s comment on Deane and Begun (2008). Journal of Human Evolution 59:150–154.

DeSilva, J. M. 2009. Functional morphology of the ankle and the likelihood of climbing in early hominins. Proceedings of the National Academy of Sciences USA 106:6567–72.

DeSilva, J. M., K. G. Holt, S. E. Churchill, et al. 2013. The lower limb and mechanics of walking in Australopithecus sediba. Science 340:1232999.

Erikson, G. E. 1963. Brachiation in New World monkeys and in anthropoid apes. Symposia of the Zoological Society of London 10:135–164.

Fleagle, J. G. 1977a. Locomotor behavior and muscular anatomy of sympatric Malaysian leaf monkeys (Presbytis obscura and Presbytis melalophos). American Journal of Physical Anthropology 46:297–308.

Fleagle, J. G. 1977b. Locomotor behavior and skeletal anatomy of sympatric Malaysian leaf-monkeys (Presbytis obscura and Presbytis melalophos). Yearbook of Physical Anthropology 20:440–453.

Fleagle, J. G. 1983. Locomotor adaptation of Oligocene and Miocene hominoids and their phyletic implications. Pp. 201–224 in R. L. Ciochon and R. Corruccini, eds., New Interpretations of Ape and Human Ancestry. New York: Plenum Press.

Fleagle, J. G. 1984a. Are there any fossil gibbons? Pp. 431–447 in H. Preuschoft, D. J. Chivers, W. Brockelman, and N. Creel, eds., The Lesser Apes: Evolutionary and Behavioral Biology. Edinburgh: Edinburgh Press.

Fleagle, J. G. 1984b. Primate locomotion and diet. Pp. 105–117 in D. J. Chivers, B. A. Wood, and A. Bolsborough, eds., Food Acquisition and Processing in Primates. New York: Plenum Press.

Fleagle, J. G. 1985. Size and adaptation in primates. Pp. 1–19 in W. L. Jungers, ed., Size and Scaling in Primate Biology. New York: Plenum Press.

Fleagle, J. G. 1999. Primate Adaptation and Evolution, Second Edition. New York: Academic Press.

Fleagle, J. G. 2013. Primate Adaptation and Evolution, Third Edition. San Diego: Elsevier.

Fleagle, J. G., and F. C. Anapol. 1992. The indriid ischium and the hominid hip. Journal of Human Evolution 22:285–305.

Fleagle, J. G., and D. J. Meldrum. 1988. Locomotor behavior and skeletal morphology of two sympatric pitheciine monkeys, Pithecia pithecia and Chiropotes santanas. American Journal of Primatology 16:227–249.

Fleagle, J. G., and R. A. Mittermeier. 1980. Locomotor behavior, body size and comparative ecology of seven Surinam monkeys. American Journal of Physical Anthropology 52:301–314.

Gebo, D. L., L. MacLatchy, R. Kityo, et al. 1997. A hominoid genus from the early Miocene of Uganda. Science 276:401–404.

Gebo, D. L., T. Smith, and M. Dagosto. 2012. New postcranial elements for the earliest Eocene fossil primate Teilhardina belgica. Journal of Human Evolution 63:205–218.

Godfrey, L. R., and W. L. Jungers. 2003. Subfossil lemurs. Pp. 1247–1252 in S. M. Goodman and J. P. Benstead, eds., The Natural History of Madagascar. Chicago: University of Chicago Press.

Grand, T. 1968. Functional Anatomy of the Upper Limb. Pp. 104–125 in M. R. Malinow, ed., Biology of the Howler Monkey (Alouatta caraya). Bibl. primato. No. 7.

Guy, F., D. E. Lieberman, D. Pilbeam, et al. 2005. Morphological affinities of the Sahelanthropus tchadensis (Late Miocene hominid from Chad) cranium. Proceedings of the National Academy of Sciences USA 102:18836–18841.

Haile-Selassie, Y., B. Z. Saylor, and A. Deino, et al. 2012. A new hominin foot from Ethiopia shows multiple Pliocene bipedal adaptations. Nature 483:565–569.

Halenar, L. B. 2011. Reconstructing the locomotor repertoire of *Protopithecus brasiliensis*. I. Body size. Anatomical Record 294:307–311.

Harrison, T. 1989. New postcranial remains of *Victoriapithecus* from the middle Miocene of Kenya. Journal of Human Evolution 18:3–54.

Hartwig, W. C., and C. Cartelle. 1996. A complete skeleton of the giant South American primate *Protopithecus*. Nature 381:307–311.

Haxton, H. A. 1947. Muscles of the pelvic limb. Anatomical Record 98:337–346.

Hunt, K. D. 1992. Positional behavior of *Pan troglodytes* in the Mahale Mountains and Gombe Stream National Parks, Tanzania. American Journal of Physical Anthropology 87:83–105.

Jenkins, F. A., Jr. 1972. Chimpanzee bipedalism: cineradiographic analysis and implications for the evolution of gait. Science 178:877–879.

Jenkins, F. A., Jr. 1973. The functional anatomy and evolution of the mammalian humero-ulnar articulation. American Journal of Anatomy 137:281–298.

Jenkins, F. A., Jr. 1974. Tree shrew locomotion and the origins of primate arborealism. Pp. 85–115 in F. A. Jenkins Jr., ed., Primate Locomotion. New York: Academic Press.

Jenkins, F. A., Jr. 1981. Wrist rotation in primates: a critical adaptation for brachiators. Symposia of the Zoological Society of London 48:429–451.

Jenkins, F. A., Jr., and S. M. Camazine. 1977. Hip structure and locomotion in ambulatory and cursorial carnivores. Journal of Zoology 181:351–370.

Jenkins, F. A., Jr., and J. G. Fleagle. 1975. Knuckle-walking and the functional anatomy of the wrists in living apes. Pp. 213–227 in R. H. Tuttle, ed., Primate Functional Morphology and Evolution. The Hague: Mouton Publishers.

Jenkins, F. A., Jr., and D. McClearn. 1984. Mechanisms of hind foot reversal in climbing mammals. Journal of Morphology 182:197–219.

Jolly, A. 1967. Lemur Behavior. Chicago: University of Chicago Press.

Jouffroy, F.-K. 1963. Contribution à la connaissance du genre *Archaeolemur* Filhol, 1895. Annales de Paléontologie 49:129–155.

Jungers, W. L., L. R. Godfrey, E. L. Simons, and P. S. Chatrath. 1997. Phalangeal curvature and positional behavior in extinct sloth lemurs (Primates, Palaeopropithecidae). Proceedings of the National Academy of Sciences USA 94:11998–12001.

Jungers, W. L., L. R. Godfrey, E. L. Simons, R. E. Wunderlich, B. G. Richmond, and P. S. Chatrath. 2002. Ecomorphology and behavior of giant extinct lemurs from Madagascar. Pp. 371–411 in J. M. Plavcan et al., eds., Reconstructing Behavior in the Primate Fossil Record. New York: Plenum Press.

Jungers, W. L., and S. E. Hartman. 1988. Relative growth of the locomotor skeleton in orangutans and other large-bodied hominoids. Pp. 380–395 in J. H. Schwartz, ed., Orangutan Biology. Oxford: Oxford University Press.

Kappelman, J., B. G. Richmond, E. R. Seiffert, A. M. Maga, and T. M. Ryan. 2003. Hominoidea (Primates). Pp. 90–124 in M. Fortelius, J. Kappelman, S. Sen, and R. L. Bernor, eds., Geology and Paleontology of the Miocene Sinap Formation, Turkey. New York: Columbia University Press.

Keith, A. 1891. Anatomical notes on Malay apes. Journal of the Straits Branch of the Royal Asiatic Society 23:77–94.

Keith, A. 1923. Man's posture: its evolution and disorders. British Medical Journal 1:451–454, 499–502, 545–548, 587–590, 624–626, 669–672.

Kibii, J. M., S. E. Churchill, P. Schmid, et al. 2011. A partial pelvis of *Australopithecus sediba*. Science 333:1407–1411.

Kinzey, W. G., A. L. Rosenberger, and M. Ramirez. 1975. Vertical clinging and leaping in a Neotropical anthropoid. Nature 255:327–328.

Kirk, E. C., P. Lemelin, and M. W. Hamrick. 2008. Intrinsic hand proportions of euarchontans and other mammals: implications for the locomotor behavior of plesiadapiforms. Journal of Human Evolution 55:278–299.

Kivell, T. L., J. M. Kibii, S. E. Churchill, et al. 2011. *Australopithecus sediba* hand demonstrates mosaic evolution of locomotor and manipulative abilities. Science 238:205–208.

Kivell, T. L., and D. Schmitt. 2009. Independent evolution of knuckle-walking in African apes shows that humans did not evolve from a knuckle-walking ancestor. Proceedings of the National Academy of Sciences USA 106:14241–14246.

Kumar, S., A. Filipski, and V. Swarna, et al. 2005. Placing confidence limits on the molecular age of the human-chimpanzee divergence. Proceedings of the National Academy of Sciences USA 102:18842–18847.

Larson, S. G. 1998. Parallel evolution in the hominoid trunk and forelimb. Evolutionary Anthropology 6:87–99.

Latimer, B. M., and C. O. Lovejoy. 1989. The calcaneus of *Australopithecus afarensis* and its implications for the evolution of bipedality. American Journal of Physical Anthropology 78:369–386.

Lieberman, D. E. 2011. The Evolution of the Human Head. Cambridge, MA: Harvard University Press.

Lieberman, D. E., D. M. Bramble, D. A. Raichlen, and J. J. Shea. 2008. Brains, brawn and the evolution of endurance running capabilities. Pp. 77–98 in F. E. Grine, J. G. Fleagle, and R. E. Leakey, eds., The First Humans: Origin of the Genus *Homo*. New York: Springer.

Lockwood, C. A., and J. G. Fleagle. 1999. The recognition and evaluation of homoplasy in primate and human evolution. Yearbook of Physical Anthropology 42:189–232.

Lovejoy, C. O. 1988. Evolution of human walking. Scientific American 259:118–125.

Lovejoy, C. O., B. Latimer, G. Suwa, et al. 2009. Combining prehension and propulsion: the foot of *Ardipithecus ramidus*. Science 326:72e1–8.

Lovejoy, C. O., S. W. Simpson, T. D. White, et al. 2009. Careful climbing in the Miocene: the forelimbs of *Ardipithecus ramidus* and humans are primitive. Science 326:70e1–8.

Lovejoy, C. O., G. Suwa, S. W. Simpson, et al. 2009. The great divides: *Ardipithecus ramidus* reveals the postcrania of our last common ancestors with African apes. Science 326:100–106.

Lovejoy, C. O., G. Suwa, L. Spurlock, et al. 2009. The pelvis and femur of *Ardipithecus ramidus*: the emergence of upright walking. Science 326:71e1–6.

MacLatchy, L. 2004. The oldest ape. Evolutionary Anthropology 13:90–103.

MacLatchy, L., D. L. Gebo, R. Kityo, and D. R. Pilbeam. 2000. Postcranial functional morphology of *Morotopithecus bishopi*, with implications for the evolution of modern ape locomotion. Journal of Human Evolution 39:159–183.

Madar, S. I., M. D. Rose, J. Kelley, et al. 2002. New *Sivapithecus* postcranial specimens from the Siwaliks of Pakistan. Journal of Human Evolution 42:705–752.

Martin, R. D. 1972. Adaptive radiation and behavior of the Malagasy lemurs. Philosophical Transactions of the Royal Society B: Biological Sciences 264:295–352.

Martin, R. D. 1990. Primate Origins and Evolution: A Phylogenetic Reconstruction. Princeton, NJ: Princeton University Press.

McArdle, J. E. 1981. Functional morphology of the hip and thigh of the lorisiforms. Contributions to Primatology 17:1–132.

McBrearty, S., and N. G. Jablonski. 2005. First fossil chimpanzee. Nature 437:105–108.

Moyà-Solà, S., M. Köhler, D. M. Alba, et al. 2004. *Pierolapithecus catalaunicus*, a new Middle Miocene great ape from Spain. Science 306:1339–1344.

Napier, J. R. 1963. Brachiation and brachiators. Symposia of the Zoological Society of London 10:183–195.

Napier, J. R. 1964. The evolution of bipedal walking in hominids. Archies de Biologies (Liège) 75:673–708.

Napier, J. R. 1967a. The antiquity of human walking. Scientific American 216:56–66.

Napier, J. R. 1967b. Evolutionary aspects of primate locomotion. American Journal of Physical Anthropology 27:333–341.

Napier, J. R., and P. H. Napier. 1967. A Handbook of Living Primates. London: Academic Press.

Napier, J. R., and A. C. Walker. 1967. Vertical clinging and leaping—a newly recognized category of locomotor behaviour of primates. Folia Primatologica 6:204–219.

Ni, X., D. L. Gebo, M. Dagosto, J. Meng, P. Tafforeau, J. J. Flynn, and K. C. Beard. 2013. The oldest known primate skeleton and early haplorhine evolution. Nature 498:60–64.

Patterson, N. D., J. Richter, S. Gnerre, E. S. Lander, and D. Reich. 2006. Genetic evidence for complex speciation of humans and chimpanzees. Nature 441:1103–1108.

Pickford, M., and B. Senut. 2001. "Millennium ancestor," a 6-million-year-old bipedal hominid from Kenya. Comptes rendus de l'Académie des Sciences de Paris, série 2a, 332:134–144.

Pilbeam, D. R. 1982. New hominoid skull material from the Miocene of Pakistan. Nature 295:232–234.

Pilbeam, D. R., and N. M. Young. 2001. *Sivapithecus* and hominoid evolution: some brief comments. Pp. 349–364 in L. de Bonis, G. Koufos, and P. Andrews, eds., Hominoid Evolution and Climatic Change in Europe, Volume Two: Phylogeny of the Neogene Hominoid Primates of Eurasia. Cambridge: Cambridge University Press.

Pilbeam, D. R., M. D. Rose, J. C. Barry, and S. M. I. Shah. 1990. New *Sivapithecus* humeri from Pakistan and the relationship of *Sivapithecus* and *Pongo*. Nature 348:237–239.

Pontzer, H., D. A. Raichlen, and M. D. Sockol. 2009. The metabolic cost of walking in humans, chimpanzees, and early hominins. Journal of Human Evolution 56:43–54.

Pook, A. G., and G. I. Pook. 1981. A field study of the socio-ecology of the Goeldi's monkey (*Callimico goeldii*) in northern Brazil. Folia Primatologica 35:288–312.

Raichlen, D. A., H. Pontzer, and M. D. Sockol. 2008. The Laetoli footprints and early hominin kinematics. Journal of Human Evolution 54:112–117.

Richmond, B. G., and W. L. Jungers. 2008. *Orrorin tugenensis* femoral morphology and the evolution of hominin bipedalism. Science 319:1662–1665.

Richmond, B. G., D. R. Begun, and D. S. Strait. 2001. Origin of human bipedalism: The knuckle-walking hypothesis revisited. Yearbook of Physical Anthropology 44:70–105.

Robinson, J. T. 1972. Early Hominid Posture and Locomotion. Chicago: University of Chicago Press.

Rolian, C., D. Lieberman, J. Hamill, J. Scott, and W. Werbel. 2009. Walking, running and the evolution of short toes in humans. Journal of Experimental Biology 212:713–721.

Rose, M. D. 1988. Another look at the anthropoid elbow. Journal of Human Evolution 17:193–224.

Rossie, J. B., M. Gutierrez, and E. Goble. 2012. Fossil forelimbs of *Simiolus* from Moruorot, Kenya. American Journal of Physical Anthropology Suppl. 54:252.

Ruvolo, M. 1997. Molecular phylogeny of the hominoids: inferences from multiple independent DNA sequence data sets. Molecular Biology and Evolution 14:248–265.

Sanders, W. J., and B. E. Bodenbender. 1994. Morphometric analysis of lumbar vertebrae UMP 67-28: implications for spinal function and phylogeny of the Miocene Moroto hominoid. Journal of Human Evolution 26:203–237.

Schmitt, D. 2003a. Insights into the evolution of human bipedalism from experimental studies of humans and other primates. Journal of Experimental Biology 206:1437–1448.

Schmitt, D. 2003b. Substrate size and primate forelimb mechanics: implications for understanding the evolution of primate locomotion. International Journal of Primatology. 24:1023–1036.

Sellers, W. I., G. M. Cain, W. Wang, and R. H. Crompton. 2005. Stride lengths, speed and energy costs in walking of *Australopithecus afarensis*: using evolutionary robotics to predict locomotion of early human ancestors. Journal of the Royal Society Interface. 2:431–441.

Shapiro, L., C. Seiffert, L. Godfrey, et al. 2005. Morphometric analysis of lumbar vertebrae in extinct Malagasy strepsirrhines. American Journal of Physical Anthropology 128:823–839.

Shea, B. T. 1985. Ontogenetic allometry and scaling: a discussion based on the growth and form of the skull in African apes. Pp. 175–205 in W. L. Jungers, ed., Size and Scaling in Primate Morphology. New York: Plenum Press.

Simpson, S. W., J. Quade, N. E. Levin, et al. 2008. A female *Homo erectus* pelvis from Gona, Ethiopia. Science 322:1089–1092.

Smith, J. M., and R. J. G. Savage. 1956. Some locomotory adaptations in mammals. Zoological Journal of the Linnean Society, London 42:603–622.

Sockol, M. D., D. A. Raichlen, and H. D. Pontzer. 2007. Chimpanzee locomotor energetics and the origin of human bipedalism. Proceedings of the National Academy of Sciences USA 104:12265–12269.

Stern, J. T., Jr. 2000. Climbing to the top: a personal memoire of *Australopithecus afarensis*. Evolutionary Anthropology 9:113–133.

Stern, J. T., Jr. 1971. Functional myology of the hip and thigh of Cebid monkeys and its implications for the evolution of erect posture. Bibliotheca Primatologica 14:1–319.

Stern, J. T., Jr., and C. E. Oxnard. 1973. Primate locomotion: some links with evolution and morphology. Primatologia 4:1–93.

Straus, W. L., Jr. 1962. Fossil evidence for the evolution of the erect bipedal posture. Clinical Orthopaedics 25:9–19.

Susman, R. L., J. T. Stern Jr., and W. L. Jungers. 1984. Arboreality and bipedality in the Hadar hominids. Folia Primatologica 43:113–156.

Suwa, G., R. T. Kono, S. Katoh, et al. 2007. A new species of great ape from the late Miocene epoch in Ethiopia. Nature 448:921–924.

Szalay, F. S. 1972. Paleobiology of the earliest primates. Pp. 3–35 in R. H. Tuttle, ed., The Functional and Evolutionary Biology Primates. Chicago: Aldine.

Szalay, F. S., and M. Dagosto. 1980. Locomotor adaptations as reflected on the humerus of Paleogene primates. Folia Primatologica 34:1–45.

Szalay, F. S., and E. Delson. 1979. Evolutionary History of the Primates. New York: Academic Press.

Tan, C. L. 1999. Group composition, home range size, and diet of three sympatric bamboo lemur species (genus *Hapalemur*) in Ranomafana National Park, Madagascar. International Journal of Primatology 20:547–566.

Vuillaume-Randriamanantena, M. 1988. The taxonomic attributions of giant subfossil lemur bones from Ampasambazimba: *Archaeoindris* and *Lemuridotherium*. Journal of Human Evolution 17:379–391.

Walker, A. C. 1974. Locomotor adaptations in past and present prosimian primates. Pp. 349–381 in F. A. Jenkins Jr., ed., Primate Locomotion. New York: Academic Press.

Ward, C. V. 2002. Interpreting the posture and locomotion of *Australopithecus afarensis*: where do we stand? Yearbook of Physical Anthropology 35:185–215.

Washburn, S. J. 1968. The Study of Human Evolution. Condon Lectures. Oregon State System of Higher Education.

Washburn, S. L. 1973. Ape into Man. Boston: Little Brown.

White, T. D., B. Asfaw, Y. Beyene, et al. 2009. *Ardipithecus ramidus* and the paleobiology of early hominids. Science 326:75–86.

Wood, B. A., and M. Collard. 1999. The human genus. Science 284:65–71.

Wood, B. A., and T. Harrison. 2011. The evolutionary context of the first hominins. Nature 470:347–352.

Young, N. M. 2003. A reassessment of living hominoid postcranial variability: implications for ape evolution. Journal of Human Evolution 45:441–464.

Zipfel, B., J. M. DeSilva, R. S. Kidd, et al. 2011. The foot and ankle of *Australopithecus sediba*. Science 333:1417–1420.

Zollikofer, C. P., M. S. Ponce de Leon, D. E. Lieberman, et al. 2005. Virtual cranial reconstruction of *Sahelanthropus tchadensis*. Nature 434:755–759.

Zuckerman, S., E. H. Ashton, R. M. Flinn, et al. 1973. Some locomotor features of the pelvic girdle in primates. Symposia of the Zoological Society of London 33:71–165.

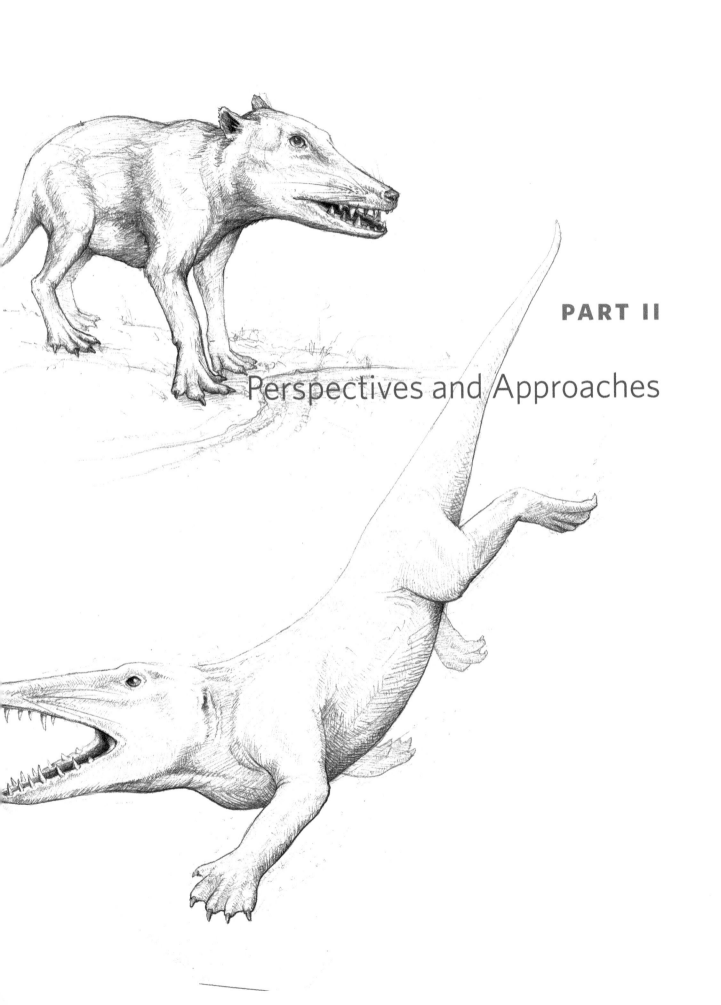

PART II

Perspectives and Approaches

16

Ontogenetic and Evolutionary Transformations: *Ecological Significance of Rudimentary Structures*

Kenneth P. Dial,* Ashley M. Heers,† and Terry R. Dial‡

Introduction

When we think about Darwin's great tree of life, or any phylogenetic tree, the images that spring to mind are usually those of adult organisms—the array of finches with varying beak morphologies, fancy pigeons exhibiting extreme artificially selected traits, or dimorphism in mature age classes throughout the great apes. Biologists and paleontologists exploring great transformations in the history of vertebrate life have typically focused on comparing adult forms, particularly when studying locomotion (but see Carrier 1996; Irschick 2000; Dial 2003; Dial et al. 2006; Dial et al. 2008; Herrel and Gibb 2006; Heers, Tobalske, and Dial 2011; Dial and Carrier 2012). Such an "adultocentric" (Minelli 2003) mindset assumes that adult forms are most visible to selection, and that juveniles are merely stepping-stones to reaching the adult condition.

Posthatching or postnatal juveniles, however, provide important insights into transformations along the tree of life, by illustrating how *transitional, morphing anatomies function in ecological settings*. First, developing young transition through distinct, sometimes dramatic morphological and physiological stages on

* Division of Biological Sciences, University of Montana
† Division of Biological Sciences, University of Montana
‡ Department of Ecology and Evolutionary Biology, Brown University

their pathway to adulthood. Individuals in many species must feed and seek protection at these immature stages (Williams 1966; Wassersug and Sperry 1977) when they are small, behaviorally naive, underdeveloped, and highly susceptible to predation and abiotic effects (e.g., Carrier 1996; Jackson et al. 2009; Heers and Dial 2012; Martin 1995; Herrel and Gibb 2006; Ricklefs 1999; Starck and Ricklefs 1998; Cheng and Martin 2012). Therefore selection on juveniles for escape behaviors and access to ecological refugia, or on parents for care that allows young to bypass vulnerable stages, is likely intense (e.g., Blob et al. 2008). Exploring how morphing juveniles negotiate their environment and reach the adult condition is thus an important but underappreciated component of understanding how species' forms are generated along extant tips in the tree of life.

Developing animals can also help elucidate deeper branching events, by providing insight into the functional attributes of fossilized predecessors. Juveniles frequently navigate their surroundings using underdeveloped anatomies that not only lack the functional specializations of adults, but that also often resemble the "transitional" features of extinct relatives (Heers and Dial 2012). By using these transitional features to locomote and survive, and by revealing how changes in form translate to changes in function, morphing juveniles help clarify the functional attributes and ecologies of fossils with similar transitional features. For instance, as metamorphosing salamanders shift from an aquatic to a terrestrial lifestyle, they progress through a series of dramatic anatomical transformations (e.g., loss of tail fins and gills) in conjunction with a shift from tail-based locomotion and gill breathing to limb-based locomotion and lung/skin breathing. Many aspects of this aquatic-to-terrestrial transformation parallel and help illuminate the origin and early history of tetrapods (for review of the fossil record, see Clack 2011). Similarly, dispersing sea squirt larvae resemble a hypothesized stage of early chordate evolution and might provide insight into the ecology of early chordates (Garstang 1951). These and other examples (Gould 1977) suggest that understanding how juveniles engage their developing features to survive and reach adulthood will help elucidate the potential functional capacities of extinct predecessors with similar morphologies. Thus, though we know very little about juvenile locomotion and ecology, quantifying developmental transitions in form and

function can offer important and novel insight into transformations along the tree of life at multiple levels.

Birds and their theropod ancestors are a particularly useful case study for illustrating how ontogenetic trajectories provide insight into evolutionary transformations. In the process of invading aerial media, developing birds and extinct theropod dinosaurs have undertaken some of the most dramatic morphological transformations in the history of vertebrates. Flight is the most physically demanding form of locomotion (Schmidt-Nielsen 1972; Alexander 2002), and flight-capable adult birds possess a number of anatomical features (e.g., keeled sternum, asymmetric feathers, specialized shoulder and forelimb joints) that are presumably adaptations or exaptations (Gould and Vrba 1982) for meeting such demands. Immature birds lack these hallmarks of flight capacity, yet in many species juveniles must feed and avoid predation in the same hazardous environment as adults. How is this accomplished? Recent work shows that flight-incapable juveniles transition to flight-capable adults through a series of morphological and behavioral stages that incrementally improve locomotor performance and culminate in full flight capacity. Though not yet capable of level flight, juveniles can reach elevated or aquatic refugia by flap-running up slopes (Dial 2003) or flap-rowing across water (Dial and Carrier 2012) and supplementing their incipient wings with their legs, until the wings can fully support body weight during flight. Many of the anatomical changes that occur during this developmental transition are similar to those observed in theropod dinosaurs during the evolutionary origin of flight (Dial et al. 2006; Heers and Dial 2012). For example, both immature birds and early winged theropods have small "proto-wings" and forelimb apparatuses that are relatively gracile and unconstrained, whereas adult birds and more derived theropods have large wings and robust, interlocking forelimb apparatuses that are associated with powerful and highly canalized flight strokes (Ostrom 1976; Vazquez 1992; Prum and Brush 2002; Norell and Xu 2005; Dial et al. 2006; Heers and Dial 2012). Given these similarities, developing birds can help illuminate both ontogenetic and evolutionary construction of the avian *bauplan*. Though we are only beginning to explore locomotor ontogeny, juveniles that reveal the functional capacities of transitional and rudimentary structures also observed in the fossil record may

FIG. 16.1 The fact that some wings are incapable of propelling their owner into the sky does not indicate that rudimentary wings are aerodynamically or functionally impotent. Protowings of developing birds (upper row) are important for a variety of locomotor behaviors (left to right: waterfowl steaming, ground birds exhibiting WAIR, arboreal forms branching). Similarly, the small wings of adults in species that have secondarily reduced their flight-capable wings (middle row: flightless cormorant, Megellanic steamer duck, kakapo [parrot from New Zealand]) are still recruited for locomotion. Finally, the current fossil record reveals that rudimentary wings were common among theropod dinosaurs (bottom row: *Caudipteryx, Anchiornis, Eosinopterx*), suggesting that these transitional (whether emerging or regressing) forelimbs may have been employed aerodynamically for important locomotor activities.

provide unique insight into many facets of life and its history.

Here, we use the theropod-avian clade to illustrate how ontogenetic transformations can help elucidate evolutionary transitions by demonstrating the ecological significance of rudimentary structures. We define "rudimentary" features as either incipient, developing structures in juveniles (e.g., protowings) or secondarily reduced structures in adults (e.g., vestigial wings) (fig. 16.1). This essay (1) provides a brief overview of theropod evolution and existing frameworks for interpreting trends in the richly expanding fossil record,

(2) discusses the importance of transitional habitats and refugia during developmental transformations, and (3) highlights three distinct, yet common, ontogenetic trajectories to (4) illustrate how the developmental acquisition of flight capacity among birds—and how adult birds with secondarily reduced wings—can provide novel insight into the locomotor capacity of extinct theropod dinosaurs. Though we focus here on the theropod-avian clade, this two-pronged ecological and ontogenetic approach can greatly enhance our understanding of transformations in many vertebrate groups.

I. Transitional Morphologies and Locomotor Evolution: A Brief Review of Extinct Feathered Theropods

So-called transitional fossils are almost never exactly intermediate between two other forms: they are mosaics of primitive and derived features, and possess unique features of their own. However, fossils often do show *features* that are transitional between other distinct forms. This is particularly apparent in theropod dinosaurs, the ancestors of birds (Gauthier 1986; Sereno 1997; Dingus and Rowe 1998; Padian and Chiappe 1998; Norell and Xu 2005; Xu 2006; Chiappe 2007; Witmer 2009; Makovicky and Zanno 2011). Initially, the theropod-avian fossil record lacked fossils with transitional, flight-related features, and consisted of non-avian dinosaurs without any preserved feathers (e.g., *Compsognathus*) and birds with full-sized wings (e.g., *Archaeopteryx*). Where did wing feathers come from? Until recently there were no transitional fossils to fill this gap. But in the 1990s our understanding of the theropod-avian transition was transformed by a series of feathered, non-avian dinosaurs discovered in China (Chen et al. 1998). These animals possessed a sequence of feather types, ranging from simple or slightly branched hair-like structures, to down-like clusters of filaments, to stalked and vaned plumes with barbs. Increasingly complex feather types correlate with the phylogenetic progression of theropods toward birds (Padian 2001; see below), and with the evolution of more bird-like wings and skeletons (Heers and Dial 2012). Newly discovered feathered dinosaurs thus record the origin and evolution of features crucial to avian flight, and demonstrate that theropods progressed through a sequence of transitional anatomi-

cal stages that became increasingly bird-like throughout the Jurassic and Cretaceous.

Transitional feathers.—With the discovery of feathered dinosaurs in 1996, it has become clear that feathers—one of the defining features of birds—actually debuted in non-avian predecessors with functions unrelated to flight. The first feathers seem to have been "downy" or plumulaceous body feathers (Norell and Xu 2005), as exemplified by basal maniraptorans such as *Sinosauropteryx* (Chen et al. 1998) (fig. 16.2). In more derived maniraptorans, such as *Caudipteryx* (Qiang et al. 1998; Zhou and Wang 2000) or *Protarchaeopteryx* (Qiang et al. 1998), these plumulaceous feathers were complemented by pennaceous (vaned) feathers that initially appeared as fans on the distal tail and as protowings on the distal forelimb (ulna and/or manus; tertials absent or not preserved). Pennaceous feathers became more widely distributed (eventually covering the entire body) and more asymmetric in the more derived paravians. For example, many paravians had frond-like tails and asymmetric feathers along their forelimbs, and some paravians, such as *Microraptor* (Xu et al. 2003) and *Anchiornis* (Hu et al. 2009), had pennaceous feathers that covered their hind limbs. However, extensive feathering of the hind limbs seems to have been reduced or lost in avialans and more derived birds (Zheng et al. 2013), and in pygostylians (e.g., *Confuciusornis*; Chiappe et al. 1999) long, frond-like tails were replaced by short tails with a pygostyle and, eventually, a rectricial bulb (~ ornithurines). Feathers thus debuted as downy-like structures relatively early in the history of theropods, and progressed through a sequence of transitional morphologies (symmetric feathers) and arrangements (protowings, frond-like tails) before reaching fully bird-like configurations in ornithurines (fig. 16.2). This progression suggests that feathers were initially associated with non-aerodynamic functions (e.g., insulation, camouflage, display, and/or brooding), and were later co-opted for flight. Behaviors that might have facilitated the acquisition of aerodynamic capacity are widely debated, though, because protowings and symmetrical primary feathers are not represented in extant flight-capable adult birds.

Transitional skeletons.—The appearance and evolution of feathers in theropods was attended by many

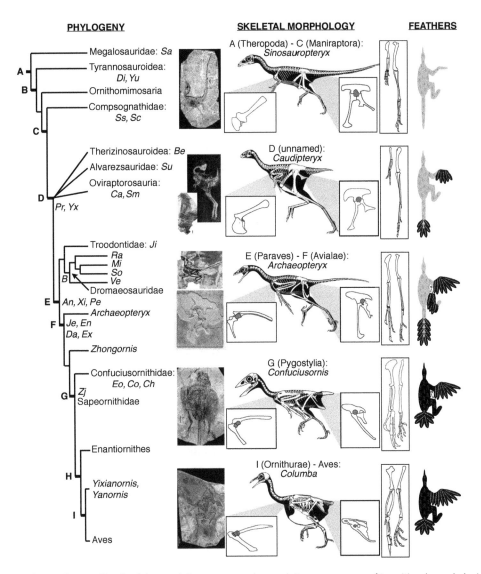

PHYLOGENY   SKELETAL MORPHOLOGY   FEATHERS

FIG. 16.2 The theropod-avian lineage. Fossils of theropod dinosaurs record an evolutionary sequence of transitional morphological stages that became more bird-like throughout the Mesozoic (see text). Clades indicated by capitalized letters: (A) Theropoda, (B) Coelurosauria, (C) Maniraptora, (D) unnamed, (E) Paraves/Eumaniraptora, (F) Avialae, (G) Pygostylia, (H) Ornithothoraces, (I) Ornithurae. Feathered theropods indicated by italicized abbreviations (all feathered theropods up through basal pygostylians included): *Sa, Sciurumimus; Di, Dilong; Yu, Yutyrannus; Ss, Sinosauropteryx; Sc, Sinocalliopteryx; Be, Beipiaosaurus; Su, Shuvuuia; Ca, Caudipteryx; Sm, Similicaudipteryx; Pr, Protarchaeopteryx; Yx, Yixianosaurus; Ji, Jinfengopteryx; B,* BPM 1 3–13; *Ra, Rahonavis; Mi, Microraptor; So, Sinornithosaurus; Ve, Velociraptor; An, Anchiornis; Xi, Xiaotingia; Pe, Pedopenna; Je, Jeholornis; Da, Dalianraptor; En, Epidendrosaurus; Ex, Epidexipteryx; Zj, Zhongjianornis; Eo, Eoconfuciusornis; Co, Confuciusornis; Ch, Changchengornis.* Feather morphology indicated by colors: gray, unbranched or "downy" feathers (types I, II, or IIIb) (Prum 2005); black, pennaceous feathers (types IIIa, IIIab, IV, or Va). Modified from Heers and Dial (2012), with additional images from Chen et al. (1998) (*Sinosauropteryx*), Arizona Museum of Natural History (*Caudipteryx*), Xu et al. (2009) (*Beipiaosaurus*), Xu et al. (2003) (*Microraptor*), Museum für Naturkunde Berlin (*Archaeopteryx*), Chiappe et al. (1999) (*Changchengornis*), Clark et al. (2006) (*Yixianornis*).

changes in the skeletal apparatus. The pectoral girdle and forelimb were enlarged, the trunk stiffened, the pelvic girdle expanded, the femur shortened, the tail reduced, and many joints modified (fig. 16.2; also refer to table S3 in Heers and Dial 2012) (Ostrom 1976; Gauthier 1986; Gatesy 1990; Jenkins 1993; Dingus and Rowe 1998; Carrano 2000; Hutchinson 2001; Hutchinson 2002; Xu and Zhang 2005). Changes in the pectoral girdle and forelimb are indicative of increases in forelimb size, strength, stroke amplitude, and joint channelization, as well as reductions in grasping capacity (Ostrom 1976; Gauthier 1986; Jenkins 1993; Dingus and Rowe 1998).

Changes in the pelvic girdle and hind limb are suggestive of increases in pelvic strength, stride length, and foot prehensility, coupled with a shift from hip- to knee-based movement (Gatesy 1990; Carrano 2000; Hutchinson 2001; Hutchinson 2002; Xu and Zhang 2005). Initially, these transformations were probably associated with "transitional," non-flight behaviors because assembly of the avian body plan began before, and continued long after, the evolutionary appearance of full-sized wings. The hind limbs, for example, have always been conspicuous among avians (with some extant exceptions) and might have facilitated the early stages of flight evolution through transitional behaviors like flap-running (Dial 2003; Dial et al. 2006; Dial et al. 2008) and climbing (Zhou and Hou 1998), and/or through launching and landing (Earls 2000). In short, evolutionary construction of the avian *bauplan* occurred through a progression of transitional anatomical stages (in both feathers and fore- and hind limb skeletons) and therefore, presumably, transitional locomotor behaviors.

Transitional locomotor behaviors.—Behaviors that might have facilitated the evolutionary acquisition of flight have sparked discussion for more than a century. Most origin-of-flight hypotheses follow a traditional assumption that there are but two scenarios to choose from: arboreal and from trees-down, or cursorial and from ground-up (table 1 in Heers and Dial 2012). However, as many researchers have pointed out, this dichotomous framework is overly simplistic and hinders research (Padian 2001; Hutchinson and Allen 2009). Bisecting extant birds into "arboreal" or "terrestrial" and "gliding" or "flapping" species grossly underrepresents extant interspecific diversity. Further, locomotor strategies and capabilities change throughout ontogeny (Jackson et al. 2009; Heers, Tobalske, et al. 2011; Dial and Jackson 2011), and thus even a single individual cannot be described by a single locomotor category. Yet, essays continue to frame the origin of flight within an ecologically simplistic framework, such that the cursorial-arboreal dichotomy has dominated discussion for over 100 years. We offer an alternative approach to exploring transitional locomotor behaviors and the origin and evolution of avian flight—an approach that replaces the arboreal-cursorial dichotomy with more integrative and realistic ecological constructs that in-

corporate interspecific and ontogenetic variation in habitat use and locomotor strategies (below). Extant birds use their wings and legs in many different ways to move through different environments, and there is no reason to expect anything different for their extinct predecessors.

II. The Importance of Transitional Habitats and Refugia during Morphological Transformations

Ecological transition zones.—If we step back from the polarized rhetoric of "trees-down" versus "ground-up" origins of avian flight, we recognize that the world is neither flat nor vertically dominated by trees. Earth is composed of three distinct media (water, land, and air) that provide a rich tapestry of three-dimensional habitats. These habitats grade into one another as ecotones—interfaces or transitional boundaries between distinct habitats, where dissimilar environments merge to create a mosaic of features. Such fine scale, environmental "stepping-stones" seem to facilitate evolutionary invasions of new habitats. Countless field studies, for example, have recorded how habitat-dependent selection generates polyphenisms or ecomorphs that grade into one another across environmental clines. Most notably, (1) small mammals, fish, and lizards have locally adapted to substrates of different colors by matching their body color to substrate color (Cott 1940; Ingles 1950; Blair 1950; Kaufman 1974; Steiner et al. 2007; Rosenblum et al. 2010; Manceau et al. 2010), (2) birds (Grant and Grant 2008) and cichlid fishes (Liem 1979) have locally adapted to varying food types and abundance through changes in beak shape and four-bar linkage systems, respectively, (3) the limbs of *Anolis* lizards are reflective of local substrates (branch size, etc.) (reviewed in Losos 2009), and (4) birds exhibiting varying degrees of vagility (sedentary/resident versus migrant/nomadic) have evolved differences in wing shape (Norberg and Rayner 1987; Mulvihill and Chandler 1991; Tornberg et al. 1999). In essence, ecological transition zones allow organisms to adapt to new environments through subtle changes in character states. When we think about theropod dinosaurs and the origin of flight, we should therefore consider how incremental adaptations to transitional habitats might have facilitated the invasion of aerial media.

The importance of avoiding predators and accessing refugia.—What stimulates organisms to invade new habitats and media, and ultimately function there? Access to new food resources is commonly viewed as a primary motivator for entering unexploited niche space (MacArthur 1967). We herein suggest that avoiding predation and securing a refuge—whether by camouflage, superior maneuverability, shear speed, or simply the ability to reach a safe place—represents an important and underappreciated stimulus for the invasion of new habitats and attending changes in locomotor design.

Whether sprinting, jumping, launching, or dodging, maximal locomotor performance is almost always associated with an animal attempting to reach a refuge or escape from a predator. For example, non-powered vertebrate fliers such as gliding frogs (e.g., *Rhacophorus*), gliding lizards (*Draco*), gliding snakes (*Chrysopelea*), gliding lemurs (Dermoptera), sugar gliders (*Petaurus*), and gliding squirrels (*Glaucomys*) routinely leap from elevated substrates (trees, cliffs, boulders) and deploy specialized anatomical structures (webbed toes, skin membranes, etc.) primarily to evade capture (Norberg 1990). Similarly, flying fish (Exocoetidae) launch out of the water and into the air to escape aquatic predators (Davenport 1994). Vertebrates that inhabit the water's edge take refuge by first evaluating their threat (is it from an aerial, terrestrial, or aquatic medium?) and then retreating quickly to the appropriate refuge by swimming, running, or if capable, flying. Finally, species that negotiate aerial environments avoid predation by retreating to a bush or tree, or to even higher altitudes that force the predator to work against gravity (Hedenström and Rosen 2001). These and many other examples demonstrate that animals often survive do-or-die encounters by moving quickly and accessing refugia in adjacent habitats or media.

Access to refugia thus presents a major selective pressure on locomotor performance in all vertebrates, including birds. However, different refuges are sought at different stages in ontogeny. As adults, most birds can reach refugia in adjacent habitats without using transitional substrates. That is, adults can leap from the ground straight into the air and fly to safety, because they are morphologically equipped to deal with both terrestrial (legs) and aerial (wings) locomotion. Even so, individuals will also access refugia by using

transitional behaviors to move through transitional habitats—for example, by flap-running up steep inclines to move from terrestrial to arboreal or aerial environments in order to reach roosting sites or leap into the air. Using transitional behaviors (where wings and legs are used cooperatively/simultaneously) and transitional habitats (that bridge aquatic, terrestrial, and/or aerial media) is facultative for most adults. In contrast, flight-incapable juveniles with underdeveloped locomotor apparatuses seem to *depend* on wing-leg cooperation for accessing refugia. Juvenile ground birds, for example, rely on wing-assisted incline running to reach otherwise inaccessible, elevated refugia (Dial 2003; Dial et al. 2006; Jackson et al. 2009). Similarly, juvenile waterfowl engage their wings and legs simultaneously to "steam" across the water and elude terrestrial predators (Dial and Carrier 2012), while juvenile hoatzins use their wings and legs to clamber through vegetation and to avoid nest predators by swimming underwater (Thomas 1996). The importance of securing a refuge, whether to avoid predation or to access a daily roosting site, thus induces animals to push the performance limits of their locomotor structures. As juvenile birds develop and their wing performance improves, they become less dependent on wing-leg cooperation, and rely more on flight to access their refuges (with some exceptions). Wing-leg cooperation and transitional habitats thus act as a developmental bridge between leg-based terrestrial locomotion and wing-based aerial locomotion, by allowing juveniles in search of refuges to move from (1) leg-based terrestrial locomotion, to (2) wing- and leg-based locomotion on/in sloped or aquatic substrates, to (3) wing-based aerial locomotion.

We argue that this developmental invasion of the air is analogous to the evolutionary invasion of new habitats. In both cases, organisms move from one set of conditions to another (terrestrial to aerial) by using ecological transition zones (sloped or aquatic substrates) that allow them to incrementally adapt (develop) to new extremes. We expand on this idea below, using three ecological case studies to illustrate how immature birds use their underdeveloped wings and legs simultaneously to access refugia during the developmental acquisition of flight, and how this may provide insight into the evolutionary acquisition of flight among theropod dinosaurs.

III. Transitional Morphologies and Locomotor Ontogeny: New Insight from Three Case Studies of Extant Feathered Theropods

Though birds are very well studied, we know extremely little about their locomotor development. The three case studies discussed here exhaust our current knowledge on the ontogeny of locomotion with respect to bird ecology. Each study has uncovered previously unappreciated locomotor behaviors that improve survivorship in juvenile birds as they acquire the ability to fly, and that may have been relevant to theropod dinosaurs during the evolution of flight. Importantly, juvenile birds employing their legs and incipient wings simultaneously perform far better than one would expect, especially given their immature musculoskeletal and feather anatomy.

Case 1. Chukar Partridge: Strategies among Precocial Ground Birds

Following Mivart's famous challenge to Darwin, "What use is half a wing during the evolution of flight?" (Mivart 1871), more than a century elapsed before the utility of incipient wings could be demonstrated in a living organism. But with the discovery of wing-assisted incline running (WAIR) and of juvenile ground birds with dinosaur-like "protowings" (Dial 2003; Dial et al. 2006), chukar partridges (*Alectoris chukar*) became an exemplary species for exploring locomotor ontogeny and the function of incipient wings among juvenile birds (http://www.youtube.com/user/UMflightlab/videos). When threatened, flight-incapable chicks reach otherwise inaccessible, elevated sites that can act as refugia by engaging their protowings and legs simultaneously

FIG. 16.3 Ontogeny of locomotion of precocial ground birds. Developing ground birds (chukar partridge) illustrate daily incremental improvement in locomotor performance as they negotiate ever-increasingly steep obstacles. When they return to lower ground they flap their protowings to generate aerodynamic forces and control the descent (Dial 2003; Dial et al. 2006; Dial et al. 2008).

to flap-run up steep surfaces. When prevented from using their wings, chukars cannot ascend slopes that are as steep (Dial 2003), because their wings generate aerodynamic forces that increase foot traction (Tobalske and Dial 2007). Developmental changes in feather structure coincide with improvements in aerodynamic performance (Heers, Tobalske, and Dial 2011) that allow birds to flap-run up steeper and steeper slopes and eventually fly, but even young birds with protowings generate useful aerodynamic forces (fig. 16.3).

Such cooperative use of wings and legs, and the corresponding ability to reach refugia, is particularly relevant to chukars and other precocial species that leave the safety of their nest early in life and receive little parental care. However, we have surveyed more than 20 avian species from different orders (spanning the super-precocial to super-altricial developmental spectrum; http://www.youtube.com/user/UMflight lab/videos), and whether precocial or altricial all surveyed juveniles flap their protowings to ascend an incline to safety (fig. 16.3). The evidence from these birds, placed in phylogenetic perspective, establishes conclusively that WAIR is neither "bizarre" nor "highly-derived" (Feduccia et al. 2005), but rather common, plesiomorphic, and ubiquitous among extant birds, both juvenile and adult. Work with chukars can therefore provide insight into the ontogeny and ecology of wing use in many species.

Chukars also provide a superb model for investigating multifunctionality of the avian hind limb, especially the foot. Birds are often categorized as arboreal perchers or terrestrial runners (e.g., Feduccia 1993; Hopson 2001), but the avian foot is an amazingly multifunctional entity. For example, most birds are capable of perching on a broad array of structures, and can negotiate diverse terrestrial substrates by adopting plantigrade, digitigrade, and unguligrade postures depending on the incline and substrate texture (fig. 16.4). During ascents, the cranially directed digits (II–IV) and sometimes the hallux hook into the substrate to pull the bird upward. In contrast, during descents digits II–IV hyperextend and contact the substrate only after the caudally directed hallux hooks in first and brakes the bird. This is evident in both juvenile and adult chukars, as well as in other species (e.g., magpies, pigeons), demonstrating that an unappreciated function of the caudally directed hallux is to act as an anchor and allow birds to descend

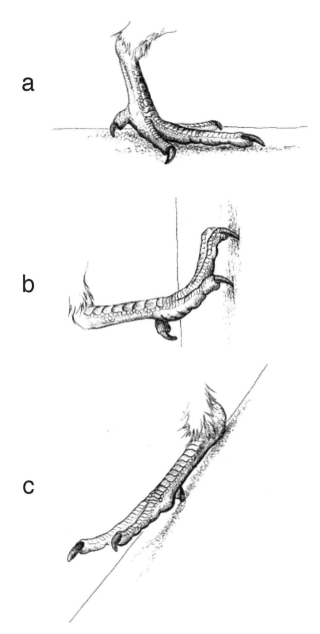

FIG. 16.4 Versatile foot postures among birds traversing varying inclines. A typical ground bird (chukar partridge) illustrates how the bird hind limb is capable of executing distinct foot postures: a, digitigrade (over near level terrain), b, unguligrade (up steep inclines), and c, plantigrade (downhill walking). Most mammals exhibit one of these foot postures whereas most birds can exhibit all three over varying three-dimensional terrestrial habitats.

steeply angled objects (fallen tree trunk, boulder, hill, ravine, etc.) without jamming their cranially directed digits (see video: http://www.youtube.com/user/UMflight lab/videos). A caudally directed hallux may be particularly useful to flight-incapable juveniles that might not be able to fly down from elevated refuges that they have reached by flap-running.

The precocial chukar has thus provided a launching point for exploring coordinated forelimb and hind limb function during bird ontogeny. Historically, attention has focused on wings because of their obvious importance for flight. However, cooperative use of wings and legs is a crucial component of bird ontogeny. Though life history strategies and the importance of wings versus legs vary widely across species, as we explore ontogenetic trajectories in other groups a common theme is beginning to emerge: wing-leg cooperation serves as an intermediate locomotor behavior that bridges obligately bipedal to flight-capable transitions, allowing vulnerable juveniles to move through an array of environmental conditions to seek food, avoid predation and reach roosting sites. As we illustrate in Cases 2 and 3 below, this common theme is manifested in different ways in species with different life history and locomotor strategies.

Case 2. Branching Owls: Strategies among Altricial Species

"Branching" is a form of wing-assisted incline running exhibited by owls (though owls walk) and, like WAIR, is used to access refugia. As a general rule, young owls hatched in open stick or ground nests leave their nests when they are still partly covered in down and not yet able to fly. Great horned (*Bubo virginianus*), great grey (*Strix nebulosa*), long-eared (*Asio otus*), short-eared (*Asio flammeus*), and marsh owls (*Asio capensis*), for example, move away from the nest by hopping or walking along adjacent branches or across the ground (Marks et al. 1999; pers. obs.). Long-eared owl siblings independently leave their platform nests about two weeks before they can fly, first flapping their diminutive, underdeveloped wings to safely descend to the ground, then walking to different trees (up to 50 m away from their nest), and finally flap-walking up the tree trunks to concealed perches in trees, where they continue to be fed by their parents (fig. 16.5). In these owls, juveniles that branch and isolate themselves from the conspicuous grouping of other, less-developed young have a higher probability of surviving to fledgling, compared to their non-branching siblings (Marks 1986). As soon as they can fly and the risk of predation is reduced, branching owlets reunite in the same

tree and their parents continue to bring them food until fully fledged (Marks 1982; Marks et al. 1994). In short, branching provides another example of flight-incapable, developing birds using rudimentary or incipient wings to access a place of safety (http://www.youtube.com/user/UMflightlab/videos.

Wing-leg cooperation therefore allows flight-incapable juveniles to access refugia in both precocial and altricial species. Though many species can be categorized along the precocial-altricial spectrum (Cases 1 and 2), a recent study shows that even within a single individual, the hind limbs and forelimbs can follow different developmental trajectories and still act cooperatively, as we describe next (Case 3).

Case 3. Mallard Ducks: Differential Maturity of Hind Limbs and Forelimbs within a Species

Not all theropod descendants answer the question, "What use is half a wing?" with wing-assisted incline running. Members of the Anseriformes—a basal lineage of extant birds (part of Galloanseriformes)—highlight an alternative use of incipient wings, and in doing so offer a compelling example of how developing birds can function in non-terrestrial environmental media. Anseriforms have a particularly unique life-history trajectory, sharing precocial aspects of hind limb development with the chukar and altricial aspects of delayed flight acquisition with the owl. Anseriforms have thus taken locomotor modularity (Gatesy and Dial 1996) to an even greater degree: in addition to segregating the forelimbs for aerial locomotion and the hind limbs for terrestrial/aquatic locomotion, they also exhibit differential timing in the development of the fore- and hind limb modules. Such sequential growth and pronounced forelimb-hind limb modularity likely contributes to these birds' ability to function in many different habitats or media.

The ecology of anseriforms like mallard ducks (*Anas platyrhynchos*) is necessarily amphibious throughout ontogeny—they routinely negotiate transitional shoreline habitats at the interface of terrestrial, aquatic, and aerial media (fig. 16.6). A newly hatched duckling forages for invertebrates along the banks of a pond, river, or lake (Collias and Collias 1963; Duttmann 1992). When startled by a land predator, it dashes for the reeds or open water; if startled in the water, the duckling dives

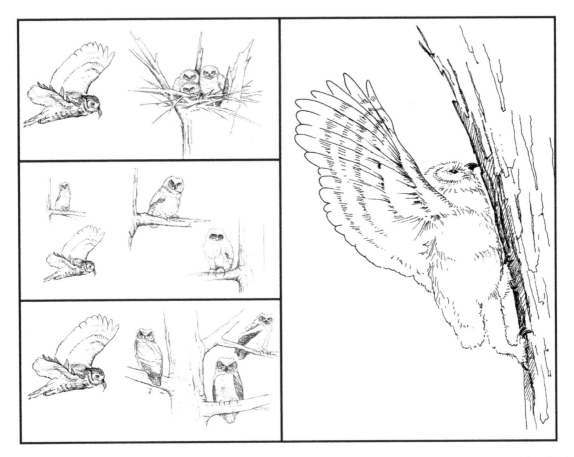

FIG. 16.5 Ontogeny of locomotion for various species of owls: "Branching." Upper left: nestlings restricted to the nest. Center left and right: branching owls with emerging wings leave the nest and disperse by flap-walking up neighboring trees. Lower left: flight-capable young converge on the same tree to be fed by parents but free to flee if threatened.

beneath the surface or scurries back to the safety of land. Older juveniles and adults spend more time grazing in the water (Collias and Collias 1963; Duttmann 1992) and may additionally respond to threats by taking to the air, once able. In transitional shoreline habitats mallards therefore have an array of foraging choices and many escape options—to aquatic, terrestrial, or aerial refugia. However, locomotor behavior changes throughout ontogeny.

Whereas chukars (Case 1) are precocial and owls (Case 2) are altricial, mallards and other anseriforms differentiate the developmental maturation of their locomotor modules: they have precocial hind limbs and altricial forelimbs. Newly hatched ducklings are thus hind limb–dependent and rely completely on running, swimming, and foot-propelled diving to escape predation and reach aquatic or terrestrial refugia. Ducklings can run and swim the day they hatch, but unlike

chukars and other galliforms (turkeys, peafowl, etc.), anseriforms as a clade do not develop flight capacity until they attain adult size, which can take over 60 days. This developmental modularity appears to allow focused growth and maturation of the hind limbs during the first half of post-hatching development, followed by rapid growth, delayed maturation, and radical allometry of the forelimbs during the latter half of development (fig. 16.6) (Dial and Carrier 2012).

Mallard juveniles, which do not begin to develop flight feathers until ~30 days after hatching, and cannot fly until ~60 days post-hatching, bridge the developmental transition from hind limb–dependent duckling to flight-capable adult by "steaming." Like wing-assisted incline running in chukars and branching in owls, steaming is a transitional locomotor behavior that involves the cooperative use of wings and legs. As soon as their developing forelimbs are long enough to reach into the

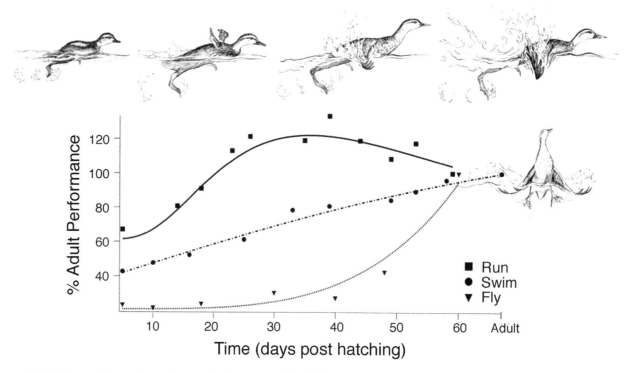

FIG. 16.6 Ontogeny of locomotor performance for Anseriformes. From left to right: hatchling using feet only with small budding wings held to side; juvenile (~20 days post hatching) with developing wing held up and unable to reach water; sub-adult using flight-incapable wing to scoop water (steaming) and supplement hind limbs; flight-capable adult using wing to paddle, supplementing hind limbs for maximum steaming performance; and below (frontal view), flight-capable adult using wings to assist takeoff from water. Graph: ontogeny of running, swimming (steaming) and flying performance.

water, ducks steam across the water surface by paddling rapidly with their feet and using their forelimbs as oars to lift the body and flap-row forward (Livezey and Humphrey 1983; Aigeldinger and Fish 1995; Dial and Carrier 2012). Ontogenetically, steaming allows continued improvement in aquatic performance until birds reach the adult stage (fig. 16.6). Steaming is thus another example of how a developing, rudimentary protowing is employed in conjunction with the hind limbs to provide useful function. Even young, nonaquatic juncos (Passeriformes) have been observed paddling with their stick-like, perching hind limbs and rowing with their developing wings to cross an aquatic barrier and avoid capture (Dial and Dial pers. obs.). Steaming, therefore, may be a fairly widespread behavior in the class Aves.

Once their wings and flight apparatus are fully developed, mallard and other "puddle-jumping" (subfamily: Anatinae) adults are capable of using their wings to take off vertically from the water (they do not need to run and flap across the water surface, and can therefore access small ponds or "puddles"). In one synchronized motion, these ducks push their wrists (ulnar/radial-carpal joint) down into the water while they simultaneously flare their feet and extend their hind limbs to launch from the water and become immediately airborne (fig. 16.6). Adult mallards are thus extremely capable fliers, and by delaying the functional development of the flight apparatus they may be able to specialize their forelimbs for high-performance migratory flight (Dial and Carrier 2012; Dial et al. 2012). Though flight capacity is not attained until the end of ontogeny, forelimb function is continually modified and enhanced during steaming. This pattern of development and habitat use is mirrored in many other waterfowl (Anseriformes), as well as other aquatic species (e.g., rails, loons, and grebes).

In summary, flap-running chukars, branching owls, and steaming mallards collectively demonstrate that wing-leg cooperation is a common technique for negotiating transitional habitats (ground → inclined substrate → air (Cases 1, 2), ground → water → air (Case 3)) to reach refugia during hind limb–dependent to flight-capable developmental transitions. Wing-leg cooperation

BOX 16.1 SOME FUTURE DIRECTIONS

Duration of development. Birds display a wide range of postnatal rates of development. Some songbirds (e.g., chipping sparrow) reach adult size and shape within 9 days posthatching (Martin 2011), whereas other avian species do not attain adult size and maturity for several years (e.g., 2 years for brush turkey, peacock, some albatross; 5–6 years for the nonflying kiwi [*Apteryx* spp.] [Bourdon et al. 2009]). Such disparate rates of development offer scientists a rich array of ecological and developmental scenarios with which to explore form and function in incipient locomotor structures. Studies on the rate of development or length of time to reach maximum size are important for understanding adaptation in ecological time, but may also provide insight into the ontogeny of extinct species (Padian et al. 2001; Horner and Padian 2004; Erickson 2005). For example, newly hatched brush turkeys (*Alectura lathami*, Megapodiidae, Galliformes) are better fliers than their parents, because brush turkeys grow so large that they essentially "outgrow" their wings and rely more on their legs (Starck and Sutter 2000; Jones and Goth 2008; Dial and Jackson 2009). Similar developmental strategies may have been used by extinct feathered dinosaurs, such that juveniles relied more on their wings and adults relied more on their legs while retaining their outgrown but still feathered forelimbs (*Caudipteryx, Velociraptor*, etc.).

Rudimentary structures in adults. This essay has focused on the ecological importance of incipient structures in developing juveniles. However, adults of many species possess similar anatomies and/or are similarly challenged to become airborne. For example, both flightless and weakly flying steamer ducks (*Tachyeres*) engage their wings and legs simultaneously to flap-row or steam rapidly across the water (Livezey and Humphrey 1986), much like juvenile ducks. Similarly, large birds with low power-to-mass ratios (Dial et al. 2009)—such as swans, loons (Norberg and Norberg 1971), and albatrosses—flap their wings while running in order to gain speed for takeoff. Shearwaters, which are awkward on their feet and cannot take off simply by leaping into the air, often flap-run up trees to take off from an elevated branch (BBC Life of Birds). In short, wing-leg cooperation is ubiquitous among birds and may be essential to incipiently flight-capable or weak-flying forms. Yet, we are just beginning to examine ontogenetic trajectories, and the locomotor capacity and performance of adult forms with secondarily reduced structures has never been empirically explored. Contrasting adult tactics with those of developing young promises to uncover novel strategies of behavior and design, and provide rich insight into the limits on anatomical function and performance.

is also particularly important to adult birds that are challenged to become airborne or that possess secondarily reduced wings (box 16.1), and may have been important during the hind limb–dependent to flight-capable evolutionary transition among theropod dinosaurs.

IV. Ontogenetic Insights into Evolutionary History

The three case studies described above illustrate a conceptual parallel between the development of flight in juvenile birds and the evolution of flight in their theropod ancestors. In each extant case, flight-incapable, hind limb–dependent organisms acquire flight capacity by transitioning through intermediate morphological, functional, and behavioral stages that incrementally improve wing performance and culminate in full flight capacity (Heers and Dial 2012). At hatching, chukars and particularly mallards are almost entirely dependent on their hind limbs, and avoid predators and seek refugia by running, jumping, climbing, or swimming; owls rely on their nests and parents for safety. Once their flight feathers begin to emerge or their forelimbs can reach the water, chukars and mallards supplement hind limb movement (incline running, aquatic paddling) by simultaneously flapping or rowing their incipient, rudimentary protowings. This improves hind limb performance and allows juveniles to ascend steeper slopes (Dial 2003) or move across water more rapidly (Aigeldinger and

Fish 1995; Dial and Carrier 2012). As the flight apparatus matures, developing birds can flap-run up increasingly steep slopes or flap-row more rapidly, and eventually fly. Juvenile owls have a more fully developed wing when they leave the nest and branch, but nevertheless rely on flap-running behaviors to reach arboreal refugia until they are flight capable. Thus, despite different life history and locomotor strategies, chukars, branching owls, mallards, and many other juveniles (video link: http://www.youtube.com/user/UMflightlab/videos) all engage their wings and legs simultaneously to negotiate transitional habitats (inclined or aquatic substrates) that link terrestrial and aerial media. Wing-leg cooperation acts as a developmental bridge between leg-based terrestrial/aquatic locomotion and wing-based aerial locomotion, because it allows juveniles to transition through intermediate morphological and functional stages that incrementally improve wing performance and culminate in full flight capacity. The development of flight among extant birds can therefore provide insight into the evolution of flight in their theropod ancestors, by revealing how animals with incipient and rudimentary anatomies behaviorally bridge obligately bipedal to flight-capable transitions. Though the functions of protowings have puzzled scientists for over a century, developing birds clearly illustrate their adaptive advantage, but only when viewed from an ecological perspective.

Developing birds can also provide quantitative insight into the potential locomotor capacities of extinct theropods. Recent work shows that juvenile birds not only have dinosaur-like protowings, but also small pectoral muscles and transitional, dinosaur-like skeletons that lack many of the hallmarks of advanced flight capacity (e.g., large keel, channelized wing joints) (Heers and Dial 2012) (fig. 16.7). Techniques such as particle image velocimetry (PIV), propeller force-plate analysis, sonomicrometry, electromyography, wind tunnels, and X-ray Reconstruction of Moving Morphology (XROMM) can be used to explore how feather development affects aerodynamic performance, and how musculoskeletal development affects power output and three-dimensional movement of wing and leg joints (fig. 16.7). By investigating form-function relationships during locomotor ontogeny, we might better understand how similar morphological changes influenced locomotor capacity and performance during the evolution

of flight. For example, preliminary work indicates that in spite of lacking many adaptations or exaptations for flight, young birds with transitional skeletons and protowings have adult-like kinematics and generate useful aerodynamic forces (Heers et al. 2011; Baier et al. 2011). This suggests that extinct theropods with similar anatomical features might also have been capable of more bird-like wing strokes and greater aerodynamic force production than implied by their transitional features. Though much remains to be done, approaches like these illustrate how quantifying form-function relationships in juveniles with transitional, dinosaur-like features can provide insight into the functional capacities of extinct theropods with similar traits. Developing birds depend on transitional anatomies for a variety of transitional flapping behaviors (Cases 1–3), and though we are only beginning to explore locomotor ontogeny, juveniles that actualize the functional capacities of transitional features are providing rich and unique insight into life and its history.

Conclusions

Though Darwin's tree of life is often viewed through an adultocentric framework, extinct animals with transitional features that record major transformations in vertebrate history are probably better understood by considering the ecology of juvenile as well as adult stages. Recent studies demonstrate that developing birds and early winged theropods share a number of unique morphological similarities, such as protowings and small or nonexistent sternal keels (Dial et al. 2006; Heers and Dial 2012). The functional significance of these rudimentary structures has intrigued the scientific community for more than a century, and juvenile birds show, for the first time, how incipient flight devices cooperate with the legs to function in ecological settings (Cases 1–3 above). Wing-leg cooperation is a common and ubiquitous phenomenon, but its relevance has been relatively unappreciated—especially in discussions of the evolution of avian flight and the ecology of locomotor development. Transitional behaviors, where wings and legs are used cooperatively, help juveniles to bridge the developmental passage from leg-based terrestrial locomotion to wing-based aerial locomotion by incrementally improving locomotor performance throughout ontogeny, and providing access to

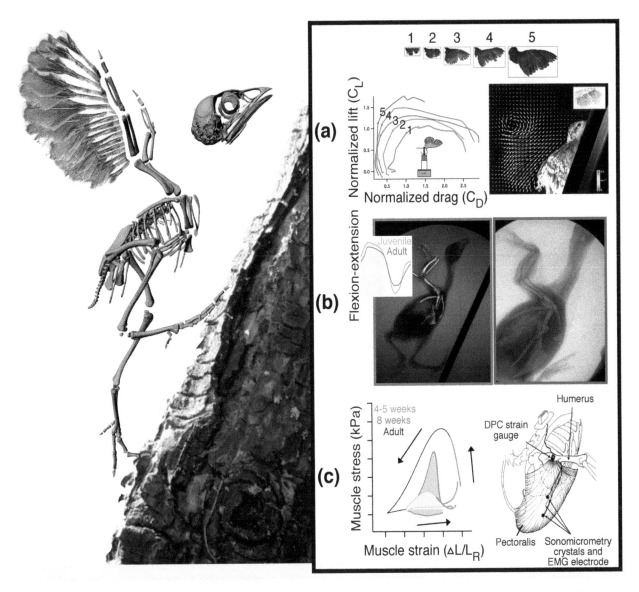

FIG. 16.7 Quantifying the ontogeny of form and function. A rich array of new experimental techniques can elucidate the ontogeny of form-function relationships in extant animals, with implications for theropod fossils. For example, wind tunnels, propeller models, and particle image velocimetry (PIV) can be used to quantify aerodynamic output (lift and drag) and thereby reveal relationships between wing or feather morphology and aerodynamic performance during development (a). Similarly, associations between three-dimensional skeletal morphology and kinematics can be explored using X-ray Reconstruction of Moving Morphology (b; www.xromm.org/), while sonomicrometry, electromyography, and assessments of muscle force can measure neuromuscular function with respect to musculoskeletal anatomy and body size (c). These are merely some of the many innovative techniques that are revolutionizing quantitative approaches to comparative anatomy and biomechanics. Our understanding of the functional capacity of transitional stages during the evolution of flight can likewise be transformed by employing these new methods (i) across extant species that display variation in a trait of interest, (ii) in models of extinct organisms, and perhaps especially (iii) in developing birds with transitional locomotor capacities and dinosaur-like morphologies. Modified from Heers and Dial (2012).

elevated or aquatic refuges (Dial et al. 2006; Tobalske and Dial 2007; Dial et al. 2008; Jackson et al. 2009; Heers, Tobalske, and Dial 2011; Dial and Carrier 2012). Just as populations adapt to new ecological extremes by using ecoclines as environmental stepping-stones, developing birds invade the aerial regime by using inclined slopes or aquatic surfaces as transitional stepping-stones to

improve wing performance until they become flight capable. In summary,

- *Transitional habitats act as stepping-stones that allow morphologically evolving/developing populations/individuals to make the transition between environmental extremes (e.g., terrestrial*

*substrates → inclined or aquatic substrates →
aerial media) through incremental adaptive stages.*

- *Organisms are stimulated to move into new or
different environments and physical media in
order to escape from predators and/or avoid harsh
abiotic conditions. Given that predator escape often
elicits maximal locomotor performance, and that
different physical conditions often require different
anatomical adaptations, access to refuges is likely
a primary trigger for the invasion of new habitats
and attending changes in locomotor design.*

- *Developing, flight-incapable birds with transitional, dinosaur-like features negotiate transitional
habitats (inclined or aquatic substrates) to reach
elevated or aquatic refugia by using their wings
and legs cooperatively (WAIR, steaming). Cooperative use of wings and legs improves locomotor
performance throughout ontogeny, and thus acts
as a developmental bridge between leg-based
terrestrial locomotion and wing-based aerial
locomotion by allowing juveniles to transition
from (1) leg-based terrestrial locomotion, to
(2) wing- and leg-based locomotion on or in
sloped or aquatic substrates, to (3) wing-based
aerial locomotion. Cooperative use of wings and
legs may have been similarly important during
the origin and evolution of flight in theropod
dinosaurs. Though we are just beginning to investigate locomotor ontogeny and the function
of rudimentary structures (box 16.1), exploring
ontogenetic trajectories can greatly improve our
understanding of evolutionary transformations,
by revealing how incipient structures function
and perform in ecological settings.*

* * *

Acknowledgments

Farish A. Jenkins Jr., dear friend and mentor, was a powerful intellectual force in evolutionary biology and profoundly influential to each person that came into his life.
He expanded our approach to science by challenging us
to integrate functional morphology and physiology with
evolutionary biology. Farish's extraordinary attention
to anatomical detail, love of discovery, and command
of the lecture hall are legendary and set the bar for his
disciples. Multiple safaris to East Africa provided the
forum to volley questions that lead to research linking evolutionary ecology to form and function. The
authors wish to thank: Thomas Martin, Bret Tobalske,
Matthew Bundle, Brandon Jackson, Kevin Padian, Neil
Shubin, and Elizabeth (Beth) Brainerd for reading and
commenting on this chapter.

References

Aigeldinger, T. L., and F. E. Fish. 1995. Hydroplaning by ducklings:
overcoming limitations to swimming at the water surface.
Journal of Experimental Biology 198:1567–1574.

Alexander, R. M. 2002. The merits and implications of travel by
swimming, flight and running for animals of different sizes.
Integrative and Comparative Biology 42:1060–1064.

Blair, W., and B. W. Frank. 1950. Ecological factors in speciation of
Peromyscus. Evolution 4:253–275.

Blob, R. W., W. C. Bridges, M. B. Ptacek, T. Maie, R. A. Cidiel,
M. M. Bertolas, M. L. Juilias, and H. L. Schoenfuss. 2008.
Morphological selection in an extreme flow environment:
body shape and waterfall-climbing success in the Hawaiian
stream fish *Sicyopterus stimpsoni*. Integrative and
Comparative Biology 48:734–749.

Carrano, M. T. 2000. Homoplasy and the evolution of dinosaur
locomotion. Paleobiology 26:489–512.

Carrier, D. R. 1996. Ontogenetic limits on locomotor performance.
Physiological Zoology 69:467–488.

Chen, P., Z. Dong, and S. Zhen. 1998. An exceptionally well-
preserved theropod dinosaur from the Yixian Formation
of China. Nature 391:147–152.

Cheng, Y.-R., and T. E. Martin. 2012. Nest predation risk and
growth strategies of passerine species: grow fast or
develop traits to escape risk? American Naturalist 180:285–
295.

Chiappe, L. M. 2007. Glorified Dinosaurs: The Origin and Early
Evolution of Birds. Hoboken NJ: Wiley-Liss.

Chiappe, L. M., S. A. Ji, Q. Ji, and M. Norell. 1999. Anatomy and
systematics of the Confuciusornithidae (Theropoda, Aves)
from the late Mesozoic of northeastern China. Bulletin of
the American Museum of Natural History 242:1–89.

Clack, J. A. 2009. The fin to limb transition: new data,
interpretations, and hypotheses from paleontology and
developmental biology. Annual Review of Earth and
Planetary Sciences 37:163–179.

Clack, J. A. 2011. The fin to limb transition: new data,
interpretations, and hypotheses from paleontology and

developmental biology. Annual Review of Earth Planetary Science 37:163-179.

Clark, J. A., Z. Zhou, and F. Zhang. 2006. Insight into the evolution of avian flight from a new clade of Early Cretaceous ornithurines from China and the morphology of *Yixianornis grabaui*. Journal of Anatomy 208:287-308.

Collias, N. E., and E. C. Collias. 1963. Selective feeding by wild ducklings of different species. Wilson Bulletin 75:6-14.

Cott, H. B. 1940. Adaptive Coloration in Animals. London: Methuen.

Davenport, J. 1994. How and why do flying fish fly? Reviews in Fish Biology and Fisheries 4:184-214.

Dial, K. P. 2003. Wing-assisted incline running and the evolution of flight. Science 299:402-404.

Dial, K. P., and B. E. Jackson. 2011. When hatchlings outperform adults: locomotor development in Australian brush turkeys (*Alectura lathami*, Galliformes). Proceedings of the Royal Society B 278:1610-1616.

Dial, K. P., B. E. Jackson, and P. Segre. 2008. A fundamental avian wing-stroke provides a new perspective on the evolution of flight. Nature 451:985-989.

Dial, K. P., R. J. Randall, and T. R. Dial. 2006. What use is half a wing in the ecology and evolution of birds? BioScience 56:437-445.

Dial, T. R., and D. R. Carrier. 2012. Precocial hindlimbs and altricial forelimbs: partitioning ontogenetic strategies in Mallard ducks (*Anas platyrhynchos*). Journal of Experimental Biology 215:3703-3710.

Dial, T. R., A. M. Heers, and B. W. Tobalske. 2012. Ontogeny of aerodynamics in mallards: comparative performance and developmental implications. Journal of Experimental Biology 215:3693-3702.

Dingus, L., and T. Rowe. 1998. The mistaken extinction: dinosaur evolution and the origin of birds. New York: W. H. Freeman.

Düttmann, H. 1992. Ontogenetische Verhaltensänderungen bei der Brandent (*Tadorna tadorna*): Schlafen, Tauchen, Nahrungserwerb. Journal of Ornithology 133:365-380.

Earls, K. D. 2000. Kinematics and mechanics of ground take-off in the starling *Sturnis vulgaris* and the quail *Coturnix coturnix*. Journal of Experimental Biology 203:725-739.

Elzanowski, A. 2002. Archaeopterygidae (Upper Jurassic of Germany). Pp. 129-159 in L. M. Chiappe and L. M. Witmer, eds., Mesozoic Birds: Above the Heads of Dinosaurs. Berkeley: University of California Press.

Frazetta, T. H. 1970. From hopeful monsters to bolyerine snakes? American Naturalist 104:55-72.

Feduccia, A. 1993. Evidence from claw geometry indicating arboreal habits of *Archaeopteryx*. Science 259:790-793.

Garstang, W. 1951. Larval Forms, and Other Zoological Verses. Oxford: Blackwell.

Gatesy, S. M. 1990. Caudofemoral musculature and the evolution of theropod locomotion. Paleobiology 16:170-186.

Gauthier, J. 1986. Saurischian monophyly and the origin of birds. Memoirs of the California Academy of Sciences 8:1-55.

Goldschmidt, R. 1940. The Material Basis of Evolution. New Haven, CT: Yale University Press.

Gould, S. 1977. Ontogeny and Phylogeny. Cambridge, MA: Harvard University Press. 640 pp.

Gould, S., and E. S. Vrba. 1982. Exaptation—a missing term in the science of form. Paleobiology 8:4-15.

Hedenström, A., and M. Rosen. 2001. Predator and prey: on aerial hunting and escape strategies in birds. Behavioral Ecology 12:150-156.

Heers, A. M., D. B. Baier, B. E. Jackson, and K. P. Dial. 2011. Developing skeletons in motion: The ontogeny of skeletal form and function in a precocial ground bird (*Alectoris chukar*). Integrative and Comparative Biology 51:e55.

Heers, A. M., and K. P. Dial. 2012. From extant to extinct: locomotor ontogeny and the evolution of avian flight. Trends in Ecology & Evolution 27:296-305.

Heers, A. M., B. W. Tobalske, and K. P. Dial. 2011. Ontogeny of lift and drag production in ground birds. Journal of Experimental Biology 214:717-725.

Herrel, A., and A. C. Gibb. 2006. Ontogeny of performance in vertebrates. Physiological and Biochemical Zoology 79: 1-6.

Hu, D., Hou, L., Zhang, and X. Xu. 2009. A pre-*Archaeopteryx* troodontid theropod from China with long feathers on the metatarsus. Nature 461:640-643.

Hopson, J. A. 2001. Ecomorphology of avian and nonavian theropod phalangeal proportions: Implications for the arboreal versus terrestrial origin of bird flight. In J. Gauthier and L. F. Gall, eds., New Perspectives on the Origin and Early Evolution of Birds: Proceedings of the International Symposium in Honor of John H. Ostrom. New Haven: Peabody Museum of Natural History, Yale University.

Hutchinson, J. R. 2001. The evolution of pelvic osteology and soft tissues on the line to extant birds (Neornithes). Zoological Journal of the Linnean Society131:123-168.

Hutchinson, J. R. 2002. The evolution of hindlimb tendons and muscles on the line to crown- group birds. Comparative Biochemistry and Physiology A 133:1051-1086.

Hutchinson, J. R., and V. Allen. 2009. The evolutionary continuum of limb function from early theropods to birds. Naturwissenschaften 96:423-448.

Irschick, D. J. 2000. Effects of behaviour and ontogeny on the locomotor performance of a West Indian lizard, *Anolis lineatopus*. Functional Ecology 14:438-444.

Jackson, B. E., P. Segre, and K. P. Dial. 2009. Precocial development of locomotor performance in a ground-dwelling bird (*Alectoris chukar*): negotiating a three-dimensional terrestrial environment. Proceedings of the Royal Society B 276:3457-3466.

Jenkins, F. A. 1993. The evolution of the avian shoulder joint. American Journal of Science 293:253-267.

Kaufman, D. W. 1974. Adaptive coloration in *Peromyscus polionotus*: experimental selection by owls. Journal of Mammalogy 55:271-283.

Ji, Q., S. Ji, H. You, J. Zhang, H. Zhang, N. Zhang, C. Yuan, and X. Ji. 2003. An early cretaceous avialan bird, *Shenzhouraptor sinensis* from Western Liaoning, China. Acta Geologica Sinica 77:21-27.

Liem, K. F. 1979. Modulatory multiplicity in the feeding mechanism in cichlid fishes, as exemplified by the invertebrate pickers of Lake Tanganyika. Journal of Zoology 189:93-125.

Livezey, B. C., and P. S. Humphrey. 1983. Mechanics of steaming in steamer-ducks. Auk 100:485-488.

Losos, J. B. 2009. Lizards in an Evolutionary Tree: Ecology and Adaptive Radiation of Anoles. Berkeley: University of California Press. 507 pp.

MacArthur, R. H. 1967. The theory of island biogeography. Vol. 1. Princeton: Princeton University Press.

Makovicky, P. J., and L. E. Zanno. 2011. Theropod diversity and the refinement of avian characteristics. Pp. 9-29 in G. Dyke and G. Kaiser, eds., Living Dinosaurs: The Evolutionary History of Modern Birds. Hoboken: John Wiley

Manceau, M., V. S. Domingues, C. R., Linnen, E. B Rosenblum, and H. E. Hoekstra. 2010. Convergence in pigmentation at multiple levels: mutations, genes and function. Philosophical Transactions of the Royal Society B: Biological Sciences 365:2439–2450.

Marks, J. S. 1982. Night stalkers along the Snake. Idaho Wildlife 3:18–21.

Marks, J. S. 1986. Nest-site characteristics and reproductive success of long-eared owls in southwestern Idaho. Wilson Bulletin 98:547–560.

Marks, J. S., D. L. Evans, and D. W. Holt. 1994. Long-eared Owl (Asio otus). In A. Poole and F. Gill, eds., The Birds of North America, No. 133. Philadelphia: Academy of Natural Sciences. 23 pp.24 pp.

Marks, J. S., R. J. Cannings, and H. Mikkola. 1999. Family Strigidae. Pp. 76–151 in J. del Hoyo, A. Elliott, and J. Sargatal, eds., Barn Owls to Hummingbirds, vol. 5 of Handbook of the Birds of the World. Barcelona: Lynx Edicions.

Martin, T. E. 1995. Avian life history evolution in relation to nest sites, nest predation, and food. Ecological Monographs 65:101–127.

Martin, T. E., P. Lloyd, C. Bosque, D. C. Barton, A. L. Biancucci, Y. R. Cheng, and R. Ton. 2011. Growth rate variation among passerine species in tropical and temperate sites: an antagonistic interaction between parental food provisioning and nest predation risk. Evolution 65:1607–1622.

Minelli, A. 2003. The Development of Animal Form: Ontogeny, Morphology, and Evolution. Cambridge: Cambridge University Press.

Mivart, S. G. J. 1871. On the Genesis of Species. New York: Appleton.

Mulvihill, R. S., and C. R. Chandler. 1991. A comparison of wing shape between migratory and sedentary dark-eyed juncos (Junco hyemalis). Condor 93:172–175.

Norberg, U. M., and J. M. Rayner, 1987. Ecological morphology and flight in bats (Mammalia; Chiroptera): wing adaptations, flight performance, foraging strategy and echolocation. Philosophical Transactions of the Royal Society of London. Series B, Biological Sciences 316:335–427.

Norberg, U. M. 1990. Vertebrate Flight: Mechanics, Physiology, Morphology, Ecology, and Evolution. Berlin: Springer-Verlag. 291pp.

Norell, M. A., and X. Xu. 2005. Feathered dinosaurs. Annual Review of Earth and Planetary Sciences 33:277–299.

Ostrom, J. H. 1973. The Ancestry of Birds. Nature 242:136.

Ostrom, J. H. 1975. The Origin of Birds. Annual Reviews of Earth and Planetary Sciences 3:55–77.

Ostrom, J. H. 1976. Some hypothetical anatomical stages in the evolution of avian flight. Smithsonian Contributions to Paleobiology 27:1–21.

Padian, K. 2001. Stages in the origin of bird flight: beyond the arboreal-cursorial dichotomy. Pp. 255–272 in J. Gauthier and L. F. Gall, eds., New Perspectives on the Origin and Early Evolution of Birds: Proceedings of the International Symposium in Honor of John H. Ostrom. New Haven: Peabody Museum of Natural History, Yale University.

Padian, K., and L. M. Chiappe. 1998. The origin and early evolution of birds. Biological Reviews 73:1–42.

Prum, R. O. 2005. Evolution of the morphological innovations of feathers. Journal of Experimental Zoology 304B(6):570–579.

Prum, R. O., and A. H. Brush. 2002. The evolutionary origin and diversification of feathers. Quarterly Review of Biology 77:261–295.

Qiang, J., P. J. Currie, M. A. Norell, and J. Shu-An. 1998. Two feathered dinosaurs from northeastern China. Nature 393: 753–761.

Rauhut, O. W. M., C. Foth, H. Tischlinger, and M. A. Norell. 2012. Exceptionally preserved juvenile megalosauroid dinosaur with filamentous integument from the Late Jurassic of Germany. Proceedings of the National Academy of Sciences 109:11746–11751.

Rosenblum, E. B., H. Römpler, T. Schöneberg, and H. E. Hoekstra. 2010. Molecular and functional basis of phenotypic convergence in white lizards at White Sands. Proceedings of the National Academy of Sciences 107:2113–2117.

Schmidt-Nielsen, K. 1972. Locomotion: energy cost of swimming, flying, and running. Science 177:222–228.

Sereno, P. C. 1997. The origin and evolution of dinosaurs. Annual Reviews of Earth and Planetary Sciences 25:435–489.

Starck, J. M., and R. E. Ricklefs. 1998. Patterns of development: the altricial-precocial spectrum. Pp. 3–30 in J. M. Starck and R. E. Ricklefs, eds., Avian Growth and Development: Evolution within the Altricial-Precocial Spectrum. London: Oxford University Press.

Starck, J. M., and E. Sutter. 2000. Patterns of growth and heterochrony in moundbuilders (Megapodiidae) and fowl (Phasianidae). Journal of Avian Biology 31:527–547.

Steiner, C. C., J. N. Weber, and H. E. Hoekstra. 2007. Adaptive variation in beach mice produced by two interacting pigmentation genes. PLoS Biology 5:e219

Thomas, B. T. 1996. Family Opisthocomidae (Hoatzin). In J. Del Hoyo, A. Elliott, and J. Sargatal, eds., Handbook of the Birds of the World, Vol. 3, Hoatzin to Auks. Barcelona: Lynx Edicions.

Tobalske, B. W., and K. P. Dial. 2007. Aerodynamics of wing-assisted incline running in birds. Journal of Experimental Biology 210:1742–1751.

Tornberg, R., M. Mönkkönen, and M. Pahkala. 1999. Changes in diet and morphology of Finnish goshawks from 1960s to 1990s. Oecologia 121:369–376.

Vazquez, R. J. 1992. Functional osteology of the avian wrist and the evolution of flapping flight. Journal of Morphology 211:259–268.

Wassersug, R. J., and D. G. Sperry. 1977. The Relationships of locomotion to differential predation on Pseudacris triseriata (Anura: Hylidae). Ecology 58:830–839.

Williams, G. C. 1966. Natural selection, the costs of reproduction, and a refinement of Lack's principle. American Naturalist 100:687–690.

Witmer, L. M. 2009. Palaeontology: feathered dinosaurs in a tangle. Nature 461:601–602.

Xu, X. 2006. Feathered dinosaurs from China and the evolution of major avian characters. Integrated Zoology 1:4–11.

Xu, X., K. Wang, K. Zhang, Q. Ma, L. Xing, C. Sullivan, D. Hu, S. Cheng, and S. Wang. 2012. A gigantic feathered dinosaur from the Lower Cretaceous of China. Nature 484:92–95.

Xu, X., H. You, K. Du, and F. Han. 2011. An Archaeopteryx-like theropod from China and the origin of Avialae. Nature 475:465–470.

Xu, X., and F. Zhang. 2005. A new maniraptoran dinosaur from China with long feathers on the metatarsus. Naturwissenschaften 92:173–177.

Xu, X., X. Zheng, and H. You. 2009. A new feather type in a nonavian theropod and the early evolution of feathers. Proceedings of the National Academy of Sciences 106:832–834.

Xu, X., Z. Zhou, X. Wang, X. Kuang, F. Zhang, and X. Du. 2003. Four-winged dinosaurs from China. Nature 421:335–340.

Zheng, X., Z. Zhou, X. Wang, F. Zhang, X. Zhang, Y. Wang, G. Wei, S. Wang, and X. Xu. 2013. Hind wings in basal birds and the evolution of leg feathers. Science 339:1309–1312.

Zhou, Z., and L. Hou. 1998. *Confuciusornis* and the early evolution of birds. Vertebrata PalAsiatica 36:136–146.

Zhou, Z., and F. Zhang. 2003. *Jeholornis* compared to *Archaeopteryx*, with a new understanding of the earliest avian evolution. Naturwissenschaften 90:220–225.

Zhou, Z. H., and X. L. Wang. 2000. A new species of *Caudipteryx* from the Yixian Formation of Liaoning, northeast China. Vertebrata PalAsiatica 38:113–130.

17

Skeletons in Motion: *An Animator's Perspective on Vertebrate Evolution*

Stephen M. Gatesy* and David B. Baier[†]

Introduction

Imagine being on a paleontological expedition to a remote part of the world in search of early dinosaurs. One day, while prospecting for fossil bones, you and your colleague stumble across a set of interesting impressions preserved on an exposed shelf of rock. After sweeping off the surface, a sequence of three-toed footprints is revealed. You walk alongside the trackway—left-right-left-right—literally following in the footsteps left by a dinosaur 200 million years ago. While photographing your discovery, a group of ravens croak loudly as they pass overhead. You watch the birds effortlessly ride the wind before flapping out of sight around a nearby cliff.

The flying ravens above and the ancient tracks below share a profound connection deeply rooted in common ancestry. As one of the classic major transformations in the history of vertebrate life, the origin of extant birds (Aves) from within a lineage of carnivorous dinosaurs (Theropoda) evokes fascinating questions. How did powered flight originate? How has bipedality changed through time? Even more generally, how can we study the evolution of ephemeral traits like locomotion across millions of years?

* Department of Ecology and Evolutionary Biology, Brown University
† Department of Biology, Providence College

This chapter is about reciprocal illumination between anatomy and movement. Our overarching theme is *animation*, both as an act and as a product. Most obviously, we strive to reanimate or "breathe life" into extinct animals known only from fragmentary evidence. An animation metaphor helps us to visualize the evolution of structures through time as well, such as a clawed predatory forelimb morphing into a feathered flapping airfoil. A perspective founded on a sequence of images also has clear connections to documenting animal movement with film and video. The history of motion analysis techniques will be briefly reviewed to provide context for a description of recent approaches founded on animation. Thus, we invoke animation as a critical conceptual and methodological link between the living animals we can study directly and the static remains we unearth as fossils.

Throughout this and many other chapters, bones are given special emphasis because body fossils dominate the paleontological record. In doing so, we must strive to avoid a paleontological conceit. Natural history museum displays are replete with original and cast specimens mounted in "lifelike" poses, but bones do not stand or walk or fly—animals do. Our goal is to understand how bony and soft tissues work together and evolve as a dynamic whole. Animation can play a vital role in this endeavor.

An Animator's Perspective

Animation—bringing life to the lifeless through the illusion of motion—is an apt metaphor for our approach to studying locomotor evolution. From a stick figure on a paper flip-book to a classic Disney film to the most sophisticated computer game, an animation fundamentally consists of nothing more than a series of static images. The act of displaying such a sequence rapidly enough to optically blend into smooth movement is extremely powerful. This power can be applied to scientific questions by maintaining flexibility in our definition of both subject matter and timescale.

One place to begin is by trying to visualize the evolutionary scope of theropod bipedalism with an animator's eye. As a starting point, consider a distant ancestor of birds, a theropod living long ago in the Late Triassic (fig. 17.1). Rather than being distracted by the entire dinosaur, try to envision just its two hind limbs. These

legs and their scaled feet have many characteristics of a bird like an emu or turkey. The long, muscular limbs are held beneath the body (unlike a more sprawled lizard or crocodile) with the ankle elevated in a digitigrade foot posture. Weight is born by three, forward-pointing toes (digits II, III, and IV), with a smaller hallux (digit I) held just above the ground. When stepping in firm mud, the plantar pads and claws would leave behind footprints very similar to those discovered on your hypothetical expedition.

Now imagine this pair of hind limbs walking steadily in place, as if on a treadmill. Right and left alternate out-of-phase, placing one foot in front of the other along a narrow path as in the fossil trackway. The limbs' morphology, posture, and kinematics (motion) throughout a stride cycle aren't identical to any bird today, but our goal is to reconnect past to present. Envision these walking legs as they transform through time along the theropod chain of descent (fig. 17.1)—morphing into ever more bird-like forms until reaching the common ancestor of all birds, then continuing onward to the most recent ancestor of living birds, and finally transitioning into a raven or some other extant species.

In theory, we could animate this lineage's bipedal history as a pair of striding legs whose anatomy and movements morph dynamically as time progresses. Changes in limb morphology (size, proportions, musculature), posture, and motion take place simultaneously in our reconstructed sequence spanning millions of years. Unlike binary approaches that treat theropods as two discrete categories (avian vs. non-avian: e.g., Gatesy 1991; Jones et al. 2000), such an animator's perspective emphasizes continuity of function and the blending of intermediates before, across, and after the origin of birds (Gatesy 1990; Hutchinson and Allen 2009; Allen et al. 2013).

Reciprocal Illumination between Anatomy and Motion

If extinct ancestors along a lineage can't be directly observed, how can such a sequence be anchored in solid evidence rather than remain purely imaginary? Software for animating 3-D skeletal models is getting cheaper and easier every day (Stevens 2002; Hutchinson and Gatesy 2006; Mallison 2011). However, we believe that simply making fossil bones move is neither a practical method

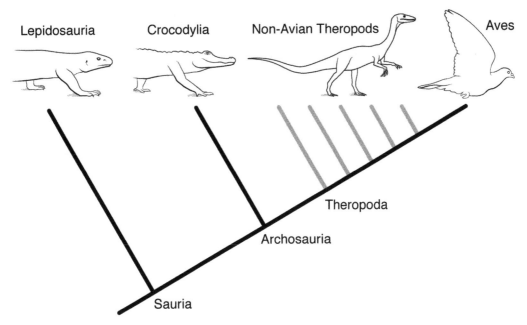

FIG. 17.1 Simplified cladogram of relationships among living saurians (black branches) showing extinct theropods (gray branches) on the line to birds (modified from Gatesy and Baier 2005; *Varanus* modified from Jenkins and Goslow 1983).

nor a desirable end in itself. Discerning a single pose, much less a sequence of poses, in an extinct animal is fraught with uncertainty. For example, the articular geometry of the femur and pelvis offers some limits on rotations about different axes at the hip, but what angles should be chosen within these ranges? When uncertainty at the hip is combined with that at the knee, ankle, and toe joints, the number of possible poses spirals catastrophically (Hutchinson and Gatesy 2006; Gatesy et al. 2009). Simply "rubbing bones together" only gets us so far.

Although starting at an ancient ancestor and working forward to the present is a compelling visual construct, we risk getting bogged down in uncertainty very quickly. In practice, going backward in time from the present is much more tractable. We and our more biomechanically minded paleontological colleagues studying dinosaurs easily devote 95% of our research time to describing and quantifying the anatomy and movement of extant birds and crocodylians. Much critical data can be acquired through dissection of cadavers. For example, a muscle's cross-section and architecture will limit maximum force output (Allen et al. 2010; Paxton et al. 2010), which together with its moment arm (Hutchinson et al. 2005; Bates et al. 2012) dictates potential torque about a joint (Hutchinson and Garcia 2002).

Likewise, joint mobility can be quantified through manipulation as ranges of motion about each rotational degree of freedom (e.g., Senter 2006; Hutson and Hutson 2012; Pierce et al. 2012).

Anatomy in action is best captured by recordings of live, locomoting animals. Sequences of real poses provide functional context—the lens through which to view skeletal form. If we can better understand why living animals move the way they do, we can discern biomechanical constraints that can be applied to extinct species. Most of these constraints will involve masses, forces, and torques that aren't directly preserved in the fossil record (Gatesy et al. 2009). Thus our approach is to exploit motion to illuminate morphology in extant forms so that morphology can help illuminate motion in extinct ones.

Recording and Quantifying Aerial Locomotion

One of the early challenges to anyone interested in how birds move was imaging. Studying a pigeon flying by at 14 meters per second is no easy feat, and even when hovering its wings still beat too rapidly for the naked eye to perceive or measure details of forelimb movement. The history of attempts to capture a series of poses of a bird "frozen" in flight predates the birth of

FIG. 17.2 A sampling of Marey's chronophotographs of bird flight. A, Lateral view of a crested heron (modified from photograph in Braun 1995; Marey 1887). B, Dorsal view of a pigeon (modified from Marey, Bibliothèque numérique Medic@).

motion pictures in the late 1800s. In the years since, dramatic technological advances have revolutionized our ability to visualize the wing beat, both inside and out. We chronicle this fascinating history by highlighting several examples.

Eadweard Muybridge is well known for taking photographs of movement in humans and other animals using a battery of cameras. Many of his image collections are still being published (e.g., Muybridge 1957); however, Muybridge's goals were largely artistic rather than scientific (Braun 1995). In contrast, Etienne-Jules Marey, a late-19th-century contemporary working at a research station outside of Paris, deserves accolades for his unusual blend of creativity and scientific rigor. Unlike Muybridge's multi-camera approach, he recorded numerous images with a single camera. Over a career spanning five decades, Marey developed innovative instruments and techniques for spatially registering pictures on photographic film (Braun 1995; Schwenk and Wagner 2010). His composite images of pigeons, gulls, pelicans, ducks, herons, and egrets capture the dynamic grace of birds in flight (fig. 17.2). These visually stunning, strobe-like photographs break motion into a

sequence of accurately timed poses—in many ways the exact opposite of animation.

For Marey, this "chronophotographic method" yielded images from which useful kinematic data could be directly extracted (Marey 1883, 1887). Rather than wanting to merge separate images into the illusion of motion, the goal was to break a continuous visual flow into separate snapshots at consistent intervals. Marey is sometimes criticized for not using his innovations to invent motion pictures, but given his desire to analyze motion by augmenting the human eye, projection of images to recreate the illusion of normal movement would have been counterproductive (Braun 1995). Well into the 20th century, authors relied on Marey's data to describe both the path of a bird's body and the phases of the wing beat.

The advent of high-speed cinematography eventually brought bird flight into focus. In a series of studies, Brown (1948, 1953) recorded freely flying birds on film at 85–95 frames per second. His exquisite images required very short exposure times to avoid blurring of the rapidly moving wingtips (fig. 17.3A). A modern counterpart is the analysis of wing movement using video (60 fields

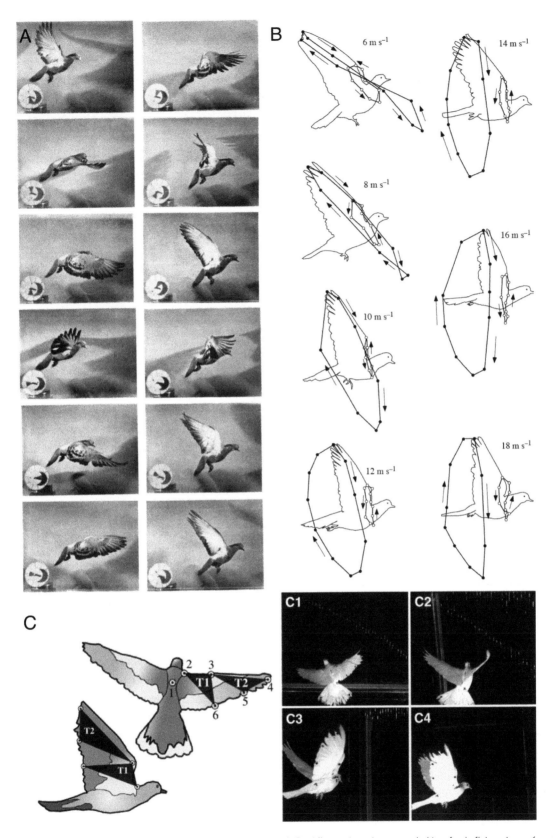

FIG. 17.3 Advances in imaging with high-speed film and video cameras. A, Rapidly moving wings recorded in a freely flying pigeon (modified from Brown 1948). B, Pigeon wing motion acquired over a range of flight speeds in a wind tunnel (modified from Tobalske and Dial 1996). C, Multiple cameras (C1-4) for reconstructing 3-D surface points and planes by stereophotogrammetry in doves (left) and cockatiels (right) (modified from Hedrick et al. 2002).

per second) and high-speed film (150–200 frames per second) by Tobalske and Dial (1996). To record multiple wing beats across a range of velocities, pigeons and magpies were trained to fly in a variable-speed wind tunnel (fig. 17.3B). The full power of high-speed video is exemplified by a dove and cockatiel study, in which four synchronized cameras (250 frames per second) were used to track marked points on the wing and body in 3-D (Hedrick et al. 2002). By connecting points on the wing surface, two triangles were created to represent motion of the proximal "arm" and distal "hand" wing for aerodynamic modeling (fig. 17.3C).

Seeing the Skeleton

Another major breakthrough took motion analysis inward from standard light imaging of the feathered airfoil to X-ray visualization of the moving skeleton (fig. 17.4). Jenkins and colleagues (Jenkins et al. 1988; Goslow et al. 1989, 1990; Dial et al. 1991) recorded starlings flying in a wind tunnel with cineradiography (X-ray movies) at 200 frames per second. Films from lateral and dorsoventral perspectives revealed motions of the pectoral girdle, shoulder, elbow, and wrist for the first time (fig. 17.4A). The difference between approximating bone position and actually seeing the changing skeletal configuration through feathers, skin, and musculature cannot be overemphasized. Despite the clarity of Brown's images (fig. 17.3A), resolving the location of a deep joint like the shoulder (glenohumeral) joint from an external image with reasonable accuracy and precision is extremely difficult, if not impossible. Similarly, tracking the motion of surface markers is not the same as tracking bones. Musculature and loose skin can prevent superficial points from maintaining a predictable relationship to underlying structures, introducing so-called errors of transformation (e.g., Zatsiorsky 1998) even if the markers' 3-D coordinates are calculated accurately.

X-rays are essentially shadows of 3-D anatomy compressed onto a 2-D film plane. For movements as complex as the avian wing beat, quantifying actual joint angles from the apparent angles projected onto a flat surface is not straightforward. Jenkins and colleagues analyzed starling cineradiographs using a technique that may seem surprising given their high-tech combination of wind tunnel and X-ray imaging system.

Following earlier studies (e.g., Jenkins and Goslow 1983), they reconstructed 3-D relationships by configuring actual starling skeletons to conform to temporally matched frames from lateral and dorsoventral perspectives (fig. 17.4B). Such sequential posing of a physical skeleton based on X-ray images is a form of "rotoscoping," a film production technique in which an animator is guided by footage of a live scene (e.g., Kerlow 1996; Dobbert 2005; Gatesy et al. 2010). Mounted skeletons were measured to yield joint angles and figured from different perspectives (fig. 17.4C, D), hinting at a new way of understanding skeletons in action.

In our studies, we have fully embraced the rotoscoping approach by adopting an entirely digital, animation-based workflow (fig. 17.5). 3-D animation software developed for the entertainment industry is used to create virtual cameras that give the viewer the same perspective on the scene as the X-ray beams (fig. 17.5A). Rather than manipulating actual osteological specimens, polygonal bone models (derived from computed tomography or laser scanning) are registered to align with the X-ray images recorded on high-speed video. We called our initial method scientific rotoscoping (Gatesy et al. 2010) to distinguish our technique from video-inspired animation done by graphic artists for nonscientific purposes. To ease animation, we articulate individual bone models into a hierarchically arranged digital marionette (fig. 17.5B), with each degree of freedom controlled independently. Once the digital skeleton is configured to match the sequence of X-ray poses as closely as possible (fig. 17.5C), its motion can be viewed from any perspective (fig. 17.5D) and quantified (see below).

Zoologists have recently begun combining X-ray systems to allow simultaneous imaging from two different perspectives, an approach pioneered by human biomechanics researchers over a decade ago (You et al. 2001). In a biplanar setup, the intersecting beams define a volume within which the 3-D bone position and orientation can be reconstructed by scientific rotoscoping much more easily than from a single view (Nyakatura and Fischer 2010; Baier et al. 2013). Moreover, if three or more radio-opaque metal markers are surgically implanted into a bone and tracked by stereophotogrammetry, a model of that bone can be animated based on the rigid body kinematics of its marker cluster (e.g., Anderst and Tashman 2003). Biplanar imaging also opens

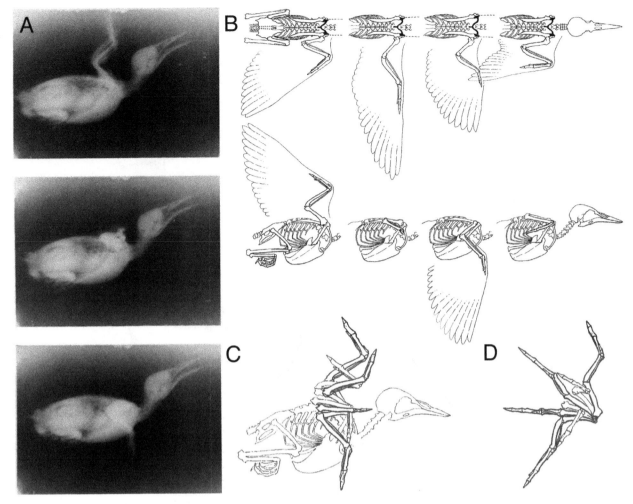

FIG. 17.4 Cineradiography of starling flight. A, Three film frames of downstroke in lateral view (modified from Goslow et al. 1990). B, Four poses in dorsal and lateral views rendered from rotoscoped anatomical specimens (modified from Jenkins et al. 1988). C, Composite rendering of six wing poses throughout the wing-beat cycle in lateral view. D, Composite wing poses from an anterolateral perspective (C and D modified from Dial et al. 1991).

up the potential for automated markerless tracking, in which a computer optimizes model registration each frame rather than a human animator (e.g., Bey et al. 2008). We now use an overarching term, X-ray Reconstruction of Moving Morphology or XROMM (Brainerd et al. 2010; Gatesy et al. 2010), to encompass all of these methods for registering bone models to X-ray video (manual markerless, automated markerless, marker-based, and variations in between).

XROMM brings video recording and animation full circle. Individual frames sampled from continuous motion as a series of discrete images are augmented with digital bone models to reconstruct 3-D relationships. A sequence of skeletal poses can be reanimated for visu-alization/exploration and then becomes the raw data for quantitative analysis.

The Power of Moving Morphology

Animating bone models with XROMM is a powerful means of measuring skeletal motion with high accuracy. Although joint angles are crucial for interpreting how the wing reconfigures within and among wing beats, the 3-D relationship of bone models provides much more than just these rotations. The conventional approach is to estimate the movement of the wing skeleton using simple line segments to symbolize forelimb elements (e.g., Brown 1948; Simpson 1983). Stick figure

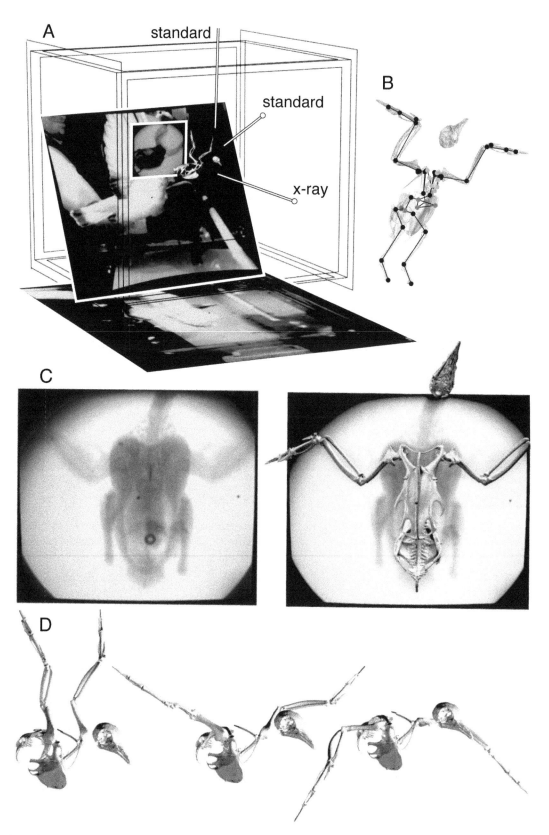

FIG. 17.5 Scientific rotoscoping of pigeon flight. A, 3-D model of the wind tunnel flight chamber, X-ray beam, standard light video cameras, and calibrated image planes forming the context for skeletal animation. B, CT-based polygonal bone models articulated into a digital marionette by joints (circles) that allow both rotation and translation. C, Frame of ventrodorsal X-ray film and registered pigeon model. D, Three downstroke poses in anterolateral view (B–D modified from Gatesy et al. 2010).

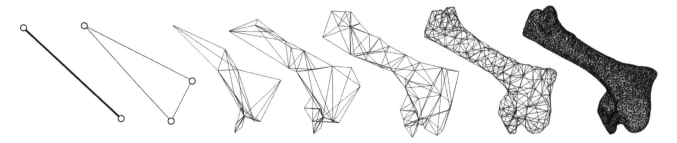

FIG. 17.6 Seven representations of a pigeon humerus segment from stick figure to detailed 3-D model. Two points spanned by a line segment provide minimal kinematic information, whereas three points form a triangle documenting six degrees of freedom. Adding polygonal faces (30, 50, 100, 500, 10,000) creates accurate articular geometry and soft tissue attachment sites at functionally significant distances from the pivot points.

representations typically treat joints as point pivots (fig. 17.6, left). Such points are approximated independently, which can cause the length of each stick to change unrealistically if joint centers are not accurately identified. Moreover, because segments are either unmarked or insufficiently marked to be treated as rigid bodies, a bone's long-axis rotation and translation cannot be calculated unambiguously. Instead of extensible sticks, animated polygonal models maintain a fixed length and embody the full complement of linear and angular movement. Thus, a distinction needs to be made between a 3-D stick figure and a set of 3-D bone models (fig. 17.6). Only from the latter can we quantify motion at each joint for all six degrees of freedom (three translations and three rotations).

Polygonal models are composed of a set of points (vertices) linked to their neighbors to form facet-like faces, typically triangular planes. Stick figures also consist of linked points, but as the complexity of a model grows, there is a shift from a primitive geometric stand-in to a more meaningful morphological representation (fig. 17.6). Very few polygonal faces are needed to perceive an animated model as a rigid segment, but such thrift is no longer required now that the graphics capabilities of most computers easily handle large polygon counts. There is great value in manipulating and registering full resolution bone models with detailed surface morphology. This choice is partly aesthetic; more accurate bone models simply look more biologically correct than stick figures or geometric primitives. But more importantly, animating morphological bone models has major implications for soft tissue motion and function.

We chose the phrase Moving Morphology for the name XROMM to emphasize the added benefits of accessing the relative movement of bone surfaces (Brainerd et al. 2010; Gatesy et al. 2010). Within a joint, we can study the contacts among elements to learn how articular geometry interacts—so-called dynamic arthrology (Anderst and Tashman 2003; Brainerd et al. 2010). Around and within the borders of joints we can measure the relative motion of attachments for passive structures such as the joint capsule and intrinsic and extrinsic ligaments. Most obviously, the motion of muscle-related features (trochanters, tubercles, tuberosities, crests, surfaces, etc.) has direct consequences for a muscle's functional relationship to a joint. The geometry of the skeletal lever system dictates how muscular tension may be converted into kinematic (translation, rotation) and kinetic (force, torque) output. XROMM's ability to visualize and measure the spatial rearrangement of these quantities in 3-D during locomotion opens up previously inaccessible variables central to muscle mechanics and the force transmission system.

Since the late 1980s, researchers' ability to instrument flying birds with ever-smaller transducers has expanded the array of quantifiable physiological attributes. Recordings of wing muscle activity (Dial et al. 1991; Dial 1992), muscle fascicle strain (Biewener et al. 1998), bone strain (Biewener and Dial 1995; Biewener et al. 1992; Dial and Biewener 1993), feather shaft strain (Corning and Biewener 1998), differential wing pressure (Usherwood et al. 2005), and whole-body acceleration (Bundle and Dial 2003; Hedrick et al. 2004) are now common in some flight laboratories. These parameters are all important to how wings operate in flight, but many would benefit from a more complete morphological context. An accurately animated skeleton offers such a 3-D environment.

Interpreting Skeletal Evolution

To be clear, we do not advocate the direct transfer of XROMM-derived flapping motion from pigeons or other species to extinct taxa. Rather, we seek to use accurately reconstructed poses in living birds and crocodylians to help reveal mechanisms at work within the complex of anatomical structures (skeletal, muscular, ligamentous, integumentary) and loads (muscular, articular, aerodynamic, gravitational) during locomotion. Each new awareness of mechanism provides a slightly, or sometimes dramatically, different outlook on morphology. Such functional insights are often additive, building upon one another into an increasingly sophisticated perspective on the significance of bone shape.

Armed with our "new set of eyes," we can approach the fossil record seeking evidence of the evolutionary persistence or change in specific functional associations. Such mechanisms, in turn, often constrain reconstruction of likely poses and movement. Unlike more traditional 2-D (e.g., Swiderski 1993; Bonnan 2004) and 3-D (e.g., Polly and MacLeod 2008) morphometric studies of skeletal shape that treat the disparity of each bone independently, animation deals with motion among multiple elements. Even sophisticated multibone analyses of covariation and integration in static morphology (Chamero et al. 2013) cannot address the dynamic geometric relationships among bone surfaces and their overlying soft tissues that we consider vital to the evolution of form.

The history of the avian coracoid bone serves as an example. An extant flying bird's coracoid differs markedly from the ancestral condition, as exemplified by the basal *Archaeopteryx*. In a now classic paper, Ostrom (1976) created a hypothetical series of possible intermediate morphologies between those of *Archaeopteryx* and a living vulture (fig. 17.7A). This compelling image is a testament to the power of an animation perspective for representing shape change over geological time. How can we better understand the implications of such a transformation? Ostrom and other paleontologists recognized that modifications in coracoid shape have important consequences for mechanics. Our approach seeks to use XROMM and 3-D models to gain a deeper awareness of coracoid features in living birds by delving into their functional associations with neighboring bones, muscles, tendons, and ligaments.

Osteologically, the two coracoids contribute to the shoulder girdle (fig. 17.7B) by connecting the sternum (breast bone) to the paired scapulae (shoulder blades), with the furcula (wishbone) spanning in between. Both the coracoid and scapula underlie the saddle-shaped glenoid (shoulder socket), forming the glenohumeral joint with the ovoid head of the humerus. This configuration makes coracoid length bear directly on muscles spanning from sternum to wing (Gray 1968). Compared to *Archaeopteryx*, the more elongate coracoid of a pigeon or vulture requires that muscles like the massive pectoralis have relatively longer fascicles. Longer fascicles are capable of longer excursions and of powering high-amplitude wing beats (Jenkins 1993), thus linking bone evolution to locomotor evolution through changes in unpreserved musculature.

A second distinct feature of extant fliers is an elevated flange known as the acrocoracoid process (right asterisk in fig. 17.7A), which is only a low bump (the so-called coracoid process or biceps tubercle; left asterisk in fig. 17.7A; Carpenter 2002; Elzanowski 2002) in *Archaeopteryx* and many other theropods. Ostrom (1976) suggested that the reorientation of two muscles might have been responsible for the dorsal extension of this process, yet he overlooked the acrocoracohumeral ligament. How might one assess the relative influence of these three components on their shared attachment site? We explored this interaction using a digital pigeon skeleton on which muscles and ligaments were mapped as 3-D vectors (fig. 17.7C; Baier et al. 2007; Baier 2012). By creating a quantitative force/torque balance model, we discovered that muscles alone cannot stabilize the shoulder during gliding (Baier et al. 2007). The acrocoracohumeral ligament is always needed to prevent dislocation of the humeral head, helps coordinate rotational degrees of freedom during flapping, and is capable of withstanding extremely high forces (over 39 times body weight) at least an order of magnitude higher than either muscle (Baier 2012). Although we cannot refute

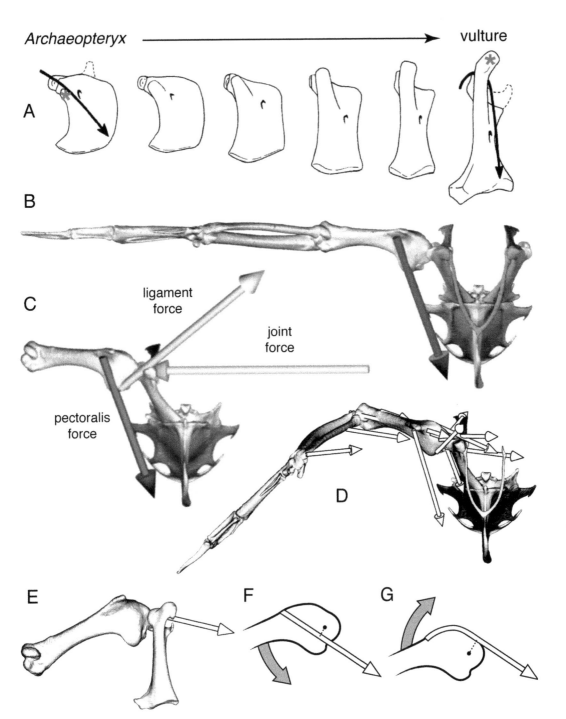

FIG. 17.7 Interpreting coracoid evolution. A, Ostrom's hypothetical sequence of coracoids transitioning from *Archaeopteryx* to a modern vulture (modified from Ostrom 1976). The biceps tubercle and homologous acrocoracoid process are marked by the red asterisk. B, Pigeon coracoid bones (purple) in osteological context articulating with the sternum (yellow), scapulae (green), furcula (red), and wing bones (gold) in anterior view. C, Results of a force-balance model of gliding showing the acrocoracohumeral ligament force preventing ventral dislocation of the humerus by the pectoralis force. Smaller muscle forces not shown (B and C modified from Baier et al. 2007). D, Muscle vectors mapped on a pigeon shoulder in early upstroke. Vector magnitudes are proportional to muscle physiological cross-sectional area on a logarithmic scale. E, Downward and inward pull of the supracoracoideus tendon (white vector) is more similar to the direction hypothesized by Ostrom for *Archaeopteryx* (left black arrow in A) than in a modern bird (right arrow in A). F, Simple, 2-D model of the supracoracoideus tendon (white vector) passing below the axis of elevation/depression (black dot) and causing depression (gray arrow), the presumed primitive condition. G, Shifting the same tendon above the axis reverses the action, yielding elevation without needing to pull upward on the humerus.

Ostrom's hypothesis, solutions from our model point to the acrocoracohumeral ligament as a major influence. Evolutionary expansion and elevation of the acrocoracoid process likely records the development of a previously unrecognized passive shoulder stabilization mechanism founded on the reorientation and increased loading of the acrocoracohumeral ligament (Baier et al. 2007). By augmenting just a single skeletal pose with 3-D force vectors, we gained new insight into wing function in a living flier, which then altered our assessment of shoulder evolution.

The acrocoracoid process is best known for its pulley-like relationship to the primary upstroke muscle, the supracoracoideus. The tendon of this ventrally located muscle passes medial to the process through a bony canal (foramen triosseum) on its way to the dorsal surface of the humerus. Ostrom (1976) postulated that in a modern flier the supracoracoideus tendon pulls dorsally (right black arrow, fig. 17.7A) to elevate and spin (supinate) the humerus (Poore et al. 1997), after undergoing a 180° turn from the muscle belly. We traced the 3-D path of the supracoracoideus tendon on animated bone models of a flapping pigeon in flight (fig. 17.7D). Surprisingly, the tendon never actually pulls dorsally as Ostrom envisioned. The pulley-like rim of the triosseal canal remains ventral (not dorsal) to the enlarged humeral head so that the supracoracoideus vector aims medially and ventrally (fig. 17.7E) akin to his reconstruction for *Archaeopteryx* (left black arrow, fig. 17.7A).

The transformation in supracoracoideus function from depression to elevation came about through a displacement in its line of action relative to the glenohumeral joint axis rather than by evolving a hairpin turn (fig. 17.7F, G). Ostrom's reconstruction of the supracoracoideus in a 140-million-year-old *Archaeopteryx* fossil may be quite correct, but his assumption about how living birds function mistakenly served as the flawed end state for his transformation scenario. Such erroneous interpretations of musculoskeletal mechanisms in extant forms are likely quite common and can easily skew our interpretation of the fossil record. As tempting as it is for paleontologists to focus on *Archaeopteryx* and other fossil forms, a more complete understanding of the dynamic geometric relationship among bones and soft tissues afforded by XROMM is essential so that extinct taxa can be illuminated with the brightest possible light.

Prospectus

We believe that animation is a fruitful path toward integrating anatomy and motion. XROMM merges X-ray videos and polygonal bone models to recreate 3-D relationships just as Jenkins and colleagues did with actual starling skeletons, but with all the power that comes with a digital format. Zoologists are only beginning to discover the potential of this method, and the next decade will undoubtedly bring a flurry of new technological advances in imaging and measuring motion. Yet we must be particularly cautious about overinterpreting results from only a few taxa or even single species. For example, the finding that the coracoids move out during downstroke and in during upstroke in starlings revealed the spring-like behavior of the wishbone (Jenkins et al. 1988; Bailey and Demont 1991). More recent analyses of partridges and pigeons confirm that their shoulders also spread, but with an in-out sequence almost completely out of phase compared to starlings (Baier et al. 2013). So much of the inner workings of living animals and their evolutionary history remains to be deciphered. For those interested in morphology, biomechanics, and the evolution of major transformations, these are exciting times to be a student of vertebrate history.

Colleagues who study fossils exclusively may question our preoccupation with living animals, but extant birds are a precious resource for students of theropod locomotion. If nothing else, direct exposure to the stunning intricacies of musculoskeletal systems in action fosters humility and tempers our desire for overly detailed explanations. Above all, we must have reasonable expectations. What we seek is not *the* answer, because one solution will rarely be an achievable goal. Rather, we must try to delineate ranges of possibility and probability that can be tested and refined by validation and sensitivity analysis (Gatesy et al. 2009; Hutchinson 2012). The challenge is to do so scientifically, proceeding with cautious optimism while being honest about shortcomings. At the same time, the flow of information need not be unidirectional. Fossils provide a historical context that enriches our understanding of living taxa and fosters new lines of inquiry. Ravens can assist in our interpretation of dinosaur tracks, but dinosaur tracks also enlighten our concept of Aves, avian bipedal locomotion, powered flight, and what it means to be a raven.

✳ ✳ ✳

Acknowledgments

We are grateful to Beth Brainerd, Ken Dial, and Neil Shubin for bringing this volume to life, and to two helpful reviewers for improving our contribution. To Farish, we are forever thankful for the inspirational vision and attention to detail that he so willingly shared as teacher, mentor, and friend. We will continue to do our best to pass his legacy on to our students.

References

Allen, V., K. T. Bates, L. Zhiheng, and J. R. Hutchinson. 2013. Linking the evolution of body shape and locomotor biomechanics in bird-line archosaurs. Nature 497:104–107.

Allen, V., R. M. Elsey, N. Jones, J. Wright, and J. R. Hutchinson. 2010. Functional specialization and ontogenetic scaling of limb anatomy in *Alligator mississippiensis*. J. Anat. 216:423–445.

Anderst, W. J., and S. Tashman. 2003. A method to estimate in vivo dynamic articular surface interaction. J. Biomech. 36:1291–1299.

Baier, D. B. 2012. Mechanical properties of the avian acrocoracohumeral ligament and its role in shoulder stabilization in flight. J. Exp. Zool. 317A:83–95.

Baier, D. B., S. M. Gatesy, and K. P. Dial. 2013. Three-dimensional, high-resolution skeletal kinematics of the avian wing and shoulder during ascending flapping flight and uphill flap-running. PLoS ONE 8:1–16.

Baier, D. B., S. M. Gatesy, and F. A. Jenkins. 2007. A critical ligamentous mechanism in the evolution of avian flight. Nature 445:307–310.

Bailey, J. P., and M. E. Demont. 1991. The function of the wishbone. Can. J. Zool. 69(11):2751–2758.

Bates, K. T., S. C. R. Maidment, V. Allen, and P. M. Barrett. 2012. Computational modeling of locomotor muscle moment arms in the basal dinosaur *Lesothosaurus diagnosticus*: assessing convergence between birds and basal ornithischians. J. Anat. 220:212–232.

Bey, M. J., S. K. Kline, R. Zauel, L. T. R., and P. A. Kolowich. 2008. Measuring dynamic *in-vivo* glenohumeral joint kinematics: techniques and preliminary results. J. Biomech. 41:711–714.

Bibliothèque numérique Medic@. www2.biusante.parisdescartes .fr/livanc/?p=32&cote=extcdf005&do=page.

Biewener, A. A., W. R. Corning, and B. W. Tobalske. 1998. *In vivo* pectoralis muscle force-length behavior during level flight in pigeons (*Columba livia*). J. Exp. Biol. 201:3293–3307.

Biewener, A. A., and K. P. Dial. 1995. *In vivo* strain in the humerus of pigeons (*Columba livia*) during flight. J. Morphol. 225:61–75.

Biewener, A. A., K. P. Dial, and G. E. J. Goslow. 1992. Pectoralis muscle force and power output during flight in the starling. J. Exp. Biol. 164:1–18.

Bonnan, M. F. 2004. Morphometric analysis of humerus and femur shape in Morison sauropods: implications for functional morphology and paleobiology. Paleobiology 30(3):444–470.

Brainerd, E. L., D. B. Baier, S. M. Gatesy, T. L. Hedrick, K. A. Metzger, S. L. Gilbert, and J. J. Crisco. 2010. X-ray reconstruction of moving morphology (XROMM): precision, accuracy and applications in comparative biomechanics research. J. Exp. Zool. Part A. 313A:262–279.

Braun, M. 1995. Picturing Time: The Work of Etienne-Jules Marey. Chicago: University of Chicago Press.

Brown, R. H. J. 1948. The flight of birds. J. Exp. Biol. 35:322–333.

Brown, R. H. J. 1953. The flight of birds II. Wing function in relation to flight speed. J. Exp. Biol. 30:90–108.

Bundle, M. W., and K. P. Dial. 2003. Mechanics of wing-assisted incline running (WAIR). J. Exp. Biol. 206:4553–4564.

Carpenter, K. 2002. Forelimb biomechanics of nonavian theropod dinosaurs in predation. Senk. Leth. 82:59–76.

Chamero, B., A. D. Buscalioni, and J. Marugan-Lobon. 2013. Pectoral girdle and forelimb variation in extant Crocodylia: the coracoid-humerus pair as an evolutionary module. Biol. J. Linnean Soc. 108:600–618.

Corning, W. R., and A. A. Biewener. 1998. In vivo strains in pigeon flight feather shafts: implications for structural design. J. Exp. Biol. 201:3057–3066.

Dial, K. P. 1992. Activity patterns of the wing muscles of the pigeon (*Columba livia*) during different modes of flight. J. Exp. Zool. 262:357–373.

Dial, K. P., and A. A. Biewener. 1993. Pectoralis muscle force and power output during different modes of flight in pigeons (*Columba livia*). J. Exp. Biol. 176:31–54.

Dial, K. P., G. E. J. Goslow, and F. A. Jenkins. 1991. The functional anatomy of the shoulder in the European starling (*Sturnus vulgaris*). J. Morphol. 207:327–344.

Dobbert, T. 2005. Matchmoving: the invisible art of camera tracking. Alameda, CA: Sybex.

Elzanowski, A. 2002. Biology of basal birds and the origin of avian flight. Pp. 211–226 in Z. Zhou and F. Zhang, eds., Proceedings of the 5th Symposium of the Society of Avian Paleontology and Evolution. Beijing: Science Press.

Gatesy, S. M. 1990. Caudofemoral musculature and the evolution of theropod locomotion. Paleobiology 16:170–186.

Gatesy, S. M. 1991. Hind limb scaling in birds and other theropods: implications for terrestrial locomotion. J. Morphol. 209:83–96.

Gatesy, S. M., and D. B. Baier. 2005. The origin of the avian flight stroke: a kinematic and kinetic perspective. Paleobiology 31:382–399.

Gatesy, S. M., D. B. Baier, F. A. Jenkins, and K. P. Dial. 2010. Scientific rotoscoping: a morphology-based method of 3-D motion analysis and visualization. J. Exp. Zool. 313A:244–261.

Gatesy, S. M., M. Baker, and J. R. Hutchinson. 2009. Constraint-based exclusion of limb poses for reconstructing theropod dinosaur locomotion. J. Vert. Paleo. 29:535–544.

Goslow, G. E., K. P. Dial, and F. A. Jenkins. 1989. The avian shoulder—an experimental approach. Am. Zool. 29:287–301.

Goslow, G. E., K. P. Dial, and F. A. Jenkins. 1990. Bird flight— insights and complications. Bioscience 40:108–115.

Gray, S. J. 1968. Animal Locomotion. London: Weidenfeld and Nicolson.

Hedrick, T. L., B. W. Tobalske, and A. A. Biewener. 2002. Estimates of circulation and gait change based on three-dimensional kinematic analysis of flight in cockatiels (*Nymphicus hollandicus*) and ringed turtle-doves (*Streptopelia risoria*). J. Exp. Biol. 205:1389–1409.

Hedrick, T. L., J. R. Usherwood, and A. A. Biewener. 2004. Wing inertia and whole-body accelerations: an analysis of instantaneous aerodynamic force production in cockatiels (*Nymphicus hollandicus*) flying across a range of speeds. J. Exp. Biol. 207:1689–1702.

Hutchinson, J. R. 2012. On the inference of function from structure using biomechanical modeling and simulation of extinct organisms. Biol. Lett. 8(1):115–118.

Hutchinson, J. R., and V. Allen. 2009. The evolutionary continuum of limb function from early theropods to birds. Naturwissenschaften 96:423–448.

Hutchinson, J. R., F. C. Anderson, S. S. Blemker, and S. L. Delp. 2005. Analysis of hindlimb muscle moment arms in *Tyrannosaurus rex* using a three-dimensional musculoskeletal computer model: implications for stance, gait, and speed. Paleobiology 31:676–701.

Hutchinson, J. R., and M. Garcia. 2002. *Tyrannosaurus* was not a fast runner. Nature 415:1018–1021.

Hutchinson, J. R., and S. M. Gatesy. 2006. Dinosaur locomotion: beyond the bones. Nature 440:292–294.

Hutson, J. D., and K. N. Hutson. 2012. A test of the validity of range of motion studies of fossil archosaur elbow mobility using repeated-measures analysis and the extant phylogenetic bracket. J. Exp. Biol. 215:2030–2038.

Jenkins, F. A. 1993. The evolution of the avian shoulder joint. Am. J. Sci. 293-a:253–267.

Jenkins, F. A., K. P. Dial, and G. E. Goslow. 1988. A cineradiographic analysis of bird flight—the wishbone in starlings is a spring. Science 241:1495–1498.

Jenkins, F. A., and G. E. J. Goslow. 1983. The functional anatomy of the shoulder of the Savannah monitor lizard (*Varanus exanthematicus*). J. Morphol. 175:195–216.

Jones, T. D., J. O. Farlow, J. A. Ruben, D. M. Henderson, and W. J. Hellenius. 2000. Cursoriality in bipedal archosaurs. Nature 406:716–718.

Kerlow, I. V. 1996. The art of 3-D computer animation and imaging. New York: Van Nostrand Reinhold.

Mallison, H. 2011. Plateosaurus in 3D: how CAD models and kinetic-dynamic modeling bring an extinct animal to life. Pp. 219–236 in N. Klein, K. Remes, C. T. Gee, and P. M. Sander, eds., Biology of the Sauropod Dinosaurs: Understanding the Life of Giants. Bloomington: Indiana University Press.

Marey, E.-J. 1883. Le vol des oiseaux. La Nature, June 16, 35–38.

Marey, E.-J. 1887. Le mécanism du vol des oiseaux éclairé par la chronophotographie. La Nature, December 5, 8–14.

Muybridge, E. 1957. Animals in Motion. Mineola NY: Dover Publications.

Nyakatura, J. A., and M. S. Fischer. 2010. Three-dimensional kinematic analysis of the pectoral girdle during upside-down locomotion of two-toed sloths (*Choloepus didactylus*, Linné 1758). Front. Zool. 7:1–21.

Ostrom, J. H. 1976. Some hypothetical stages in the evolution of avian flight. Smithson. Contrib. Paleobiol. 27:1–21.

Paxton, H., N. B. Anthony, S. A. Corr, and J. R. Hutchinson. 2010. The effects of selective breeding on the architectural properties of the pelvic limb in broiler chickens: a comparative study across modern and ancestral populations. J. Anat. 217:153–166.

Pierce, S. E., J. A. Clack, and J. R. Hutchinson. 2012. Three-dimensional limb joint mobility in the early tetrapod *Ichthyostega*. Nature 486:523–526.

Polly, P. D., and N. MacLeod. 2008. Locomotion in fossil Carnivora: an application of eigensurface analysis for morphometric comparison of 3D surface. Palaeontol. Electron. 11(2):13p.

Poore, S. O., A. Ashcroft, A. Sanchez-Haiman, and G. E. Goslow. 1997. The contractile properties of the M. supracoracoideus in the pigeon and starling: a case for long-axis rotation of the humerus. J. Exp. Biol. 200:2987–3002.

Schwenk, K., and G. P. Wagner 2010. Visualizing vertebrates: new methods in functional morphology. J. Exp. Zool. 313A:241–243.

Senter, P. 2006. Comparison of forelimb function between *Deinonychus* and *Bambiraptro* (Theropoda: Dromaeosauridae). J. Vert. Paleontol. 26:897–906.

Simpson, S. F. 1983. The flight mechanism of the pigeon *Columba livia* during take-off. J. Zool. 200:425–443.

Stevens, K. A. 2002. DinoMorph: parametric modeling of skeletal structures. Senck. Leth. 82:23–34.

Swiderski, D. L. 1993. Morphological evolution of the scapula in tree squirrels, chipmunks, and ground squirrels (Sciuridae): an analysis using thin-plate splines. Evolution 47(6):1854–1873.

Tobalske, B. W., and K. P. Dial. 1996. Flight kinematics of black-billed magpies and pigeons over a wide range of speeds. J. Exp. Biol. 199:263–280.

Usherwood, J. R., T. L. Hedrick, C. P. McGowan, and A. A. Biewener. 2005. Dynamic pressure maps for wings and tails of pigeons in slow, flapping flight, and their energetic implications. J. Exp. Biol. 208:355–369.

You, B. M., P. Siy, W. Anderst, and S. Tashman. 2001. In vivo measurement of 3-D skeletal kinematics from sequences of biplane radiographs: application to knee kinematics. IEEE T. Med. Imaging 20:514–525.

Zatsiorsky, V. M. 1998. Kinematics of Human Motion. Champaign, IL: Human Kinetics.

18

Developmental Mechanisms of Morphological Transitions:
Examples from Archosaurian Evolution

Arhat Abzhanov*

Introduction

Few vertebrate groups can rival the Archosauria (the "ruling reptiles") in terms of the number of species described and the extent of their morphological disparity and diversity. Since their explosive diversification in the early Triassic period, many archosaurs evolved to specialize on large prey and came to dominate many ecological systems that they inhabited since the Mesozoic era (Sereno 1991; Brusatte et al. 2010, 2011; Nesbitt 2011). Described archosaur species collectively number well over 12,000 species (both extinct and extant) and their modern representatives, such as birds, are still among the most numerous members of the terrestrial fauna on all continents (Wang and Dodson 2006; Brusatte et al. 2011; Duncan et al. 2012). This biological success was accompanied by a number of important morphological transitions and some remarkable evolutionary novelties. Multiple morphological alterations allowed the archosaurs to survive, adapt, and prosper in the ever-changing environment over the last 250 million years. This review will discuss some of the recent breakthroughs in our understanding of several important morphological transitions in Archosauria, especially those that occurred in the theropod dinosaur and avian lineage, at both large- and small-scale phylogenetic levels.

* Department of Organismic and Evolutionary Biology, Harvard University

Two of the main branches within class Reptilia are Archosauromorpha and its sister group Lepidosauromorpha (lizards, snakes, tuataras, and plesiosaurs) (fig. 18.1A; Benton 2004). The node Archosauriformes shows the appearance of many characteristic archosaurian traits, including the very large premaxilla and the antorbital fenestra. These are represented by the superficially crocodile-like Permian reptiles *Archosaurus rossicus* and *Proterosuchus sp.* (Broom 1946; Sennikov 1990, 1999; Gower and Sennikov 2003; Borsuk-Białynicka and Sennikov 2009; Borsuk-Białynicka and Evans 2009). Some of the most basally branching groups within the division Archosauria (reptiles featuring antorbital and mandibular fenestra and laterally compressed, thecodont teeth) included *Prolacerta broomi* (Prolacertiformes), a monitor-looking reptile from the Early Triassic (Parrington 1935; Benton 1985; Modesto and Sues 2002). None of the early members of the stem Archosauria survived to the present but, fortuitously, one of the relatively basally branching lineages, the Crocodylia (clade Crurotarsi), persists to the modern time and is represented by a number of species (fig. 18.1A). The crocodylians also retained a number of important morphological characteristics of the more basal Archosauria, including many primitive cranial and postcranial traits, such as the rib-bearing neck and the more complete composition of the skull (fig. 18.1B; Brent 2005). Therefore, for workers in evolutionary developmental biology ("evo-devo"), the Crocodylia represent a very useful reference for studying the more morphologically derived groups, such as birds. Birds represent the only other surviving archosaurian branch that evolved within the Avemetatarsalia (Dinosauria and allies) clade, historically one of the most dominant groups of land vertebrates (fig. 18.1A; Gauthier and de Queiroz 2001; Benton 2004, 2005; Brusatte et al. 2010, 2011). Modern birds are one of the most specious and diverse groups of vertebrates, and they arguably have the most modified vertebrate body plan (fig. 18.1C; Benton 2004).

Advances in modern molecular biology and genetics produced a number of breakthroughs in our understanding of how multicellular organisms are constructed during development. In particular, knowledge of developmental genes and pathways as well as their specific functions at the cellular, tissue, and organismal levels, allow us to more meaningfully describe and compare mechanisms involved in morphological evolution. From the earliest stages, each step of embryonic development is orchestrated by a number of DNA-binding molecules, called transcription factors, that directly and indirectly influence the behavior of cells as these proliferate, differentiate, assemble into tissues, tissues form organs, and organs undergo morphogenesis (Carroll 2008; Gilbert 2010). Multiple types of signaling molecules used for cell-to-cell communication are produced by secreting cells, diffuse to target tissues, bind receptors on/in the receiving cells, and activate specific signal transduction pathways. The signal transduction pathways, in turn, regulate expression and/or function of downstream genes, such as the transcription factors. Much of what we know about the "developmental mechanics" of multicellular organisms comes from studying only a few species of laboratory animals, which are amenable to genetic and/or biochemical analyses used to represent large animal groups. Major invertebrate and vertebrate "model" species include the fruit fly *Drosophila melanogaster,* sea urchins, zebrafish *Danio rerio,* clawed frog *Xenopus laevis*, mouse *Mus musculus*, and chicken *Gallus gallus.* Chicken embryos were among the first animal systems used to study principles of embryonic development in general and amniotic tetrapods in particular (Gilbert 2010). For instance, the discovery of the three germ layers that give rise to all of a vertebrate embryo's organs, ectoderm, mesoderm, and endoderm was made by Christian Pander and Karl Ernst von Baer about 200 years ago in the 1820s when they were analyzing development of early chicken embryos (von Baer 1860). It was a fortunate and auspicious incident that one of the archosaur species, such as the chicken, had been chosen by early developmental biologists to study the principles of embryonic development. In addition to providing us with a wealth of information about various general aspects of vertebrate development, it also proved to be a very useful experimental system for analyzing morphological evolution specifically in archosaurs and other birds.

This review is concerned primarily with recently published reports that reveal developmental mechanisms for evolutionary transitions and novelties in skeletal morphology, which produced the modern birds. Particular attention will be paid to the changes in their axial (vertebral column) and cranial (skull and jaws) skeletons. There are two major challenges to studying the evolution of archosaurs along the avian lineage. First, much of the archosaur radiation became extinct and is only accessible in the fossil record (fig. 18.1A). Second, the two extant archosaur lineages, birds and crocodylians, which we can subject to developmental analyses, both represent derived conditions as compared to the more basally branching groups. Given these restrictions, one has to collect and interpret developmental genetic data from the existing archosaur species with a proper attention to both phylogeny and the known pattern of morphological change.

Overview

Interior Design Improvements: Modifications of the Archosaurian Axial Skeleton

Modern birds are one of the most profoundly modified groups of vertebrates in terms of skeletal morphology. The avian axial skeleton has undergone multiple modifications associated with adaptation to flight and bipedal walking (Benton 2005). These include expansion of the neck and repression of cervical ribs, fusion of thoracic vertebrae, and formation of the synsacrum and pygostyle via fusion of lumbar-sacral and caudal vertebrae, respectively (fig. 18.1C). Such changes in the axial skeleton support various novel avian modes of locomotion, feeding, and other behaviors.

The axial skeleton variation, termed the "axial formula," consists of both differences in the morphologies of segments within each region of the vertebral column (cervical, thoracic, lumbar, sacral, caudal), and also of differences in the number of vertebrae in each region. The avian axial formula is vastly altered when compared with more basal archosaurs. Birds are characterized by long and flexible necks containing an expanded number of cervical rib-less vertebrae, as the ribs were lost via fusion of cervical rib precursors with the vertebral centra (fig. 18.1C; Goodrich 1930). In contrast, thoracic, lumbar, and caudal regions display reduced numbers of vertebrae and show widespread vertebral fusions: fusions between thoracic vertebrae form a fulcrum for the wings, fusion of thoracic, lumbar, and sacral segments with the pelvic girdle form the synsacrum, and posterior caudal segments fuse to form the pygostyle (Gill 2007). In the chicken, all of these features form during embryogenesis and can be compared directly with

development of corresponding homologous structures in crocodylian embryos (Bellairs 2005; Mansfield and Abzhanov 2010).

The crocodylian axial formula lacks the avian-specific modifications described above. In comparison, modern crocodylians have a conservative axial formula; most of them display 8–9 cervical, 13 dorsal (thoracic and lumbar), and 2 sacral vertebrae (Hofstetter and Gasc 1969). Moreover, the crocodylian body plan closely resembles that of basally branching archosaurs, sister groups to archosaurs, and even basal amniotes, which generally had 8 cervical vertebrae with long and free (unfused) ribs, approximately 20–25 rib-bearing dorsal vertebrae, and 2 sacral vertebrae (fig. 18.1B; Ewer 1965; Cruickshank 1972; Gow 1975; Gauthier et al. 1988; Berman et al. 2004). Figure 18.1D compares the avian and crocodylian axial formulae to those of *Orobates pabstii* (Diadectomorpha), a diadectomorph representative of the basal amniotes, and *Prolacerta broomi* (Prolacertiformes), from a close sister group to Archosauria (Benton 1985; Brusatte et al. 2010, 2011; Mansfield and Abzhanov 2010; Nesbitt 2011).

During development both the overall axial formula and morphology of individual vertebrae are regulated, at least in part, by Hox genes. Hox genes (from an abbreviation of homeobox DNA binding domain) are a group of related transcription factors that control the body plan of the embryo along the anterior-posterior (head-tail) axis (Gilbert 2010). After the embryonic segments (somites) have formed, the Hox proteins determine the type of segment structures that will form on a given segment. Thus, Hox proteins confer segmental identity, but do not form the actual segments themselves (Gilbert 2010). These genes are expressed in restricted and partially overlapping domains along the anterior-posterior axis, including in the somites and lateral plate mesoderm that gives rise to most of the axial skeleton, musculature, and connective tissue. Mutations in Hox genes can produce dramatic homeotic transformations in identities of structures along the main body axis (Krumlauf 1994). The exact identity of axial structures, such as vertebrae and ribs, is established by unique combinations of Hox genes, also termed the "Hox code" (Kieny et al. 1972; Kessel and Gruss 1991; Yueh et al. 1998; Nowicki and Burke 2000; Carapuco et al. 2005). Therefore, avian-specific modifications in axial skeleton may be generated by changes in the embryonic Hox

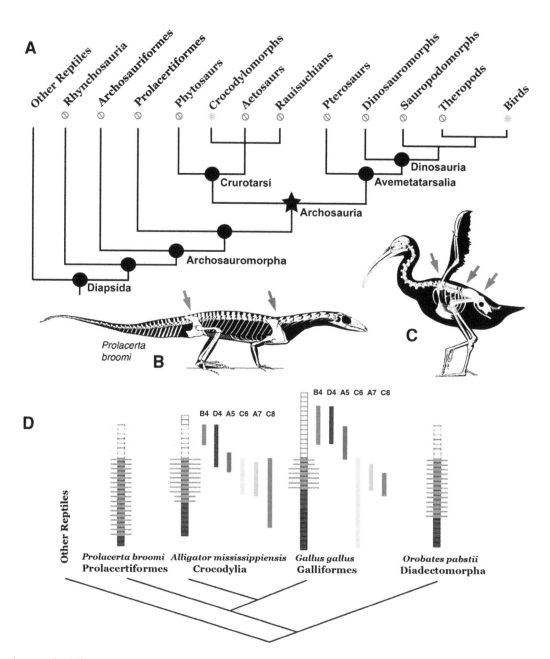

FIG. 18.1 Large-scale phylogeny of archosaurs and evolution of their body plan. (A) Archosauria clade consists of two major groups, the Cruro-tarsi and Avemetatarsalia. Each group is represented by only one surviving lineage (green star), the Crocodylia and Aves, respectively. Modified after Brusatte et al. (2010, 2011) and Nesbitt (2011). (B) Skeleton of *Prolacerta broomi* (Prolacertiformes), which shows a neck bearing ribs (neck-thorax boundary is indicated with a red arrow) and no morphological thoraco-lumbar boundary (lumbar-sacral border is indicated with a blue arrow). *P. broomi* image is modified after Parrington (1935). (C) Skeleton of a bird, which has a long, flexible rib-less neck as well as a clear thoraco-lumbar boundary (purple arrow). The ibis *Xenicibis xympithecus* image is modified after Longrich and Olson (2011). (D) Expression borders of some Hox genes correlates with morphological boundaries and modifications as revealed by comparative studies on alligator and chicken embryos. Modified after Mansfield and Abzhanov (2010).

expression code. Recently, expression patterns of multiple key Hox genes have been comparatively analyzed in chicken (Galliformes) and alligator (Crocodylia) embryos (Mansfield and Abzhanov 2010). A summary of the antero-posterior Hox expression boundaries in alligator and chicken embryos is illustrated in figure 18.1D. The somitic Hox expression boundaries are largely conserved between crocodylians and birds, revealing overall similarity of amniotic design. However, several differences were detected and two of these, in Hoxc-8 and Hoxa-5 expression domains, correlate with skeletal morphologies particular to each lineage (Mansfield and Abzhanov 2010).

In most amniotes studied, including chickens and alligators, Hoxc-8 has an anterior somitic anterior expression boundary at the mid-thoracic region, with strongest expression in the anterior-most 4–5 segments of its domain (Burke et al. 1995; Ohya et al. 2005). In mice, Hoxc-8 is required to pattern the thoracic neural arches and spines of the vertebrae, and mutant defects include formation of ectopic sternebrae and malformation of the xiphoid process (Le Mouellic et al. 1992; Ohya et al. 2005). Unlike other amniotes, alligator embryos do not show concentrated Hoxc-8 expression in the anterior-most 4–5 somites of its domain, and the anterior boundary is shifted forward, to the cervical-thoracic transition (fig. 18.1D; Mansfield and Abzhanov 2010). This is in striking correlation with crocodylian thoracic segments, which have proportionally larger dorsal neural spines supporting larger epaxial musculature (Organ 2006). Such skeletal modification is believed to be required with alligators' mode of feeding, which involves underwater rolling and tearing prey. Given the dose-dependent Hoxc-8 function in regulation of shape and size of thoracic neural arches and dorsal neural spines in mice, loss of its strong expression domain is expected to promote larger spines in alligators (Pollock et al. 1995; Yueh et al. 1998).

Another conspicuous difference between alligator and chicken embryos is in Hoxa-5 expression. Chicken Hoxa-5 is expressed at the highest level in cervical somites, whereas alligators show mostly thoracic expression (fig. 18.1D; Mansfield and Abzhanov 2010). Curiously, Hoxa-5 is also expressed in a similar pattern in cervical vertebrae in mouse embryos (Nowicki et al. 2003; Durland et al. 2008). Thus, strong cervical expression of Hoxa-5 correlates with flexible and rib-less vertebral morphology (fig. 18.1D). Importantly, Hoxa-5 has rib-repressing activity in mice, as loss-of-function mutants form ribs on the last cervical vertebra (Jeannotte et al. 1993; Tabaries et al. 2007). Moreover, chicken Hoxa-5 is highly expressed in the dorsal and ventral lateral sclerotome of cervical segments, a region that contains rib precursors (Aoyama et al. 2005). Thus, it is likely that in the avian lineage cervical rib repression occurred, at least in part, through recruitment of Hoxa-5 to these cells to alter rib cartilage development. Inter-estingly, although mammals and birds evolved their rib-less necks independently, the exact developmental mechanisms of cervical rib repression are quite comparable, as rib loss in both birds and mammals occurs through fusion of cervical rib progenitors with vertebral centra to form transverse foramen (Goodrich 1930; Gauthier et al. 1988). From the evo-devo perspective, it will be exciting to determine whether these lineage-specific morphological transitions in crocodylian and avian axial skeletons were caused by changes in the coding sequences of Hoxc-8 and Hoxa-5 homologs, respectively, or by mutations affecting regulatory elements both upstream and downstream of Hox genes or both.

Like (Juvenile) Father Like Son: Birds Have Paedomorphic Dinosaur Skulls

Birds have extremely modified skulls that display a number of unique features as compared with the more basal archosaurs (fig. 18.2). In particular, birds have relatively very large eyes, and the bird brain has expanded to contain sophisticated visual and neuromuscular coordination systems for surviving in complex environments and for flight coordination (Benton 2004). The bird face is also highly distinct and develops a toothless and pointed bill formed by the upper and lowers beaks. Many birds evolved diverse and highly specialized beak shapes for a variety of adaptive reasons. Ontogenetically, the basally branching members of the dinosaur clade Eumaniraptora, which includes *Archaeopteryx* and modern birds, appear to undergo little change from juvenile to adult states. For example, the skull of the young Eichstätt specimen of *Archaeopteryx lithographica* (Archaeopterygidae, Aves) is about half size of the Berlin specimen's skull but has a nearly identical cranial morphology, and both of them resemble the juvenile theropod skulls (fig. 18.2B; Bhullar et al. 2012). Likewise, juvenile modern extant birds are mostly morphologically similar to their adults (Gill 2007). In contrast, skulls of crocodylians and the more basal dinosaurs like *Coelophysis bauri* (Saurischia) undergo extensive change during ontogeny with juveniles displaying the typical juvenile amniote features, such as the relatively shorter facial (antorbital) regions as well as the larger brains and eyes (fig. 18.3). Thus, it appears that birds and their close relatives are paedomorphic in terms of

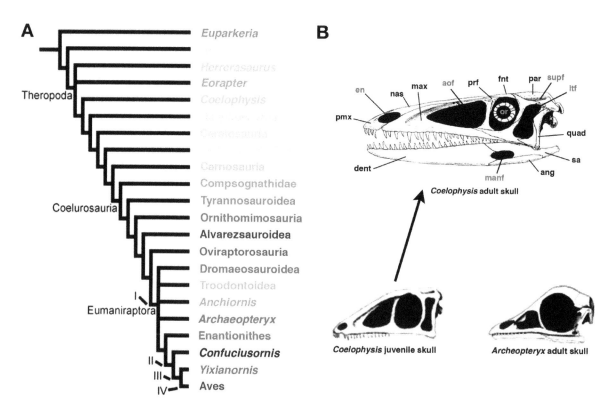

FIG. 18.2 Patterns of skull evolution in archosaurs and birds. (A) Phylogeny of archosaurs in the avian lineage. Heterochronic transformations are indicated with numerals I, II, III and IV. (B) Theropod dinosaurs, such as *Coelophysis*, as adults had elongated skulls with extended snout and face regions. By comparison, theropod juveniles had a proportionally shorter snout, but larger eyes and brain. Likewise, adult avialans (modern birds and immediate relatives), such as *Archaeopteryx*, had a skull morphology similar to that of non-avian archosaur juveniles. Abbreviations for (B): bones—ang (angular), dent (dentary), par (parietal), fnt (frontal), max (maxillary), nas (nasal), pmx (premaxillary), prf (prefrontal), quad (quadrate); fenestrae and other openings—aof (antorbital fenestra), en (external naris), ltf (lateral temporal fenestra), manf (mandibular fenestra), or (orbit), supf (supratemporal fenestra). Skulls in (B) are modified after Colbert (1989). All other images are from Bhullar et al. (2012).

their cranial morphology in retaining a morphology as adults that resembles that of the juveniles or embryos of other archosaurs, including theropod dinosaurs.

To test whether bird skulls are indeed paedomorphic, geometric morphometric and principal component analyses (PCA) of shape variation were recently performed that sampled broadly across theropods using detailed photographs and CT scans of skulls (fig. 18.3A; Bhullar et al. 2012). Unlike traditional morphometrics, which is concerned with absolute measurements, the "landmark"-based geometric morphometrics uses spatial information contained in the data as "landmark" coordinates, the discrete anatomical loci considered homologous in all individuals under analysis. This analysis included the archosaur stem representative *Euparkeria capensis* (Archosauromorpha), crocodylian *Alligator mississippiensis* (Crocodylia), and a number of non-avian and selected avian theropods, such as comp-

sognathids and therizinosaurs (Bhullar et al. 2012). To incorporate ontogeny into the analysis, multiple juvenile/adult pairs were also included where available. Over 40 landmarks provided a comprehensive coverage of the lateral view of the cranium taking advantage of the fact that most of the archosaur skulls are mediolaterally compressed. The first two principal components (PCs) of the PCA explained most of the skull variation in archosaurs (43% and 14%, respectively; fig. 18.3). The second principal axis (PC2) accounts for phylogenetic transformations from the primitive archosaurs toward the coelurosaurian theropods and shows dorso-ventral narrowing of the face, changes in shapes of orbital and premaxillary bones, enlargement of neurocranium, and the characteristic postero-ventral rotation of the braincase (fig. 18.3A, B). The first principal axis (PC1), on the other hand, captures changes during ontogenetic growth, as it describes the extension of the face, resulting in the relative reduction of orbit size and neurocranium, and a constriction in the lower temporal fenestra (fig. 18.3). PC1 also reveals a highly conserved pattern of extensive ontogenetic change across archosaurs with similar ontogenetic trajectories (fig. 18.3A).

The combined morphometric and PCA analyses provide clear evidence for the role of a particular type of heterochrony, called paedomorphosis, in avian evolution. While non-eumaniraptoran archosaur adults cluster together, birds and their immediate eumaniraptoran relatives cluster with the embryos and very young juveniles of other non-avian archosaurs. Remarkably, the more advanced avialan *Confuciusornis sanctus* (Confuciusornithiformes) features a skull nearly identical to those of the more basal archosaur embryos, for example,

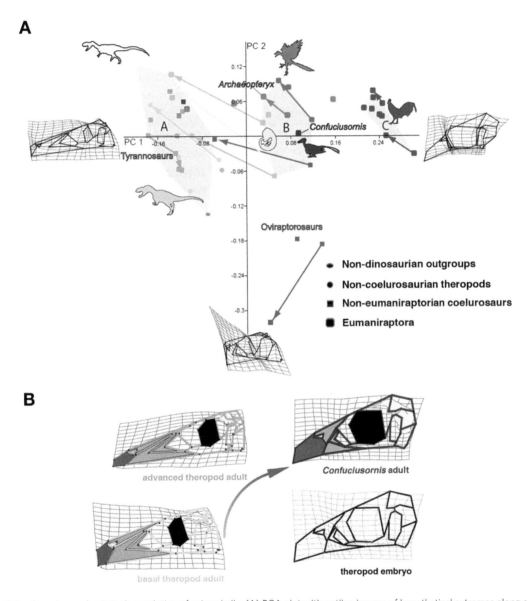

FIG. 18.3 Role of paedomorphosis in the evolution of avian skulls. (A) PCA plot with outline images of hypothetical extremes along each axis. Taxonomic names are color coded the same way as the data points in figure 18.2A. Arrows indicate ontogenies. Major groupings are outlined, shaded, and labeled. (B) During dinosaur evolution from the basal theropod *Herrerasaurus* (green) to the more advanced theropod *Guanlong* (orange), they evolved more elongated snouts with small premaxillary bones (shaded dark gray) and large nasal and maxillary bones (shaded light gray), as well as relatively small eyes (shaded black) and brains (shaded yellow). In the bird lineage these proportions changed dramatically as the face collapses (nasal and maxillary bones become much shorter), while the eyes, brain, and premaxillary bones all expand. The skull of the basal bird *Confuciusornis* (brown) closely resembles that of the perinate of the more basal theropod *Compsognathus* (unshaded wireframe image in the right lower corner). All images are from Bhullar et al. (2012).

the perinate alligator or basal theropod *Compsognathus* (fig. 18.3B; Bhullar et al. 2012). The cluster of adult bird and early eumaniraptoran with non-avian embryonic skulls is widely separated from the adult non-avian skulls cluster. At least four heterochronic transformations could be detected (figs. 18.2, 18.3). Transition I describes paedomorphosis between most theropods and Eumaniraptora; this transition was accompanied by a drastic reduction of the overall body size that was not observed for the subsequent transitions. Transition II is paedomorphosis between Eumaniraptora and *Confuciusornis* where the skull of the latter very closely resembles the skull shape of the previous theropod groups. The ornithurine *Yixianornis grabaui* (Yanornithiformes) spans the gap between these groups and modern birds, as might be expected phylogenetically. Paedomorphic transitions III and IV are between *Confuciusornis* and *Yixianornis*, and between *Yixianornis* and more advanced Aves. Overall, the morphological changes in skull structure move largely along the PC2 axis and then take an abrupt 90 degree turn to move in the reverse-ontogenetic direction along PC 1. Modern extant birds appear to be extremely paedomorphic, moving the farthest in the direction toward the embryos of other archosaurs. Curiously, transitions III and IV are also accompanied by the increasing heterometric change through localized peramorphosis (growth beyond the state of adult ancestors)/heterotopy (novel expansion) of the fused premaxillary bone, which forms a small capping element in basal archosaurs and most dinosaurs but expands to structurally and functionally dominate the upper beak in the bird lineage (Benton 2004, 2005; Bhullar et al. 2012).

There are two ways by which paedomorphosis can arise. One is called neoteny (juvenilization), when somatic development of an animal is slowed or delayed so at the onset of sexual maturity it still retains juvenile characteristics. One of the best-known examples of neoteny is certain species of urodele amphibians, which sexually mature despite preserving a number of larval features, such as a finned tail and external gills. The second type of paedomorphosis is called progenesis when sexual development occurs faster than in the ancestral lineage and results in early somatic maturation and juvenile-looking adults. Paedomorphosis in the bird lineage is probably one of the most striking examples of

progenesis reported to date. The PCA graph shows that birds and the closely related eumaniraptorans have significantly shorter ontogenetic trajectories than all other studied archosaurs (fig. 18.3). Importantly, these short trajectories also match histological data indicating that sexual and somatic maturation times were truncated during each of the paedomorphic transformations (Bhullar et al. 2012). For modern crocodylians and primitive non-avian dinosaurs it takes 6–10 years to reach sexual maturity and 20+ years for somatic maturity (Ferguson 1985; Erickson et al. 2007; Lee and Werning 2008). Even for the smallest crocodiles, for example, the dwarf crocodile *Osteolaemus tetraspis* (Crocodylia), which is about 0.7–1 meters long at maturity (and quite comparable with large birds), it takes about 5 years to become reproductively active (Kofron and Steiner 1994). In contrast, the paleognath bird tinamou (Tinamiformes) and neognath chicken (Galliformes) become sexually mature after only 0.8 and 0.4 years, respectively (Crawford 1990). Thus, ontogenetic trajectories of some modern birds are 8–20 times faster than those in more basal archosaurs. The avian paedomorphic trend is largely independent of size differences (except for transition I). The theropod *Compsognathus longipes* (Saurischia) is quite small (size of a modern turkey) yet as a non-eumaniraptoran it clusters, as expected, with an ancestral adult cluster (fig. 18.3A). On the other hand, the ostrich *Struthio camelus* (Struthioniformes) and emu *Dromaius novaehollandiae* (Casuariiformes) from the analysis are both large-bodied birds, but they group with other birds (fig. 18.3).

Staring into the Face of a Bird: Molecular Mechanisms for Beak Shape Diversity

Comprising 30 orders, 193 families, 2,099 genera and close to 10,000 species, extant birds are the most successful land vertebrates, and much of their success can be attributed to variation in beak shape (Gill 2007). Birds use beaks of a bewildering variety of different designs to obtain food items (hooked beaks of predatory birds, long and straight beaks of herons and hummingbirds, deep beaks of finches and sparrows, serrated beaks of mergansers and flamingos for catching fish and filter feeding, respectively), to build or weave nests (most birds), to dig burrows in the ground (puffins and swallows), to hammer or excavate wood to build a hollow enclosure (woodpeckers and chickadees), to dissipate heat (toucans) and to communicate (song birds, storks, owls). Above all, the bird beak is a precision grasping tool, and its exact shape is often critical to its function.

Understanding genetic and developmental mechanisms responsible for differences in shapes of avian beaks is an important task for students of archosaur evolution. Ideally, comparisons should be made first among species that have a fairly recent common ancestor but which have evolved significant morphological differences. In this way, meaningful differences in molecular patterning are more easily identified. A series of recent reports focused on the natural diversity of beak shapes in the Galápagos (Darwin's) finches (Thraupidae, Passeriformes). These birds inhabit Galápagos and Cocos Islands and comprise a monophyletic group of 14 closely related species that represent a classic example of adaptive radiation, niche partitioning, and rapid morphological evolution (fig. 18.4A). Darwin's finches, in fact, occupy ecological niches normally occupied by different families of birds on the mainland, such as warblers, finches, thrushes, grosbeaks, woodpeckers, etc. (Bowman 1961; Grant and Grant 2008). Several congruent molecular phylogenies based on microsatellite, mitochondrial DNA and nuclear gene sequences exist for Darwin's finches that can be used for comparative analyses (fig. 18.4A; Grant and Grant 2008). The sharp-billed finch *G. difficilis,* the most basal species in the monophyletic genus *Geospiza,* has a small symmetrical beak, while the ground finches have more specialized deep and broad bills adapted to crack hard seeds, and cactus finches evolved elongated, narrow and pointed bills for probing cactus flowers and fruit (fig. 18.4B1–3).

Bill morphology in Darwin's finches was first analyzed using primarily linear measurements when calipers were applied to measure various aspects of the beak length, depth, and width. Such analyses performed on hundreds of birds over many generations revealed important clues on how these birds continuously adapt to their environment (Grant and Grant 2008). More recently the Darwin's finches' bills were analyzed using geometric morphometrics, and the principal component analysis, based on a combination of "landmarks" and "semilandmarks" (points on an outline determined by extrinsic criteria), revealed quantitative contribution of relative size, position and curvature of the upper and lower mandibles in each of the species (Foster et al. 2008). Finally, to understand the nature of the changing geometry of the finch bills, a novel mathematical method was applied that uses digitized outlines of bill shapes to uncover whether the curvatures representing beak shapes could be collapsed using affine transformations, such as scaling or shear (fig. 18.4C; Campàs et al. 2010). The entire diversity of beak morphologies in Darwin's finches could be classified into three groups by the scaling transformations. These distinct "group shapes" can belong either to a single species, to a group of species within a single genus, or even to species that reside in multiple genera. Curiously, applying the more parameter-rich shear transformation accounted for the variation among "group shapes" and, therefore, for all the key morphological differences of beak shapes in Darwin's finches (Campàs et al. 2010).

Morphological observations indicated that beak morphology in Darwin's finches was a trait that was highly genetically regulated with heritability index ranging from 0.7 to 0.9, one of the highest recorded in nature (Grant 1986, Grant and Grant 2008). It was also reported that the hatchlings of these species already displayed the species-specific features, which were, thus, developmentally regulated. To understand the developmental basis for beak morphological variation in a group not amenable to artificial breeding and direct genetic experimentation, comparative analysis methods were employed, such as the candidate gene approach, cDNA microarray screens, and qRT-PCR (Abzhanov et al. 2004, 2006; Mallarino et al. 2011). Several developmental candidate pathways were successfully discovered using a combination of the comparative approaches. First, a candidate gene approach was employed with in situ hybridization probes to reveal expression patterns of many developmental genes on thin cranial sections of embryonic heads from members of *Geospiza* with three different beak shapes. *Bmp4 (Bone morphogenetic protein 4)* was identified as a candidate factor whose strong and early expression correlated with deeper and wider beaks of ground finch species (Abzhanov et al. 2004). Next, a microarray screen using chips featuring several thousand mRNA transcripts expressed in embryonic beak primordia uncovered several developmental

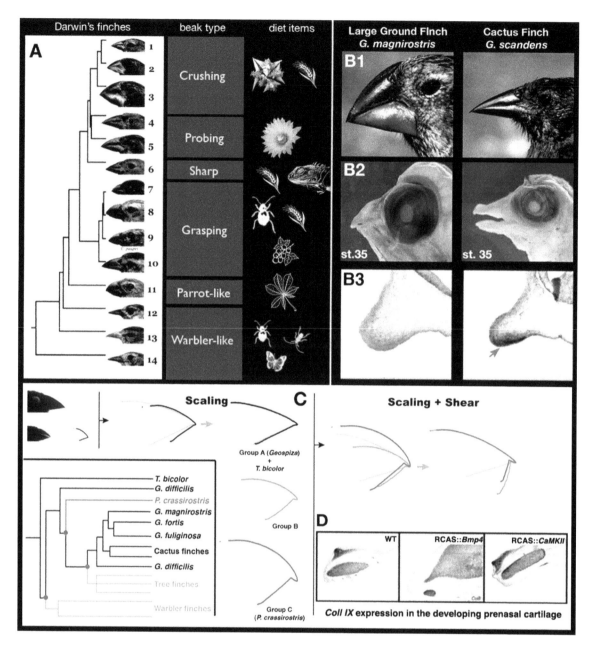

FIG. 18.4 Beak shape evolution in Darwin's finches (Thraupidae, Passeriformes). (A) Molecular phylogeny of 14 species of Darwin's finches illustrating a range of beak shapes in this group of birds. These species have beaks that function in different ways to feed on many different diets: insects, seeds, berries, and even young leaves (vegetarian finch). Species are numbered as follows: 1—small ground finch *G. fuliginosa*, 2—medium ground finch *G. fortis*, 3—large ground finch *G. magnirostris*, 4—cactus finch *G. scandens*, 5—large cactus finch *G. conirostris*, 6—sharp-billed finch *G. difficilis*, 7—small tree finch *C. parvulus*, 8—large tree finch *C. psittacula*, 9—medium tree finch *C. pauper*, 10—woodpecker finch *C. pallidus*, 11—vegetarian finch *P. crassirostris*, 12—Cocos finch *P. inornata*, 13—warbler finch *C. fusca*, 14—warbler finch *C. olivacea* (phylogeny from Grant and Grant 2008). (B1) Large ground finch (left) has a very deep and wide bill adapted to crack hard and large seeds, while the cactus finch (right) has an elongated and pointy beak for probing cactus flowers and fruits (photo courtesy of Dr. Peter Grant). (B2) Bills, especially upper beaks, develop their distinct shapes during embryogenesis and are apparent upon hatching (stage 35 embryos are shown from Abzhanov et al. 2004). (B3) The cactus finch–specific expression of *CaM* was validated by in situ hybridization after it was identified as a candidate by the microarray screen (Abzhanov et al. 2006). (C) The entire diversity of beaks in Darwin's finches can be collapsed into three "group shapes" by the scaling affine transformations. Scaling and shear transformations relate all of the beak shapes (Campàs et al. 2010). (D) Functional assays using retroviruses either misexpressing *Bmp4* (middle) or upregulating *CaM*-dependent pathway (right) in chick embryos, led to increase of embryonic beak depth or length, respectively (Abzhanov et al. 2004, 2006).

regulatory genes that were expressed with high specificity and at high levels in elongated and pointy beaks of cactus finches (fig. 18.4D; Abzhanov et al. 2006). The microarray approach identified *CaM* (*Calmodulin*), a molecule involved in mediating Ca^{2+} signaling and *CaM*-dependent signaling pathway, as a potential developmental regulator of the cactus finch beak morphology (figs. 18.4D, 18.5A). Both *Bmp4* and *CaM* were shown to be associated with developing prenasal cartilage, which forms the beak shapes in early and mid-stage embryos. Another recent microarray screen was used to identify *TGFβIIr*, *β-catenin*, and *Dickkopf-3* (*Dkk3*) as genes differentially expressed in skeletal tissue at later embryonic stages during formation of premaxillary beak bone, which structurally and functionally dominates the upper beak skeleton, especially in species with deep and broad beaks (Mallarino et al. 2011).

Comparative developmental analyses on embryos of *Geospiza* Darwin's finches, which have distinct beak shapes as adults, suggested that the level and/or timing of expression of several regulatory genes have been altered (fig. 18.5; Abzhanov et al. 2004, 2006; Mallarino et al. 2011). In order to understand if such heterochronic and heterotopic changes in gene expression were responsible for the observed morphological variation, functional experiments were performed that aimed to mimic these molecular changes. These experiments were performed using chicken embryos as a model avian system and expression vectors based on the replication-competent avian sarcoma retrovirus (RCAS) that allow strong and stable tissue transgenesis. The earlier and stronger expression of *Bmp4* observed in ground finch embryos was reproduced in chick embryonic beaks with the RCAS::*Bmp4* virus. The infected beaks grew on average about 2.5 times as wide and 1.5 times as deep as the uninfected control beaks (fig. 18.4D; Abzhanov et al. 2004). The change in the *Bmp4*-infected beak morphology occurred, as expected, due to an increase in the size of the prenasal cartilage. Similarly, upregulation of the *CaM*-dependent pathway using a virus carrying the constitutively active version of *CaM*-dependent kinase II (RCAS::CA-*CaMKII*) resulted in a significant elongation of the infected chicken beaks relative to control embryos, but the beak width and depth were unaffected (figs. 18.4, 18.5; Abzhanov et al. 2006). To mimic the ground finch–specific expression patterns of *TGFβIIr*, *β-catenin*, and *Dkk3* in the developing premaxillary bones, infections

were performed with constitutively-active viruses that either increased the activity of these molecules or interfered with their normal functions (Mallarino et al. 2011). All experiments that upregulated activities of the three candidate molecules led to a significant expansion of beaks along the depth and length dimensions but beak width remained unaltered. Thus these phenotypes were also distinct from those caused by either RCAS:*Bmp4* or RCAS::CA-*CaMKII* and helped better explain the existing beak diversity within the *Geospiza* where all three dimensions of beak growth have been altered in one way or another (fig. 18.5A; Mallarino et al. 2011). Beak morphology in birds is highly "modular," that is, different parameters of the beak shape, such as depth, width, and length, can be regulated quite independently as in pelicans, toucans, and storks. Such "modules" are possible due to the fact that these dimensions are controlled by independent regulatory molecules and pathways (fig. 18.5A). Moreover, a comprehensive investigation combining the morphometric analysis of beak shapes, the comparative developmental analysis of candidate genes, the microarray-based screens, and functional assays revealed a highly modular and flexible program for beak development and morphogenesis controlled by multiple regulatory networks. Future research efforts aimed both to provide a wider phylogenetic sampling of passerines and other birds and to pinpoint specific genetic alterations at the DNA level will give us an unprecedented understanding of developmental mechanisms for archosaurian morphological evolution.

Discussion

Since their emergence as a group shortly after the Permian-Triassic extinction event, many archosaur lineages have undergone dramatic evolutionary changes and generated an astonishing array of diverse and disparate body plans (Benton 2004; Brusatte et al. 2011). Over the last 250 million years the archosaurs evolved into animals as small as the bee hummingbird *Mellisuga helenae* (Trochiliformes), which weighs only 1.8 grams (0.063 oz.) and has a length of only 5 cm (1.5 in.) and as large as the massive sauropod *Argentinosaurus huinculensis* (Saurischia), which probably weighed up to 80,000–100,000 kg (the largest ever land animal) or as long as *Supersaurus ivianae* (Saurischia), which measured up to 35 meters (112 ft.), the longest ever land

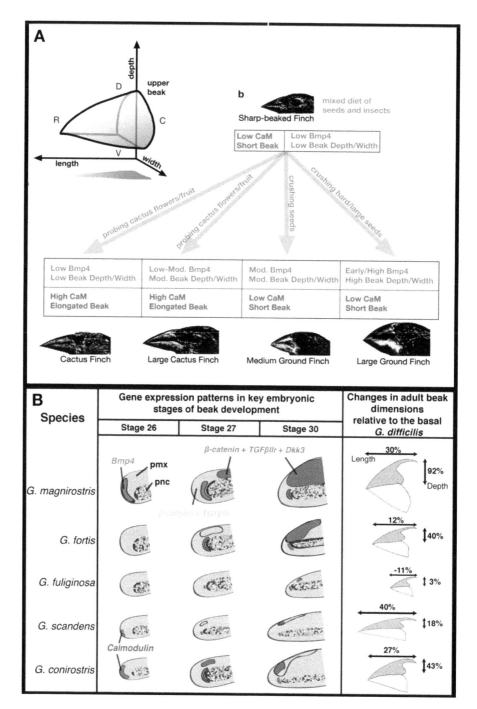

FIG. 18.5 Developmental mechanisms for evolution of beak shape diversity. (A) Developmental program that generates beak shape is highly modular in Darwin's finches, for example, different parameters of beak morphology (depth, width, and length) are controlled by independent regulatory pathways (e.g., *Bmp4* and *Calmodulin*). Such modularity allows for great flexibility in exploring novel beak shapes during evolution (Abzhanov et al. 2004, 2006). (B) Overall summary of beak developmental program that shows multiple independent molecular regulatory networks (*Bmp4/CaM* and *TGFβIIr/Dkk3/β-catenin*) regulate morphogenesis by controlling two tissue modules, prenasal cartilage (gray) and premaxillary bone (brown) (Mallarino et al. 2011).

vertebrate (Paul 1997; Mazzetta et al. 2004; Lovelace et al. 2007; Piper 2007). Some archosaurs, in particular pterosaurs and birds, evolved to be skilled fliers, for example, the giant pterosaur *Quetzalcoatlus northropi* (Pterosauria) was probably the largest flying animal that ever existed. This species is believed to have weighed up to 250 kg (550 lb.) and had a wingspan of up to 11 m (36 ft.) (Henderson 2010). The peregrine falcon *Falco peregrinus* (Falconiformes) is considered to be the fastest animal on the planet in its hunting dive, the stoop, and it was recorded to achieve a top speed of 389 km/h (242 mph) (Harpole 2005). The archosaurs, both extinct and extant, can be found on all continents and in all climatic zones. Members of this clade developed many adaptations for inhabiting the air, land, and water, and they were historically an extremely influential component of the fauna, often dominating the landscapes that they occupied. Thus, it is very important to understand the processes that promoted morphological diversification of such a fascinating group of animals at both large- and small-scale taxonomic levels.

A variety of approaches and techniques that are currently available to study the developmental origins of morphological change can be generally divided into three categories (Mallarino and Abzhanov 2012). First, the pattern of morphological change should be investigated using morphometric techniques, which range from linear measurements of entire organisms or sub-organismal structures, often defined as "modules" (such as limbs, tails, and beaks), to the more sophisticated geometric morphometrics methods, which take into account the size parameters and, when combined with the principal component analysis, reveal the major vectors of morphological change (Gatesy and Dial 1996). Such morphometric analysis should ideally cover multiple ontogenetic stages, include both males and females, and should be analyzed in the context of a strong, preferably molecular phylogeny. This was the approach that allowed recognizing the paedomorphic nature of the avian skulls as compared to the more basal archosaurs, including their theropod dinosaur ancestors (Bhullar et al. 2012). Likewise, the novel mathematical analysis specifically designed to test whether different beak shapes of all Darwin's finches could be collapsed under affine transformations showed that the entire diversity of their beak shapes could be reduced to only three "group shapes" (Campàs et al. 2010). Second, candidate developmental mechanisms, such as particular transcription factors or signaling molecule transduction pathway members, can be uncovered using a comparative genomic, genetic, and/or molecular screens, such as the QTL mapping of the morphological traits of interest, RNA transcript sequencing (RNA-Seq), or analysis of candidate genes using immunohistochemistry (cross-reacting antibodies or species-specific in situ hybridization probes). In this way, strong correlations can be found between the changes in expression of certain molecules and alterations in morphology (Mallarino and Abzhanov 2012). For example, both candidate gene and microarray approaches were very successful in finding developmental pathways associated with the highly specialized beak shapes in *Geospiza*, such as those in ground and cactus finches (Abzhanov et al. 2004, 2006; Mallarino et al. 2011). These associations between molecules and morphology can be directly tested with functional assays, which are aimed to show the clear causative relationship (Abzhanov et al. 2004, 2006; Mallarino and Abzhanov 2012). The potential roles for *Bmp4*, *TGFβIIr*, *β-catenin*, and *Dickkopf-3* (*Dkk3*) as well as *CaM* in beak development were tested with infections by retroviruses carrying these genes or their functionally altered versions. These experiments showed that mimicking the heterochronic and heterotopic changes in these genes' expression patterns could reproduce the morphological modifications that occurred during evolution, such as deep and broad bills in ground finches and narrow, elongated bills in cactus finches (figs. 18.4, 18.5; Abzhanov et al. 2004, 2006). Similar functional experiments could, in principle, be performed on Hox genes in both the trunks and limbs of crocodylians and birds to test the sufficiency of detected expression changes on their morphology.

Archosaurs have a spectacular quarter-billion year record of evolution, which is continued today by two extant and thriving clades, crocodylians and birds. Both developmental and genetic mechanisms underlying archosaur morphological transitions can and should be studied using a combination of morphometric, molecular, and functional analyses using paleontological, embryonic, juvenile, and adult material. Such comprehensive inspection is likely to produce fascinating new insights into how Archosauria came to rule the Earth.

Acknowledgments

I thank Bhart-Anjan Bhullar, Drs. James Hanken, Clifford Tabin, Jonathan Losos, Kenneth Dial, Elizabeth Brainerd, and Neil Shubin for their deeply insightful and tremendously useful comments on the manuscript. A.A. is supported by NIH 1R21DE021535-01, NSF 1257122, and Templeton Foundation Grant RFP-12-01.

References

Abzhanov, A., W. P. Kuo, C. Hartmann, B. R. Grant, P. R. Grant, and C. J. Tabin. 2006. The Calmodulin pathway and the evolution of elongated beak morphology in Darwin's finches. Nature 442:563–567.

Abzhanov, A., M. Protas, B. R. Grant, P. R. Grant, and C. J. Tabin. 2004. *Bmp4* and morphological variation of beaks in Darwin's finches. Science 305:1462–1465.

Aoyama, H., S. Mizutani-Koseki, and H. Koseki. 2005. Three developmental compartments involved in rib formation. Int. J. Dev. Biol. 49:325–333.

Bellairs, R. 2005. Atlas of Chick Development, Second Edition. San Diego, CA: Academic Press.

Benton, M. 1985. Classification and phylogeny of the diapsid reptiles. Zoological Journal of the Linnean Society 84:97–165.

Benton, M. J. 2004. Origin and relationships of Dinosauria. Pp. 7–24 in D. B. Weishampel, P. Dodson,. and H. Osmolska, eds., The Dinosauria. 2nd ed. Berkeley: University of California Press.

Benton, M. J. 2005. Vertebrate Palaeontology. 3rd ed. Oxford: Blackwell.

Berman, D. S., A. C. Henrici, R. A. Kissel, S. S. Sumida, and T. Martens. 2004. A new diadectid (Diadectomorpha), *Orobates pabsti*, from the Early Permian of Central Germany. Bull. Carnegie Mus. Nat.Hist. 35:1–36.

Bever G. S., J. A. Gauthier, and G. P. Wagner. 2011. Finding the frame shift: digit loss, developmental variability, and the origin of the avian hand. Evolution & Development 13:269–279.

Bhullar, B., J. Marugán-Lobón, F. Racimo, G. S. Bever, M. A. Norell, T. B. Rowe, and A. Abzhanov. 2012. Birds have paedomorphic dinosaur skulls. Nature *487:223–226.*

Borsuk-Białynicka, M., and S. E. Evans. 2009. A long-necked archosauromorph from the Early Triassic of Poland. Palaeontologia Polonica 65:203–234.

Borsuk-Białynicka, M., and A. G. Sennikov. 2009. Archosauriform postcranial remains from the Early Triassic karst deposits of southern Poland. Palaeontologica Polonica 65:283–328.

Bowman, R. I. 1961. Morphological Differentiation and Adaptation in the Galápagos Finches. Berkeley: University of California Press.

Broom, R. 1946. A new primitive proterosuchid reptile. Annals of the Transvaal Museum 20: 343–346.

Brusatte, S. L., M. J. Benton, J. B. Desojo, and M. C. Langer. 2010. The higher-level phylogeny of Archosauria (Tetrapoda: Diapsida). Journal of Systematic Palaeontology 8:3–47.

Brusatte, S. L., S. L. Brusatte, M. J. Benton, G. T. Lloyd, M. Ruta, and S. C. Wang. 2011. Macroevolutionary patterns in the evolutionary radiation of archosaurs (Tetrapoda: Diapsida). Earth and Environmental Science Transactions of the Royal Society of Edinburgh 101:367–382.

Burke, A. C., C. E. Nelson, B. A. Morgan, and C. J. Tabin. 1995. Hox genes and the evolution of vertebrate axial morphology. Development 121:333–346.

Campàs, O., R. Mallarino, A. Herrel, A. Abzhanov, and M. P. Brenner. 2010. Scaling and shear transformations capture beak shape variation in Darwin's finches. Proceedings of the National Academy of Sciences 107:3356–3360.

Carapuco, M., A. Nóvoa, N. Bobola, and M. Mallo. 2005. Hox genes specify vertebral types in the presomitic mesoderm. Genes Dev. 19:2116–2121.

Carroll, S. B. 2008. Evo-devo and an expanding evolutionary synthesis: a genetic theory of morphological evolution. Cell 134:25–36.

Colbert, E. H. 1989. The Triassic dinosaur *Coelophysis*. Bulletin of the Museum of Northern Arizona 57:1–160.

Crawford, R. D. 1990. Poultry Breeding and Genetics. San Diego: Elsevier Academic Press.

Cruickshank, A. R. I. 1972. The protesuchian thecodonts. In K. A. Joysey and T. S. Kemp, eds., Studies in Vertebrate Evolution. Edinburgh: Oliver and Boyd.

Duncan, R. P., A. G. Boyer, and T. M. Blackburn. 2012. Magnitude and variation of prehistoric bird extinctions in the Pacific. Proceedings of the National Academy of Sciences USA 110:6436–6441.

Durland, J. L., M. Sferlazzo, M. Logan, and A. C. Burke. 2008. Visualizing the lateral somitic frontier in the Prx1Cre transgenic mouse. J. Anat. 212:590–602.

Erickson, G. M., et al. 2009. Was dinosaurian physiology inherited by birds? reconciling slow growth in Archaeopteryx. PLoS ONE 4, e7390.

Ewer, R. F. 1965. The anatomy of the thecodont reptile *Euparkeria capensis* Broom. Phil. Trans. R. Soc. Lond. B 751:379–435.

Ferguson, M. W. J. 1985. Reproductive biology and embryology of the crocodilians. Pp. 330–491 in A. C. Gans, F. S. Billet, and P. Maderson, eds., Biology of Reptilia: Volume 14, Development Wiley and Sons, New York.

Gatesy, S. M., and K. P. Dial. 1996. From frond to fan: *Archaeopteryx* and the evolution of short-tailed birds. Evolution 50:2037–2048.

Gauthier, J., and K. de Queiroz. 2001. Feathered dinosaurs, flying dinosaurs, crown dinosaurs, and the name "Aves." Pp. 7–41 in J. Gauthier and L. F. Gall, eds., New Perspectives on the Origin and Early Evolution of Birds: Proceedings of the International Symposium in Honor of John H. Ostrom. New Haven: Peabody Museum of Natural History, Yale University.

Gauthier, J., A. Kluge, and T. Rowe. 1988. Amniote phylogeny and the importance of fossils. Cladistics 4:105–209.

Gilbert, S. F. 2010. Developmental Biology. 9th ed. Sunderland: Sinauer Associates.

Gill, F. B. 2007. Ornithology. London: W. H. Freeman and Company.

Goodrich, E. S. 1930. Studies on the structure and development of Vertebrates. London: Macmillan.

Gow, C. E. 1975. A new heterodontosaurid from the Redbeds of South Africa showing clear evidence of tooth replacement. Zoological Journal of the Linnean Society 57:335-339.

Gower, D. J., and A. G. Sennikov. 2003. Early archosaurs from Russia. Pp. 140-159 in M. J. Benton, M. A. Shishkin, and D. M. Unwin, eds., The Age of Dinosaurs in Russia and Mongolia. Cambridge: Cambridge University Press.

Grant, P. R. 1986. Ecology and Evolution of Darwin's Finches. Princeton: Princeton University Press.

Grant, P. R., B. R. Grant, and A. Abzhanov. 2006. A developing paradigm for the development of bird beaks. Biological Journal of the Linnean Society 88:17-23.

Grant, P. R., and B. R. Grant. 2008. How and Why Species Multiply: The Radiation of Darwin's Finches. Princeton: Princeton University Press.

Harpole, T. 2005. Falling with the falcon. Smithsonian Air & Space Magazine, March 1.

Henderson, D. M. 2010. Pterosaur body mass estimates from three-dimensional mathematical slicing. Journal of Vertebrate Paleontology 30:768-785.

Hofstetter, R., and J. P. Gasc. Vertebrae and ribs of modern reptiles. In C. Gans, A.d'A Bellairs, T. S. Parsons, eds., Biology of the Reptilia. Morphology A. Vol. 1. London: Academic Press.

Jeannotte, L., M. Lemieux, J. Charron, F. Poirier, and E. J. Robertson. 1993. Specification of axial identity in the mouse: role of the Hoxa-5 (Hox1-3) gene. Gene Dev. 7:2085-2096.

Kessel M., and P. Gruss. 1991. Homeotic transformations of murine vertebrae and concomitant alteration of Hox codes induced by retinoic acid. Cell 67:89-104.

Kieny, M., A. Mauger, and P. Sengel. 1972. Early regionalization of somitic mesoderm as studied by the development of axial skeleton of the chick embryo. Dev Biol. 28:142-161.

Kofron, P., and C. Steiner. 1994. Observations on the African dwarf crocodile, Osteolaemus tetraspis. Copeia 2:533-535.

Krumlauf, R. 1994. Hox genes in vertebrate development. Cell 78:191-201.

Kumazawa, Y., and M. Nishida. 1999. Complete mitochondrial DNA sequences of the green turtle and blue-tailed mole skink: statistical evidence for Archosaurian affinity of turtles. Molecular Biology and Evolution 16:784-792.

Le Mouellic, H., Y. Lallemand, and P. Brulet. 1992. Homeosis in the mouse induced by a null mutation in the Hox-3.1 gene. Cell 69:251-264.

Lee, A. H., and S. Werning. 2008. Sexual maturity in growing dinosaurs does not fit reptilian growth models. Proceedings of the National Academy of Sciences 105:582-587.

Longrich, N. R., and S. L. Olson 2011. The bizarre wing of the Jamaican flightless ibis Xenicibis xympithecus: a unique vertebrate adaptation. Proceedings of the Royal Society B 278:2333-2337.

Lovelace, D. M., S. A. Hartman, and W. R. Wahl. 2007. Morphology of a specimen of Supersaurus (Dinosauria, Sauropoda) from the Morrison Formation of Wyoming, and a re-evaluation of diplodocid phylogeny. Arquivos do Museu Nacional 65:527-544.

Mallarino, R. M., and A. Abzhanov. 2012. Paths less travelled: evo-devo approaches to investigating animal morphological evolution. Annual Review in Cell and Developmental Biology 28:743-763.

Mallarino, R. M., A. Herrel, W. P. Kuo, B. R. Grant, P. R. Grant, and A. Abzhanov. 2011. Two developmental modules establish 3D beak-shape variation in Darwin's finches. Proceedings of the National Academy of Sciences 108:4057-4062.

Mansfield, J. H., and A. Abzhanov. 2010. Hox expression in the American alligator and evolution of archosaurian axial patterning. Journal of Experimental Zoology Part B: Molecular and Developmental Evolution 314:629-644.

Mazzetta, G. V., P. Christiansen, and R. A. Farina. 2004. Giants and bizarres: body size of some southern South American Cretaceous dinosaurs. Historical Biology 16:71-83.

Modesto, S. P., and H.-D. Sues. 2002. The skull of the Early Triassic archosauromorph reptile Prolacerta broomi and its phylogenetic significance. Zoological Journal of the Linnean Society 140:335-351.

Nesbitt, S. J. 2011. The early evolution of archosaurs: relationships and the origin of major clades. Bulletin of the American Museum of Natural History 352:1-292.

Nowicki, J. L., and A. C. Burke. 2000. Hox genes and morphological identity: axial versus lateral patterning in the vertebrate mesoderm. Development 127:4265-4275.

Nowicki, J. L., R. Takimoto, and A. C. Burke. 2003. The lateral somitic frontier: dorso-ventral aspects of anterio-posterior regionalization in avian embryos. Mech. Dev. 120:227-240.

Organ, C. L. 2006. Thoracic epaxial muscles in living archosaurs and ornithopod dinosaurs. Anat. Rec. A Discov. Mol. Cell Evol. Biol. 288: 782-793.

Ohya, Y., S. Kuraku, and S. Kuratani. 2005. Hox code in embryos of Chinese soft-shelled turtle Pelodiscus sinensis correlates with the evolutionary innovation in the turtle. J. Exp. Zool. (Mol. Dev. Evol.) 304B:107-118.

Parrington, F. R. 1935. On Prolacerta broomi gen. et sp. nov. and the origin of lizards. Annals and Magazine of Natural History 16:197-205.

Paul, G. S. 1997. Dinosaur models: the good, the bad, and using them to estimate the mass of dinosaurs. Pp. 129-154 in D. L. Wolberg, E. Stump, G. D. Rosenberg, eds., DinoFest International Proceedings. Philadelphia: Academy of Natural Sciences.

Piper, R. 2007. Extraordinary Animals: An Encyclopedia of Curious and Unusual Animals. Greenwood Press.

Pollock, R. A., T. Sreenath, L. Ngo, and C. J. Bieberich. 1995. Gain of function mutations for paralogous Hox genes—implications for the evolution of Hox gene function. P. Natl. Acad. Sci. USA 92: 4492-4496.

Sennikov, A. G. 1990. New data on the rauisuchids of Eastern Europe. Paleontological Journal 3:3-16.

Sennikov, A. G. 1999. The evolution of the postcranial skeleton in archosaurs in connection with new finds of the Rauisuchidae in the Early Triassic of Russia. Paleontological Journal 6:44-56.

Senter, P. 2007. A new look at the phylogeny of Coelurosauria (Dinosauria: Theropoda). Journal of Systematic Palaeontology 5:429-463.

Sereno, P. C. 1991. Basal archosaurs: phylogenetic relationships and functional implications. Memoir (Society of Vertebrate Paleontology) 2:1-5.

Tabaries, S., M. Lemieux, J. Aubin, and L. Jeannotte. 2007. Comparative analysis of Hoxa5 allelic series. Genesis 45:218–228.

von Baer, K. E. 1860. Über ein allgemeines Gesetz in der Gestaltung der Flußbetten. Kaspische Studien 8, S., 1–6.

Wang, S. C., and P. Dodson. 2006. Estimating the diversity of dinosaurs. PNAS 103:13601–13605.

Wang, Z., R. L. Young, H. Xue, and G. P. Wagner. 2011. Transcriptomic analysis of avian digits reveals conserved and derived digit identities in birds. Nature 477:583–586.

Yueh, Y. G., D. P. Gardner, and C. Kappen. 1998. Evidence for regulation of cartilage differentiation by the homeobox gene Hoxc-8. Proc. Natl. Acad. Sci. USA. 95:9956–9961.

Zardoya, R., and A. Meyer. 1998. Complete mitochondrial genome suggests diapsid affinities of turtles. Proceedings of the National Academy of Sciences USA 95:14226–14231.

19

Microevolution and the Genetic Basis of Vertebrate Diversity:
Examples from Teleost Fishes

Sydney A. Stringham* and Michael D. Shapiro†

If it could be demonstrated that any complex organ existed, which could not possibly have been formed by numerous, successive, slight modifications, my theory would absolutely break down.
—CHARLES DARWIN (1859)

Nature does make jumps now and then, and a recognition of the fact is of no small importance in disposing of many minor objections to the doctrine of transmutation.
—THOMAS HUXLEY (1860)

Great transformations among the vertebrates can only be appreciated and understood by elucidating the micro-transformational mechanisms responsible for form and function. However, when studying major transformations that occurred many millions of years ago, we have limited access to the molecular mechanisms underlying these changes. For example, evolutionary biologists can only dream of using controlled genetic crosses between birds and non-avian theropod dinosaurs to map the key genetic changes in the evolution of flight, or crossing a fish and a tetrapod to identify the genes that matter in fin versus limb development and function. Even among extant vertebrates, anatomically divergent species are typically too distantly related to allow traditional genetic approaches, which require the production of fertile offspring. Moreover, although the complete sequences of many vertebrate genomes are now available, determining which of the millions of DNA

* Department of Biology, University of Utah
† Department of Biology, University of Utah

sequence and structural differences among species are actually responsible for particular trait differences remains a major challenge.

Organismal diversity, and morphological diversity in particular, is rooted in changes to developmental programs. That is, major anatomical changes among adults of different populations and species must manifest sometime between fertilization of an egg and sexual maturity. Developmental differences, in turn, are regulated largely (but by no means exclusively) by changes in genetic programs. Much of what we know about the molecular genetic basis of vertebrate development comes from mechanistic studies of traditional laboratory models such as the mouse, chicken, African claw-toed frog, and zebrafish. Despite major advances in our understanding of organismal construction from normal and mutant inbred laboratory populations, we know considerably less about the genetic and developmental basis of *natural variation* among vertebrates. Evolutionary developmental genetics (often referred to as "evo-devo") takes advantage of variation in the wild to directly address the link between genotype and phenotype among species, which will lead to a better understanding of the molecular origins of diversity.

In contrast to most other chapters in this volume, we focus on variation and transformations among populations and closely related species. This scale of investigation has the advantage of using traditional genetic approaches to understand vertebrate diversity, a strategy that typically is not available when studying major transformations among lineages with distant common ancestors. Fortunately, in a limited number of extant species, different populations have evolved anatomical, physiological, or behavioral changes of a magnitude that typically characterizes different species. Not many species meet this criterion, but the ones that do are emerging as important models in evolutionary genetics and developmental biology.

By understanding the genetic changes that underlie phenotypic changes in these special cases, we can begin to address central questions about the mechanisms underlying morphological transformations within and among species. For example, how many genetic changes underlie substantial morphological changes? Where do these changes occur, in the coding or regulatory regions of genes? Finally, do the same genetic changes underlie

the repeated evolution of similar traits in different populations and species?

We focus here on examples of particularly striking variation in teleost fishes. With nearly 29,000 extant species (Santini et al. 2009), teleosts are among the most successful radiations of vertebrates. In some cases, changes among populations *within* a species are so pronounced that they resemble in magnitude the differences *among* species. These cases of intraspecific variation in extant taxa are especially important to our understanding of the mechanisms that give rise to phenotypic transformations, and perhaps ultimately to new species and adaptive radiations. Within teleosts, we discuss examples of genetic mechanisms of diversification in sticklebacks, Mexican cavefish, and African cichlids. Each of these groups evolved dramatic—and repeated—phenotypic transformations in response to novel habitats, and each provides an ideal framework to examine the genetic basis of organismal diversity. These are not the only teleost groups in which the genetic basis of variation has been studied; however, the traits and transformations we highlight below introduce important themes and trends in the evolution of teleosts and other vertebrates.

Each of these groups of teleosts also offers important advantages as a model system in evolutionary genetics. First, different populations or closely related species within each group can be interbred to produce fertile offspring. This important characteristic facilitates traditional genetic mapping of traits of interest. Second, all three groups have been studied for many decades from the perspectives of ecology, natural history, and to a lesser extent, classical genetics and developmental biology. This foundation provides an important entry point to dissect the molecular genetic changes that control organismal diversity. Below, we consider micro-evolutionary transformations in each group, then discuss their impact on our understanding of broader trends of the genetic basis of vertebrate diversity.

Sticklebacks (Family Gasterosteidae)

Sticklebacks comprise seven species of small teleost fish that are widespread and often locally abundant across the Northern Hemisphere. A subset of these species exhibits tremendous intraspecific variation in

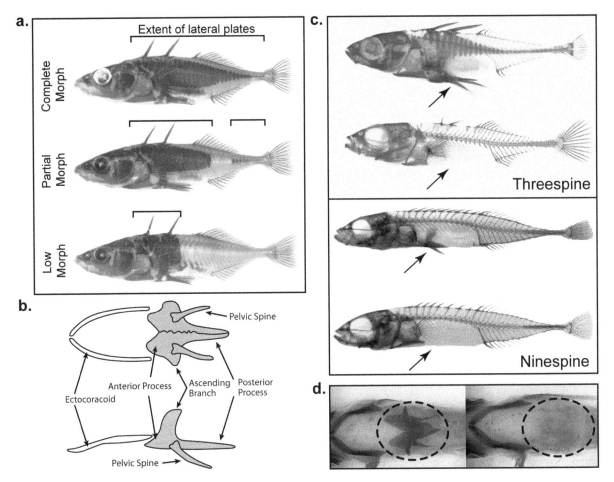

FIG. 19.1 (a) Variation in lateral plate number in marine threespine sticklebacks: complete morph (top), partial morph (center), and low morph (bottom). Bony structures in all panels were visualized by staining with alizarin red. Fish found in marine habitats nearly always possess 30 or more plates per side (a phenotype referred to as the "complete morph"). In freshwater, fish typically have less than 10 plates per side ("low morph"), or, less frequently, have an intermediate number of plates ("partial morph") (Hagen and Gilbertson 1972). Partial morphs exhibit a stereotypical pattern of plate loss, with plates at the most anterior and most posterior regions of the body and a mid-body gap in between. Images courtesy of Jun Kitano, modified after Kitano et al. (2008). (b) Ventral (top) and lateral (bottom) illustrations of the stickleback pelvis and ectocoracoid. The ectocoracoid is located anterior to the pelvis. (c) Pelvic loss has evolved in multiple populations of freshwater threespine sticklebacks (G. aculeatus) (top) and ninespine sticklebacks (P. pungitius) (bottom). In both species, the ancestral marine populations possess a complete pelvis; therefore, this trait has evolved independently in each species. (d) Ventral view of ninespine sticklebacks with a complete pelvis (left) and a missing pelvis (right).

skeletal morphology, body shape, color, behavior, and physiological adaptations. The most recent adaptive radiation of the threespine stickleback began with the retreat of glacial ice less than 20,000 years ago (Bernatchez and Wilson 1998; Hewitt 2000). This retreat created new inland freshwater habitats, which were subsequently colonized by marine stickleback populations. The transition to resident freshwater environments presented novel trophic, predatory, and physiological challenges. For example, freshwater habitats vary dramatically from marine habitats in temperature, topological complexity, water chemistry, and predator loads (Heuts 1947; Hagen and Gilbertson 1973b; Moodie et al. 1973; Hagen and Moodie 1982; Coad 1983; Giles 1983; Reimchen 1992, 1995; Kitano et al. 2008).

Geographically and phylogenetically distant populations of threespine sticklebacks have evolved strikingly similar suites of characteristics in response to the shift to freshwater habitats. For example, many populations have lost major components of their bony armor, including the lateral plates and pelvic girdle, in response to new predator loads and other factors (Bell and Foster 1994) (fig. 19.1). Furthermore, parallel phenotypic changes occur not only among populations of threespine sticklebacks (*Gasterosteus aculeatus*), the focus of most recent genetic and genomic studies, but also across species that diverged millions of years ago (e.g., the ninespine stickleback *Pungitius pungitius*, and the brook stickleback *Culaea inconstans*) (Nelson and Atton 1971; Wootton 1976; Blouw and Boyd 1992; Bell and Foster 1994; Ziuganov and Zotin 1995). Thus, this multispecies system provides an excellent model to examine the genetics of adaptive traits on both micro- and macroevolutionary levels.

Armor Plate Variation

Armor plates are composed of thin dermal bone and almost completely cover the lateral sides of marine threespine sticklebacks ("complete morph"; fig. 19.1a, top). In contrast, the number and size of these plates is reduced in most freshwater populations ("low morph"; fig. 19.1a, bottom) in response to strong selection in freshwater habitats (discussed below), and the genetic basis of this variation has been the subject of classical genetic studies for decades (Hagen and Gilbertson 1973a; Avise 1976; Ziuganov 1983; Banbura 1994). Laboratory crosses between different morphs showed that probably only a few genes control most of the variation in plate number (Hagen and Gilbertson 1973a; Avise 1976; Ziuganov 1983; Banbura 1994).

More recently, Colosimo et al. (2004) used a molecular genetic approach to identify the major locus controlling plate reduction. To do this, they crossed a complete-morph marine fish (Hokkaido Island, Japan) to a low-morph freshwater fish (Paxton Lake, British Columbia); the grandchildren (F_2 progeny) of this cross showed a wide range of plate morphologies, including fish that had high or low numbers of plates like their grandparents. By looking for associations between plate phenotypes and segments of chromosomes inherited from either the complete- or low-morph grandparent, Colosimo et al. (2004) found a single position in the genome (a quantitative trait locus, or QTL) on linkage group (LG) 4 that largely determined whether fish had the complete, partial, or low-plate morph (see fig. 19.2). Other studies suggested that LG4 controls plate phenotypes in multiple populations of threespine sticklebacks (Cresko et al. 2004; Schluter et al. 2004). However, key questions remained: which gene(s) in the major QTL region controlled armor variation, and were the muta-

tions the same or different among the many populations with low plates?

Further genetic mapping studies showed that variation in the gene *Ectodysplasin* (*Eda*) was the most likely cause of armor diversity (Colosimo et al. 2005). In vertebrates, *Eda* plays a key role in the development of several tissues derived from the ectoderm, including hair, teeth, sweat glands, and scales (Thesleff and Mikkola 2002; Kangas et al. 2004; Harris et al. 2008). The external armor of sticklebacks is also derived from ectoderm. Importantly, Colosimo et al. (2005) showed that, by injecting low-plated embryos with an engineered DNA construct containing a functional version of *Eda*, they could partially restore plate formation in low-plated fish. This provided functional evidence that *Eda* plays a critical role in plate development.

Strikingly, nearly every low-plated population throughout the range of the species appears to have the *same* chromosome segment containing the *Eda* gene (Colosimo et al. 2005). This indicates that the repeated evolution of low plates probably resulted from selection on the same mutant version of *Eda*, rather than by independent mutations in *Eda* in each population. The key to the spread of the low-plate allele resides in the marine populations that colonize new freshwater habitats: the low-plate version of *Eda* typically found in freshwater populations is also found in a small proportion of marine fish, suggesting that high-plated ocean populations are a "genetic reservoir" for the low-plate allele (Colosimo et al. 2005). Once the allele enters a freshwater habitat with the arrival of new marine colonists, selection drives it to high frequency. Transition from high to low plates can happen very quickly. In one Alaskan lake population, for example, Bell et al. (2004) observed a dramatic shift from predominantly high-plated to low-plated in less than 12 years (also see Kitano et al. 2008). Paradoxically, while the genetic basis for this trait is well understood and there is strong evidence for selection on plate phenotypes and the *Eda* locus, the ecological mechanism driving selection is less clear (reviewed in Barrett 2010).

Reduction and Loss of the Pelvic Fin Complex

In addition to variation in lateral armor, at least 20 freshwater populations of threespine stickleback also exhibit reduction or loss of the pelvis (Bell 1974; Moodie

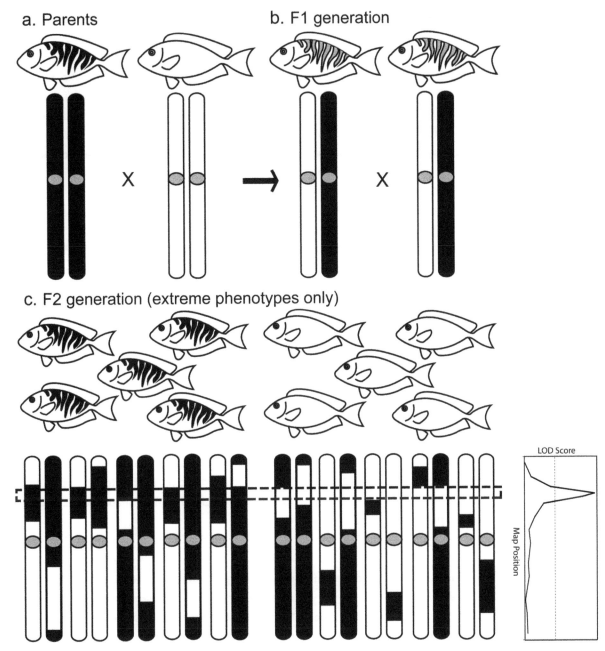

FIG. 19.2 Schematic of quantitative trait locus (QTL) mapping in a laboratory cross. (a) Individuals or populations that show variation in a trait of interest (in this case stripes) are crossed to produce F_1 offspring (b), which exhibit a phenotype intermediate to the phenotypes of the parental generation. F_1 individuals are crossed to produce an F_2 generation (c), which will now show segregation of the trait of interest if the number of genes controlling the trait is small. In this case, only the phenotypic extremes (dark stripes or no stripes, but not intermediate stripes) are shown. The genomes of the F_2 individuals are then analyzed with a set of genomic markers to detect statistical associations between genotypes and phenotypes. These associations define QTL, which are chromosome regions that are linked to phenotypes of interest. The identity of the specific genes that underlie phenotypic variation might not immediately be known because QTL associations often span many genes. Chromosome segments inherited from the striped and unstriped founders of the cross are indicated by black and white, respectively, and only the chromosome containing the causative mutation is depicted here. If individuals that inherit one version of the chromosome segment (black) nearly always exhibit one phenotype (dark stripes) and individuals inheriting the alternative version (white) nearly always exhibit the alternative phenotype (no stripes), then that segment is probably linked (physically close on a chromosome) to the causative mutation. In this example, a dashed box indicates the chromosome region associated with the stripe trait. The different versions of the chromosomes can be detected using markers such as polymorphic microsatellite markers (short repeat sequences that often differ in length among individuals) and single nucleotide polymorphisms (SNPs). These markers are assembled into linkage groups, and relative marker positions are determined based on recombination rates. Ideally, each chromosome in the genome will be represented by a single linkage group, and together the groups comprise a linkage map. Likelihood of odds (LOD) scores provide a statistical test of associations between genotypes and traits. The LOD plot at bottom right shows a region of a chromosome that exceeds a significance threshold (dashed line) and is therefore associated with variation in the trait of interest.

and Reimchen 1976; Campbell and Williamson 1979; Edge and Coad 1983; Bell 1987). The stickleback pelvis is homologous to the pelvic fin skeleton of other teleosts as well as to the tetrapod hind limb. It is composed of a pelvic girdle and serrated pelvic spines that provide protection from gape-limited predators such as large piscivorous fish (Hoogland et al. 1957; Hagen and Gilbertson 1972; Moodie 1972; Gross 1978; Lescak and von Hippel 2011) (fig. 19.1b–d). However, reduction of pelvic structures is advantageous in some populations where grasping predators such as aquatic invertebrates are a greater threat, especially to juvenile fish (Hoogland et al. 1957; Reimchen 1980, 1983; Bell et al. 1993; Bell and Orti 1994; Bourgeois et al. 1994). Large pelvic skeletons could be disadvantageous in these habitats because spines provide an additional surface for insects to capture and hold their prey (Reimchen 1980; Reist 1980; Ziuganov and Zotin 1995; Marchinko 2009).

Using a QTL mapping approach similar to the armor plate study, Shapiro et al. (2004) identified the gene *Pitx1* as a major influence on pelvic morphology. *Pitx1* contributes to hind limb identity and development in vertebrates, and mice with an inactive form (knockout) of the gene exhibit reduced and malformed hind limbs but normal forelimbs (Lanctôt et al. 1999; Marcil et al. 2003). Furthermore, sticklebacks from the genetic mapping cross that retained pelvic spines showed a marked asymmetry with larger spines on the left side, a feature also seen in the limbs of mice with an inactive version of *Pitx1* and humans with a *Pitx1* mutation (Lanctôt et al. 1999; Gurnett et al. 2008).

Unlike in the mouse *Pitx1* knockout, mutations were not found in the coding region of *Pitx1* in pelvisless freshwater stickleback populations compared to marine fish (Shapiro et al. 2004). Consequently, the Pitx1 proteins encoded by the marine and freshwater populations were the same. However, the location of the gene's expression was drastically different between populations. As in other vertebrates, *Pitx1* was expressed in the developing pelvis of marine larvae. In contrast, expression was greatly reduced or absent in the pelvic region of freshwater stickleback larvae, yet other regions of normal expression, such as the jaws, were not affected (Shapiro et al. 2004; Shapiro, Marks, et al. 2006). Therefore, the change in *Pitx1* was predicted to affect a DNA sequence that regulates when and where the gene is expressed. Chan et al. (2010) confirmed this

hypothesis by finding DNA deletions near the *Pitx1* gene in several pelvic-reduced populations. When attached to the protein-coding sequence of *Pitx1* and injected into embryos from pelvisless sticklebacks, this regulatory region (also known as an enhancer) was capable of restoring pelvic development, thus verifying that the deletion was critical in the evolution and development of pelvic reduction. In contrast to repeated selection on the same low-plate version of *Eda*, Chan et al. detected *different* deletions near *Pitx1* in different populations, suggesting that pelvic reduction in threespine sticklebacks arose repeatedly by independent mutations in different populations.

A likely factor in the repeated involvement of the *Pitx1* regulatory element, as opposed to mutations in the coding sequence of the gene, is pleiotropy; that is, selection on one trait, such as pelvic reduction, has the potential to affect development of other traits controlled by the same gene. In mice, the pleiotropic effects of *Pitx1* mutations are especially pronounced: complete inactivation of the gene leads not only to hind limb anomalies, but also jaw and brain deformities (Lanctôt et al. 1999). In contrast, the pelvis-specific *regulatory* mutation in sticklebacks yields an adaptive phenotype that is specific to one trait, while leaving other developmental roles of *Pitx1* intact (Shapiro et al. 2004; Chan et al. 2010).

Pelvic reduction is not limited to a single species of stickleback. The ninespine stickleback (*Pungitius pungitius*) diverged from the threespine stickleback at least 10 million years ago, yet these two species have a similar history of postglacial freshwater colonization and repeated evolution of pelvic reduction (Aldenhoven et al. 2010). Based on studies of the ninespine stickleback from two localities (Canada and Finland), *Pitx1* appears to play a role in pelvic reduction in this species as well (Shapiro, Bell, and Kingsley 2006; Shikano et al. 2013). These results in extant, genetically tractable stickleback species might hold clues about mechanisms of pelvic reduction in other species as well. For example, the extensive fossil record of *Gasterosteus doryssus*, an extinct relative of the threespine stickleback, documents the repeated evolution of pelvic reduction in a Miocene population (Bell 1974b; Bell et al. 1985; Bell 1988). As in modern threespine sticklebacks, pelvic reduction in *G. doryssus* shows a pronounced left-side bias, a morphological signature of *Pitx1*-mediated changes (Shapiro et al. 2004; Shapiro, Bell, and Kingsley 2006). This morphological trend extends beyond sticklebacks, as pelvic remnants in manatees also show a left-side bias (Shapiro, Bell, and Kingsley 2006). The genetic basis of hind limb reduction in manatees is not known, but this shared morphological signature of *Pitx1*-mediated reduction provides clues about the molecular mechanisms involved. Together, these examples show that genetics in one species can potentially generate hypotheses for study in other, less genetically tractable species.

Pitx1 probably does not universally play a major role in pelvic reduction, however. In another population of ninespine sticklebacks (Point MacKenzie, Alaska), the major QTL for pelvic reduction is clearly not *Pitx1* (Shapiro et al. 2009). This result suggests that ninespine stickleback populations use both the same and different genetic mechanisms as threespine sticklebacks to converge on the same pelvic phenotype.

Body Shape Variation

Sticklebacks from a variety of habitats exhibit enormous variation in overall body shape. The ancestral marine form is generally large and streamlined with a deep body and head, long fins, and a narrow caudal region. These adaptations are thought to be optimal for navigating open water (Walker 1997; Walker and Bell 2000; Spoljaric and Reimchen 2007; Albert et al. 2008). Freshwater populations, particularly those that inhabit littoral regions and feed on macroinvertebrates, generally have bodies that are short and deep, with shorter fins and a wider caudal region, resulting in a more maneuverable body that is better suited to foraging and evading predators in a complex habitat (Webb 1982; Walker 1997; Walker and Bell 2000; Spoljaric and Reimchen 2007).

While many studies have highlighted recurring trends in body shape and their link to particular habitats, less is known about the genetic architecture of these changes (reviewed in Reid and Peichel 2010). To address this shortcoming, Albert et al. (2008) used a cross between marine and freshwater fish to conduct QTL mapping for body and head shape. Perhaps not surprisingly, they found that the genetic architecture of body shape is more complex than discrete traits such as plate variation and pelvic reduction. However, similar to discrete traits, the same genomic regions underlie similar body shape traits in different populations. For example, some of the same chromosome regions influence differences not only between marine and freshwater populations, but also between semi-isolated benthic and limnetic populations that occur within several lakes (Gow et al. 2006; Reid and Peichel 2010).

Collectively, these studies suggest that similar suites of shape changes are key transformations in adaptation to new freshwater habitats, and similar suites of genes might govern these repeated changes species-wide (also see Hohenlohe et al. 2010; Jones, Chan, et al. 2012; Jones, Grabherr, et al. 2012).

Summary

Molecular genetic studies of microevolutionary transformations in sticklebacks provide important insights into general trends underlying the molecular basis of a classic adaptive radiation. First, dramatic phenotypic changes such as pelvis and armor reduction can result largely from changes at a few genetic loci (e.g., *Pitx1* and *Eda*, respectively, plus a modest number of loci of small effect). Furthermore, repeated evolution of the same trait can result from repeated selection on a common ancestral chromosome segment (lateral armor evolution and *Eda*) or independent mutations in the same gene (pelvic evolution and *Pitx1*). However, comparisons across stickleback species suggest that these mechanisms are not necessarily universal. Other adaptive changes, such as body shape modifications that characterize populations in different habitats, have a more complex genetic architecture, yet still repeatedly involve a similar suite of genomic regions.

Mexican Cavefish (Family Characidae, *Astyanax mexicanus*)

Introduction

As with freshwater habitat specialization in sticklebacks, cave specialization has resulted in the repeated evolution of similar traits across diverse lineages of metazoans, including teleost fishes. Constructive traits that are common in cave-dwelling animals include increased numbers of taste buds, increased fat storage, larger egg size, and more sensitive nonvisual sensory systems (Culver 1982); regressive traits, such as loss of eyes and pigmentation, have evolved repeatedly across phyla as well.

The Mexican cavefish (*Astyanax mexicanus*) is an ideal model to study the genetic basis of cave phenotypes in vertebrates. Multiple populations within this species have converged on similar phenotypes, providing another opportunity to test whether the same or different genetic mechanisms underlie repeated morphological changes. At least 30 populations of *A. mexicanus* are distributed across northeastern Mexico (Hubbs and Innis 1936; Wilkens and Burns 1972; Mitchell et al. 1977; Espinasa et al. 2001), and phylogenetic analyses suggest that the cave form does not have a single evolutionary origin (Espinasa and Borowsky 2001; Dowling et al. 2002; Strecker et al. 2003; Strecker et al. 2004).

Pigmentation Variation

In the darkness of a cave environment, the usual roles of pigmentation (camouflage, mate selection, etc.) are no longer relevant and the loss of pigmentation has occurred in cave-dwelling species across phyla. However, the adaptive significance (if any) of this phenotype in cavefish and other cave animals is still unclear. Pigmentation variation in cavefish encompasses a number of distinct phenotypes, including complete albinism, pigmentation reduction, and decreased melanophore number, each with a distinct genetic architecture.

Albinism was long known to be controlled by a single major locus, and possibly the same gene in multiple populations (Sadoglu 1957; Sadoglu and McKee 1969; Wilkens 1988). More recently, QTL mapping in cavefish led to the discovery of a deletion in the *Oca2* gene that underlies albinism in the Pachón population (Protas et al. 2006) (fig. 19.3a–c). *Oca2* encodes a key protein in melanin synthesis, and mutations in this gene also cause albinism in both humans and mice (Rinchik et al. 1993; Yi et al. 2003). Albinism in a second cavefish population, Molino, is also due to a deletion in *Oca2*, but this deletion is distinct from the Pachón version and therefore must have arisen independently (Protas et al. 2006). Albinism in a third population, Japonés, probably results from a regulatory mutation in the same gene as no coding changes were identified (Protas et al. 2006). Hence, as with *Pitx1* and pelvic reduction in

sticklebacks, different mutations in the same gene led to similar phenotypes in different populations.

Another pigment-reduction phenotype, *brown* (characterized by brown instead of black eyes and reduced melanophore number), results from mutations in the *Melanocortin-1 receptor* (*Mc1r*) gene (fig. 19.3d–f). *Mc1r* encodes a receptor protein expressed in pigment-producing cells, and its activity can regulate melanin content and melanocyte dispersal in fish (Richardson et al. 2008; Tezuka et al. 2011). Like *Oca2* and albinism, the *brown* phenotype results from more than one mutation in different cavefish populations, although at least one of these mutations has probably spread to several populations (Gross et al. 2009).

Together, these examples of pigment variation illustrate that convergent phenotypes can occur by independent mutations in the same genes (similar to the repeated evolution of pelvic reduction in sticklebacks), and by selection on standing genetic variants (similar to repeated evolution of armor phenotypes in sticklebacks). In cavefish, independent deletions in the coding region of *Oca2*, as well as a possible regulatory mutation, have both been implicated in albinism. Likewise, independent mutations in *Mc1r* led to repeated evolution of the *brown* phenotype, perhaps by a combination of selection on mutant alleles that originated in the surface population, and new mutations in different cave populations (Gross et al. 2009).

Eye Loss

One of the most dramatic changes in cavefish compared to their surface-dwelling relatives is severe eye reduction (fig. 19.3a–c). During embryonic development in cavefish, eyes begin to form but eventually stall and degenerate, beginning with the lens (Cahn 1958; Yamamoto et al. 2004). However, transplanting a surface fish lens into a developing cavefish eye can halt degeneration, demonstrating that this structure is a critical signaling center in eye development (Jeffery and Martasian 1998; Yamamoto and Jeffery 2000; Strickler, Yamamoto, and Jeffery 2007).

Genetic and developmental experiments suggest that between 6 and 12 genes contribute to eye regression in cavefish (Wilkens 1988; Protas et al. 2007), and that the same genetic mechanisms do not underlie regression in all cave populations (Wilkens 1971; Wilkens

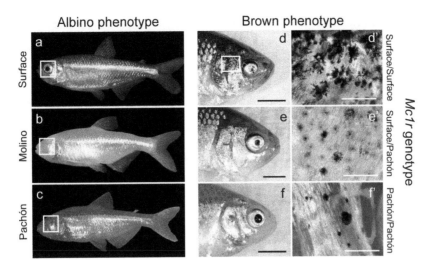

FIG. 19.3 Surface morph of *Astyanax mexicanus* (a) compared to cavefish populations from the Molino (b) and Pachón (c) populations. Each of these cave populations exhibits pigment loss mediated by *Oca2* and eye reduction (white boxes). In some populations, these changes have probably evolved independently. (d-f') The partially pigmented "brown" phenotype results from a decrease in melanin content and number of melanophores (pigment-containing cells). The severity of the phenotype depends on the number of cave alleles of *Mc1r* in an individual. In this example, two copies of the Pachón allele yield the most severe phenotype. Boxed area in (d) indicates area of magnification in (d'-f'). (a-c) Images courtesy of Richard Borowsky; (d-f') images courtesy of Josh Gross, modified after Gross et al. (2009).

and Strecker 2003; Borowsky 2008). This complex trait probably entails genetic pathways that control cell death and proliferation (Protas et al. 2007; Strickler, Byerly, and Jeffery 2007; Gross et al. 2008), response to environmental stress (Hooven et al. 2004), photoreceptor development (Kozmik 2008; Strickler and Jeffery 2009), and morphogenesis (Jeffery and Martasian 1998; Yamamoto et al. 2004; Strickler and Jeffery 2009). In summary, eye degeneration in cavefish is probably not under simple genetic control. Although several specific genes have been shown to affect eye development in this species, no specific mutations have yet been identified that correlate with the eyeless phenotype in any cave population.

Selection, Neutral Mutation, and Pleiotropy

While it is intuitive to envision natural selection driving the acquisition of heightened sensory traits such as increased taste bud number and increased sensitivity to vibrations in a cave environment, the adaptive consequences of eye and pigment loss are less clear. Perhaps unnecessary structures in a dark environment, such as the eye, are a liability; for example, eyes could be targets for predators, injury, or infection (Poulson 1963; Poulson and White 1969; Culver 1982; Jeffery 2005).

Alternatively, neutral mutation could explain eye and pigment loss (Kimura and Ohta 1971; Culver 1982; Wilkens 1988). In a dark environment, otherwise deleterious mutations in pigment and eye developmental pathways might not be selected against, as long as they do not result in other disadvantageous phenotypes. Therefore, given sufficient time, pathways involved in eye and pigment development could accumulate enough mutations for the associated structures to be lost. Interestingly, in genetic crossing experiments, cave alleles tend only to contribute to decreases in eye size, consistent with selection on eye regression, while cave alleles contribute to both increases and decreases in number of melanophores, suggesting drift might play a central role in pigmentation traits (Protas et al. 2007).

The loss of eyes and pigmentation in cavefish might also result from pleiotropy. Genetic and experimental evidence suggest that eye reduction might be a secondary effect of selection on alleles that are advantageous in the cave environment for increased gustatory or mechanical sensitivity (Yamamoto et al. 2004, 2009; Yoshizawa et al. 2010, 2013; Borowsky 2013). For example, in hybrid crosses between cave and surface fish, the number of taste buds is inversely correlated with eye size (Yamamoto et al. 2009). A compelling example of this effect on the developmental level comes from the gene *Sonic hedgehog* (*Shh*), which is expressed in the oral-pharyngeal region and the developing taste buds of both cave and surface forms. When this gene is experimentally overexpressed in both forms, embryos develop wider jaws and more taste buds, as well as smaller eyes (Yamamoto et al. 2004, 2009).

Summary

As in sticklebacks, genetic dissection of derived traits in cavefish demonstrates that dramatic phenotypes can potentially fall under the control of a modest number of genomic regions of large effect. Furthermore, these studies also show that similar phenotypes can arise through independent mutations in the same genes: *Oca2* and *Mc1r* underlie pigmentation variation in several populations, but different populations carry different mutations. Derived pigmentation traits in cavefish can also result from either coding or regulatory mutations: at least one population of albino cavefish probably harbors a regulatory mutation in *Oca2*, while most other albino populations have coding changes that lead to a decrease or loss of function. Other phenotypes, such as eye loss, are genetically more complicated and are probably the result of changes in multiple genes.

Although great strides are being made to identify the genetic basis of derived traits, these data do not necessarily lead directly to an understanding of the adaptive significance of phenotypes. Both pigment and eye reduction might result from positive selection for these traits, neutral mutation, or pleiotropy as the result of selection on other, as yet unknown, adaptive phenotypes.

Cichlids (Family Cichlidae)

Background

Cichlids, a third example of a morphologically diverse and species-rich group of teleosts, inhabit lakes throughout Central and South America, Madagascar, India, and Africa. Several lakes throughout this range include classic examples of rapid adaptive radiations. Two especially notable cases occur in the African rift lakes, where more than 500 species in Lake Victoria and over 700 species in Lake Malawi arose within the last

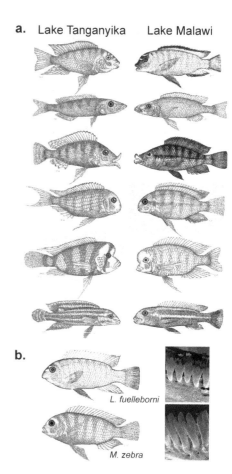

a. Lake Tanganyika   Lake Malawi

b.

L. fuelleborni

M. zebra

FIG. 19.4 (a) A sample of the cichlid diversity in Lake Tanganyika (left) and Lake Malawi (right), highlighting the convergent phenotypes that have evolved independently in these two lakes. (b) *Labeotropheus fuelleborni* (top left) feeds by biting algae from rock surfaces. This species has a shorter lower jaw and tricuspid teeth (top right). In contrast, *Metriaclima zebra* (bottom left) is a suction feeder with a long lower jaw and bicuspid dentition (bottom right). Images courtesy of Craig Albertson, modified after Albertson and Kocher (2006).

1 million years after multiple colonization events and hybridization (Banister and Clarke 1980; Meyer et al. 1990; Owen et al. 1990; Meyer 1993; Kocher et al. 1995; Turner et al. 2001; Joyce et al. 2011). Within a single lake, these species occupy habitats from shallow water to depths of over 100 meters. Different species also have diverse feeding strategies from generalist fish, zooplankton, and algae feeders to specialized crab, snail, and scale eaters (reviewed in Turner 2007). Furthermore, similar feeding strategies have arisen multiple times, providing another opportunity to examine the genetic basis of convergence in adaptively relevant phenotypes (Kocher et al. 1993) (fig. 19.4a). Like sticklebacks and cavefish,

genetic mapping of derived traits in cichlids is greatly facilitated by the ability of many distinct forms to interbreed and produce fertile offspring in a laboratory setting.

Feeding Morphology

Some of the best-studied adaptive traits in cichlids involve craniofacial structures. Different cichlid species have evolved to feed on an enormous variety of food types, and this diversification has produced a wide range of specialized head, jaw, and tooth morphologies (Albertson and Kocher 2006) (fig. 19.4). Genetic control of jaw and head morphology is highly complex and involves at least 40 chromosome regions, many of them affecting multiple elements of the feeding apparatus (Albertson and Kocher 2001; Albertson et al. 2003a, 2003b).

To reduce this complexity, Albertson et al. (2005) specifically examined functionally relevant aspects of jaw morphology in two divergent species. The first species, *Metriaclima zebra*, feeds on algae, diatoms, and plankton from the water column, and has a narrow, forward-directed mouth optimized for suction feeding (Ribbink et al. 1983). In contrast, the jaw of *Labeotropheus fuelleborni* is short and square with a downward orientation that allows it to bite algae from rocks while remaining horizontal (Ribbink et al. 1983). One QTL identified in the Albertson et al. study included *Bone Morphogenetic Protein 4* (*Bmp4*), a member of a large gene family that also regulates growth and differentiation during craniofacial development in other vertebrates (Abzhanov et al. 2004; Wu et al. 2004). At early developmental stages, the jaws of the suction-feeder *M. zebra* had much lower *Bmp4* expression than the biting-feeder *L. fuelleborni* (Albertson et al. 2005). Interestingly, when Albertson et al. overexpressed *Bmp4* in the embryos of zebrafish (suction-feeders, like *M. zebra*), the lower jaw shape shifted to a shape more suited for biting (like *L. fuelleborni*). Therefore, the results of experimental developmental studies in the zebrafish model system were consistent with genetic findings in wild cichlid species.

In another study using the same two species, Roberts et al. (2011) implicated the gene *Patched 1* (*Ptch1*)—a receptor in the hedgehog pathway that contributes to dermal bone development (Abzhanov et al. 2007)—in morphological differences in the lower jaw. Beyond *M. zebra* and *L. fuelleborni*, additional species-specific alleles of *Ptch1* were found in other cichlids with divergent feeding strategies, suggesting that this gene might affect jaw morphology in multiple lineages (Roberts et al. 2011).

Summary

The search for molecular changes that contribute to adaptive changes among cichlid species has thus far identified a small number of genes that contribute to diversity in feeding morphology, a key feature of this group's radiation. However, the genetic basis of variation in feeding structures is complex, with numerous chromosome regions contributing to differences in morphology. As with body shape variation in sticklebacks and eye reduction in cavefish, feeding morphology in cichlids involves several genomic regions that contribute to variation in multiple structures.

Discussion

Genetic Architecture of Derived Traits

The examples outlined above show that the genetic architecture of some major morphological changes can be relatively simple, with large effects produced by changes in only a few genes or genomic regions. Plate and pelvic reduction in sticklebacks, as well as albinism in cavefish, are largely controlled by single major genes. However, some derived traits have a more complex genetic architecture, including changes in stickleback body shape, variation in cichlid jaw morphology, and reduction of the cavefish eye. These contrasting degrees of complexity might represent different temporal stages of morphological transformations. Theoretical models of adaptation by new mutations (as opposed to selection on standing genetic variation) suggest that a small number of initial mutations lead to large fitness effects, so early adaptive stages can have a simple genetic architecture; subsequently, "modifier" mutations of smaller effect accumulate over time (Orr 1998; Orr 2002). By this model, several examples of genetically simple changes discussed above might reflect very recent transformations, while a more complex architecture could potentially reflect a longer period of trait evolution or selection on a large number of preexisting genetic variants.

We also note that, in all three teleost examples, several QTL regions control more than one trait. For instance, in sticklebacks, LG4 appears to be a "hotspot" of variation in body shape, lateral plates, and pelvic phenotypes (Colosimo et al. 2004; Shapiro et al. 2004; Albert et al. 2008; Shapiro et al. 2009). In cavefish, 13 genomic regions are known to influence multiple traits (Protas et al. 2008); these regions could contain multiple genes that affect a suite of traits beneficial to cave-dwellers, or single genes that have pleiotropic effects. Finally, in cichlids, LG5 influences tooth morphology, female sex determination, pigmentation, and also contains genes important for color perception (Carleton and Kocher 2001; Albertson et al. 2003a; Streelman et al. 2003; Kocher 2004; Streelman and Albertson 2006). This trend is by no means limited to loci that underlie diversity in fishes; the genetic clustering of QTL that control ecologically relevant traits could allow rapid evolutionary change through linkage of advantageous alleles in many different organisms (e.g., Garber and Quisenberry 1927; Mather 1950; Sheppard 1953; Murray and Clarke 1973; Joron et al. 2006; Joron et al. 2011).

Coding versus Regulatory Mutations

Among the teleost examples we discuss above, some of the genetic changes are (or are predicted to be) in noncoding regulatory regions of genes, while others directly affect protein-coding sequences, which in turn can affect protein function. This dichotomy, and relative contributions of each type of mutation to evolutionary change in general, has sparked considerable interest in the recent evolutionary genetics literature (e.g., Hoekstra and Coyne 2007; Wray 2007; Carroll 2008; Stern and Orgogozo 2008). While it is clear that not all evolutionary change results from *cis*-regulatory mutations, a number of hypotheses have been put forth to explain why these noncoding mutations might be a primary driver of evolutionary change, especially morphological change. One compelling argument centers on the modularity of regulatory regions (reviewed in Carroll 2008). Modularity refers to the semi-independent function of each *cis*-regulatory element with respect to other *cis*-regulatory elements. Therefore, a mutation in one of several regulatory regions of a gene can affect gene expression in only a subset of tissues or developmental time points, thereby avoiding potentially detrimental side effects on other developmental processes (pleiotropy). The potential importance of regulatory changes has been appreciated since the description of bacterial operons by Jacob and Monod (1961), and *cis*-regulatory changes are clearly important in morphological, physiological, and behavioral evolution (reviewed in Wray 2007).

An argument against the dominance of *cis*-regulatory changes in evolutionary change is that there are currently more confirmed examples of coding changes, but this could simply be because coding mutations are much easier to identify than regulatory mutations (reviewed in Stern and Orgogozo 2008). However, the pace of discovery (or implication) of *cis*-regulatory changes has recently begun to closely track the discovery of coding changes (Stern and Orgogozo 2008). In summary, both coding and regulatory mutations have the potential to contribute to significant evolutionary transformations, and ongoing work in fishes and other organisms will further elucidate general trends, if any exist.

Convergent Evolution

Teleosts exhibit repeated evolution of similar phenotypes among different populations within a species, and in some cases, between species. In many populations of threespine sticklebacks, lateral armor reduction evolved by repeated selection on a standing variant of the *Eda* locus. In contrast, other convergent evolutionary changes are the products of different mutations in the same genes. For example, different mutations in *Pitx1* underlie pelvic reduction in several populations of threespine sticklebacks, and *Oca2* and *Mc1r* mutations differ among cavefish populations with similar pigmentation phenotypes.

Comparisons *between* stickleback species also yield novel insights about convergent phenotypes. For example, pelvic reduction in at least two populations of ninespine sticklebacks probably results from changes to *Pitx1*, just as in threespine sticklebacks (Shapiro, Bell, and Kingsley 2006; Shikano et al. 2013). However, in another population of ninespine sticklebacks, pelvic reduction is controlled by a genomic region distinct from *Pitx1*; QTL for other skeletal traits (including lateral armor) and sex determination also differ between the two species (Shapiro et al. 2009). Therefore, a multispecies approach can be particularly informative in dissecting a broad range of genetic mechanisms underlying similar phenotypes.

Future Directions

Biologists are intensely interested in how vertebrates undergo transformations both great and small, yet we know remarkably little about the genetic basis of phenotypic change. In several examples above, QTL results were leveraged to fine-map and functionally test specific candidate genes for the evolution of derived traits. While these cases are exciting, it is important to note that they are also currently the exceptions—mapping traits to the gene level and demonstrating functional consequences of mutations is still uncommon.

Traits with a simple genetic architecture are easier to analyze than those with more genetic complexity, and many traits that have been examined in natural populations of teleosts and other organisms are ones that are relatively easy to see and quantify. Therefore, observable and relatively simple traits are preferentially studied, and we have a poorer understanding of complex anatomical, physiological, and behavioral traits that are undoubtedly important for evolutionary transformations (Rockman 2012).

New genomic tools, and the ability to compare dozens of genomes simultaneously, can help identify signatures of selection in suites of genes that affect traits that are not easily visualized. Recent studies, perhaps most notably in sticklebacks (Hohenlohe et al. 2010; Jones, Chan, et al. 2012; Jones, Grabherr, et al. 2012), have taken this "bottom-up" approach to identify genomic regions under selection in marine versus freshwater environments, as well as in benthic versus limnetic freshwater habitats. With precipitous drops in the cost of DNA sequencing and generation of new genetic resources, we expect that techniques pioneered for a limited number of species will become widely available to investigate important evolutionary transformations in other vertebrates as well.

* * *

Acknowledgments

We thank Eric Domyan, Josh Gross, Craig Miller, and the editors for comments on earlier drafts of this chapter. M. D. S. is forever grateful for mentorship by Farish Jenkins in the lab, field, and classroom. This work was supported by NIH training grant T32-GM007464 (S.A.S.), NSF grants IOS-0744974 and DEB-1149160 (M.D.S.), and the Burroughs Wellcome Fund (M.D.S.).

Glossary

Allele: Variant of a given gene or marker.

Coding mutation: A change in DNA sequence that occurs in a part of a gene that codes for a protein.

Genetic architecture: A general description of how traits are controlled by genotypes. For example, genetic architecture includes the number and location of genes that underlie a trait, as well as the number of alleles at these loci and the interactions among them.

Genetic marker: A DNA sequence that shows variability among individuals, and thus the inheritance of different alleles can be traced from one generation to the next. Examples include single nucleotide polymorphisms (SNPs) and microsatellites (simple DNA sequence repeats).

Genotype: The genetic makeup of an organism.

Linkage group: A group of genes or genetic markers that reside on the same chromosome. Genes or markers that are physically close to one another tend to be inherited together; as a result, markers can be ordered by tracking transmission from one generation to the next (also called genetic mapping). The sum of linkage groups comprises a linkage map.

Locus (plural: loci): The location of a gene or DNA sequence on a chromosome or linkage group.

Phenotype: The observable characteristics of an organism.

Pleiotropy: When one gene affects more than one trait or developmental process.

QTL (quantitative trait locus): A genomic region that contributes to variation in a trait. Quantitative traits are typically controlled by multiple loci.

QTL mapping: An experimental approach that often begins by crossing strains of organisms that differ in a trait or traits of interest. Molecular markers across the genome are used to track the co-inheritance of genotypes and phenotypes of offspring. Correlations between the trait(s) of interest and molecular markers are assessed (see fig. 19.2, and Miles and Wayne 2008).

Regulatory (*cis-*) mutation: A change in DNA sequence that affects a region controlling the level or location of expression of a gene, but (typically) does not affect the protein encoded by the gene (see also Wray 2007; Carroll 2008).

References

Abzhanov, A., M. Protas, B. R. Grant, P. R. Grant, and C. J. Tabin. 2004. Bmp4 and morphological variation of beaks in Darwin's finches. Science 305:1462–1465.

Abzhanov, A., S. J. Rodda, A. P. McMahon, and C. J. Tabin. 2007. Regulation of skeletogenic differentiation in cranial dermal bone. Development 134:3133–3144.

Albert, A. Y., S. Sawaya, T. H. Vines, A. K. Knecht, C. T. Miller, B. R. Summers, S. Balabhadra, D. M. Kingsley, and D. Schluter. 2008. The genetics of adaptive shape shift in stickleback: pleiotropy and effect size. Evolution 62:76–85.

Albertson, R. C., and T. D. Kocher. 2001. Assessing morphological differences in an adaptive trait: a landmark-based morphometric approach. J. Exp. Zool. 289:385–403.

Albertson, R. C., and T. D. Kocher. 2006. Genetic and developmental basis of cichlid trophic diversity. Heredity (Edinb.) 97:211–221.

Albertson, R. C., J. T. Streelman, and T. D. Kocher. 2003a. Directional selection has shaped the oral jaws of Lake Malawi cichlid fishes. Proc. Natl. Acad. Sci. USA 100:5252–5257.

Albertson, R. C., J. T. Streelman, and T. D. Kocher. 2003b. Genetic basis of adaptive shape differences in the cichlid head. J. Hered. 94:291–301.

Albertson, R. C., J. T. Streelman, T. D. Kocher, and P. C. Yelick. 2005. Integration and evolution of the cichlid mandible: the molecular basis of alternate feeding strategies. Proc. Natl. Acad. Sci. USA 102:16287–16292.

Aldenhoven, J. T., M. A. Miller, P. S. Corneli, and M. D. Shapiro. 2010. Phylogeography of ninespine sticklebacks (*Pungitius pungitius*) in North America: glacial refugia and the origins of adaptive traits. Mol. Eco.l 19:4061–4076.

Avise, J. C. 1976. Genetics of plate morphology in an unusual population of threespine sticklebacks (*Gasterostus aculeatus*). Genet. Res. 27:33–46.

Banbura, J. 1994. A new model of lateral plate morph inheritance in the threespine stickleback, *Gasterosteus aculeatus*. Theor. Appl. Genet. 88:871–876.

Banister, K. E., and M. A. Clarke. 1980. A revision of the large *Barbus* (Pisces, Cyprinidae) of Lake Malawi with a reconstruction of the history of the southern African Rift Valley lakes. J. Nat. Hist. 14:483.

Barrett, R. D. 2010. Adaptive evolution of lateral plates in three-spined stickleback *Gasterosteus aculeatus*: a case study in functional analysis of natural variation. J. Fish. Biol. 77:311–328.

Bell, A. M., G. Orti, J. A. Walker, and J. P. Koenings. 1993. Evolution of pelvic reduction in threespine stickleback fish—a test of competing hypotheses. Evolution 47:906–914.

Bell, M. A. 1974. Reduction and loss of the pelvic girdle in *Gasterosteus* (Pisces): a case of parallel evolution. Nat. Hist. Mus. L.A. Contrib. Sci. 257:1–36.

Bell, M. A. 1987. Interacting evolutionary constrains in pelvic reduction of threespine sticklebacks, *Gasterosteus aculeatus* (Pisces, Gasterosteidae). Biol. J. Linn. Soc. 31:347–382.

Bell, M. A. 1988. Stickleback fishes: bridging the gap between population biology and paleobiology. Trends Ecol. Evol. 3:320–325.

Bell, M. A., W. E. Aguirre, and N. J. Buck. 2004. Twelve years of contemporary armor evolution in a threespine stickleback population. Evolution 58:814–824.

Bell, M. A., J. V. Baumgartner, and E. C. Olson. 1985. Patterns of temporal change in single morphological characters of a Miocene stickleback fish. Paleobiology 11:258–271.

Bell, M. A., and S. A. Foster. 1994. The Evolutionary Biology of the Threespine Stickleback. Oxford: Oxford University Press.

Bell, M. A., and G. Orti. 1994. Pelvic reduction in threespine stickleback from Cook Inlet lakes: geographic distribution and intrapopulation variation. Copeia 1994:314–325.

Bernatchez, L., and C. C. Wilson. 1998. Comparative phylogeography of Nearctic and Palearctic fishes. Mol. Ecol. 7:431–452.

Blouw, D. M., and G. J. Boyd. 1992. Inheritance of reduction, loss, and asymmetry of the pelvis of *Pungitius pungitius* (ninespine stickleback). Heredity 68:33–42.

Borowsky, R. 2008. Restoring sight in blind cavefish. Curr. Biol. 18:R23–24.

Borowsky, R. 2013. Eye regression in blind *Astyanax* cavefish may facilitate the evolution of an adaptive behavior and its sensory receptors. BMC Biol. 11:81.

Bourgeois, J. F., D. M. Blouw, and M. A. Bell. 1994. Multivariate analysis of geographic covariance between phenotypes and environments in the threespine stickleback, *Gasterosteus aculeatus*. Can. J. Zool. 72:1497–1509.

Cahn, P. H. 1958. Comparative optic development in *Astyanax mexicanus* and two of its blind cave derivatives. Bull. Am. Mus. Nat. Hist. 115:75–112.

Campbell, R. N., and R. B. Williamson. 1979. The fishes of inland waters in the Outer Hebrides. Proc. R. Soc. Edinb. 77B:377–393.

Carleton, K. L., and T. D. Kocher. 2001. Cone opsin genes of african cichlid fishes: tuning spectral sensitivity by differential gene expression. Mol. Biol. Evol. 18:1540–1550.

Carroll, S. B. 2008. Evo-devo and an expanding evolutionary synthesis: a genetic theory of morphological evolution. Cell 134:25–36.

Chan, Y. F., M. E. Marks, F. C. Jones, G. Villarreal Jr., M. D. Shapiro, S. D. Brady, A. M. Southwick, D. M. Absher, J. Grimwood, J. Schmutz, R. M. Myers, D. Petrov, B. Jonsson, D. Schluter, M. A. Bell, and D. M. Kingsley. 2010. Adaptive evolution of pelvic reduction in sticklebacks by recurrent deletion of a Pitx1 enhancer. Science 327:302–305.

Coad, B. W. 1983. Plate morphs in freshwater samples of *Gasterosteus aculeatus* from Arctic and Atlantic Canada: complementary comments on a recent contribution. Can. J. Zool. 61:1174–1177.

Colosimo, P. F., K. E. Hosemann, S. Balabhadra, G. Villarreal Jr., M. Dickson, J. Grimwood, J. Schmutz, R. M. Myers, D. Schluter, and D. M. Kingsley. 2005. Widespread parallel evolution in sticklebacks by repeated fixation of Ectodysplasin alleles. Science 307:1928–1933.

Colosimo, P. F., C. L. Peichel, K. Nereng, B. K. Blackman, M. D. Shapiro, D. Schluter, and D. M. Kingsley. 2004. The genetic architecture of parallel armor plate reduction in threespine sticklebacks. PLoS Biol. 2:E109.

Cresko, W. A., A. Amores, C. Wilson, J. Murphy, M. Currey, P. Phillips, M. A. Bell, C. B. Kimmel, and J. H. Postlethwait. 2004. Parallel genetic basis for repeated evolution of armor loss in Alaskan threespine stickleback populations. Proc. Natl. Acad. Sci. USA 101:6050–6055.

Culver, D. C. 1982. Cave Life : Evolution and Ecology. Cambridge, MA: Harvard University Press.

Darwin, C. 1859. On the Origin of Species by Means of Natural Selection. London: John Murray.

Dowling, T. E., D. P. Martasian, and W. R. Jeffery. 2002. Evidence for multiple genetic forms with similar eyeless phenotypes in the blind cavefish, *Astyanax mexicanus*. Mol. Biol. Evol. 19:446–455.

Edge, T. A., and B. W. Coad. 1983. Reduction of the pelvic skeleton in the three-spined stickleback *Gasterosteus aculeatus* in 2 lakes of Quebec Canada. Can. Field-Nat. 97:334–336.

Espinasa, L., and R. B. Borowsky. 2001. Origin and relationships of cave populations of the blind Mexican tetra *Astyanax fasciatus*, in the Sierra de El Abra. Environ. Biol. Fishes 62:233–237.

Espinasa, L., P. Rivas-Manzano, and H. Espinosa Pérez. 2001. A new blind cave fish population of the genus *Astyanax*: geography, morphology and behavior. Environ. Biol. Fishes 62:329–344.

Garber, R. J., and K. S. Quisenberry. 1927. The inheritance of length of syle in buckwheat. J. Agric. Res. 34:181–183.

Giles, N. 1983. The possible role of environmental calcium levels during the evolution of phenotypic diversity in Outer-Hebridean populations of the three-spined stickleback, *Gasterosteus aculeatus*. J. Zool. 199:535.

Gow, J. L., C. L. Peichel, and E. B. Taylor. 2006. Contrasting hybridization rates between sympatric three-spined sticklebacks highlight the fragility of reproductive barriers between evolutionarily young species. Mol. Ecol. 15:739–752.

Gross, H. P. 1978. Natural selection by predators on the defensive apparatus of the three-spined stickleback, *Gasterosteus aculeatus* L. Can. J. Zool. 56:398–413.

Gross, J. B., R. Borowsky, and C. J. Tabin. 2009. A novel role for Mc1r in the parallel evolution of depigmentation in independent populations of the cavefish *Astyanax mexicanus*. PLoS Genet. 5:e1000326.

Gross, J. B., M. Protas, M. Conrad, P. E. Scheid, O. Vidal, W. R. Jeffery, R. Borowsky, and C. J. Tabin. 2008. Synteny and candidate gene prediction using an anchored linkage map of *Astyanax mexicanus*. Proc. Natl. Acad. Sci. USA 105:20106–20111.

Gurnett, C. A., F. Alaee, L. M. Kruse, D. M. Desruisseau, J. T. Hecht, C. A. Wise, A. M. Bowcock, and M. B. Dobbs. 2008. Asymmetric lower-limb malformations in individuals with homeobox PITX1 gene mutation. Am. J. Hum. Genet. 83:616–622.

Hagen, D. W., and L. G. Gilbertson. 1972. Geographic variation and environmental selection in *Gasterosteus aculeaus* L. in the Pacific northwest, America. Evolution 26:32–51.

Hagen, D. W., and L. G. Gilbertson. 1973a. The genetics of plate morphs in freshwater threespine sticklebacks. Heredity 31:75–84.

Hagen, D. W., and L. G. Gilbertson. 1973b. Selective predation and the intensity of selection acting upon the lateral plates of threespine stickleacks. Heredity 30:75–84.

Hagen, D. W., and G. E. E. Moodie. 1982. Polymorphism for plate morphs in *Gasterosteus aculeatus* on the east coast of Canada and an hypothesis for their global distribution. Can. J. Zool. 60:1032–1042.

Harris, M. P., N. Rohner, H. Schwarz, S. Perathoner, P. Konstantinidis, and C. Nusslein-Volhard. 2008. Zebrafish eda and edar mutants reveal conserved and ancestral roles of ectodysplasin signaling in vertebrates. PLoS Genet. 4:e1000206.

Heuts, M. C. 1947. The phenotypical variability of *Gasterosteus aculeatus* (L.) populations in Belgium; its bearing on the general geographical variability of the species. Antwerpen: Standarrd-boekhandel.

Hewitt, G. 2000. The genetic legacy of the Quaternary ice ages. Nature 405:907–913.

Hoekstra, H. E., and J. A. Coyne. 2007. The locus of evolution: evo devo and the genetics of adaptation. Evolution Int. J. Org. Evolution 61:995–1016.

Hohenlohe, P. A., S. Bassham, P. D. Etter, N. Stiffler, E. A. Johnson, and W. A. Cresko. 2010. Population genomics of parallel adaptation in threespine stickleback using sequenced RAD tags. PLoS Genet. 6:e1000862.

Hoogland, R. D., D. Morris, and N. Tinbergen. 1957. The spines of sticklebacks (*Gasterosteus* and *Pygosteus*) as means of defense against predators (*Perca* and *Esox*). Behaviour 10:205–230.

Hooven, T. A., Y. Yamamoto, and W. R. Jeffery. 2004. Blind cavefish and heat shock protein chaperones: a novel role for hsp90alpha in lens apoptosis. Int. J. Dev. Biol. 48:731–738.

Hubbs, C. L., and W. T. Innis. 1936. The first known blind fish of the family Characidae: a new genus from Mexico. Occas. Papers Mus. Zool. Univ. Michigan 342:1–7.

Huxley, T. H. H. 1860. The origin of species. Westminster Review 17:541–570.

Jacob, F., and J. Monod. 1961. Genetic regulatory mechanisms in the synthesis of proteins. J. Mol. Biol. 3:318–356.

Jeffery, W. R. 2005. Adaptive evolution of eye degeneration in the Mexican blind cavefish. J. Hered. 96:185–196.

Jeffery, W. R., and D. P. Martasian. 1998. Evolution of eye regression in the cavefish *Astyanax*: apoptosis and the *Pax-6* gene. Amer. Zool. 38:685–696.

Jones, F. C., Y. F. Chan, J. Schmutz, J. Grimwood, S. D. Brady, A. M. Southwick, D. M. Absher, R. M. Myers, T. E. Reimchen, B. E. Deagle, D. Schluter, and D. M. Kingsley. 2012. A genome-wide SNP genotyping array reveals patterns of global and repeated species-pair divergence in sticklebacks. Curr. Biol. 22:83–90.

Jones, F. C., M. G. Grabherr, Y. F. Chan, P. Russell, E. Mauceli, J. Johnson, R. Swofford, M. Pirun, M. C. Zody, S. White, et al. 2012. The genomic basis of adaptive evolution in threespine sticklebacks. Nature 484:55–61.

Joron, M., L. Frezal, R. T. Jones, N. L. Chamberlain, S. F. Lee, C. R. Haag, A. Whibley, M. Becuwe, S. W. Baxter, L. Ferguson, et al. 2011. Chromosomal rearrangements maintain a polymorphic supergene controlling butterfly mimicry. Nature 477:203–206.

Joron, M., R. Papa, M. Beltran, N. Chamberlain, J. Mavarez, S. Baxter, M. Abanto, E. Bermingham, S. J. Humphray, J. Rogers, et al. 2006. A conserved supergene locus controls colour pattern diversity in *Heliconius* butterflies. PLoS Biol. 4:e303.

Joyce, D. A., D. H. Lunt, M. J. Genner, G. F. Turner, R. Bills, and O. Seehausen. 2011. Repeated colonization and hybridization in Lake Malawi cichlids. Curr. Biol. 21:R108–109.

Kangas, A. T., A. R. Evans, I. Thesleff, and J. Jernvall. 2004. Nonindependence of mammalian dental characters. Nature 432:211–214.

Kimura, M., and T. Ohta. 1971. Theoretical Aspects of Population Genetics. Princeton, NJ: Princeton University Press.

Kitano, J., D. I. Bolnick, D. A. Beauchamp, M. M. Mazur, S. Mori, T. Nakano, and C. L. Peichel. 2008. Reverse evolution of armor plates in the threespine stickleback. Curr. Biol. 18:769–774.

Kocher, T. D. 2004. Adaptive evolution and explosive speciation: the cichlid fish model. Nat. Rev. Genet. 5:288–298.

Kocher, T. D., J. A. Conroy, K. R. McKaye, and J. R. Stauffer. 1993. Similar morphologies of cichlid fish in Lakes Tanganyika and Malawi are due to convergence. Mol. Phylogenet. Evol. 2:158–165.

Kocher, T. D., J. A. Conroy, K. R. McKaye, J. R. Stauffer, and S. F. Lockwood. 1995. Evolution of NADH dehydrogenase subunit 2 in east African cichlid fish. Mol. Phylogenet. Evol. 4:420–432.

Kozmik, Z. 2008. The role of Pax genes in eye evolution. Brain Res. Bull. 75:335–339.

Lanctôt, C., A. Moreau, M. Chamberland, M. L. Tremblay, and J. Drouin. 1999. Hindlimb patterning and mandible development require the Ptx1 gene. Development 126:1805–1810.

Lescak, E. A., and F. A. von Hippel. 2011. Selective predation of threespine stickleback by rainbow trout. Ecol. Freshwat. Fish. 20:308–314.

Marchinko, K. B. 2009. Predation's role in repeated phenotypic and genetic divergence of armor in threespine stickleback. Evolution 63:127–138.

Marcil, A., E. Dumontier, M. Chamberland, S. A. Camper, and J. Drouin. 2003. Pitx1 and Pitx2 are required for development of hindlimb buds. Development 130:45–55.

Mather, K. 1950. The genetical architecture of heterostyly in *Primula sinensis*. Evolution 4:340–352.

Meyer, A. 1993. Phylogenetic relationships and evolutionary processes in East African cichlid fishes. Trends Ecol. Evol. 8:279–284.

Meyer, A., T. D. Kocher, P. Basasibwaki, and A. C. Wilson. 1990. Monophyletic origin of Lake Victoria cichlid fishes suggested by mitochondrial DNA sequences. Nature 347:550–553.

Miles, C. M., and M. Wayne. 2008. Quantitative Trait Locus (QTL) Analysis. Nat. Educ. 1:1–8.

Mitchell, R. W., W. H. Russell, and W. R. Elliot. 1977. Mexican eyeless characin fishes, genus *Astyanax*: environment, distribution and evolution. Spec. Publ. Mus. Texas Tech Univ. 12:1–89.

Moodie, G. E. E. 1972. Predation, natural selection and adaptation in an unusual threespine stickleback. Heredity 28:155–167.

Moodie, G. E. E., J. D. McPhail, and D. W. Hagen. 1973. Experimental demonstration of selective predation on *Gasterosteus aculeatus*. Behaviour 47:95–105.

Moodie, G. E. E., and T. Reimchen. 1976. Phenetic variation and habitat differences in *Gasterosteus* populations of the Queen Charlotte Islands. Syst. Zool. 25:49–61.

Murray, J., and B. Clarke. 1973. Supergenes in polymorphic land snails—examples from genus *Partula*. Genetics 74:S188–S189.

Nelson, J. S., and F. M. Atton. 1971. Geographic and morphological variation in the presence and absence of the pelvic skeleton in the brook stickleback, *Culaea inconstans* (Kirtland), in Alberta and Saskatchewan. Can. J. Zool. 49:343–352.

Orr, H. A. 1998. The population genetics of adaptation: the distribution of factors fixed during adaptive evolution. Evolution 52:935–949.

Orr, H. A. 2002. The population genetics of adaptation: the adaptation of DNA sequences. Evolution 56:1317–1330.

Owen, R. B., R. Crossley, T. C. Johnson, D. Tweddle, I. Kornfield, S. Davison, D. H. Eccles, and D. E. Engstrom. 1990. Major low levels of Lake Malawi and their implications for speciation rates in cichlid fishes. Proc. R. Soc. London B 240:519–553.

Poulson, T. L. 1963. Cave adaptation in amblyopsid fishes. Am. Mid. Nat. 70:257–290.

Poulson, T. L., and W. B. White. 1969. The cave environment. Science 165:971–981.

Protas, M., M. Conrad, J. B. Gross, C. Tabin, and R. Borowsky. 2007. Regressive evolution in the Mexican cave tetra, *Astyanax mexicanus*. Curr. Biol. 17:452–454.

Protas, M., I. Tabansky, M. Conrad, J. B. Gross, O. Vidal, C. J. Tabin, and R. Borowsky. 2008. Multi-trait evolution in a cave fish, *Astyanax mexicanus*. Evol. Dev. 10:196–209.

Protas, M. E., C. Hersey, D. Kochanek, Y. Zhou, H. Wilkens, W. R. Jeffery, L. I. Zon, R. Borowsky, and C. J. Tabin. 2006. Genetic analysis of cavefish reveals molecular convergence in the evolution of albinism. Nat. Genet. 38:107–111.

Reid, D. T., and C. L. Peichel. 2010. Perspectives on the genetic architecture of divergence in body shape in sticklebacks. Integr. Comp. Biol. 50:1057–1066.

Reimchen, T. E. 1980. Spine deficiency and polymorphism in a population of *Gasterosteus aculeatus*—an adaptation to predators. Can. J. Zool. 58:1232–1244.

Reimchen, T. E. 1983. Structural relationships between spines and lateral plates in threespine stickleback (*Gasterosteus aculeatus*). Evolution 37:931–946.

Reimchen, T. E. 1992. Injuries on stickleback from attacks by a toothed predator (*Oncorhynchus*) and implications for the evolution of lateral plates. Evolution 46:1224.

Reimchen, T. E. 1995. Predator-induced cyclical changes in lateral plate frequencies of *Gasterosteus*. Behaviour 132:1079.

Reist, J. D. 1980. Predation upon pelvic phenotypes of brook stickleback, *Culaea inconstans*, by selected invertebrates. Can. J. Zool. 58:1253–1258.

Ribbink, A. J., A. C. Marsh, C. C. Ribbink, and B. J. Sharp. 1983. A preliminary survey of the cichlid fishes of rocky habitats in Lake Malawi. S. Afr. J. Zool. 18:149–310.

Richardson, J., P. R. Lundegaard, N. L. Reynolds, J. R. Dorin, D. J. Porteous, I. J. Jackson, and E. E. Patton. 2008. mc1r Pathway regulation of zebrafish melanosome dispersion. Zebrafish 5:289–295.

Rinchik, E. M., S. J. Bultman, B. Horsthemke, S. T. Lee, K. M. Strunk, R. A. Spritz, K. M. Avidano, M. T. Jong, and R. D. Nicholls. 1993. A gene for the mouse pink-eyed dilution locus and for human type II oculocutaneous albinism. Nature 361:72–76.

Roberts, R. B., Y. Hu, R. C. Albertson, and T. D. Kocher. 2011. Craniofacial divergence and ongoing adaptation via the hedgehog pathway. Proc. Natl. Acad. Sci. USA 108:13194–13199.

Rockman, M. V. 2012. The QTN program and the alleles that matter for evolution: all that's gold does not glitter. Evolution 66:1–17.

Sadoglu, P. 1957. A Mendelian gene for albinism in natural cave fish. Experientia 13:394.

Sadoglu, P., and A. McKee. 1969. A second gene that affects eye and body color in Mexican blind cavefish. J. Hered. 60:10–14.

Santini, F., L. J. Harmon, G. Carnevale, and M. E. Alfaro. 2009. Did genome duplication drive the origin of teleosts? A comparative study of diversification in ray-finned fishes. BMC Evol. Biol. 9:194.

Schluter, D., E. A. Clifford, M. Nemethy, and J. S. McKinnon. 2004. Parallel evolution and inheritance of quantitative traits. Am. Nat. 163:809–822.

Shapiro, M. D., M. A. Bell, and D. M. Kingsley. 2006. Parallel genetic origins of pelvic reduction in vertebrates. Proc. Natl. Acad. Sci. USA 103:13753–13758.

Shapiro, M. D., M. E. Marks, C. L. Peichel, B. K. Blackman, K. S. Nereng, B. Jonsson, D. Schluter, and D. M. Kingsley. 2004. Genetic and developmental basis of evolutionary pelvic reduction in threespine sticklebacks. Nature 428:717–723.

Shapiro, M. D., M. E. Marks, C. L. Peichel, B. K. Blackman, K. S. Nereng, B. Jónsson, D. Schluter, and D. M. Kingsley. 2006. Corrigendum: genetic and developmental basis of evolutionary pelvic reduction in threespine sticklebacks. Nature 439.

Shapiro, M. D., B. R. Summers, S. Balabhadra, J. T. Aldenhoven, A. L. Miller, C. B. Cunningham, M. A. Bell, and D. M. Kingsley. 2009. The genetic architecture of skeletal convergence and sex determination in ninespine sticklebacks. Curr. Biol. 19:1140–1145.

Sheppard, P. M. 1953. Polymorphism, linkage and the blood groups. Am. Nat. 87:283–294.

Shikano, T., V. N. Laine, G. Herczeg, J. Vilkki, and J. Merila. 2013. Genetic architecture of parallel pelvic reduction in ninespine sticklebacks. G3 Genes Genom. Genet. 3:1833–1842.

Spoljaric, M. A., and T. E. Reimchen. 2007. 10,000 years later: evolution of body shape in Haida Gwaii three-spined stickleback. J. Fish Biol. 70:1484.

Stern, D. L., and V. Orgogozo. 2008. The loci of evolution: how predictable is genetic evolution? Evolution 62:2155–2177.

Strecker, U., L. Bernatchez, and H. Wilkens. 2003. Genetic divergence between cave and surface populations of *Astyanax* in Mexico (Characidae, Teleostei). Mol. Ecol. 12:699–710.

Strecker, U., V. H. Faundez, and H. Wilkens. 2004. Phylogeography of surface and cave *Astyanax* (Teleostei) from Central and North America based on cytochrome b sequence data. Mol. Phylogenet. Evol. 33:469–481.

Streelman, J. T., and R. C. Albertson. 2006. Evolution of novelty in the cichlid dentition. J. Exp. Zool. B Mol. Dev. Evol. 306:216–226.

Streelman, J. T., R. C. Albertson, and T. D. Kocher. 2003. Genome mapping of the orange blotch colour pattern in cichlid fishes. Mol. Ecol. 12:2465–2471.

Strickler, A. G., M. S. Byerly, and W. R. Jeffery. 2007. Lens gene expression analysis reveals downregulation of the anti-apoptotic chaperone αA-crystallin during cavefish eye degeneration. Dev. Genes Evol. 217:771–782.

Strickler, A. G., and W. R. Jeffery. 2009. Differentially expressed genes identified by cross-species microarray in the blind cavefish *Astyanax*. Int. Zool. 4:31.

Strickler, A. G., Y. Yamamoto, and W. R. Jeffery. 2007. The lens controls cell survival in the retina: evidence from the blind cavefish *Astyanax*. Dev. Biol. 311:512–523.

Tezuka, A., H. Yamamoto, J. Yokoyama, C. van Oosterhout, and M. Kawata. 2011. The MC1R gene in the guppy (*Poecilia reticulata*): genotypic and phenotypic polymorphisms. BMC Res. Notes 4:31.

Thesleff, I., and M. L. Mikkola. 2002. Death receptor signaling giving life to ectodermal organs. Sci. STKE 2002:pe22.

Turner, G. F. 2007. Adaptive radiation of cichlid fish. Curr. Biol. 17:R827–831.

Turner, G. F., O. Seehausen, M. E. Knight, C. J. Allender, and R. L. Robinson. 2001. How many species of cichlid fishes are there in African lakes? Mol. Ecol. 10:793–806.

Walker, J. A. 1997. Ecological morphology of lacustrine three-spine stickleback *Gasterosteus aculeatus* L. (Gasterosteidae) body shape. Biol. J. Linn. Soc. 61:3–50.

Walker, J. A., and M. A. Bell. 2000. Net evolutionary trajectories of body shape evolution within a microgeographic radiation of threespine sticklebacks. J. Zool. Lond. 252:293–302.

Webb, P. W. 1982. Locomotor patterns in the evolution of actinopterygian fishes. Am. Zool. 22:329–342.

Wilkens, H. 1971. Genetic interpretation of regressive evolutionary processes: studies of hybrid eyes of two *Astyanax* cave populations (Characidae, Pisces). Evolution 25:530–544.

Wilkens, H. 1988. Evolution and genetics of epigean and cave *Astyanax fasciatus* (Characidae, Pisces). Evol. Biol. 23:271–367.

Wilkens, H., and R. L. Burns. 1972. A new *Anoptichthys* cave population (Characidae, Pisces). Ann. Spéléol. 27:263–270.

Wilkens, H., and U. Strecker. 2003. Convergent evolution of the cavefish *Astyanax* (Charcidae, Telesotei): genetic evidence from reduced eye-size and pigmentation. Biol. J. Linn. Soc. 80:545–554.

Wootton, R. J. 1976. The Biology of the Sticklebacks. London: Academic.

Wray, G. A. 2007. The evolutionary significance of cis-regulatory mutations. Nat. Rev. Genet. 8:206–216.

Wu, P., T. X. Jiang, S. Suksaweang, R. B. Widelitz, and C. M. Chuong. 2004. Molecular shaping of the beak. Science 305:1465–1466.

Yamamoto, Y., M. S. Byerly, W. R. Jackman, and W. R. Jeffery. 2009. Pleiotropic functions of embryonic sonic hedgehog expression link jaw and taste bud amplification with eye loss during cavefish evolution. Dev. Biol. 330:200–211.

Yamamoto, Y., and W. R. Jeffery. 2000. Central role for the lens in cavefish eye degeneration. Science 289:631–633.

Yamamoto, Y., D. W. Stock, and W. R. Jeffery. 2004. Hedgehog signalling controls eye degeneration in blind cavefish. Nature 431:844–847.

Yi, Z., N. Garrison, O. Cohen-Barak, T. M. Karafet, R. A. King, R. P. Erickson, M. F. Hammer, and M. H. Brilliant. 2003. A 122.5-kilobase deletion of the P gene underlies the high prevalence of oculocutaneous albinism type 2 in the Navajo population. Am. J. Hum. Genet. 72:62–72.

Yoshizawa, M., S. Goricki, D. Soares, and W. R. Jeffery. 2010. Evolution of a behavioral shift mediated by superficial neuromasts helps cavefish find food in darkness. Curr. Biol. 20:1631–1636.

Yoshizawa, M., K. E. O'Quin, and W. R. Jeffery. 2013. Evolution of an adaptive behavior and its sensory receptors promotes eye regression in blind cavefish: response to Borowsky (2013). BMC Biol. 11:82.

Ziuganov, V. V. 1983. Genetics of osteal plate polymorphism and microevolution of threespine stickleback (*Gasterosteus aculeatus* L). Theor. Appl. Genet. 65.

Ziuganov, V. V., and A. A. Zotin. 1995. Pelvic girdle polymorphism and reproductive barriers in the ninespine stickleback *Pungitius pungitius* (L.) from northwest Russia. Behaviour 132:1095–1105.

20

The Age of Transformation: *The Triassic Period and the Rise of Today's Land Vertebrate Fauna*

Kevin Padian* and Hans-Dieter Sues[†]

Introduction

The Triassic Period (about 251 to 200 million years ago on the most recent geological timescales) was revolutionary in two senses. First, faunal turnover among terrestrial tetrapods was greater, by some measures, than at any other time in history. In addition, independent advances in locomotion, growth rates, and associated physiological features in two major lineages of tetrapods, derived nonmammalian synapsids (therapsids) and the ornithodiran archosaurs, were more profound than at any other time during the evolutionary history of continental tetrapods. (Only the emergence of tetrapods onto land during the Devonian Period is comparable.)

Here we describe the principal changes in communities of continental tetrapods during the Triassic and how they unfolded over some 50 million years. Different groups dominated different global regions, changes in the tetrapod faunas through time were not in lockstep, and several waves of faunal replacement took place (Padian 1986, 2013; Fraser 2006; Sues and Fraser 2010; Irmis 2011; Irmis et al. 2007). Although we want to avoid overgeneralizing, we stress that these observations and inferences are based on a fossil record that is necessarily incomplete, but is surprisingly good in many respects.

* Department of Integrative Biology and Museum of Paleontology, University of California, Berkeley
† Department of Paleobiology, National Museum of Natural History, Smithsonian Institution

The first taxonomic shift happened after the diversity crisis at the end of the Permian. Late Paleozoic terrestrial communities had been dominated by stem-amphibians (temnospondyls) and nonmammalian synapsids (Kemp 1982, 2005; Ruta and Benton 2008). Reptiles remained a minor component of many Permian faunas. The best-known groups were the aquatic mesosaurs, the vaguely lizard-like millerosaurs, bolosaurs, and procolophonoids, and the large and robust pareiasaurs, along with some still poorly known groups of diapsid reptiles. All but the pareiasaurs were small animals, rarely exceeding 50 cm in total length. They were considerably smaller than most therapsids of the Late Permian.

During the Triassic, both continental and marine ecosystems changed in many ways. Reptiles rapidly diversified and by the Middle Triassic became dominant in many communities of continental tetrapods, increasingly relegating synapsids to rather minor roles for the remainder of the Mesozoic Era. Reptilian groups diversified into a range of functional and ecological roles unmatched in the history of tetrapods on land, and only rivaled by Cenozoic mammals. Although crown-group mammals themselves did not evolve until the Late Triassic, many key cranial, dental, locomotory, and physiological features of the mammalian lineage appeared earlier (see below). With the exception of the birds, all major extant tetrapod groups, as well as many others that have since become extinct, first appeared during the Triassic. Even so, some characteristic avian features had already appeared in theropod dinosaurs during the Late Triassic.

In this paper we want to tell the story of Triassic tetrapods from two perspectives. First, we review briefly the macroevolutionary patterns and phylogenetic relationships of the groups that took over the terrestrial realm during this period. We then want to recast these groups in functional and ecological terms, to show the extent of convergence and divergence in major community roles, and to emphasize how critical the environment of the Triassic was (in both abiotic and biotic senses) in fostering continental vertebrate diversity. The important take-away message is that the present-day continental vertebrate biotas are difficult to understand fully without reference to the Triassic, the most functionally and ecologically diverse period in the history of terrestrial tetrapods, when most of these revolutionary changes began.

Phylogenetic Relationships of Triassic Tetrapods

Triassic tetrapod diversity was far too great to encompass in a single diagram. The most fundamental division is between Amphibia *sensu lato*, represented in the Triassic by various groups of stem-amphibians (including stem-frogs), and Amniota. Amniota (fig. 20.1) comprises two principal lineages, Reptilia and Synapsida. Reptilia comprises a few Paleozoic "holdovers" plus Diapsida, which includes the great reptilian radiations of the Mesozoic Era. Synapsida comprises mammals and their relatives, which (contrary to the traditional parlance of "mammal-like reptiles") were never part of the reptiles. The two major lineages of Triassic Synapsida are Anomodontia and Cynodontia (including mammals). Within Diapsida there are two major groups: one leading to extant lizards, snakes, and rhynchocephalians (Lepidosauria) and their various relatives (Lepidosauromorpha), and the other leading to the crocodylian and dinosaur-bird lineages (Archosauria) and their diverse, mostly Triassic relatives (Archosauromorpha). During the Triassic, Archosauromorpha (including Archosauria) was the taxonomically and ecologically most diverse clade of tetrapods on land, and they will provide most of the examples that we cite in the following sections. Space is too limited for us to describe all these animals in detail here, but readers are referred to Benton (2014), Fraser (2006), and Sues and Fraser (2010) for more detailed introductions to most of them. In the following section we show how these groups can be sorted into three global "faunas" that replaced each other successively during the Triassic. Then we will explore the ecological roles that they played.

Triassic Faunal Turnover

Three general continental tetrapod faunas dominated the Triassic. These faunas were not cohesive phylogenetically, but represent successive collections of groups that dominated at particular times (fig. 20.2). The first group comprises "Paleozoic holdovers," taxa that were diverse during the Permian and survived the end-Permian extinction to proliferate further during

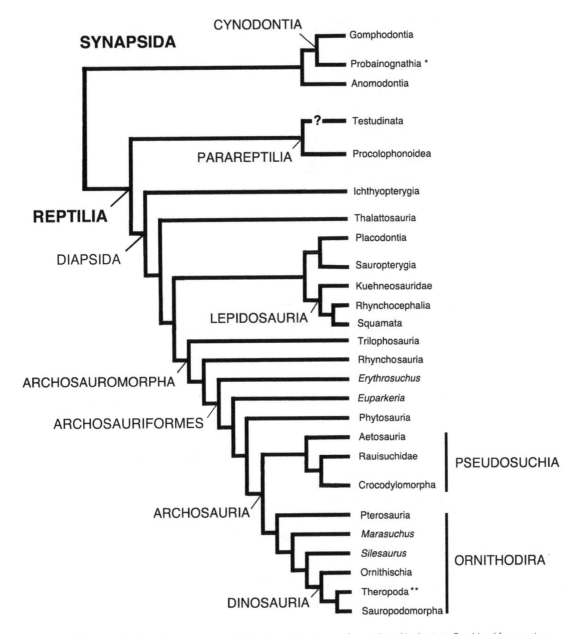

FIG. 20.1 Hypothesis of interrelationships of major groups of Triassic amniote tetrapods mentioned in the text. Combined from various sources. The placement of turtles (Testudinata) remains highly controversial. Probainognathia (*) includes mammals and their closest relatives (Mammaliaformes), and Theropoda (**) includes birds and their closest relatives (Avialae).

the Triassic. The second is called "indigenous" Triassic taxa, so named because they originated and became extinct during the Triassic. The third is the "living" fauna (we avoid the term "modern"), which comprises groups that first appeared during the Triassic but radiated much more extensively thereafter. Some representative types are pictured in figures 20.3 and 20.4.

The Late Paleozoic "Holdover" Fauna

The end-Permian diversity crisis included the disappearance of some temnospondyl stem-amphibians and gorgonopsian and dinocephalian synapsids. Some groups of reptiles (bolosaurs, millerosaurs, mesosaurs) and basal synapsids (sphenacodontids) had become

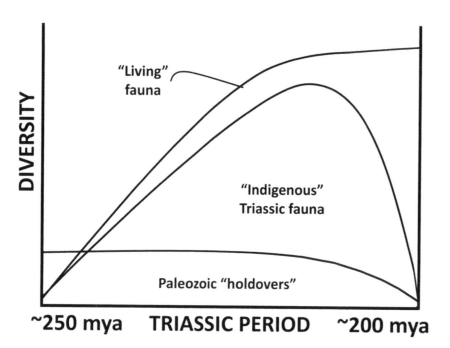

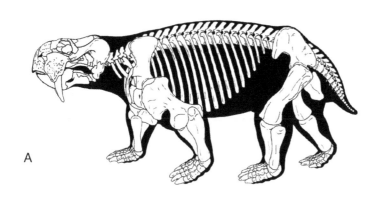

FIG. 20.2 Schematic depiction of three "evolutionary faunas" of Triassic terrestrial vertebrates: the Paleozoic "holdovers," the "Indigenous" Triassic fauna, and the "Living" fauna. Estimated diversity correlates with morphological and ecological disparity rather than with Linnean taxonomic categories.

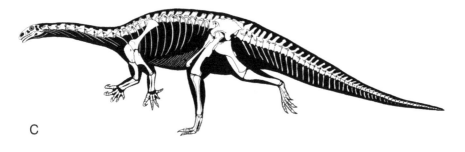

FIG. 20.3 Reconstructed skeletons (with body outline in black) of large Triassic herbivores. Top: Paleozoic "holdover"—dicynodont therapsid *Dinodontosaurus* (courtesy of Leonardo Morato); total length up to 3 m. Center: "Indigenous"—aetosaurian pseudosuchian *Stagonolepis* (modified from Walker 1961); total length up to 2.1 m. Bottom: "Living fauna"—sauropodomorph dinosaur *Plateosaurus* (modified from Weishampel and Westphal 1986); total length up to 9 m.

FIG. 20.4 Reconstructed skeletons (with body outline in black) of Triassic carnivores. Top: "Living fauna"—theropod dinosaur *Coelophysis* (from Paul 1993); total length up to 3 m. Bottom: "Indigenous"—rauisuchid pseudosuchian *Postosuchus* (courtesy of Jonathan Weinbaum), shown in bipedal pose; total length up to 5 m.

extinct before the end of the Permian. Other synapsid groups, such as dicynodonts and therocephalians, suffered major losses in diversity although some lineages (in the case of dicynodonts, at least four: Kemp 2005) did survive into the Triassic. Among temnospondyls, several mostly aquatic lineages persisted (Ruta and Benton 2008). Chroniosuchians, a lineage of armored non-amniote anthracosaurs, survived into the late Middle Triassic (Witzmann et al. 2008). Procolophonoids and archosauromorph reptiles survived the end-Permian crisis, although they are poorly known in the Permian. Both archosauromorph (bird-crocodylian) and lepidosauromorph (sphenodontian-squamate) lineages must have arisen in the Permian, but they are all but absent from the fossil record, so these "ghost lineages" must be inferred by phylogenetic analysis (e.g., Modesto et al. 2001). The effects of the end-Permian biotic crisis on other groups are still difficult to elucidate because much remains to be learned about reptilian diversity during the later part of the Permian and the timing of the originations and extinctions of various groups.

The "Indigenous" Triassic Fauna

By "indigenous" we mean groups that appeared and died out within the Triassic, and mostly within even a part of it. In the interest of space we refer readers to the summaries by Kemp (2005), Fraser (2006), and Sues and Fraser (2010) for details about the diversity and anatomy of particular groups.

We noted above that some temnospondyl and therapsid lineages survived the end-Permian crisis. These lineages evolved into new groups during the Triassic, although most of them became extinct by the end of the Triassic. Among temnospondyls, Stereospondyli comprised a wide range of sizes and body plans, ranging from superficially salamander- or crocodile-like stem-amphibians with long and narrow snouts to the peculiar plagiosaurs, whose heads were much wider than long. Among therapsids, the kannemeyeriiform dicynodonts became typically large (up to at least 3 m long) herbivores in many Middle and Late Triassic tetrapod communities. Cynodont therapsids, on the other hand, included two major groups, the omnivorous or herbivorous gomphodonts and the carnivorous/insectivorous probainognathians, which likely included the ultimate precursors of mammals (Hopson and Kitching 2001).

The Triassic Period was the apex in the evolution of major body types of archosauromorph reptiles, particularly the more basal forms. Archosauromorpha comprises Archosauria, which includes crocodylians, dinosaurs, and pterosaurs, and a host of other non-archosaurian groups that died out by the end of the Triassic (Gauthier 1986; Nesbitt 2011). Some 34 known archosauromorph lineages first diversified during the Triassic (although some may have first appeared in the Permian), but all except crocodyliforms, dinosaurs, and pterosaurs (unless turtles are also included in this clade) became extinct by the end of the Triassic (Nesbitt 2011, fig. 58). Examples include two clades of presumably plant-eating forms, the rhynchosaurs and trilophosaurs, and a host of semiaquatic to fully aquatic archosauriform reptiles. Among archosaurs themselves, recent discoveries from late Early and early Middle Triassic strata have established that the division between pseudosuchians (crocodile-line archosaurs) and ornithodirans (bird-line archosaurs) had already taken place by the early Middle Triassic (Nesbitt et al. 2010; Nesbitt 2011). Pseudosuchians comprised mostly medium-sized to large carnivores and herbivores (2–9 m in length).

The Triassic also witnessed the appearance of several reptilian groups that adapted to life in the sea (discussed below). Of these, only ichthyosaurs and plesiosaurs survived through the end of the Triassic and diversified during the Jurassic Period.

The relationships of all these groups are shown in fig. 20.1, but this barely scratches the surface of the morphological and ecological diversity (including substantial convergence) of these "indigenous" Triassic tetrapod groups.

The Living (Triassic to Recent) Fauna

Almost every major group of living terrestrial tetrapods (or their immediate stem-forms) evolved by the end of the Triassic. (A "major group" is one with considerable diversity and a distinct ecological role through some extent of geological time.) These groups comprise dinosaurs (including birds) and pterosaurs, as well as crocodylians, turtles, lepidosaurs, the groups of extant amphibians (frogs, salamanders, and caecilians), and mammals. Some groups, such as lizards and snakes, still have no recognized Triassic fossil record but can be safely inferred as "ghost lineages" because their sister taxa (as implied by phylogenetic analyses) were already present. (The oldest sphenodontians, the sister group of squamates, date from the late Middle Triassic; Jones et al. 2013.) If these missing groups were present, we can perhaps not recognize them because they had not yet evolved the characteristics that we use to diagnose them. It is also important to stress that the first, earliest members of extant tetrapod groups often looked quite different from their extant relatives and performed different ecological roles. For example, based on extant taxa, we think of crocodylians as aquatic or amphibious

ambush predators. However, Triassic crocodylomorphs were mostly small to medium-sized (often less than a meter in length), lightly built terrestrial forms such as *Terrestrisuchus* and *Hesperosuchus*. They had an erect stance and a parasagittal gait, and their limb proportions suggest cursoriality. Crocodylomorphs apparently did not become aquatic until the Early Jurassic, when they first invaded both freshwater and marine environments.

In this evolutionary fauna we also include groups such as nonavian dinosaurs and pterosaurs, which survived the Triassic and flourished during the remainder of the Mesozoic (nearly 140 million years). The marine ichthyosaurs and plesiosaurs and their relatives followed the same pattern. All of them evolved by the early Middle Triassic, some 240 million years ago, and plesiosaurs survived until about 66 million years ago.

Triassic Functional-Ecological Revolutions

Two Roads to Erect Stance and Parasagittal Gait: Synapsids and Reptiles

The locomotion of an animal says much about its way of life, its surroundings, and its physiology. A habitually erect stance implies that a considerable level of energy is going into maintaining that position, compared to a sprawling posture in which the animal is often resting its body on the ground. Animals with long, parasagittally oriented limb segments are generally capable of fast running, which implies an energy budget for short bursts of speed or sustained chases at variable speeds.

During the Triassic, the two principal clades of amniotes, the synapsids and the reptiles, independently evolved erect stance and parasagittal gait in some lineages. They did so in somewhat different ways, though they began from similar starting points. These advances in the synapsids led to the mammalian condition—which primitively was not cursorial but a more generalized gait capable of walking and climbing (Jenkins 1971a). Among reptiles these advances are manifested in Ornithodira, the bird-line archosaurian clade that includes dinosaurs and pterosaurs. As we will see, the crocodile-line archosaurian clade also evolved a kind of upright posture and possibly parasagittal gait, but in a different way.

Synapsids

Kemp (2005) summarized the evolutionary history of synapsid gait from the earliest amniotes of the Late Carboniferous to the basal mammaliaforms of the Late Triassic. During the Permian Period, early ("pelycosaurian") synapsids such as *Dimetrodon* greatly reduced or lost the lateral undulation of the vertebral column that characterized all basal tetrapods. The loss of axial flexibility accompanied the evolution of long, ventrally directed ribs that anchored muscles that aided in resisting axial sagging. The shoulder socket had a wide, complex shape that restricted movement of the humerus to the anteroposterior axis (Jenkins 1971a). As the humerus was drawn backward in the step cycle, the radius and ulna twisted to accommodate this motion while the hand remained planted; the radius rotated against the humerus while the ulna rotated on the wrist.

The pelvic girdle and hind limb were similarly primitive. The acetabulum was shallow and wide, and the femur, like the humerus, mostly moved in an anteroposterior direction. The femur could also be raised, lowered, and rotated. The lower leg seems to have been oriented ventrolaterally with respect to the knee. The tibia had a broad articular surface that contacted the distal end of the femur, whereas the fibula contacted only the posterolateral corner of this end. Conversely, the tibia articulated with the ankle only on the medial side of the astragalus, whereas the fibula broadly contacted both the astragalus and calcaneum (Kemp 2005). Thus, the posture was essentially sprawling, and the gait was rotatory (Padian et al. 2010).

The emergence of therapsids during the Permian brought important cranial and dental changes, but also substantial changes in posture and gait shared by today's mammals (Kemp 2005). The shoulder girdle became more lightly built and less intimately connected to the ribs, allowing longer strides. The glenoid fossa was altered from a "corkscrew shape" to a dorsoventrally bifaceted notch that faced posteroventrally, much as in dinosaurs and certain other archosaurs. Kemp (1982) related these changes to the greater functional separation of the shoulder girdle from the ribcage, and the need to evolve separate muscular systems for connecting the anterior girdle to the body and moving the forelimbs in locomotion. The articular surface of the humeral head was no longer a spiral but a long hemicylinder. As it rolled anteriorly over the glenoid, while the humerus was retracted, the bone also pronated nearly 90°. At the elbow, the radius rotated against the axis formed by the humerus and ulna, whereas, at the wrist, the ulna rotated and the radius did not. This combination, as in basal synapsids, produced an effective anchoring of the manus, which was probably plantigrade because both the metacarpals and phalanges were short.

The hind limb of early therapsids brought a new innovation in the ability to bring about both a rotatory and a more or less parasagittal gait (Kemp 1978, 1982; Sues 1986; for terminology see Padian et al. 2010). The latter style of gait approximates the mammalian condition in having the knees face more or less forward, but the elbows still faced laterally rather than posteriorly. This is a function of the orientation of the femur, which could be adducted for parasagittal gait and abducted for a more rotatory gait. The flexibility was made possible by a distinct, medially offset femoral head and a femoral shaft with a sigmoid curvature, such that the femoral head was offset approximately 90° to the distal condyles of the femur (Padian 1983, 1986). The femoral head articulated with the acetabulum so that both kinds of gaits could be accommodated, much as in extant crocodylians (Brinkman 1980) and perhaps other pseudosuchians (Bonaparte 1984). The difference is that archosaurian reptiles evolved these features in the Triassic, whereas synapsids had evolved them by the Late Permian.

However, the Triassic is when most of the evolution of Cynodontia, which includes mammals and their closest relatives, took place (Hopson and Kitching 2001). Therefore we have to ask what changes in posture and locomotion occurred then, and whether they were important to the diversification and success of the immediate precursors of mammals. In fact, the overall architecture of the girdles and limbs had not changed substantially from the condition in early therapsids, regardless of the numerous changes in the skull and dentition. Larger muscles and modified articulations seem to have facilitated more powerful and maneuverable limb movements, but the hind limb seems to have reduced its ability to produce a rotatory gait, as evidenced by the strongly offset femoral head and the evolution of the mammal-like greater trochanter, as well as modification of the geometry of the pelvic bones (Kemp 1982, 2005). The closest lineages to mammals (tritheledontid and tritylodontid cynodonts: Sues and Jenkins 2006) evolved pelves that were even more "mammalian" with the reduction and even loss of the posterior portion of the iliac blade and development of the characteristic anterodorsal prong. At the same time, the pubis and ischium formed a more extensive ventral connection, and the obturator foramen became enlarged. But just as important were the changes in the pectoral girdle and forelimb. The coracoid was reduced, and the humeral head became bulbous, forming with the glenoid a ball-and-socket joint capable of a range of motions. Jenkins (1971a) showed that the humerus was more adducted, bringing the forelimb more underneath the body like the hind limb, with the elbow now directed posteriorly.

By the end of the Triassic, therefore, many mammalian-grade structures and functions had arisen. However, Jenkins (1971, 1971b) showed that truly "cursorial" abilities did not appear until well into the history of crown-group mammals, and that early Mesozoic mammaliaforms did not have all features of the posture and gait present in crown-group mammals. Bramble and Jenkins (1989) charted the assembly of basal and derived cynodont locomotory features, noting that the hind limb and girdle took on "mammalian" characteristics before the forelimb and girdle did. They inferred that, as the limbs adopted a more parasagittally oriented stance, the possibility of developing asynchronous gaits (such as galloping) increased. The reduction of the posterior dorsal ribs has been linked with the evolution of a mammal-like diaphragm, which would have been critical to the efficiency of certain gaits (Carrier 1987).

Reptiles

Late Paleozoic reptiles were generally small, and had a sprawling posture and rotatory gait in both forelimbs and hind limbs. As with synapsids and basal tetrapods in general, the power for terrestrial locomotion mainly came from the hind limbs, while the forelimbs mainly supported the front end of the body and kept it off the ground and moving.

The fossil record of amniote trackways from the late Paleozoic (e.g., Haubold 1971) reflects trackmakers with widely spread limbs engaging in rotatory locomotion.

Narrow trackways with footprints close to the body midline are rare (i.e., some examples of *Rotodactylus*). However, most reptilian trackways from the Early Triassic are not close to the midline, their pace angulations are high, and their toes point anterolaterally instead of anteriorly, all of which suggests that their makers retained the basal tetrapod rotatory locomotion pattern.

Functional studies of Triassic archosauriforms suggest that the basal condition for the clade was a sprawling stance and rotatory gait. However, among Archosauria, two independent ways of reaching relatively erect stance and parasagittal gait evolved, at least in the hind limb (Sullivan, this volume), within Pseudosuchia and Ornithosuchia.

Bonaparte (1984) and Parrish (1986, 1987) showed that in some pseudosuchians an erect stance and parasagittal gait was achieved much as it was in the birdline archosaurs; others accomplished it by having a slightly inturned head of the femur nested in the acetabulum underneath a laterally deflected iliac blade. This allowed the femur to be adducted so that the knee faced nearly forward. But the slightly inturned femoral head allowed the femur to be abducted as well, so that the more sprawling posture typical of basal archosauromorphs could be retained. Present-day crocodylians can execute both sprawling and relatively erect postures (Brinkman 1980). Parrish (1986) argued on the basis of careful reconstruction of the possible excursions at the limb joints that the ability to walk parasagittally first appeared in aetosaurs among pseudosuchians. This was reasonable because the joints of phytosaurs appeared to be too generalized to support a parasagittal gait. However, if the Triassic trackways known as *Apatopus* were made by phytosaurs, as Baird (1957) first suggested, the trackmaker most likely had an erect stance and parasagittal gait, because the limbs were brought in close to the body (Padian et al. 2010). If this interpretation holds, then parasagittal gait is a basal feature of Pseudosuchia. Nesbitt (2011) recently reevaluated the phylogeny of Archosauriformes and recovered phytosaurs outside Archosauria proper, in which case parasagittal gait would be primitive for the whole group.

This makes even more sense if we consider Ornithodira (the group within Ornithosuchia that includes most bird-line archosaurs), the sister taxon to Pseudosuchia, which had a very different functional-evolutionary history. All ornithodirans appear to have had an erect stance and parasagittal gait, and they were most likely bipedal from their inception (Padian 1983, 2008; Nesbitt 2011). Traits that support this interpretation include a femur with a head that is distinctly offset from the shaft, a femoral shaft with a double (rather than simply sigmoid) curvature that reflected anteroposterior excursion, a reduced fibula (reflecting a lack of the torsion seen in animals with a rotatory gait), a mesotarsal ankle joint, an elongated metatarsus (reflecting habitual digitigrady, another sign of extended limb excursion common to cursorial animals using a parasagittal gait), and long digits. These features certainly appeared by the Middle Triassic with the earliest known ornithodirans (Sereno 1991). It is notable that dinosaurs and pterosaurs survived the end-Triassic extinctions and diversified extensively throughout the Mesozoic Era, whereas all pseudosuchians except crocodylomorphs became extinct. The pseudosuchian innovation of the buttressed ilium overhanging the hip joint can be seen as promoting versatility in gait (presumably the parasagittal gait allowed longer strides and more rapid progression) without requiring a fundamental change in physiology (they could still sprawl when not progressing quickly). However, the ornithodiran condition of mandatory erect posture and parasagittal gait seems to have required a commitment not just to rapid locomotion but to higher supporting metabolic levels, which we will explore below.

Two Roads to High Metabolic Rates: Synapsids and Reptiles

Although the term "warm-blooded" is widely used in professional and popular scientific works, it has no precise meaning, and has been used to refer to a variety of physiological syndromes that have different underlying causes and are not necessarily correlated. Present-day birds and mammals are considered "warm-blooded," but their physiological systems are very different. It is likely that many features commonly considered characteristic of "warm-bloodedness" evolved in archosaurian reptiles (specifically ornithodirans) during the Triassic, but whether many of those features evolved in synapsids before the Jurassic is less clear.

Rather than using "warm-blooded" versus "cold-blooded," we define and contrast three sets of terms.

Endothermic denotes that an animal generates most of its body heat itself, usually by burning calories. *Ectothermic* denotes that it gets most of the heat needed to run metabolic processes from external sources such as the sun. The body temperature of a *poikilothermic* tetrapod fluctuates significantly, even over the course of a single day whereas a *homeothermic* animal has a relatively constant body temperature (although many homeotherms can actively lower their body temperatures for daily periods, a syndrome called "heterothermy"). A *tachymetabolic* animal has relatively high metabolic rates, because its biochemical processes work at higher rates than those of *bradymetabolic* animals. There is no dichotomous distinction between any of these contrasting terms, nor are there absolute quantitative ranges into which the categories fall. Historically, birds and mammals have been considered "warm-blooded" (i.e., with a relatively high body temperature), endothermic, homeothermic, and tachymetabolic. Whereas this characterization is simplistic and there are many deviations and combinations, animals that do not fit these descriptions have generally been labeled by their antonyms. And taxonomic terms have come to bring their own typological baggage: "reptiles" are held to be "cold-blooded," but birds evolved from and are themselves reptiles, so the generalization of "cold-blooded" is meaningless in evolutionary terms.

Among extinct animals it is difficult to assess endothermy, because we have no direct indication of the source of metabolic heat, although isotopic ratios of carbon and oxygen (as well as other elements such as calcium and phosphorus) in fossilized hard tissues may provide indirect evidence of body temperature (e.g., Eagle et al. 2010). The stability of body temperature (homeothermy) is also difficult to specify: although birds are generally considered "warm-blooded," many of them actively lower their temperature for parts of the day. We cannot tell whether extinct animals did the same. Metabolic rate, however, is a slightly different story, because it is a partial determinant of growth rate: generally speaking, to grow quickly, an animal needs a high metabolic rate to sustain the processes of growth. However, although most slowly growing animals have lower metabolic rates, a slowly growing animal does not have to have low metabolic rates: among mammals, primates grow relatively slowly yet they are not bradymetabolic. Generally, in studying the growth rates

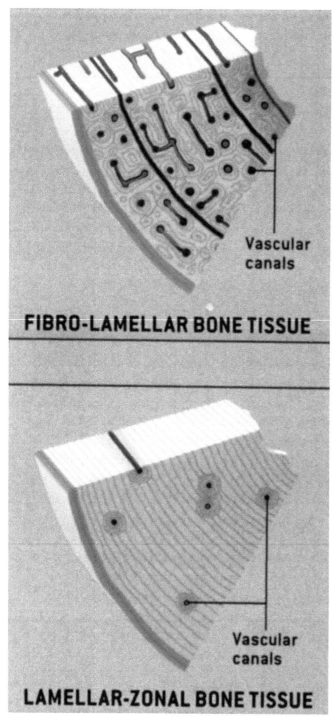

FIBRO-LAMELLAR BONE TISSUE

Vascular
canals

LAMELLAR-ZONAL BONE TISSUE

Vascular
canals

FIG. 20.5 Contrast between rapidly growing "fibro-lamellar" bone tissue found in most dinosaurs and mammals, and "lamellar-zonal" tissue that is more prevalent in other reptiles and amphibians. The red and blue circles and lines represent vascular canals, which are more numerous in fibro-lamellar bone and generally reflect higher growth rates and underlying metabolic rates. From Horner et al. (2005), reprinted by courtesy of Scientific American.

of extinct tetrapods, paleobiologists have been able to say only that if an animal is growing rapidly, its metabolic rates were probably relatively high (Padian et al. 2001).

Vertebrate bone preserves a direct indication of growth rate by the expression of certain tissue types that reflect the rate of growth (e.g., Castanet et al. 1996). These rates are normalized on the basis of experimental measurements taken from present-day species, and are usually expressed as microns of new bone deposition per day. In most extant non-avian reptiles and amphibians, cortical bone is deposited at a rate of 2 μm or less per day. In present-day birds and mammals, that rate can be five to ten times higher or more. Even the fabric of the bone tissue looks different between these two general groups (fig. 20.5). Compared to the bone tissue of mammals and birds, amphibian and reptilian bone typically has fewer vascular canals, the tiny spaces where the osteocytes resided are generally smaller, fewer, and circumferentially aligned, and the circumferential rest lines that mark annual growth periods are closer together and more distinct, and may even manifest themselves in annuli of bone tissue with few or no canals and osteocytes. The faster-growing bone of mammals and birds not only has more vascular canals; in contrast to the sparser canals that generally run longitudinally in amphibian and reptile bone, the canals in mammal and bird bone can also extend circumferentially and radially, and can often connect with each other (Padian and Lamm 2012).

The high growth rates found in extant birds and mammals are absent in the fossil record of most tetrapods, with some exceptions. Among reptiles, the ornithodirans are most conspicuous (Padian et al. 2001, 2004). Among synapsids, somewhat higher growth rates seem to evolve in the therapsids, but mammals grow at even higher rates, especially the placentals (de Ricqlès 1969, 1972, 1974, 1980; Chinsamy-Turan 2011). Also in the Triassic, archosauromorph reptiles began to show higher rates of bone deposition, more osteocytes, vascular canals with more complex structure, and a greater preponderance of rapidly growing fibrolamellar bone than in other reptiles (Werning et al. 2011). Among these animals, dinosaurs and pterosaurs reached rates comparable to those of mammals and birds, and the largest species grew most rapidly, a common feature of tetrapods (Padian et al. 2001). And these histological advances tend to be correlated with the advent of advances in posture and gait that were detailed above. In the Triassic, therefore, a revolution of sorts occurred in both the posture and gait and the growth and metabolic regimes of advanced synapsids and archosaurs, particularly ornithodirans. These functional and metabolic changes signaled a commitment to "living fast," as it were—establishing energy budgets that required a constant supply of high-quality food sources, whether plant or animal. Ecologists would see these commitments as risky, and perhaps they are. But the groups of animals that made these commitments went on to dominate all later communities of tetrapods on land.

Ecological Roles and Diversification during the Triassic

Given the large number of tetrapod clades that evolved during the Triassic Period, and given some of their functional and metabolic innovations, it was inevitable that ecological diversification would follow. The interesting thing about continental tetrapods in the Triassic is that so many independent groups repeatedly exploited the same adaptive zones—and often for the first time. For example, before the Triassic there had been no active fliers among tetrapods and few undisputed marine reptiles. What follows is a brief catalog of some of the major adaptive types, with remarks on the peculiarities of adaptations and also structural differences among the groups. For details see Fraser (2006) and Sues and Fraser (2010).

We begin with three generalizations. First, many or most taxa mentioned here are likely to have been trophic generalists, and even omnivores; we have no way of telling. Both biologists and paleobiologists often make sharp distinctions among (and even within) dietary categories, but countless studies on extant tetrapods show that these categories represent a continuum. Many herbivores consume animal protein, and many predominantly carnivorous animals occasionally feed on plants. Second, carnivory (with an emphasis on small prey consumed whole) or omnivory represented the plesiomorphic condition among tetrapods in general and tetrapods in particular, and the clades that populated the Triassic are no exception. There were at least 10 independent evolutionary forays into herbivory

among Triassic tetrapods (Reisz and Sues 2000). Until the reign of the dinosaurs during the Jurassic and Cretaceous, when carnivorous and herbivorous taxa were closely similar in diversity, and, in terms of numbers of individuals, herbivores must have greatly outnumbered carnivores in terrestrial tetrapod communities (as they do today), carnivory was more the rule and herbivory the exception. Some late Paleozoic basal synapsids (caseids and edaphosaurs) predominantly subsisted on plants, and pareiasaurs, at least some procolophonids, and larger captorhinids took this route among reptiles. Dicynodont therapsids were the most taxonomically diverse group of late Paleozoic herbivores. But the Triassic is the first period in Earth's history when herbivores really diversified morphologically in communities of terrestrial tetrapods across a variety of major clades. And they included both synapsids and reptiles. Third, these ecological innovations not only occurred in different clades but at different times during the Triassic (which lasted about 50 million years). For example, some macrocarnivores evolved in the Early Triassic, others in the Middle Triassic, and yet others in the Late Triassic.

What we learn from the Triassic diversification of trophic groups among tetrapods is that the simple conical, often labiolingually flattened teeth that we usually interpret as indicative of carnivory turn out to be both the generalized condition for most groups and likely quite versatile for dietary preferences. Examples of some of these varied feeding types are shown in figures 20.6 and 20.7 and summarized in table 20.1.

Generalized Smaller Carnivores

We use this term broadly to include small to medium-sized tetrapods with a (presumably) mostly carnivorous diet, probably subsisting on arthropods and small tetrapods (including juveniles of larger species) that could be dispatched using the jaws and swallowed more or less whole. Our best clues to this inference come from the form of the teeth and shape of the skull. Among many smaller taxa presumed to qualify here (such as *Marasuchus*, *Lewisuchus*, and *Dromomeron*), little is as yet known of the skull and jaws. Where known, the skulls tend to be "long and low," to use a common descriptor; this means that they are two to three times as long in total as they are high at the orbit, and they tend

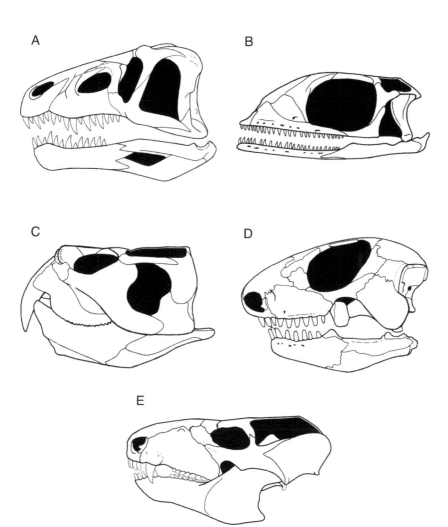

FIG. 20.6 Skulls of Triassic land-dwelling carnivorous and herbivorous tetrapods in left lateral view. (A) Large carnivore: archosauriform *Erythrosuchus* (modified from Gower 2003). (B) Small carnivore/insectivore: lepidosauromorph *"Kuehneosaurus"* (modified from Robinson 1962). (C) Large herbivore: rhynchosaurian archosauromorph *Hyperodapedon* (modified from Benton 1983). (D) Small herbivore: procolophonid parareptile *Procolophon* (modified from Carroll and Lindsay 1985). (E) Large herbivore: cynodont therapsid *Traversodon* (modified from Barberena 1981).

to taper steadily in height toward the tip of the snout. The tooth crowns are more or less conical to labiolingually flattened, often recurved, and they may have cutting edges on the mesial and (or) distal edges, which often are serrated. The teeth are relatively small and numerous in the jaws. Examples of this trophic morph include the Late Triassic ornithodiran *Scleromochlus* (20 cm long), the early crocodylomorph *Terrestrisuchus* (50 cm long), the earliest well-known archosauriform *Proterosuchus* (about 1.5 m long), and the theropod dinosaur *Coelophysis* (up to 3 m long). Among synapsids, many cynodonts (e.g., *Thrinaxodon*) have dentitions suggesting that they ate arthropods and small tetrapods, and this may have been the plesiomorphic diet for the Triassic-Jurassic lineages that were the closest relatives to true mammals (tritheledontids and morganucodontids).

Terrestrial Macrocarnivores

We use the term "macrocarnivore" to denote a predator that takes prey of a substantial percentage of its own body size, prey that would have to be killed and bitten into smaller pieces, rather than simply swallowed whole, as predators that hunt smaller game would do with their food. These are mostly large animals (with a length of 2 m or more), often with a proportionately large head relative to body size. Skull length tends to be about twice its height at the orbit. Furthermore, skull height stays relatively high throughout the length of the skull. The teeth are large, typically labiolingually flattened and with serrated cutting edges, deeply implanted, and rather few in number in the jaw. The early Middle Triassic cynodont *Cynognathus*, which attained a skull length of more than 40 cm and has blade-like

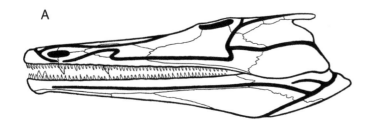

A

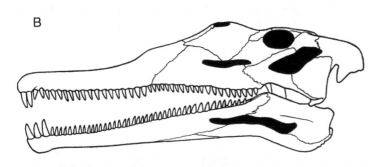

B

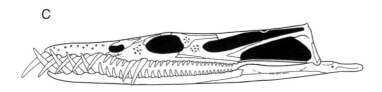

C

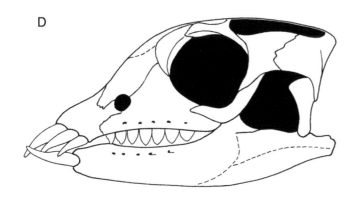

D

FIG. 20.7 Skulls of Triassic freshwater and marine tetrapods in left lateral view. (A) Freshwater carnivore: temnospondyl *Benthosuchus* (from Bystrov and Efremov 1940). (B) Freshwater carnivore: phytosaurian archosauriform *Machaeroprosopus* (modified from Colbert 1947). (C) Marine carnivore: nothosaurid sauropterygian *Nothosaurus* (modified from Rieppel 2000b). (D) Marine durophage: placodont *Paraplacodus* (modified from Rieppel 2000a).

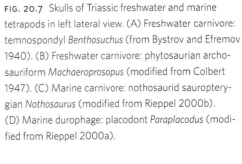

"cheek" teeth with recurved, serrated cusps, was a macrocarnivore. In archosauriforms with this skull shape the antorbital and (or) narial opening may be relatively large. Examples of archosauromorph macrocarnivores include erythrosuchids (up to 5 m), ornithosuchids (up to 3 m long), the basal dinosaur *Herrerasaurus* (up to 4 m long), as well as poposaurids and rauisuchids (as defined by Nesbitt 2011; 5 m or more in length). Of these, *Herrerasaurus* and *Poposaurus* were obligate bipeds. Most of the other forms mentioned (with the exception of the erythrosuchids) are often considered facultatively bipedal, largely because their hind limbs are considerably longer and more robust than their forelimbs. This is not unusual for reptiles, however, regardless of size. During rapid locomotion the forelimbs may have been unable to keep pace with the powerful hind limbs and may have simply been lifted off the ground. By itself, however, limb disparity is insufficient to infer bipedality. *Herrerasaurus*, like all ornithodirans, had an erect stance and parasagittal gait

TABLE 20.1 Partial listing of Triassic tetrapod lineages, grouped by generalized trophic/ecological categories

Generalized smaller carnivores	Some proterosuchids, protorosaurs, small pseudosuchians, crocodylomorphs, lagerpetids, lagosuchids, small theropods, miscellaneous ornithodirans, turtles, chiniquodontid and possibly tritheledontid cynodonts
Terrestrial macrocarnivores	Some erythrosuchids, ornithosuchids, larger pseudosuchians, poposaurids, large theropods, cynognathid therapsids
Freshwater carnivores	Various temnospondyl stem-amphibians, proterosuchids, proterochampsids, phytosaurs, choristoderans, some protorosaurs, somedrepanosaurids, *Vancleavea*, doswelliids
Marine reptiles	Sauropterygians (placodonts, plesiosaurs, nothosaurs, pistosaurs, pachypleurosaurs), ichthyosaurs, thalattosaurs, saurosphargids, some protorosaurs
Herbivores	Rhynchosaurs, trilophosaurs, aetosaurs, silesaurids, revueltosaurids, some sauropodomorphs, [ornithischians,] procolophonids, dicynodonts, gomphodont cynodonts
Aerial reptiles	Pterosaurs (flyers), kuehneosaurids, *Mecistotrachelos*, *Sharovipteryx* (gliders)

Note: These taxa are not of equivalent "Linnean" ranks or of comparable taxonomic diversity. Names in brackets represent ghost lineages.

because the structure of its hip joint restricted femoral excursion to a fore-and-aft plane, and the other joints of the hind limb acted as hinges (Sullivan, this volume).

Aquatic Carnivores

During the Triassic at least six lineages of reptiles adopted a semiaquatic or fully aquatic existence in freshwater. Some (proterosuchids, proterochampsids, phytosaurs) were relatively large (up to 7 m in length) carnivores with peg-like or conical teeth. Their skulls were typically long, particularly in the antorbital region. The snouts tended to be attenuated, sometimes gavial-like, which suggests a diet of fish. The jaws often held a large number of teeth: phytosaurs, for example, have up to 40 functional teeth in each jaw quadrant. Phytosaurs looked much like crocodylians superficially, and were either the most basal pseudosuchians or outside that entire clade. Their snouts were narrow and long, and the external narial openings in front of the eyes were on a raised prominence of the nasal bones. The back of the skull was wide transversely. Choristoderans, which probably evolved in the Triassic but became common only during the later Mesozoic and Paleogene, shared many of these features and also had a wide posterior region of the skull.

A range of generally smaller (less than 1 m body length) reptiles also invaded the freshwater environments of lakes, streams, and ponds. They generally lacked the long snouts of the larger fish-eaters mentioned above, and had smaller skulls and simpler teeth. A group of archosauromorph reptiles called protorosaurs diversified into freshwater (and marine) environments. *Tanytrachelos* (20 cm long) was one of them. Other aquatic archosauromorphs include drepanosaurids such as the deep-tailed *Hypuronector* (about 15 cm long) and the archosauriform *Vancleavea* (about 1 m long), which had extensive dermal armor and a deep tail. Unusual for this trophic category, *Vancleavea* has a short skull with large, somewhat recurved teeth of varying size. *Vancleavea* and perhaps doswelliids adopted at least semiaquatic modes of life. There were doubtless many other forms of small aquatic tetrapods, because those mentioned in this paragraph only came to light within the past few decades. Triassic microvertebrate assemblages (e.g., Kaye and Padian 1994) preserve teeth and skeletal elements from taxa that are so far unidentified, suggesting a further unrecognized diversity of aquatic tetrapods.

In addition to these reptiles, it should be noted that most Triassic temnospondyl stem-amphibians were probably aquatic or mostly so. They range in size from less than 1 m to more than 5 m in length, and skull shapes range from long and attenuated to wider than long. Typical Triassic members include metoposaurids, which are found in many Late Triassic freshwater deposits in North America and Europe. Metoposaurids have large, flat skulls with rather small, anteriorly placed, and dorsally facing eyes. The first stem-frogs (*Triadobatrachus, Czatkobatrachus*) are known from the Early Triassic, and it may be presumed that they were at least partly aquatic.

The Return to the Sea

Not only did at least four reptilian groups return to the sea during the Triassic; they did so in completely different ways. In each case, we have as yet no evidence that the groups first ventured into freshwater and then moved on to marine environments. It seems rather that freshwater forms evolved inland in lakes and streams, and marine taxa in brackish or marine settings, because in most cases their earliest members are found in nearshore marine environments.

Placodonts have short, broad, and robust skulls, often with procumbent teeth at the front of the jaws and massive crushing teeth at the back of the jaws and especially on the palate. *Placodus* and related forms probably swam using undulation of the tail. Some derived placodonts had turtle-like dermal armor and probably swam by paddling. Most placodonts probably subsisted on hard-shelled marine invertebrates that were attached to the substrate. A poorly known lineage superficially similar in some respects to both placodonts and sauropterygians, Saurosphargidae (Li et al. 2011), represents yet another lineage of marine Triassic reptiles.

Although they were very different from other marine reptiles, placodonts may be most closely related to the long-necked plesiosaurs and their relatives, collectively known as Sauropterygia (Rieppel 2000a, 2000b, 2002). The Triassic members of this group include pachypleurosaurs, nothosaurs, and pistosaurs (the sister group to plesiosaurs). The small (usually less than 1 m long) pachypleurosaurs have small skulls with broad, rounded snouts and uniform dentitions. *Nothosaurus* was larger (up to 4 m long); its skull could attain a length of 50 cm. The snout is elongated with "pincer" jaws and a heterodont "fish-trap" dentition (Rieppel 2000b). Although their limbs became increasingly flipper-like, pachypleurosaurs and nothosaurs probably still relied on the tail as their principal means of propulsion, in contrast to plesiosaurs, which seem to have had a more or less rigid body and swam using both fore- and hind limbs that were modified as flippers (Robinson

1975; Braun and Reif 1985). Specialized sauropterygians extended their necks by increasing the number of cervical vertebrae: some Cretaceous plesiosaurs had more than 70.

Thalattosaurs were superficially similar to and probably related to sauropterygians. Generally rather lizard-like in body form, they have somewhat modified limbs and laterally flattened, deep tails, which, together with their occurrence in marine deposits, suggest an aquatic existence. The skulls were robust and had tapering, often deflected snouts.

Here then is an example of the dynamic tension between phylogenetic unity and morphological and ecological diversification: the thalattosaurs, placodonts, saurosphargids, pachypleurosaurs, nothosaurs, pistosaurs, and plesiosaurs (even only considering Triassic forms) were far more different from each other morphologically and ecologically than the "adaptive radiations" one commonly hears labeling extant cichlid fishes or dung beetles; yet these bizarre marine reptiles may have all had a single phylogenetic origin. Their rapid diversification exemplifies how unusual the Triassic Period was.

Ichthyosaurs are the most widely known Mesozoic marine reptiles, but their familiar dolphin- or tuna-like body shape did not become widely established until the Jurassic (Motani 2005). The earliest known ichthyosaurs were small, long-bodied forms such as *Utatsusaurus*, which swam using undulations of the body and tail (Braun and Reif 1985); they are known from nearshore marine environments. A variety of later Triassic ichthyosaurs such as *Shonisaurus* and *Cymbospondylus* attained large size (10 m or longer) and have long heads, deep bodies with strange, distally expanded ribs, and long, narrow forefins. These were open-water forms. Early ichthyosaurs were ecologically diverse; they are generally thought to have been active swimmers preying on fishes or cephalopods, but the more basal forms were probably not pursuit predators and may have favored a varied diet of invertebrates.

In addition to these major groups, several other reptilian clades invaded the seas. The protorosaurs that were mentioned above as aquatic foragers had several members that became large and lived in nearshore environments, including *Macrocnemus* and *Tanystropheus*, which had a neck that made up half its body length (up to 6 m) but still comprised the same basic number of vertebrae shared by most other reptiles. Rather unexpectedly, the oldest known stem-turtle, *Odontochelys* from early Late Triassic marine deposits in China, has a shell comprising dorsal neural plates (and expanded dorsal ribs) and a fully developed ventral plastron (Li et al. 2008). Its shell may represent an intermediate stage in the evolution of the typical turtle shell (Lyson et al. 2013). Geologically slightly younger turtles such as *Proganochelys* appear to have been terrestrial, with both upper and lower shells (Joyce and Gauthier 2004).

As noted above, stem-crocodylians do not seem to have become aquatic until the Jurassic. However, the basal crocodile-line archosaur *Qianosuchus* from early Middle Triassic marine deposits in China already has a laterally flattened and dorsoventrally deep tail, which is suitable for swimming, although its skull appears to be that of a terrestrial carnivore.

All told, Triassic reptiles diversified quickly to exploit a range of food sources in several kinds of marine environments, evolving a diversity of body plans not only among, but even within, major lineages such as ichthyosaurs and sauropterygians.

Herbivores

As the phylogeny (fig. 20.1) shows, herbivory is a secondary specialization from carnivory/insectivory that independently evolved numerous times, and the transition usually seems to have involved an omnivorous phase, although we have little evidence of the actual transitions in many groups. Basal sauropodomorph ("prosauropod") dinosaurs, which together with the carnivorous theropods make up the clade Saurischia, are a good example. It now appears that the earliest sauropodomorphs such as the aptly named *Panphagia* ("eats everything") still had teeth suggestive of omnivory (Martínez and Alcober 2009). The teeth of *Panphagia* have triangular crowns like those of many herbivorous dinosaurs, but they also have finely serrated mesial and distal cutting edges, as well as labial and lingual keels that suggest puncturing as well as slicing. More derived sauropodomorphs such as *Plateosaurus* have more typical lanceolate, labiolingually flattened tooth crowns with coarser serrations, reminiscent of those in present-day plant-eating iguanid lizards.

The dentitions of some early ornithischian dinosaurs such as *Lesothosaurus* (an Early Jurassic form) have simpler conical or recurved teeth at the front of the jaws (again suggestive of carnivorous or omnivorous ancestry). However, recent work indicates that ornithischian dinosaurs were very rare during the Triassic (Irmis et al. 2007). *Pisanosaurus*, from the early Late Triassic of Argentina, has long been considered the earliest ornithischian, but the only known specimen is poorly preserved and its phylogenetic status must be considered uncertain. Irmis et al. (2007, 5) reviewed the case of the missing Triassic ornithischians, and pointed out that "supposed ornithischian dental synapomorphies such as low, triangular tooth crowns, the separation of the crown and root by a distinct neck, and the presence of asymmetrical teeth with serrated denticles, also occur in other Late Triassic archosaurs, including aetosaurs (Walker 1961), *Silesaurus* (Dzik 2003) and now R[*evueltosaurus*] *callenderi* (Parker et al. 2005). Thus, these character-states cannot be used to assign isolated teeth to the Ornithischia." This poses a major problem, first because most alleged Triassic ornithischian specimens (now almost all eliminated as ornithischian) are teeth, and, second, because ornithischians are considered the sister taxon to saurischians, the two groups must be equally ancient. Good material of one taxon, *Eocursor*, was described from strata of possibly Late Triassic age in South Africa (Butler et al. 2007), but the age of this record is poorly constrained.

Other taxa with more or less ornithischian-like teeth have been considered herbivores, including aetosaurs (and their relative *Revueltosaurus*) and silesaurids. Aetosaurs were basal crocodile-line archosaurs (Nesbitt 2011). Their bodies were protected by extensive dermal armor dorsally and ventrally, occasionally with prominent spikes in the shoulder region. The jaws were typically edentulous in front. Some forms have a beveled snout, which suggested rooting to some authors. Teeth assigned to *Revueltosaurus* were originally thought to belong to ornithischian dinosaurs, but when associated skeletal remains were eventually found, the tooth-bearer turned out to be a crocodile-line archosaur closely related to aetosaurs (Parker et al. 2005). Silesaurids were first thought to be ornithischians, but they were instead close relatives of dinosaurs, quadrupedal and apparently herbivorous (Dzik 2003; Nesbitt et al. 2010).

Other Triassic lineages that apparently evolved herbivory include two basal archosauromorph groups

(Reisz and Sues 2000). Rhynchosaurs were among the most unusual Triassic tetrapods. The skull has a down-turned upper beak formed by the premaxillae, and the anterior ends of the dentaries are turned up. The maxillae each bear two or more rows of teeth bordering a groove into which the mandibular tooth row fits. The back of the skull is transversely wide, presumably to accommodate powerful jaw-closing musculature. On the other hand, the jaws of *Trilophosaurus* and *Teraterpeton* are edentulous at the front and have transversely broad molar-like teeth with cusps. The cheek region of the skull is deep and lacks a lower temporal opening.

Procolophonoid parareptiles were diverse during the Triassic. *Procolophon* and its Triassic relatives (Procolophonidae) have transversely broad teeth at the back of the jaws and incisor-like teeth at the front. The teeth at the back of the jaws interdigitated during jaw closure and have prominent apical crests and often cusps.

This makes at least eight independent major lineages of herbivorous reptiles known from the Triassic Period, and we can add at least two more lineages of herbivorous therapsids to this total. Gomphodont cynodonts have transversely wide, molar-like teeth at the back of the jaws. In many Triassic forms, the tooth crowns had several cusps and the upper and lower "cheek teeth" met in precise occlusion: the lower teeth met the upper ones and moved upward and backward. Gomphodonts were mostly smaller forms but a few Late Triassic forms such as *Exaeretodon* attained a length of at least 2 m. Triassic dicynodonts for the most part lack teeth. In a turtle-like manner, a keratinous beak covered the front of the snout. Food was sliced between the occluding halves of the beak by a posteriorly directed jaw motion, which was facilitated by a sliding jaw joint. The kannemeyeriiform dicynodonts became important elements in communities of Triassic terrestrial tetrapods. They often attained large body size, with a length of up to 3 m and a weight of at least one metric ton (e.g., *Stahleckeria*).

Perhaps the most striking thing about these 10 or more independent lineages of herbivorous Triassic tetrapods is that the teeth and jaws were different in each, and likely the occlusal mechanics were also different according to the shape of the food-processing apparatus and the type of plants being consumed. This diversity pales by comparison in extant herbivorous reptiles and mammals.

Aerial Reptiles

We distinguish gliders (which are mostly passive fliers) from true flyers, who power themselves through the air by means of a flight stroke. There is no evidence that any group of active flyers among tetrapods is related to gliders or passed through a gliding stage during the evolution of active flight (Padian 1985; pace Dudley et al. 2007). Pterosaurs were the first group of tetrapods to evolve active flight. To date, at least seven lineages are known from the Late Triassic (Dalla Vecchia 2013). Early pterosaurs typically have long jaws, often with multicusped teeth at the back of the jaws and large fangs near the front. Fish remains have been found in the gut regions of some Triassic pterosaur skeletons. Even the most basal pterosaurs are already fully winged and capable of active flight; they apparently descended from bipedal ornithodirans (Padian 2008). Late Triassic forms such as *Eudimorphodon* and *Peteinosaurus* still have wingspans of well under 1 m. Jurassic pterosaurs had simpler teeth but otherwise resembled their Triassic precursors in wing shape and skeletal structure.

In contrast, gliding seems to have evolved independently in at least three groups of small Triassic reptiles. The Late Triassic kuehneosaurids such as *Icarosaurus* were distantly related to lepidosaurs (Evans and Jones 2010). They have greatly elongated thoracic ribs upon which a gliding membrane of skin was stretched, much like in the present-day lizards of the genus *Draco*. The Late Triassic *Mecistotrachelos* was a long-necked (possible) archosauromorph with elongated ribs for gliding (Fraser et al. 2007). The Middle or Late Triassic *Sharovipteryx* has an extensive skin fold stretched between its hind limbs, which some workers have interpreted as a feature for parachuting or gliding. (Its forelimbs are small but poorly known.) Another small diapsid, *Longisquama*, which lived alongside *Sharovipteryx*, has long, recurved, blade-like epidermal structures along its back that have reminded some workers of bird feathers. However, there is nothing really feather-like about these structures, which form a single row along the midline of the back and had nothing to do with gliding or active flight. Nevertheless, this degree of diversification of aerial tetrapods into one flying lineage and at least three gliding ones has only been exceeded during the Cenozoic, with the great diversification of birds and the evolution of the flying bats and a great variety of gliding marsupials, rodents, and dermopterans (Dudley et al. 2007).

Smaller Tetrapods and Their Varied Roles

As in present-day continental ecosystems, small tetrapods played important ecological roles in the Triassic, including small carnivores/insectivores (generally a size-dependent category), omnivores, and herbivores, but the players were different. Based on the presence of Early Triassic stem-frogs (such as *Triadobatrachus*), frogs must have existed during the Triassic but have no known fossil record until the Early Jurassic. Thus, their sister taxon, comprising newts and salamanders (caudates) must have been present as well, but, with the possible exception of the Middle or Late Triassic *Triassurus*, their fossil record begins in the Middle Jurassic.

Among Lepidosauria, squamates (lizards and snakes) are so far unknown from the Triassic, but their sister group, the sphenodontians (Rhynchocephalia, restricted today to the genus *Sphenodon*), dates back to the Middle Triassic and diversified during the Late Triassic (Jones et al. 2013). Some sphenodontians were probably insectivores/carnivores, but others fed on plants (Jones 2008).

Triassic mammaliaforms and haramiyid mammals were small (5–10 g) and presumably insectivorous, as were the closely related tritheledontid and brasilodontid cynodonts (Kemp 2005).

Among now-extinct groups of small tetrapods, procolophonids and gomphodont cynodonts appear to have been predominantly herbivorous (Fraser 2006; Sues and Fraser 2010).

Carnivores included a variety of small ornithodiran archosaurs, such as *Marasuchus*, as well as others more closely related to crocodylians, such as *Gracilisuchus*. All these animals are known from at least partial skeletons. But much remains to be learned about the diversity and structure of many other Triassic small tetrapods. Recovery of their skeletal remains usually requires breaking up and sieving bulk samples of bone-bearing sedimentary rock. In one example, the *Placerias* Quarry in the Upper Triassic of Arizona, the known diversity of vertebrates was tripled by analysis of microvertebrate remains (Kaye and Padian 1994). The problem is that the processing of the fossil-bearing matrix dissociates and often damages bones and teeth to

a point where anatomical and taxonomic identification of the elements can become difficult if not impossible.

Functional Ecology and the Structure of Triassic Communities

Given the morphological and taxonomic diversity of Triassic land-dwelling tetrapods, what can we learn about the diversity and evolution of their communities and how they were structured? The reconstruction of "food webs" and other diagrams of trophic flow in communities is difficult because we cannot directly observe diets (except when food remains are preserved in a digestive tract); diets may change with size through the lifetime of an animal or even seasonally; and we cannot presume that a given animal lived in a particular environment of deposition merely because it was preserved there (it could have washed or been carried in, for example).

Generally speaking, carnivores feed on all members of a community that they encounter, depending on body size and ontogenetic stage. Small carnivores feed on small vertebrates and insects, again dependent in part on body size. Herbivores flourish and diversify depending on availability of plants, but there is little evidence of specific herbivores eating specific plants. Aquatic and semiaquatic animals presumably fed on fishes and invertebrates, but unless their jaws (e.g., placodonts) or body forms (e.g., thunniform ichthyosaurs) are obviously specialized for feeding or locomotory strategies, it is difficult to go beyond generalizations.

Size Increase and Ecological Diversification

The median body size of Triassic tetrapods was substantially greater than those of their Permian predecessors, partly because there were more large taxa in the Triassic. The largest known Permian tetrapods included the dinocephalian and some dicynodont synapsids and, among reptiles, pareiasaurs (up to 3 m in length). During the Triassic certain stereospondyl stem-amphibians, kannemeyeriiform dicynodonts, the archosauriform *Erythrosuchus*, phytosaurs, crocodile-line archosaurs (aetosaurs, ornithosuchids, rauisuchids), and saurischian dinosaurs attained lengths of up to 5 m and sometimes more (particularly the largest phytosaurs, rauisuchids, and sauropodomorphs). Most of these taxa did not reach large body size by growing quickly through high metabolic levels; exceptions were the sauropodomorphs and possibly the erythrosuchians. Most of these large taxa grew slowly and took many years to reach full size (de Ricqlès et al. 2008). Apart from the large stem-amphibians and dicynodonts, whose lineages persisted with new members from the Permian to the Triassic, the vast majority of new large animals in the Triassic were reptiles. They include members of proterosuchids, erythrosuchids, trilophosaurs, rhynchosaurs, protorosaurs, phytosaurs, aetosaurs, paracrocodylomorphs, and sauropodomorphs. It can be presumed that these animals became larger because their food sources were abundant enough throughout the year to enable them do so. Larger size enables resistance to smaller predators for prey species, and access to a greater range of prey (including larger prey) for predators. Expanding the upper limits of size in communities has other important effects on community diversity: both smaller species and the young of larger forms can play similar roles.

How Stable Were Triassic Continental Vertebrate Communities?

"Stability" is a difficult question for community ecologists who study diversity in the short term. Throughout the Triassic, some notable changes took place. The overriding theme is that whereas ecological roles diversified in the Triassic, from one temporal interval to the next (and also geographically), ecological roles may have been filled by members of either the same or different clades (or both) as those from the previous interval.

As detailed above, Triassic communities of continental tetrapods had three major components: late Paleozoic "holdovers," "indigenous" Triassic taxa that diversified and became extinct during that period, and "living" groups that came to dominate more recent communities. During the Triassic, one clade of predominantly aquatic stem-amphibians, Stereospondyli, diversified, but most other temnospondyl lineages had vanished by the end of the Permian. Derived synapsids (therapsids), however, had a relatively smaller role in these ecosystems, because the reptiles (which had been restricted to mostly small forms during the Permian) greatly diversified and because therapsids declined in diversity during the Triassic. However, two groups flourished during the Triassic: cynodonts (ranging from *Thrinaxodon* to *Exaeretodon* to *Morganucodon*) and certain dicynodonts (especially *Lystrosaurus* and kannemeyeriiforms) (Kemp 2005).

Although Early Triassic tetrapod communities were still taxonomically dominated by therapsids and temnospondyls, reptiles were diversifying even then, and, by the early Middle Triassic, they clearly were in ascendancy. By the Late Triassic, reptiles had become the dominant tetrapods in continental ecosystems. Archosauromorph reptiles, whose phylogeny is detailed in figure 20.1, were the principal component of this Late Triassic diversification. In other words, there was a turnover of taxa not only in terms of species within clades, but also of clades that dominated ecological roles. The latter did not include dinosaurs: there were only a few small to medium-sized carnivorous theropods and some medium-sized to large, predominantly herbivorous sauropodomorphs. Had life ended after the Triassic, dinosaurs would not have appeared as interesting as many other groups of the time.

Despite the apparent absence of major topographic barriers, Pangaea was not a homogenous place. Many tetrapod taxa were restricted to some regions, more diverse in certain regions, and existed for longer periods of time in some regions than in others. Some of this differentiation may have been latitudinal: for example, Whiteside et al. (2011) argued for latitudinal differences in the distribution of procolophonid parareptiles and gomphodont cynodonts from the Late Triassic in eastern North America. Dinosaurs did not replace their relatives as quickly or at the same times in low latitudes as in higher ones (Irmis et al. 2007). As interesting, and just as problematic, is that although theropod dinosaurs ranged widely across Pangaea during the Late Triassic, sauropodomorph dinosaurs were abundant in Europe, southern Africa, and South America during that time but are unknown from North America. When sauropodomorphs finally did appear in North America during the Early Jurassic, they seem to represent lineages that came from three different geographic areas (Rowe et al. 2011). The question of stability in Triassic continental ecosystems requires understanding that different groups occupied different roles in time and space, and that phylogeny and biogeography are important components of the mix.

FIG. 20.8 Reconstruction of a Late Triassic habitat with various tetrapods in what is now the American Southwest. The dicynodont therapsid *Placerias* (in the upper left corner, drinking) represents one of the geologically youngest members of this group of "Paleozoic holdovers." The heavily armored aetosaur *Desmatosuchus* (near the center of the scene, raising itself up on a tree stump) and the crocodile-like phytosaur *Machaeroprosopus* (partially concealed under a fallen tree) both represent "indigenous" groups of reptiles that did not survive the end-Triassic extinction event. Three early theropod dinosaurs, representing the "living fauna," scurry across the scene. Image courtesy of and copyright by Mary A. Parrish (National Museum of Natural History).

What Caused the Triassic Faunal Changes?

In brief, there is no satisfactory causal explanation for the taxonomic and ecological revolutions that occurred during the Triassic, and perhaps we should not expect to be able to divine them (Padian 2013). Climate change has often been invoked as a driver of biotic change in Triassic tetrapod communities. Earlier studies attempted to explain the evolutionary success of diapsid reptiles over synapsids by climatic factors. Following the late Paleozoic "Icehouse" world, global climates during the Triassic Period were for the most part warm and became increasingly dry, especially toward the end of this period. Robinson (1971) hypothesized that Triassic climates favored diapsids over synapsids in part because the former were able to excrete nitrogen with little loss of water. Extant reptiles and birds excrete nitrogen in the form of uric acid, either as a nearly dry pellet (in lizards) or as a paste (in birds). By contrast, present-day mammals (and presumably their synapsid precursors) almost exclusively excrete nitrogen as urea, which requires copious amounts of water to be removed from the body. Given increasingly drier climatic conditions during the early Mesozoic, the water-saving disposal of nitrogen waste in archosaurs could have conferred a competitive advantage on these reptiles.

Irmis (2011) argued that climatic changes as currently understood during the Triassic could not account for the observed patterns of taxonomic change, although there were regional and latitudinal differences in which particular taxa survived and coexisted (Irmis et al. 2007). Throughout the Triassic, differential extinction favored some groups over others, though we do not know the ecological and environmental causes of these changes. We know that nonmammalian synapsids and temnospondyls declined in taxonomic and ecological diversity compared to reptiles during the Triassic. We have no way to test hypotheses of competition among these groups, nor their implication that the surviving taxa must have been somehow competitively superior. The notion that dinosaurs simply got a "lucky break" (Brusatte et al. 2008) is intriguing but untestable. The traditional idea of "competitive superiority" of archosaurian reptiles, particularly dinosaurs, does not hold water because early dinosaurs coexisted with their relatives (Irmis et al. 2007) and with likely pseudosuchian competitors for more than 20 million years in some regions.

The Triassic—a Time of Functional and Ecological Innovation

The singular importance of the Triassic Period for the evolution of continental tetrapods consists not simply of the explosion of lineages, but of the diversifications of functional form and ecological exploitation. The taxonomic component of that diversity mainly comprised a burgeoning of archosauromorph reptiles, culminating in the loss at the end of the Triassic of most groups of "indigenous" Triassic tetrapod groups and the survival of dinosaurs, pterosaurs, and the reptilian groups that would dominate the rest of the Mesozoic and Cenozoic eras, along with the other components of living tetrapod communities.

The ecological diversity of Triassic reptiles can best be expressed by considering the number of independent evolutionary iterations of particular ecomorphs, notably terrestrial macrocarnivores and herbivores. Marine reptiles comprised entirely separate evolutionary radiations into varied body plans. With the exception of the "Cambrian explosion" and its effects on the diversification of multicellular animals, it is difficult to think of another period in the history of life that witnessed so much rapid diversification. Even the Cenozoic radiations of mammals, although different in many respects, did not exceed the Triassic diversification of terrestrial tetrapods. The replacement of late Paleozoic "holdovers," first by "indigenous" Triassic groups and then by representatives of the "living fauna" that has dominated since the early Mesozoic, resembles the waves of mammalian groups that replaced each other from the Paleocene to the Neogene.

* * *

Acknowledgments

We salute the late Farish Jenkins for having been the very model of a scientist, teacher, colleague, and friend for more than four decades. As much as he distinguished himself through elegant experimental research on vertebrate locomotion, his annual forays into the field greatly enriched the record of early Mesozoic vertebrate life and encouraged others to follow his lead. Every paper written by Farish tells a compelling story, weaving an exquisite narrative of functional evolution that integrates morphology, paleontology, and physiology in an innovative manner. He explored the Triassic for 30 years because he knew it held the secrets to the early evolution and assembly of the mammalian body plan and its physiological components, and he was wildly successful. Few can claim such a diverse legacy or set such a fine example as both a researcher and teacher.

We are grateful to Sterling Nesbitt and Corwin Sullivan for helpful reviews of the manuscript, and to J. Michael Parrish, William Parker, Jeffrey Martz, Nick Fraser, Rainer Schoch, Randall Irmis, Sarah Werning, and Paul E. Olsen for helpful discussions. We thank Mary Parrish for the painting reproduced as fig. 20.8. This is UCMP Contribution No. 2049.

References

Baird, D. 1957. Triassic reptile footprint faunules from Milford, New Jersey. Bull. Mus. Comp. Zool., Harvard Coll. 117:447–520.

Barberena, M. C. 1981. Novos materiais de *Traversodon stahleckeri* da Formação Santa Maria (Triássico do Rio Grande do Sul). Pesquisas 14:149–162.

Benton, M. J. 1983. The Triassic reptile *Hyperodapedon* from Elgin: functional morphology and relationships. Phil. Trans. R. Soc. Lond. B 302:605–720.

Benton, M. J. 2014. Vertebrate Palaeontology. 4th ed. Oxford: Wiley-Blackwell.

Bonaparte, J. F. 1984. Locomotion in rauisuchid thecodonts. J. Vert. Paleont. 3:210–218.

Bramble, D. M., and F. A. Jenkins Jr. 1989. Structural and functional integration across the reptile-mammal boundary: the locomotor system. Pp. 133–144 in D. B. Wake and G. Roth, eds., Complex Organismal Functions: Integration and Evolution in Tetrapods. Chichester: John Wiley and Sons.

Braun, J., and W.-E. Reif. 1985. A survey of aquatic locomotion in fishes and tetrapods. N. Jb. Geol. Paläont., Abh. 169:307–332.

Brinkman, D. 1980. The hind limb step cycle of *Caiman sclerops* and the mechanics of the crocodile tarsus and metatarsus. Can. J. Zool. 58:2187–2200.

Brusatte, S., M. J. Benton, M. Ruta, and G. T. Lloyd. 2008. Superiority, competition, and opportunism in the evolutionary radiation of dinosaurs. Science 321:1485–1488.

Butler, R. J., R. M. H. Smith, and D. B. Norman. 2007. A primitive ornithischian dinosaur from the Late Triassic of South Africa and the early evolution and diversification of Ornithischia. Proc. R. Soc. B 274:2041–2046.

Bystrov, A. P., and I. A. Efremov. 1940. [*Benthosuchus sushkini* Efr.—a labyrinthodont from the Eotriassic of Sharzhenga River.] Trudy Paleont. Inst. Akad. Nauk SSSR 10:1–152. [Russian with English translation.]

Carrier, D. R. 1987. The evolution of locomotor stamina in tetrapods: circumventing a mechanical constraint. Paleobiology 13:326–341.

Carroll, R. L., and W. Lindsay. 1985. Cranial anatomy of the primitive reptile *Procolophon*. Can. J. Earth Sci. 22:1571–1587.

Castanet, J., A. Grandin, A. Abourachid, and A. de Ricqlès. 1996. Expression de la dynamique de croissance dans la structure de l'os périostique chez *Anas platyrhynchos*. Compt. Rend. Acad. Sci. Paris, Sci. de la Vie 319:301–308.

Chinsamy-Turan, A., ed. 2011. The Biology of Non-Mammalian Therapsids: Insights from Bone Microstructure. Bloomington: Indiana University Press.

Colbert, E. H. 1947. Studies of the phytosaurs *Machaeroprosopus* and *Rutiodon*. Bull. Amer. Mus. Nat. Hist. 88:57–96.

Dalla Vecchia, F. M. 2013. Triassic pterosaurs. Pp. 119–155 in S. J. Nesbitt, J. B. Desojo, and R. B. Irmis, eds., Anatomy, Phylogeny and Palaeobiology of Early Archosaurs and Their Kin. Geological Society, London, Special Publications 379. Bath: Geological Society Publishing House.

de Ricqlès, A. 1969. Recherches paléohistologiques sur les os longs des tétrapodes II.—Quelques observations sur la structure des os longs des thériodontes. Ann. Paléont. 55:3–52.

de Ricqlès, A. 1972. Recherches paléohistologiques sur les os longs des tétrapodes III.—Titanosuchiens, dinocéphales et dicynodontes. Ann. Paléont. 58:17–60.

de Ricqlès, A. 1974. Recherches paléohistologiques sur les os longs des tétrapodes IV.—Eothériodontes et pélycosaures. Ann. Paléont. 60:1–39.

de Ricqlès, A. 1980. Tissue structures of dinosaur bone—functional significance and possible relation to dinosaur physiology. Pp. 103–139 in R. D. K. Thomas and E. C. Olson, eds., A Cold Look at the Warm-Blooded Dinosaurs. Boulder, CO: Westview Press.

de Ricqlès, A., K. Padian, F. Knoll, and J. R. Horner. 2008. On the origin of rapid growth rates in archosaurs and their ancient relatives: complementary histological studies on Triassic archosauriforms and the problem of a "phylogenetic signal" in bone histology. Ann. Paléont. 94:57–76.

Dudley, R., G. Byrnes, S. P. Yanoviak, B. Borrell, R. M. Brown, and J. A. McGuire. 2007. Gliding and the functional origins of flight: biomechanical novelty or necessity? Ann. Rev. Ecol. Evol. Syst. 38:179–201.

Dzik, J. 2003. A beaked herbivorous archosaur with dinosaur affinities from the early Late Triassic of Poland. J. Vert. Paleont. 23:556–574.

Eagle, R. A., E. A. Schauble, A. K. Tripati, T. Tütken, R. C. Hulbert, and J. M. Eiler. 2010. Body temperatures of modern and extinct vertebrates from ^{13}C-^{18}O bond abundances in bioapatite. Proc. Nat. Acad. Sci. USA 107:10377–10382.

Evans, S. E., and M. E. H. Jones. 2010. The origin, early history and diversification of lepidosauromorph reptiles. Pp. 27–44 in S. Bandyopadhyay, ed., New Aspects of Mesozoic Biodiversity. (Lecture Notes in Earth Sciences 132.) Berlin and Heidelberg: Springer-Verlag.

Fraser, N. 2006. Dawn of the Dinosaurs: Life in the Triassic. Bloomington: Indiana University Press.

Fraser, N. C., P. E. Olsen, A. C. Dooley Jr., and T. R. Ryan. 2007. A new gliding tetrapod (Diapsida: ?Archosauromorpha) from the Upper Triassic (Carnian) of Virginia. J. Vert. Paleont. 27:261–265.

Gauthier, J. A. 1986. Saurischian monophyly and the origin of birds. Pp. 1–55 in K. Padian, ed., The Origin of Birds and the Evolution of Flight. Memoirs of the California Academy of Sciences 8. San Francisco: California Academy of Sciences.

Gower, D. J. 2003. Osteology of the early archosaurian reptile *Erythrosuchus africanus*. Ann. S. Afr. Mus. 110:1–84.

Haubold, H. 1971. Ichnia amphibiorum et reptiliorum fossilium. Pp. 1–124 in O. Kuhn, ed., Handbuch der Paläoherpetologie, Part 18. Stuttgart: Gustav Fischer Verlag.

Hopson, J. A., and J. W. Kitching. 2001. A probainognathian cynodont from South Africa and the phylogeny of nonmammalian cynodonts. Bull. Mus. Comp. Zool., Harvard Univ. 156:5–35.

Horner, J. R., K. Padian, and A. de Ricqlès. 2005. How dinosaurs grew so large—and so small. Sci. Am. 293(1):46–53.

Irmis, R. B. 2011. Evaluating hypotheses for the early diversification of dinosaurs. Earth Environm. Sci. Trans. R. Soc. Edinb. 101:397–426.

Irmis, R. B., S. J. Nesbitt, K. Padian, N. D. Smith, A. H. Turner, D. Woody, and A. Downs. 2007. A Late Triassic dinosauromorph assemblage from New Mexico and the rise of dinosaurs. Science 317:358–361.

Irmis, R. B., W. G. Parker, S. J. Nesbitt, and J. Liu. 2007. Early ornithischian dinosaurs: the Triassic record. Hist. Biol. 19:3–22.

Jenkins, F. A., Jr. 1971a. Limb posture and locomotion in the Virginia opossum (*Didelphis marsupialis*) and in other non-cursorial mammals. J. Zool. (Lond.) 165:303–315.

Jenkins, F. A., Jr. 1971b. The postcranial skeleton of African cynodonts. Bull. Peabody Mus. Nat. Hist., Yale Univ. 36:1–216.

Jones, M. E. H. 2008. Skull shape and feeding strategy in *Sphenodon* and other rhynchocephalians (Diapsida: Lepidosauria). J. Morph. 269:945–966.

Jones, M. E. H., C. L. Anderson, C. A. Hipsley, J. Müller, S. E. Evans, and R. R. Schoch. 2013. Integration of molecules and new fossils supports a Triassic origin for Lepidosauria (lizards, snakes, and tuatara). BMC Evol. Biol. 13.

Joyce, W. G., and J. A. Gauthier. 2004. Palaeoecology of Triassic stem turtles sheds new light on turtle origins. Proc. R. Soc. Lond. B 271:1–5.

Kaye, F. T., and K. Padian. 1994. Microvertebrates from the Placerias Quarry: a window on Late Triassic diversity. Pp. 171–196 in N. C. Fraser and H.-D. Sues, eds., In the Shadow of the Dinosaurs. New York: Cambridge University Press.

Kemp, T. S. 1978. Stance and gait in the hindlimb of a therocephalian mammal-like reptile. J. Zool. (Lond.) 186:143–161.

Kemp, T. S. 1982. Mammal-like Reptiles and the Origin of Mammals. London: Academic Press.

Kemp, T. S. 2005. The Origin and Evolution of Mammals. Oxford: Oxford University Press.

Li, C., X. Wu, O. Rieppel, L. Wang, and L. Zhao. 2008. An ancestral turtle from the Late Triassic of southwestern China. Nature 456:497–501.

Li, C., X. Wu, O. Rieppel, L. Zhao, and L. Wang. 2011. A new Triassic marine reptile from southwestern China. Vert. Paleont. 31:303–312.

Lyson, T. R., G. S. Bever, T. M. Scheyer, A. Y. Hsiang, and J. A. Gauthier. 2013. Evolutionary origin of the turtle shell. Curr. Biol. http://dx.doi.org/10.1016/j.cub.2013.05.003.

Martínez, R. N., and O. A. Alcober. 2009. A basal sauropodomorph (Dinosauria: Saurischia) from the Ischigualasto Formation (Triassic: Carnian) and the early evolution of Sauropodomorpha. PLoS One 4(2):e4397.

Modesto, S. P., H.-D. Sues, and R. J. Damiani. 2001. A new Triassic procolophonoid reptile and its implications for procolophonoid survivorship during the Permo-Triassic extinction event. Proc. R. Soc. Lond. B 268:2047–2052.

Motani, R. 2005. Ichthyosauria: evolution and physical constraints of fish-shaped reptiles. Ann. Rev. Earth Planet. Sci. 33:395–420.

Nesbitt, S. J. 2011. The early evolution of archosaurs: relationships and the origin of major clades. Bull. Amer. Nat. Hist. 352:1–292.

Nesbitt, S. J., C. A. Sidor, R. B. Irmis, K. D. Angielczyk, R. M. H. Smith, and L. A. Tsuji. 2010. Ecologically distinct dinosaurian sister group shows early diversification of Ornithodira. Nature 464:95–98.

Padian, K. 1983. A functional analysis of flying and walking in pterosaurs. Paleobiology 9:218–239.

Padian, K. 1985. The origins and aerodynamics of flight in extinct tetrapods. Palaeontology 28:423–433.

Padian, K. 1986. Introduction. Pp. 1–8 in K. Padian, ed., The Beginning of the Age of Dinosaurs: Faunal Change across the Triassic-Jurassic Boundary. New York: Cambridge University Press.

Padian, K. 2008. Were pterosaur ancestors bipedal or quadrupedal? Morphometric, functional, and phylogenetic considerations. Zitteliana 28B:21–28.

Padian, K. 2013. The problem of dinosaur origins: integrating three approaches to the rise of Dinosauria. Earth Environm. Sci. Trans. R. Soc. Edinb. 103:1–20.

Padian, K., and E.-T. Lamm, eds. 2012. Bone Histology of Fossil Tetrapods. Berkeley: University of California Press.

Padian, K., A. de Ricqlès, and J. R. Horner. 2001. Dinosaurian growth rates and bird origins. Nature 412:405–408.

Padian, K., J. R. Horner, and A. de Ricqlès. 2004. Growth in small dinosaurs and pterosaurs: the evolution of archosaurian growth strategies. J. Vert. Paleont. 24:555–571.

Padian, K., C. Li, and J. Pchelnikova. 2010. The trackmaker of Apatopus: early diversification of archosaur stance and gait. Palaeontology 53:175–189.

Parker, W. G., R. B. Irmis, S. J. Nesbitt, J. W. Martz, and L. S. Browne. 2005. The Late Triassic pseudosuchian Revueltosaurus callenderi and its implications for the diversity of early ornithischian dinosaurs. Proc. R. Soc. Lond. B 272:963–969.

Parrish, J. M. 1986. Locomotor adaptations in the hindlimb and pelvis of the Thecodontia. Hunteria 1:1–35.

Parrish, J. M. 1987. The origin of crocodilian locomotion. Paleobiology 13:396–414.

Paul, G. 1993. Are Syntarsus and the Whitaker Quarry theropod the same genus? Pp. 397–402 in S. G. Lucas and M. Morales, eds., The Nonmarine Triassic. New Mexico Mus. Nat. Hist. Sci. Bull. 3.

Reisz, R. R., and H.-D. Sues. 2000. Herbivory in late Paleozoic and Triassic terrestrial tetrapods. Pp. 9–41 in H.-D. Sues, ed., Evolution of Herbivory in Terrestrial Tetrapods: Perspectives from the Fossil Record. New York: Cambridge University Press.

Rieppel, O. 2000a. Paraplacodus and the phylogeny of the Placodontia. Zool. J. Linn. Soc. 130:635–659.

Rieppel, O. 2000b. Sauropterygia I. Pp. 1–134 in P. Wellnhofer, ed., Handbuch der Paläoherpetologie, Part 12A. Munich: Verlag Dr. Friedrich Pfeil.

Rieppel, O. 2002. Feeding mechanics in Triassic stem-group sauropterygians: the anatomy of a successful invasion of Mesozoic seas. Zool. J. Linn. Soc. 135:33–63.

Robinson, J. A. 1975. The locomotion of plesiosaurs. N. Jb. Geol. Paläont., Abh. 149:286–332.

Robinson, P. L. 1962. Gliding lizards from the Upper Keuper of Great Britain. Proc. Geol. Soc. Lond. 1601:137–146.

Robinson, P. L. 1971. A problem of faunal replacement on Permo-Triassic continents. Palaeontology 14:131–153.

Rowe, T. B., H.-D. Sues, and R. R. Reisz. 2011. Dispersal and diversity in the earliest North American sauropodomorph dinosaurs, with a description of a new taxon. Proc. R. Soc. B 278:1044–1053.

Ruta, M., and M. J. Benton. 2008. Calibrated diversity, tree topology and the mother of mass extinctions: the lesson of temnospondyls. Palaeontology 51:1261–1288.

Ruta, M., D. Pisani, G. T. Lloyd, and M. J. Benton. 2007. A supertree of Temnospondyli: cladogenetic patterns in the most species-rich group of early tetrapods. Proc. R. Soc. B 274:3087–3095.

Sereno, P. C. 1991. Basal archosaurs: phylogenetic relationships and functional implications. Soc. Vert. Paleontol. Mem. 2:1–53.

Sues, H.-D. 1986. Locomotion and body form in early therapsids (Dinocephalia, Gorgonopsia, and Therocephalia). Pp. 61–70 in N. Hotton III, P. D. MacLean, J. J. Roth, and E. C. Roth, eds., The Ecology and Biology of Mammal-Like Reptiles. Washington, DC: Smithsonian Institution Press.

Sues, H.-D., and N. C. Fraser. 2010. Triassic Life on Land: The Great Transition. New York: Columbia University Press.

Sues, H.-D., and F. A. Jenkins Jr. 2006. The postcranial skeleton of *Kayentatherium wellesi* from the Lower Jurassic Kayenta Formation of Arizona and the phylogenetic significance of postcranial features in tritylodontid cynodonts. Pp. 114–152 in M. T. Carrano, R. W. Blob, T. J. Gaudin, and J. R. Wible, eds., Amniote Paleobiology: Perspectives on the Evolution of Mammals, Birds, and Reptiles. Chicago: University of Chicago Press.

Walker, A. D. 1961. Triassic reptiles from the Elgin area: *Stagonolepis*, *Dasygnathus* and their allies. Phil. Trans. R. Soc. Lond. B 244:103–204.

Weishampel, D. B., and F. Westphal. 1986. Die Plateosaurier von Trossingen im Geologischen Institut der Eberhard-Karls-Universität Tübingen. Tübingen: ATTEMPTO Verlag.

Werning, S., R. Irmis, N. Smith, A. Turner, and K. Padian. 2011. Archosauromorph bone histology reveals early evolution of elevated growth and metabolic rates. P. 213 in Abstracts of the 71st Annual Meeting of the Society of Vertebrate Paleontology.

Whiteside, J. H., D. S. Grogan, P. E. Olsen, and D. V. Kemp. 2011. Climatically driven biotic provinces of Late Triassic tropical Pangea. Proc. Nat. Acad. Sci. USA 108:8972–8977.

Witzmann, F., R. R. Schoch, and M. W. Maisch. 2008. A relict basal tetrapod from Germany: first evidence of a Triassic chroniosuchian outside Russia. Naturwissenschaften 95:67–72.

21

How Do Homoplasies Arise?
Origin and Maintenance of Reproductive Modes in Amphibians

Marvalee H. Wake*

One of the most significant of the several "Great Transformations in Vertebrate Evolution" is the evolution of live-bearing. It has developed in all of the major clades of vertebrates except birds, and it has a number of common attributes despite independent origins in diverse clades. Among tetrapods, live-bearing is presumed to facilitate increased terrestriality, such that, in more derived cases (of amphibians), a free-living larval period in water is obviated, and the young are born ready to live like adults as they reach maturity. Furthermore, the developing embryos and young are not abandoned, but maintained in or on the body of the mother, so they are carried, "protected," and often provided with nutrients after they have resorbed their yolk and throughout their developmental period. Thereby the developing larvae are not subject to specific predation or to food limitations. This is thought to increase the fitness of the young. A trade-off, though, is that if the pregnant mother dies before the birth of the young, her entire "investment" is lost, and fitness is zero. (See reviews by Packard et al. 1989 and Wake 1989.) Modification of reproductive modes is among the most complex of the "Great Transitions" that vertebrates have achieved in their morphology, function, and ecology.

Concomitantly, one of the "great unanswered questions" in evolutionary biology is how "similarity" evolves, so that members of distantly related lineages

* Department of Integrative Biology and Museum of Vertebrate Zoology, University of California, Berkeley

achieve virtually the same features—such as being live-bearers (this chapter) or having tongue-projecting feeding mechanisms (D. Wake et al., this volume). That is, how does homoplasy arise, and what maintains it? Why is "similarity" a dominant theme in vertebrate evolution? What are the biological reasons homoplasy is as common as it appears to be? If we want to understand the "great transformations" in the history of life, homoplasious situations present both opportunities (to understand mechanism) and challenges (difficulties in reconstructing the pattern of evolution). I address these questions by examining the evolution of viviparity (live-bearing) from a mechanistic perspective. I first consider current ideas about homoplasy—what it is and how to study it—then examine two cases of homoplasious evolution of viviparity to try to assess factors that mediate its evolution (largely based on morphology, and including new information), and conclude with some ideas about the nature of homoplasy based on my research in progress.

Several definitions of homoplasy occur in the literature:

Lankester (1870; founder of the term): "close agreement in *form* which may be attained in the course of evolutionary changes by organs or parts in two animals which have been subjected to similar moulding conditions of the environment, but have not close genetic community of origin . . . although they have a certain identity in primitive quality which is accountable for the agreement of their response to similar moulding conditions."

Sanderson and Donoghue (1989): "Homoplasy in cladistic analyses results when features hypothesized at the outset of an analysis to be homologous are found to arise more than once on a cladogram, or to originate and then be lost."

Wake, D. (1991): "Homoplasy is the *appearance* of "sameness" that results from independent evolution." (1999): "Homoplasy is derived similarity that is not the result of common ancestry . . . roles for functionalist . . . and structuralist . . . approaches to the . . . problem, which always must have historical (phylogenetic) rooting."

Hall (2007): "Homoplasy and homology are not dichotomies but the extremes of a continuum, reflecting deep or more recent shared ancestry

based on shared cellular mechanisms and processes and shared genes and gene pathways and networks."

Wake et al. (2011): "Homoplasy is the diametric opposite of homology—underlying similarity that does not result from inheritance at the hierarchical level . . . being considered."

Nixon and Carpenter (2012): "Homoplasy is error, in coding . . ."

Three "kinds" of homoplasy have long been characterized:

1. convergence—similar features develop in distantly related lineages, not derived from a shared ancestral condition, and via different generative programs.
2. reversal—features thought lost in lineages recur (genetic basis is retained and recalled).
3. parallelism—similar features develop from a common (identifiable?) substrate/program, elicited independently in different lineages by the same or different "signals."

Homoplasy—what it is, how it comes to be, and even whether it is "real"—has become a focus of new attention by evolutionary and developmental biologists. The development of molecular and genetic techniques for analyzing various patterns of comparative ontogenies and the relationships of taxa within lineages has prompted new attention to the nature of homology and homoplasy. Homoplasies are often identified by mapping characters on a tree that presents a phylogenetic hypothesis of relationships of taxa. If the features occur in lineages that do not have recent common ancestors, they are considered homoplasious. Because the research goal for such studies is usually either the phylogenetic hypothesis or an assessment of how much homoplasy exists in the character set, only rarely to date have attempts been made to assess the mechanistic basis for the evolution of homoplasious conditions. However, with new tools, new approaches to questions of homology and homoplasy are being undertaken.

Recently, several researchers have questioned whether the "kinds" of similarity mentioned above overlap conceptually and pragmatically, and whether they all are different "kinds" of homology, or whether homoplasy really is conceptually different from homology. The development of molecular and genetic tools has facilitated major changes in two arenas that dramatically affect the analysis of homology and especially homoplasy—the provision of molecular phylogenies based on a diversity of genes and gene products (sometimes used in a "total evidence" mode in combination with morphological characters), and the analysis of the genes and control mechanisms that underlie the development of morphological features, organism- and population-level life history characters, and even those of ecology and behavior. (e.g., Hall 1999, 2002, 2003, 2007; Sanderson and Hufford 1996, Wake 1991, 1999, 2003; Powell 2007; and Shubin et al. 2009 for different perspectives on this advance.) The ways that we think about pattern and process of evolution are changing and expanding rapidly. For example, Abouheif (2008) has proposed that parallelism is the basis for what he terms "mesoevolution," and that new tools and information about the genetics of development reveal pattern and process of mesoevolution. Arendt and Reznick (2008a, 2008b) consider parallelism and convergence a continuum, that parallelism is simply a kind of convergence, and that it is appropriate to recognize a broad class of such phenomena. Leander (2008) responded that a distinction between the two is important for assessing patterns and processes, such as the genetics of adaptation. Scotland (2011) opted for a direct dichotomy, claiming that all homoplasy is effectively a broad view of convergence, with phenotypic homoplasy as convergence and genotypic homoplasy being parallelism—the convergence of molecular data. He also stated, "Parallel changes of molecular traits may or may not be associated with convergent phenotypes but if so describe homoplasy at two biological levels—genotype and phenotype. Therefore convergence and parallelism are not alternative terms, but rather the genetics of homoplasy can explain, at the molecular level, how convergent phenotypes evolve." Wake et al. (2011) discussed these issues and pointed out that convergence generally results from different genetic mechanisms, but parallelisms have similar genetic bases. They illustrate how new "tools" in genetics, genomics, development, and phylogenetics can expand the analysis of homoplasy and the way that such studies will take our science from describing patterns to understanding processes and mechanisms of evolution. They also suggested that by using hierarchical approaches, we will reveal that "similarity" has multiple dimensions, such that rigid definitions of "kinds" of homoplasies might not serve full understanding of the evolution of like features, whether we have complete knowledge of the genetic basis for the mechanisms or not.

One of the two major foci of my research is the comparative biology of the evolution of derived modes of reproduction, especially viviparity (live-bearing) in the Lissamphibia—frogs, salamanders, and caecilians (the other is the biology and evolution of members of the caecilian clade). It occurred to me that the evolution of viviparity provided an excellent example of the evolution of homoplasious conditions, especially parallelisms (because of strong evidence of common substrates [e.g., morphologies, hormones] but with responses elicited independently). Such study is amenable to examination of features that would be either *common to,* or *different in,* the lineages that share the purported parallelisms, and this potentially allows assessment of the *mechanisms* by which parallel conditions evolve.

A research approach that examines the "similarities" at several different but inherently related levels of the hierarchy of biological organization is essential to the analysis of homoplasy. In my experience, several aspects of the biology of the study organisms must be explored in order to assess mechanistic processes and their effects on multiple aspects of the feature(s) under study. Therefore, in order to assess the evolution of similarity in derived modes of amphibian reproduction, I examine both maternal and embryonic properties of live-bearing development: the condition of the maternal parent (e.g., ovaries, oviducts, and skin) in order to evaluate physiology, endocrinology, and, to the degree possible, the underlying genetics, the trajectory and characteristics of development of the embryos/ fetuses, and aspects of ecology and behavior (especially courtship and fertilization), all looking for downward and upward causation. I use a diversity of techniques (developmental, morphological, analytical). I examine several taxa within each of the comparator clades, rather than a single typological example, in order to assess both within-clade and across-clade similarity and variation, in as robust a phylogenetic context as is available. I use the evidence drawn from these studies to generate hypotheses about the mechanisms by which the evolution of live-bearing has occurred.

I focus this account on the evolution of viviparity as an examination of mechanisms by which homoplasies arise in two cases of the evolution of "similarity" in live-bearing reproductive modes:

1. Back-brooding, often through metamorphosis of the young, in all species of the frog genera *Gastrotheca* (Hemiphractidae) and *Pipa* (one genus in the family Pipidae [also includes *Xenopus, Hymenochirus, Pseudhymenochirus,* and *Silurana*]); and

2. Intra-oviductal development through metamorphosis with provision of nutrients after yolk resorption in caecilians (Dermophiidae [presumably all members of the four genera included in the family], Indotyphlidae [one species of one genus of seven genera; mostly direct developers, so far as is known], Scolecomorphidae [presumably all members of one genus of the two in the family], Typhlonectidae [presumably all members of the five genera], the salamander *Salamandra atra* [Salamandridae], and the toad *Nimbaphrynoides* [Bufonidae, both of the species in the genus]).

I provide a brief background for the research on these cases, and present new information and my analyses in a hierarchical framework. I emphasize the proximal interaction of endocrine mediation and its correlation with seasonal/ecological cues as fundamental to the evolution of live-bearing for both mothers and young. How does "similarity" evolve? What aspects of mechanisms of live-bearing are "similar" and "different" in different lineages that have evolved "similarity" at presumed end-point levels of study? This is work in progress, and it opens more new questions than it provides hard conclusions.

Case 1: Brooding the Clutch of Fertilized Eggs in the Skin of the Back of the Maternal Frog

I first briefly compare and contrast the biology of the two distantly related back-breeding lineages, *Gastrotheca* and *Pipa*.

Gastrotheca are terrestrial frogs; courtship and mating are not dependent on nearby water. At sunset, males emerge and establish calling territories. Females emerge subsequently, and if they encounter a male, it usually results in courtship and amplexus. The courtship terminates with the female in a head-down position as the male clasping her back fertilizes her eggs as they emerge singly from her cloaca. He guides the eggs into the entrance of a permanent dorsal skin pouch (fig. 21.1A) on the back of the female (see Elinson et al. 1990 for a summary). So far as is known, of the 62 species of *Gastrotheca* currently recognized, the majority give birth to fully metamorphosed froglets, the rest to advanced tadpoles. In some species, the female is reported to use her hind feet to help open the pouch to aid her young to emerge (Duellman and Maness 1980).

The reproductive biology and development of one species, *Gastrotheca riobambae,* is relatively well known, largely owing to the seminal work of Eugenia del Pino. Jones et al. (1973) set the stage for del Pino by finding that estrogen administered at 12 weeks postfertilization causes pouch formation; the skin on both sides of the dorsal midline proliferates, becomes hypervascular and hyperplasic, folds, and fuses to form the pouch and the crypts in the epidermal internal lining. Epidermal keratin, dermal poison glands, and dermal chromatophore units (fig. 21.2A) are reduced or lost during pouch formation. During the nonbreeding season, the skin of the pouch is similar to that of the rest of the body. Del Pino rigorously examined the morphology of the pouch in *G. riobambae* and several of its relatives, the ovary and oogenesis and other aspects of the ovarian cycle, and the unique pattern of development of the embryos in a series of publications from 1975 to the present. She found that during the follicular phase of the endocrine cycle approximately 100 ova become mature (3 mm dia; del Pino and Sanchez 1977; de Albuja et al. 1983), and the skin of the pouch proliferates and vascularizes (del Pino et al. 1975; del Pino 1980, 1983). Whether the pouch is open or closed is correlated with the ovarian cycle; progesterone conditions the pouch and maintains hyperemia and vascularization and initiates pouch closure well before ovum maturation and oviposition (del Pino 1983; de Albuja et al. 1983). Del Pino was reluctant to identify the evacuated egg follicles as corpora lutea (del Pino and Sanchez

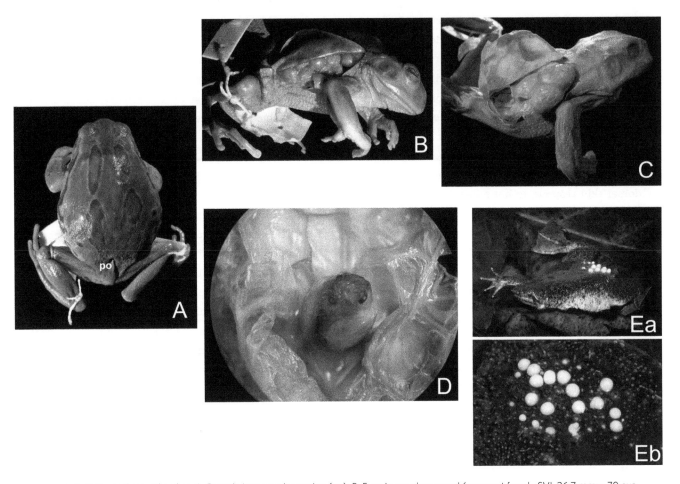

FIG. 21.1 A–B, *Gastrotheca riobambae*: A, Dorsal view, pouch opening (po). B, Eggs in pouch exposed (pregnant female SVL 36.7 mm; ~ 70 ova, each 3.2 mm dia; two layers). C, *Gastrotheca longipes*: Eggs in pouch exposed (pregnant female SVL 73. 5 mm; 17 ova, each 11. 5 mm dia). D, Ventral pouch skin of *G. testudinea* with an unborn froglet on it. Ea, *Pipa parva* pregnant female with eggs on her back. Eb, Close-up of eggs in Ea. Photos A–D by David Buckley/M. H. Wake; photos Ea and Eb by Danté Fenolio, used with his permission.

1977), but described their incomplete invasion by cells, and their regression 30–50 days later (in *G. riobambae* advanced tadpoles are born from the pouch approximately 100 days after the insertion of the fertilized ova), and measured progesterone levels and their effects on the pouch, oogenesis, and development. Eggs/embryos are necessary for pouch maintenance, as del Pino (1983) elegantly showed with an experiment in which she removed eggs from the pouch and substituted inert beads the size of the eggs. She then administered either progesterone or human chorionic gonadotropin to assess effects on behavior and on maintenance of the beads. Frogs with beads but not progesterone did not show the normal postmating behavior; frogs that received only progesterone did, leading to the conclusion that the behavior is hormonally mediated. The

progesterone-enhanced beads were incubated for about one week, then released, covered with pouch epithelium; del Pino (1983) concluded that the physical presence of embryos or beads, as well as progesterone-like hormone, is necessary for formation of the chambers for embryos in the pouch, and continuous presence of the hormonal stimulus is necessary for maintenance and incubation. The pouch entrance relaxes before birth. The pouch is a permanent structure, but it regresses after birth, then recrudesces with the next seasonal/hormonal cycle. Del Pino's extensive research on the endocrinology of back-brooding, including embryonic development and bell gills (the expanded, sac-like gills that cover the embryo's body) morphology (e.g., del Pino 1989; del Pino and Escobar 1981; del Pino and Elinson 1983; Elinson and del Pino 1985; Alcocer et al.

1992), has effectively established a research field on the cell biology of the evolution of derived modes of reproduction—which needs much more work!

My students and I have examined the histology of the pouch, ovaries, oviducts, and embryos in pregnant females of several species of *Gastrotheca* (fig. 21.2A–C); our work on eight additional species largely corroborates del Pino's observations of *G. riobambae*. We found in all species examined that the oviducts as well as the dorsal skin respond to hormones by proliferating the mucosa, and that the tissues become more vascularized and their cells larger and more fluid-filled. The two layers of the pouch fuse; the top has typical skin morphology; the bottom reduces pigmentation and keratinization, and increases mucous secretion. The ovary includes ova at several stages of maturation, as well as definite corpora lutea and corpora atretica. Del Pino found that the "corpora lutea" in *G. riobambae* regress well before birth and that tadpoles are born. We observed that the bell gills covering the body form a thin layer that encloses fluid (fig. 21.1D); we also noted that the tail is highly vascular in all the species with advanced tadpoles that we examined. Both have been suggested to be involved in gaseous exchange while appressed to the pouch epithelium. Egg and clutch size vary considerably across the species we examined; del Pino considered *G. riobambae* eggs large at 3 mm dia (fig. 21.1B); we found that in a *G. longipes* (which gives birth to fully metamorphosed froglets) the egg diameter is 11.5 mm (fig. 21.1C). Furthermore, del Pino considered the clutch of approximately 100 eggs "small"; the clutch of the *G. longipes* we examined is 17. This suggests that life history characters such as egg and clutch size and length of retention continue to evolve in various directions in the lineage, despite the plesiomorphic pouch deposition potentially constraining the evolution of those features in terms of size and physiological accommodation of the embryos.

We have not yet been able to accumulate full ontogenetic sequences of development for the species we examined, so our observations are dependent on samples of early, mid-, and late development in several species. Without complete sequences for each species, we cannot adequately assess whether there is significant variation in developmental patterns among the species, but we can at least begin to generalize about some similarities and differences in developmental features. I have

been particularly interested in examining late stages in order to evaluate any mechanistic basis for birth as tadpoles or as metamorphosed froglets. More research is necessary, but accumulating evidence suggests that corpus luteum maintenance and activity mediates the stage at which birth takes place. If corpora lutea regress at mid- to late tadpole development, tadpoles will be born shortly afterward. If the corpora lutea maintain function for a longer time, metamorphosis ensues, and froglets are born.

Members of the genus *Pipa* (fig. 21.1E) are aquatic northern South American frogs; far less is known about their reproductive biology than that of *G. riobambae*. They have a complex courtship, with acrobatic turns in the water and amplexus ending with the female head-down and the male clasping her back inguinally, positions that facilitate the male's fertilizing the eggs as they emerge from the female's cloaca and guiding the eggs onto the back of the female (Rabb and Rabb 1960; Rabb and Snedigar 1960; Weygoldt 1976). Some eggs do not embed adequately and fall off (Trueb and Massemin 2000), but most develop in the skin of the female's back, which overgrows the eggs, leaving a small aperture in nearly all species. The skin thickens and becomes fluid-filled and vascularized, forming a cup that holds the young. *Pipa arribali* is apparently the only species in which the skin completely covers the embryos (fig. 21.2D), thought by Buchacher (1993) to protect the embryos as the pregnant females migrate on land among ponds. Members of the [*P. aspera* + [*P. arrabali* + [*P. snethlageae* + *P. pipa*]]] clade give birth to fully metamorphosed froglets; [*P. parva* + *P. myersi*] + [*P. carvalhoi*]], the more basal taxa, have free-living tadpoles (Trueb and Massemin 2000). *P. carvalhoi* tadpoles exhaust their yolk supply 2/3 through gestation, and are born as advanced tadpoles. Eggs are larger and fewer in *Pipa* species that give birth to metamorphosed froglets (Trueb and Massemin 2000). Fernandes et al. (2011) hypothesized that there is nutrient transport over vascularized maternal and tadpole epithelia, but did not characterize the purportedly nutrient material.

Greven and Richter (2009) examined the morphology of the skin of the dorsum of *Pipa carvalhoi* in females with embryos and young at different stages of development. They carefully documented histological modifications during pregnancy. In *P. carvalhoi*, the skin overgrows each egg, such that they lie in cups with

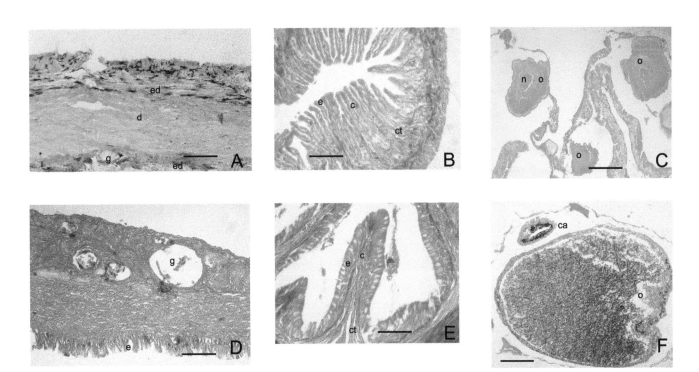

FIG. 21.2 Histological sections of ovaries, skin, oviducts of pregnant back-brooding frogs. A, *Gastrotheca riobambae* pouch skin; note the pigmentation of the dorsal layer, the depth of the fused skin layers that form the pouch, and the absence of pigment, the presence of glands, etc., in the inner (ventral) pouch layer. B, Hyperemic, well-vascularized oviductal mucosa of a pregnant *G. guentheri*. C, Ovary of a pregnant *G. riobambae*. D, *Pipa arrabali* skin covering the developing tadpole; note the complexity of the ventral layer adjacent to the tadpole. E, Hyperemic, well-vascularized oviduct of a pregnant *P. arrabali*. F, Postovulatory ovary of *P. snethlageae* with a "mature" ovum and a corpus atreticum. Photos by David Buckley/M. H. Wake. Scale bar in A, B, D, E = 10 μm, in C = 0.2 mm, and in F = 0.5 mm. Abbreviations: ca = corpus atreticum; c = capillary; ct = connective tissue (blue); d = dermis; e = epithelium; ed = epidermis; f = follicle; g = gland; n = nucleus; o = ovum.

open mouths. The epidermis becomes bilayered, thin, and lacks a stratum corneum; the dermis is highly vascularized and loose. Tadpoles are head-down before birth, with their tails waving through the aperture of the cup. They are born as well-developed, pre–limb extension, tadpoles. Greven and Richter examined the morphology of the ovaries, but only in terms of oogenesis, and they did not comment on the presence/absence of corpora lutea and corpora atretica. They did not assess the endocrinology of the system, but used morphology to hypothesize about hormonal mediation, noting that (1) follicles in early stages of development are numerous in pregnant females, but late-stage follicles are very few, perhaps implying the developmental delay characteristic of the influence of progesterone, and (2) the presence of corpora lutea have not been demonstrated in any species of *Pipa*. They also compared *P. carvalhoi* skin with that of *G. riobambae* as reported in the literature, and commented on similarities.

Our preliminary data for five species of *Pipa* suggest that the ovarian cycle is similar to that of *Gastrotheca*, with the presence of corpora lutea (observed in *P. parva* and *carvalhoi*) and corpora atretica in *P. snethlageae* (fig. 21.2F); however, because we do not have ontogenetic series, I cannot say how long corpora lutea are maintained. I predict that the same cycle of follicular estrogen stimulation of skin cup formation followed by luteal secretion of progesterone to maintain the pregnancy and inhibit ovum development occurs. As in *Gastrotheca*, the oviducts retain responsiveness to progesterone because they are hyperemic, highly vascularized (fig. 21.2E), have thickened mucosae, and are convoluted whenever young are embedded in the dorsal skin. Of the species we have examined to date, *P. arrabali* has the most extreme skin response with extensive proliferation, vascularization, etc., and the complete skin covering over the developing embryo/tadpole/froglet likely correlated with its tendency to be semiterrestrial. Gaseous exchange via the dilated tadpole tail is another aspect of maintenance that should be investigated in both lineages of back-brooding frogs.

An examination of such data with regard to the mechanisms by which "similarities" arise, and differences persist, can lead to alternative interpretations of the evolution of live-bearing. For example, Wiens et al. (2007) looked carefully at character evolution in *Gastrotheca* and its relatives, using a molecular phylogeny to estimate patterns of life-history evolution (fig. 21.3). They examined two specific life history characters: complex life cycles, which they defined as those with egg, free-living tadpole, and adult stages, and direct development, those with egg and adult stages only, thus evaluating whether or not there was a free-living larval stage. Their phylogeny included 43 of the 86 species in the family; they mapped the life history mode for each species on the phylogenetic reconstruction and found that *Gastrotheca* species with "complex life cycles" are nested within lineages of species with direct development. To their surprise, what they considered to be misleading inferences about character evolution resulted from their analysis, given that two hypotheses both received high statistical support—one being that the tadpole stage was lost early in the phylogeny but reappeared many times in *Gastrotheca*, the other that the tadpole stage was lost multiple times and never regained. One of the two hypotheses cannot be correct, but it is not clear which in their analysis.

I have examined the Wiens et al. data from a different perspective, that of the state of development of the tadpoles when born and what might mediate the birth process. In so doing, I do not see dichotomous modes, but a potential continuum. Wiens et al. considered complex life cycles with birth of tadpoles and direct development with birth of froglets as two distinct evolutionary end points. Information on the basis for these differences was not included in their analysis. For me, a major arena of missing information lies at other, but essential, hierarchical levels—how long are corpora lutea actively producing progesterone, which would maintain the pregnancy such that longer development ensues before birth? At what stage of development were the tadpoles when they were born? I predict, based on my preliminary data, that species vary considerably in the timing of corpus luteum regression, such that

"young" may be born at a range of different stages and sizes of development from mid-larvae to fully metamorphosed froglets. I suspect that this might also vary *within* species, depending on the specificity of regulation interaction with seasonal and/or other ecological effects. Elinson et al. (1990) reported that developmental rates and successes vary in *G. riobambae* when they are raised in labs in cities at different altitudes! And see discussions of *Nimbaphrynoides, Salamandra atra,* and some caecilians (below). Another question opened by my analysis is whether the skin of the dorsum in *Pipa* and *Gastrotheca* has significantly more estrogen and progesterone receptors than the skin of other frogs, how skin receptors correlate with the numbers and timing of activity of the oviductal receptors, and whether there is extensive skin receptor up-regulation and maintenance. Understanding the evolution of the phenomenon of back-brooding in frogs, especially the homoplasy represented by the two lineages, requires a hierarchical and integrative approach that considers molecular, cellular, physiological/endocrinological, behavioral, and ecological parameters of the parallelism, and it is only beginning to be explored.

The following points of similarity and of distinction in these two cases of the evolution of back-brooding require examination and delineation in order to understand the mechanisms of origin and the extents of the homoplasy in the two lineages:

1. In both clades, estrogen apparently prepares the dorsal skin for "pregnancy," and progesterone conditions and maintains gestation. However, the skin response differs between and even within clades. The oviducts maintain their responses to the hormones throughout gestation. The presence of the eggs/embryos is apparently necessary to maintain feedback to the hypothalamic/pituitary/gonadal axis to effect ovarian hormonal action.

2. Eggs apparently are larger and fewer in species that give birth to metamorphosed froglets, probably a maternal investment correlated with a longer brooding/developmental period. Selection on life-history parameters could affect offspring survival in both back-brooding clades via similar mechanisms.

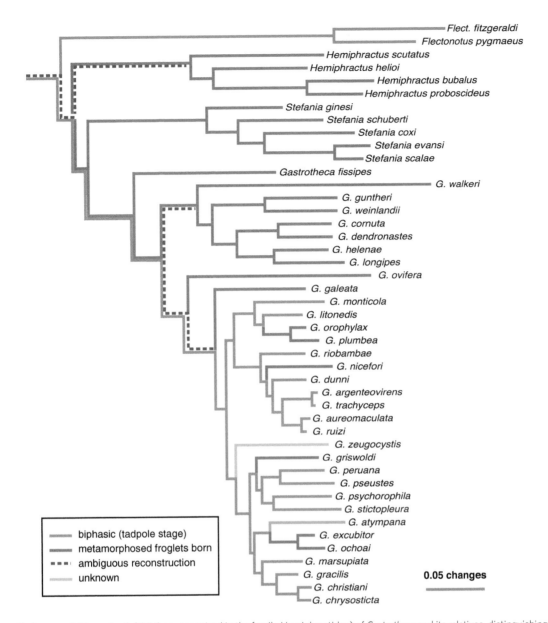

FIG. 21.3 Cladogram of 43 species (of 86 then recognized in the family Hemiphractidae) of *Gastrotheca* and its relatives, distinguishing species that give birth to tadpoles from those that give birth to metamorphosed froglets. Redrawn from Wiens et al. (2007), with permission from *Evolution*/John Wiley Publ. See text for discussion.

The cladogram shows species and a legend with the following categories:
- biphasic (tadpole stage)
- metamorphosed froglets born
- ambiguous reconstruction
- unknown

Scale: 0.05 changes

Species listed:
Flect. fitzgeraldi
Flectonotus pygmaeus
Hemiphractus scutatus
Hemiphractus helioi
Hemiphractus bubalus
Hemiphractus proboscideus
Stefania ginesi
Stefania schuberti
Stefania coxi
Stefania evansi
Stefania scalae
Gastrotheca fissipes
G. walkeri
G. guntheri
G. weinlandii
G. cornuta
G. dendronastes
G. helenae
G. longipes
G. ovifera
G. galeata
G. monticola
G. litonedis
G. orophylax
G. plumbea
G. riobambae
G. nicefori
G. dunni
G. argenteovirens
G. trachyceps
G. aureomaculata
G. ruizi
G. zeugocystis
G. griswoldi
G. peruana
G. pseustes
G. psychorophila
G. stictopleura
G. atympana
G. excubitor
G. ochoai
G. marsupiata
G. gracilis
G. christiani
G. chrysosticta

3. Courtship and amplexus differ dramatically between members of the aquatic and terrestrial clades; however, in both cases, behavior has evolved so that females assume postures that allow the males to guide fertilized ova to their backs.

4. Embryos develop modifications for respiration/gaseous exchange during gestation—bell gills in *Gastrotheca* and expanded, highly vascularized tails in *Pipa*.

5. The mechanism of control of timing of parturition is not known. However, corpora lutea resorb well before "birth" of tadpoles in *G. riobambae*, so loss of progesterone may be involved; consequently duration of corpora lutea and their function may determine stage of development at birth, probably in both clades. Our data suggest that the length of time the pregnancy can be maintained determines whether or not metamorphosis will be completed before birth.

Case 2: Intraoviductal Retention of Internally Fertilized Ova through Metamorphosis of the Embryos, with Provision of Maternal Nutrition Following Yolk Resorption

Oviductal viviparity in amphibians is relatively rare, particularly so in frogs and salamanders (Wake 1989). Internal fertilization, required for oviductal retention of fertilized eggs, is accomplished by distinctly different means in each of the three orders. The very few frogs that have internal fertilization use cloacal apposition or insertion of the vestigial tail (*Ascaphus*: see Sever et al. 2001); conversely, 90% of salamanders fertilize the ovulated eggs by the female's picking up a spermatophore deposited on the substrate by the male during courtship and holding it in her cloaca, from which the sperm swim up the oviducts to the ova. All caecilians apparently have internal fertilization via the male inserting the extruded rear part of his cloaca into the vent of the female, effecting direct transfer of the sperm. Many salamanders lay their eggs after a brief period of retention, such that free-living larvae develop in water. A few internally fertilizing frogs lay their eggs and guard them as they develop; *Eleutherodactylus coqui* is an example of a direct-developer that has internal fertilization (Townsend et al. 1981). Several anuran species retain yolky fertilized eggs in the oviducts and give birth to fully metamorphosed froglets (e.g., *Nectophrynoides* [see Wake 1980c; Channing et al. 2006] and the extinct *E. jasperi* [Wake 1978]). Among salamanders, prolonged retention of fertilized eggs is constrained largely to a few members of the family Salamandridae, with additional reports for a plethodontid (see below). Several subspecies of *Salamandra salamandra* have oviductal retention of internally fertilized ova; the embryos are yolk dependent and birth typically occurs as larvae, such that an aquatic period is required during which metamorphosis takes place. Two other subspecies of *S. salamandra* have a form of viviparity that involves siblicide, with birth following metamorphosis (Buckley et al. 2009). There is little information for caecilians, but even those that lay their clutch may retain the fertilized ova for a time; *Ichthyophis glutinosus* lays its eggs at neurulation, and the female guards them until the larvae hatch and wriggle to streams to continue developing through metamorphosis (Sarasin and Sarasin 1887–90). However, perhaps a third of caecilian species retain the developing young through metamorphosis in their oviducts, and yolk dependence without maternal nutrition after yolk resorption is not yet known in caecilians because all of the species that retain their young beyond neurula appear to provide a nutrient material secreted by the oviductal epithelium. A feature of such viviparity in all three orders is that egg size (yolk provisioning) is reduced, concomitant with the longer gestation periods and extensive maternal nutrient secretion following yolk resorbtion (summarized in Wake 1977a, 1977b, 1993; Wake and Dickie 1998).

In frogs, oviductal retention with maternal nutrition provided after yolk is resorbed is documented only for *Nimbaphrynoides occidentalis* (formerly included in the genus *Nectophrynoides*). Françoise Xavier, in a series of exquisite papers, integrated information on the ecology, reproductive cycle, development of the young, and the state of the ovaries and oviducts, including the basic endocrinology (see Xavier 1977, 1986, and fig. 21.4 for summaries). She found that the species has internal fertilization by cloacal apposition; the gestation period is nine months, directly correlated with the cold and dry and warm and rainy seasons; the maternal female provides nutrient material to the maturing "tadpoles" after yolk is fully resorbed, and that it is secreted by a region of the anterior end of the oviduct. Ovulation and fertilization occur in late October–early November, following an intergravidic period during which the ovarian follicles mature and secrete estrogens. The oocytes mature and take up yolk. Following courtship and mating, the female then goes underground during the dry season. Following ovulation, corpora lutea form and secrete progesterone. This inhibits oocyte growth, and slows embryonic development, but maintains some secretory activity of the oviductal mucosa. The female emerges in April with the inception of the rainy season, and commences active life above ground for six or seven months. At the time of her emergence, the corpora lutea begin to degenerate, progesterone secretion decreases, her uterine mucosa becomes more secretory and hyperemic, and the embryos undergo rapid growth through metamorphosis. The next season's oocytes increase in size. Parturition occurs in June, in mid–rainy season, so that prey are abundant and ~7.5 mm total length, 45 mg froglets are born and able to feed immediately.

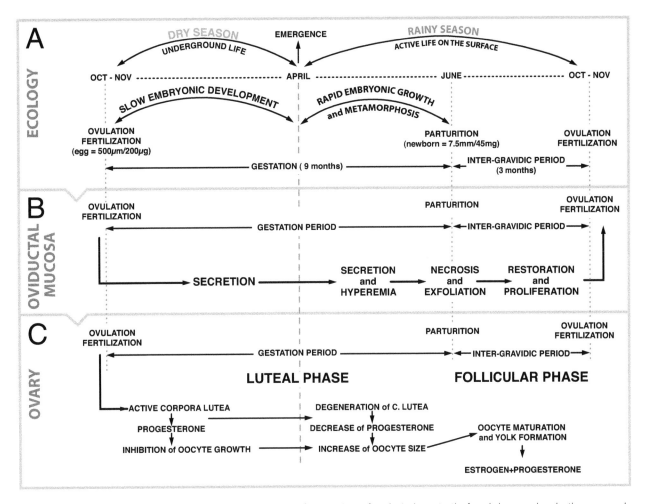

FIG. 21.4 Françoise Xavier's diagram summarizing the courses and interactions of ecological events, the female hormonal cycle, the ovary and oviduct cycles, and intraoviductal development of tadpoles in *N. occidentalis*. Redrawn from Xavier (1977) with permission from Springer Publ.

During the three-month intergravidic period, the oviductal epithelium first becomes necrotic and exfoliates, then proliferates under the influence of the new crop of maturing follicles secreting estrogens. Xavier clearly demonstrated that estrogen conditions the pregnancy and progesterone maintains it, and the ovarian and uterine cycles are highly correlated with habitat conditions and seasonality.

Obligate viviparity in salamanders is well documented for only one species, *Salamandra atra*, which has a 2–5 year gestation period and provision of maternal nutrition by ingestion of modified oviductal epithelial mucosal cells in an anterior region of the oviduct, the zona trophica. Viviparity also has been reported for the closely related *Lyciasalamandra luschani*, which gives birth to two fully metamorphosed young, one from

each oviduct, but it lacks a zona trophica and therefore presumably does not supply maternal nutrition (Ozeti 1979; Olgun et al. 2001; Polymeni and Greven 1992); *Salamandra algira* of Morocco is also reported to have viviparous populations (Donaire-Barroso et al. 2001). It is not clear whether either species is entirely yolk dependent or not; the authors defined "viviparity" solely in terms of giving birth to fully metamorphosed young, rather than the two characters (provision of maternal nutrition; birth of metamorphosed young) that I prefer to see expressed. Similarly, females of the distantly related European plethodontid *Speleomantes sarrabusensis* are reported to give birth to metamorphosed young, but nothing is known of the mechanism by which this occurs (Lanza and Leo 2001). Most, if not all, of these taxa likely will prove to be ovoviviparous—yolk

FIG. 21.5 *Salamandra atra* and her near-term fetus.

dependent—when their reproductive biology is better known. *Proteus anguinus* had been considered viviparous, but recent observations in captivity of courtship, egg-laying, and hatching of larvae have essentially rejected that idea (Juberthie et al. 1996).

Research on viviparity in *S. atra* (fig. 21.5) has been extensive. Vilter and Vilter (1960, 1964) found that corpora lutea are present throughout pregnancy, but that their numbers, sizes, and activity decrease late in gestation (probably regulating growth and initiating oogenesis). Joly and Picheral (1972), however, reported that removal of the ovaries has no effect on the course of gestation, and suggested that while the hypervascularization of the "uterus" is a response to estrogen, prostaglandins and/or local stimuli may also be involved. Greven and his colleagues (Greven 1977, 1980, 1998, 2002, 2003a; Greven and Guex 1994; Greven and Rütterbories 1984; Guex and Greven 1994) and Joly (Lostanlen et al. 1976; summarized in Joly et al. 1994) have rigorously examined oviduct morphology and function in pregnant and nonpregnant females of the viviparous *S. atra* and its sister species, the ovoviviparous *S. salamandra*, especially the oviductal epithelium and its secretions, including the egg jelly and the nutrient material secreted by the mother (Greven 2003b) and the specific region, the zona trophica, which produces the nutritive cells (Greven 1984). They also examined the development of the embryos and the way they obtain the nutrient material (Greven 1984, 2003a; Guex and Chen 1986). They report that when the embryos first hatch, they practice oophagy, eating disintegrated, perhaps unfertilized ova (and possibly less advanced embryos, because only one young is born from each oviduct). The more advanced embryos develop in a posterior, aglandular part of the

oviduct but roam anteriorly, using their specialized fetal dentition to scrape the epithelium of the zona trophica, the cells detaching from the connective tissue below as a consequence of partial necrosis. Stripping and regeneration of the epithelium continue to cycle in different parts of the zona trophica throughout the rest of the gestation period so that the young have abundant nutrition via epitheliophagy. Greven (1998) commented that the factors that induce and maintain the zona trophica are largely unknown, but he claimed that development of the zone "could not be induced by estrogen and/or progesterone," apparently because of the decline in number and function of corpora lutea during the period of active feeding. He also states that direct stimulation of the zone by the larval dentition is improbable, and that fetal secretion of stimuli, "perhaps prostaglandins, may trigger the induction of the trophic epithelium." Indeed, much research remains to be done, but I suspect that hormonal mediation will be found important.

In contrast to the situation in frogs and salamanders, viviparity has evolved independently four times in caecilians (determined by mapping the feature on the phylogenetic hypothesis of Wilkinson et al. 2011); it characterizes almost one-third of the species for which we know, or infer, reproductive modes. All caecilians, so far as is known, have internal fertilization via the male's insertion of his everted posterior cloaca into the vent of the female and transferring sperm directly, so retention of fertilized ova is facilitated. The gestation period of live-bearers, known for only a few species, ranges from 6 to 11 months, depending on the species. Ova are reduced in size, ~2 mm dia or less; clutch number also appears to be reduced relative to species with free-living larvae (Wake 1977a, 1977b, 1980b; Exbrayat

2006a). Ovoviviparity is not (yet) known for caecilians; all viviparous species apparently provide maternal nutrition after yolk is resorbed by the embryos early in the gestation period (Wake 1985, 1993; Wake and Dickie 1998), and all give birth to fully metamorphosed young. In *Dermophis mexicanus* (fig. 21.6A), the gestation period is eleven months, and in the population investigated, females are synchronous in having ova fertilized in June–July, an 11-month gestation period, and birth in May–June at the inception of the rains and increased prey abundance (Wake 1980b). Exbrayat and colleagues (summarized by Exbrayat 2006a, 2006b) observed a shorter gestation period in *Typhlonectes compressicaudus*, with events synchronized but to different months; nevertheless, they too concluded that ecological parameters are correlates of gestation and timing of birth. Hatching from the egg membrane in *D. mexicanus* occurs concomitantly with yolk depletion and the development of the jaws and their musculature and a species-specific "fetal" dentition at approximately 3 months into the 11-month gestation period (Wake and

Hanken 1982). The oviductal mucosa just at the time of hatching proliferates and becomes hyperemic, highly vascularized, and secretory (Wake 1993). The dentition (the shape of which changes over the gestation period [fig. 21.6B] [Wake 1980a]) is apparently used to scrape the oviductal mucosa of its secretions as the developing fetuses (fig. 21.6C) move about the lengths of the oviducts. The fetal teeth are shed very shortly after birth, with the adult dentition (characterized by tooth crown shape and position) erupting within ~48 hours after birth (Wake 1980a). There apparently is not a specialized region for nutrient secretion; virtually the entire mucosa is secretory (fig. 21.6D) (Wake 1970, 1972; Wake and Lai unpubl.). The ovaries of pregnant females are characterized by the presence of several large corpora lutea (fig. 21.6E), as well as maturing ova (Wake 1968). The corpora lutea appear to be present during the entire gestation period, at least in *D. mexicanus*, and fully metamorphosed young are born. Wake (1993 and unpubl. data) measured circulating progesterone during late pregnancy, and found it to be five times that

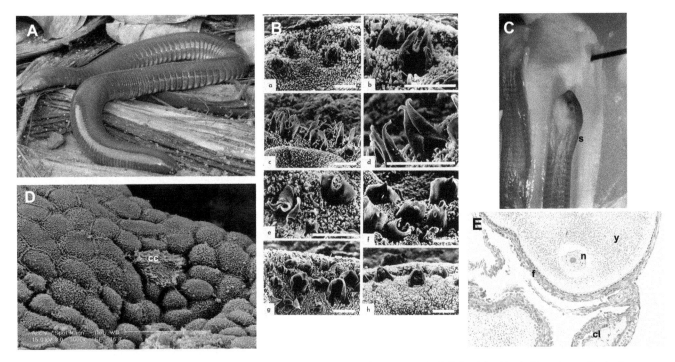

FIG. 21.6 A–D, *Dermophis mexicanus*. A, Adult (photo by Sean Rovito, used with his permission). B, Fetal dentition in *D. mexicanus*, illustrating tooth crown shape changes during pre-birth ontogeny (a = 35 mm total length [TL]; h = 60 mm TL). Scale bars = 10 μm (from Wake 1980a). Compare with fetal teeth of *S. atra*, figure 11b in Greven (1998). C, Intraoviductal fetuses of a pregnant *D. mexicanus* (photo by M. H. Wake). D, SEM of oviduct of early pregnant female *D. mexicanus* (from Wake and Lai, unpubl.). E, Histological section of ovary of *Scolecomorphus vittatus* including a maturing ovum and a corpus luteum (from Wake 1968). Abbreviations: cc = ciliated cell; cl = corpus luteum; f = follicle; n = nucleus with nucleolus; s = gill slit; y = yolk.

present in nonpregnant females. Because of the activity of the fetuses, the proliferation of secretory mucosa becomes confined to crypts in the epithelium lining the oviducts. Mature females appear to reproduce every other year, pregnant in one, yolking up during the next. The endocrine cycle is consistent with the typical vertebrate/amphibian pattern, but the phases may be prolonged. The endocrinology of several aspects is not known: (a) the inception of the fetal dentition and the fact that in *D. mexicanus* the tooth crown shapes vary with fetal "age" (Wake 1980a), (b) the mediation of the changes in composition of the nutrient material secreted by the maternal oviductal mucosa (initially free amino acids and simple carbohydrates, complex amino acids and carbohydrates in mid-pregnancy, and proteins, carbohydrates, and lipids during the last months of gestation: Welsch et al. 1977; Wake 1993 and unpubl.; Wake and Dickie 1998), and (c) the regulation of the disjunct metamorphosis (Wake unpubl.; Exbrayat 2006a, 2006b). Exbrayat and his colleagues made very similar observations of pregnancy, ovarian and oviductal morphology, the fetal dentition, the biennial female cycle, and the lengthy metamorphosis, etc., in their study species, *T. compressicaudus.* See particularly the work on ovarian and oviductal morphology and cycles (Exbrayat 1984a, 1984b, 1988; Exbrayat and Collenot 1983; Hraoui-Bloquet et al. 1994), development (Exbrayat and Delsol 1988; Sammouri et al. 1990), and their summarized work (Exbrayat 2000, 2006a, 2006b; Exbrayat and Estabel 2006). Their research has added considerable information about the ultrastructure of various structures, and they have initiated extensive studies of neurendocrinology in the species, including that influencing the reproductive cycle. Their work, especially on the effects of prolactin, will extend our understanding of the correlation of the hypophyseal-pituitary-gonadal interactions with development and the environment, as the research to date on the viviparous frog, *N. occidentalis,* the salamander, *S. atra,* and viviparous caecilians (*D. mexicanus, T. compressicauda*) has suggested.

It appears undeniable that the endocrine cycle mediates intraoviductal pregnancy and maternal provision of nutrients in the frogs, salamanders, and caecilians as well as back-breeding in frogs; corpus luteum longevity with progesterone secretion mediates the gestation period and stage of development of young at birth; and there is a strong influence of environmental cues on the endocrine cycle, implying integration of perception, the hypothalamic-pituitary axis, and the physiology. Browning (1973), in his exquisite but overlooked discussion of the evolutionary history of the corpus luteum, summarized research on vertebrate corpora atretica and lutea, including discussion of those of *Salamandra atra* and *Nectophrynoides occidentalis*. He made a number of general points about the independent co-option of those bodies in different lineages for different functions, including the support of gestation in live-bearers. He noted that "follicle structure *per se* dictates its potential economical utilization as a hormone source" and "It is not so much the structures that have evolved but rather the uses to which they are put and the controls that are exerted over them." Browning's ideas presage current work on the evolution of viviparity. How does one assess the homoplasy of such systems that have similarities and differences at multiple hierarchical levels of examination?

Preliminary Conclusions

Several general and preliminary conclusions can be drawn from comparisons of back-brooding and intraoviductal pregnancy across amphibians as cases of "homoplasy":

1. Internal fertilization is necessary but not sufficient for intraoviductal pregnancy. Among amphibians, mechanisms of internal fertilization evolved independently and differ dramatically (cloacal apposition or vestigial tail insertion in frogs, spermatophore deposition and pickup in salamanders, insertion of the male's intromittent organ into the vent of the female in caecilians).

2. Both oviducts and skin respond morphologically and physiologically to ovarian hormones by proliferation, hyperemia, and increased vascularity. Estrogen prepares the skin or oviduct, and progesterone maintains the pregnancy. The skin and oviductal morphologies are similar across the relevant taxa, although in the oviduct the secretory regions may be general or restricted, and the secretions of the cells vary in composition; i.e., morphology can be "the same," but what it produces biochemically and physiologically (and behaviorally) can differ considerably across taxa.

3. The influences of other maternal hormones and how they interact with embryonic-fetal hormones remain virtually unstudied in these cases. Furthermore, relatively little is known about hormones and their effects in gestation, development, metamorphosis, and birth for all but a very few live-bearing taxa; several hormones and their interactions have been virtually ignored to date. Consequently, direct comparisons remain to be made.

4. Clutch size is reduced in live-bearing taxa, compared with direct-developers and species with free-living larvae. Egg size is relatively larger in back-brooding frogs, especially those that give birth to metamorphs, and is significantly reduced in amphibians that provide maternal nutrition after yolk is resorbed, which also give birth to metamorphs. Both phenomena likely have evolved to provide an adequate nutrient investment for full development.

5. Embryos/fetuses develop clade-specific means of facilitating gaseous exchange (across modified gills, expanded tails, and body surfaces and the mucosa of the skin or oviduct, given their increased vascularity, (nutrient uptake by gills has been proposed [see Exbrayat 2000], but labeled uptake studies are needed to demonstrate this). Similarly new structures for obtaining nutrients (fetal dentitions, tufts of skin around the mouth) have evolved in intraoviductal embryos and fetuses. In those species, the fetal modifications are lost near birth, presumably under hormonal control.

6. Metamorphosis is protracted; little is known about its regulation.

7. Young maintained intraoviductally are born either as advanced larvae or fully metamorphosed, as are those of back-brooders; species with obligate oviductal maternal nutrition typically give birth to fully metamorphosed young.

8. The exposure to maternal hormones throughout the lengthy gestation periods appears to regulate development and growth, the female responding to internal and especially external environmental conditions, whether the gestation period is either loosely or tightly correlated. When progesterone exposure declines, birth is imminent.

9. Maintenance of development in or on the body of a parent is a complex interaction of morphology, development, endocrinology, and ecology, as well as historical contingency, that is underappreciated, understudied, and deserving of extensive attention. It is especially important to rectify this, given that the existence of many species with derived modes of reproduction is threatened.

10. The mechanistic bases of the evolution of similarity can be dependent on different deployment of common substrates (e.g., morphologies, hormones) or have unique attributes that achieve a common end point, depending on the levels of the hierarchy of organization of the phenomenon under study. It is the integration of the elements that allow us to determine pattern, process, divergence, and similarity.

What Is Homoplasy?

Given this approach to analysis of the evolution of viviparity in diverse amphibian clades, what, then, is homoplasy? Is it pattern, process, or the end result—the phenotypic expression? Is it a combination of these? Is it merely "deep homology," as Hall (2007) asserts? My analyses suggest that it is "all of these." Any attempt to understand homoplasious features beyond their identification by mapping "summary" features (e.g., viviparity) on a cladogram requires a hierarchical analysis of the probable mechanisms that underlie development and maintenance. Only then can the features exhibited by the distantly related taxa for which homoplasy is imputed be evaluated. Phylogenies can reveal patterns, but likely not mechanisms.

Homoplasy may include "deep homology," but that is not all that it is. For example, all amphibians and reptiles have the "equipment" (estrogen, progesterone, skin, oviducts) to evolve live-bearing, but they do not. How and why selection operates in the evolution of these modes has been suggested (the "cold hypothesis," competition for resources, reduction of larval predation, etc.), but rarely demonstrated, and never in the context of the complexity of the molecular, cellular, organismal, and ecological interactions involved. Similarly, homoplasy should not focus entirely on the "end product" or phenotypic expression either. For example, is "viviparity" in all its forms a singular state, or does the term cover a diversity of processes and "end products" at several different levels of biological organization? Which components are homologous, and which are homoplasious? Two sets of organisms are involved, maternal and fetal, each with a suite of characters whose functional attributes have different regulatory systems. Multiple traits require investigation at multiple hierarchical levels.

In my opinion, now that we can generate robust phylogenetic hypotheses, a new approach should be taken to the examination of homoplasy: (1) At a minimum, clear identification of the level of study and the definition of features must be stated, so that those features can be explicitly compared. At the same time, we must recognize that phylogenies can aid in assessment of pattern, but rarely of process or mechanism. (2) As exemplified by this research on the evolution of different modes of viviparity, homoplasy and "deep homology" are suites of features, and we may err when we focus on the presumed end points of evolution. (3) We should keep in mind that the species, and the characters and their associations, are likely still evolving, so any current examination is of one point in "real time," and especially in evolutionary time. (4) It is important to examine homoplasious traits hierarchically, so that "similarity" of the "end points" can be evaluated in terms of both similarity and differences in mechanisms of evolution and development. In fact, as the maintenance of pregnancy in a diversity of live-bearing modes proximally regulated by progesterone levels exemplifies, the basis for similarity, and of variation, may rest at levels below and above that of the feature itself.

These notions are not particularly new; they are also elements in the conclusions of Hall, D. Wake, Sanderson, Donoghue, and others, but I try to frame them in a more broadly inclusive context. Indeed, Lankester (1870), when he coined the term "homoplasy," considered it a second kind of homology, the other being what he called "homogeny," which differed in that the most recent common ancestor presumably had the feature in question and its descendent lineages retained it. The conditions for his two "kinds" were not antithetic. Homoplasy illustrates both constraints—an organism can only use the material it has—and opportunities—there are innovative ways to use that material in response to selection, and they may co-occur in distantly related lineages.

The study of the mechanistic basis of homoplasy opens new ways to explore the selection pressures that elicit responses that are mechanistically and phenotypically similar. The understanding of homoplasious evolution requires a hierarchical and integrative research framework that examines features and mechanisms at a diversity of levels of organization and analyzes their interactions, as the study of the evolution of viviparity illustrates. The bases for similarities (and differences) in clades in which the most recent common ancestor did *not* have the feature(s) in question provides systems that allow hypotheses of presumed mechanisms of origin and maintenance of derived features to be proposed and tested. Hierarchies in biology have long been recognized, but not often examined (Dullemeijer 1974, 1981; Eldredge 1985; Eldredge and Salthe 1984; Grene 1987). Most discussions of hierarchical approaches, in addition, restrict themselves to subsets—macroevolution (Eldredge 1982; Gregory 2004), ecology (MacMahon et al. 1978; Wainwright and Reilly 1994), and others. However, this is changing—the search for "a more complete evolutionary theory" has prompted broader examination (e.g., Gould 1980; Vrba and Eldredge 1984), and "integrative biologists" are responding (e.g., Wake and Larson 1987; Wake 1990, 2003, 2008a, 2008b; Gilbert and Epel 2009). With the integration of the tools provided by genetics, genomics, development, epigenetics, and phylogenetics, new understanding of the basis and evolution of homoplasious conditions is under way. Strict cladists consider homoplasy to be "error, in coding" (Nixon and Carpenter 2012). No—homoplasy is "real," and deserves better understanding. The vocabulary, including the classic terms "convergence," "parallelism," and "reversals," needs explication and clarification so that communication is more effective. A concept of homoplasy that incorporates both pattern and process at different levels of biological organization would serve evolutionary biologists well. Understanding the *biology* that is the basis for the development of similarity in distantly related lineages is fundamental to understanding patterns and processes of evolution overall. The "Great Transformations" in vertebrate biology all are complex events, fraught with barely explored information about the way that evolution works.

Acknowledgments

I am honored to contribute this account to the celebration of the career of my revered colleague, Professor Farish Jenkins. I thank David Blackburn and Jim Stewart for enthusiastic and constructive reviews of the manuscript. I appreciate many years' worth of discussions, research collaborations, and fieldwork with colleagues and students who share concerns about how (and why) to try to understand evolutionary biology. Jacob Chin and David Buckley prepared some of the photos, and I also thank curators and colleagues at the California Academy of Sciences, the University of Kansas Museum of Natural History, the Museum of Comparative Zoology at Harvard University, the Field Museum of Natural History, and the Museum of Vertebrate Zoology, University of California, Berkeley, for the loan of specimens. I especially appreciate the long-term support by the National Science Foundation for my research program.

References

Abouheif, E. 2008. Parellelism as the pattern and process of mesoevolution. Evol. Dev't. 10:3–5.

Alcocer, I., X. Santacruz, H. Steinbeisser, K.-H. Thierauch, and E. M. del Pino. 1992. Ureotelism as the prevailing mode of nitrogen excretion in larvae of the marsupial frog *Gastrotheca riobambae* (Fowler) (Anura, Hylidae). Comp. Biochem. Physiol. 101A:229–231.

Arendt, J., and D. N. Reznick. 2008a. Convergence and parallelism reconsidered: what have we learned about the genetics of adaptation? Trends Ecol. Evol. 23:26–32.

Arendt, J., and D. N. Reznick. 2008b. Moving beyond phylogenetic assumptions about evolutionary convergence: response to Leander. Trends Ecol. Evol. 23:483–484.

Browning, H. C. 1973. The evolutionary history of the corpus luteum. Biol. Repro. 8:128–157.

Buchacher, C. O. 1993. Field studies on the small Surinam toad, *Pipa arrabali*, near Manaus, Brazil. Amphib.-Rept. 14:59–69.

Buckley, D., M. Alcobendas, M. García-París, and M. H. Wake. 2007. Heterochrony, cannibalism, and the evolution of viviparity in *Salamandra salamandra*. Evol. Dev't. 9:105–115.

Channing, A., K. S. Finlow-Bates, S. E. Haarklau, and P. G. Hawkes. 2006. The biology and recent history of the critically endangered Kihansi Spray Toad, *Nectophrynoides asperginis* in Tanzania. J. East African Nat. Hist. 95:117–138.

de Albuja, C. M., M. Campos, and E. M. del Pino. 1983. Role of progesterone in oocyte maturation in the eggbrooding hylid frog *Gastrotheca riobambae* (Fowler). J. Expt. Zool. 227:271–276.

del Pino, E. M. 1980. Morphology of the pouch and incubatory integument in marsupial frogs (Hylidae). Copeia 1980:10–17.

del Pino, E. M. 1983. Progesterone induces incubatory changes in the brooding pouch of the frog *Gastrotheca riobambae* (Fowler). J. Expt. Zool. 227:159–163.

del Pino, E. M. 1989. Modifications of oogenesis and development in marsupial frogs. Development 107:169–187.

del Pino, E. M., and R. P. Elinson. 1983. A novel development pattern for frogs: gastrulation produces an embryonic disk. Nature 306:589–591.

del Pino, E. M., and B. Escobar. 1981. Embryonic stages of *Gastrotheca riobambae* (Fowler) during maternal incubation and comparison of development with that of other eggbrooding hylid frogs. J. Morphol. 167:277–295.

del Pino, E. M., M. L. Galarza, C. M. de Albuja, and A. A. Humphries Jr. 1975. The maternal pouch and development in the marsupial frog *Gastrotheca riobambae* (Fowler). Biol. Bull. 149:480–491.

del Pino, E. M., and G. Sánchez. 1977. Ovarian structure of the marsupial frog *Gastrotheca riobambae* (Fowler). J. Morphol. 153:153–162.

Donaire-Barroso, D., S. Bogaerts, and D. Herbert. 2001. Confirmación de desarrollo larvario completo intrauterino en *Salamandra algira* (Bedriaga, 1883) del Noroeste de Marruecos. Bul. Soc. Catalana d'Herpetologia, 5:107–110.

Duellman, W. E., and S. J. Maness. 1980. The reproductive behavior of some hylid marsupial frogs. J. Herpet. 14:213–222.

Dullemeijer, P. 1974. Concepts and approaches in animal morphology. Assen, The Netherlands: Van Gorcum.

Dullemeijer, P. 1981. Functional morphology and evolutionary biology. Acta Theor. 29: 151–250.

Eldredge, N. 1982. Hierarchies in macroevolution. Pp. 42–61 in J. W. Valentine, D. Jablonski, D. H. Erwin, and J. H. Lipps, eds., Evolutionary Paleobiology: In Honor of James W. Valentine. Chicago: University of Chicago Press,.

Eldredge, N. 1985. Unfinished synthesis: Biological hierarchies and modern evolutionary thought. New York: Oxford University Press.

Eldredge, N., and S. N. Salthe. 1984. Hierarchy and evolution. Oxford Surv. Evol. Biol. 1:182–206.

Elinson, R. P., and E. M. del Pino. 1985. Cleavage and gastrulation in the egg-brooding, marsupial frog, *Gastrotheca riobambae*. J. Embryol. Exp. Morph. 90:223–232.

Elinson, R. P., E. M. del Pino, D. S. Townsend, F. C. Cuesta, and P. Eichhorn. 1990. A practical guide to the developmental biology of terrestrial-breeding frogs. Biol. Bull.179:163–177.

Exbrayat, J. M. 1984a. Cycle séxuel et réproduction chez un Amphibien Apode: *Typhlonectes compressicaudus* (Duméril et Bibron, 1841). Bull. Soc. Herpel. France 32:31–35.

Exbrayat, J. M. 1984b. Quelques observations sur l'évolution des voies génitales femelles de *Typhlonectes compressicaudus* (Duméril et Bibron, 1841), Amphibien Apode vivipare, au cours du cycle de réproduction. C.R.S. Acad. Sci. Paris 298:13–18.

Exbrayat, J. M. 1988. Croissance et cycle des voies génitales femelles de *Typhlonectes compressicaudus* (Duméril et

Bibron, 1841), amphibien apode vivipare. Amphib.-Rept. 9:117–134.

Exbrayat, J. M. 2000. Les Gymnophiones ces curieux amphibiens. Paris: Soc. Nouvelle des Editions Boubée.

Exbrayat, J. M. 2006a. Endocrinology of reproduction in Gymnophiona. Pp. 183–229 in J. M. Exbrayat and B. G. M. Jamieson, eds., Reproductive Biology and Phylogeny of Gymnophiona (Caecilians). Enfield, NH: Science Publishers.

Exbrayat, J. M. 2006b. Modes of parity and oviposition. Pp. 303–323 in J. M. Exbrayat and B. G. M. Jamieson, eds., Reproductive Biology and Phylogeny of Gymnophiona (Caecilians). Enfield, NH: Science Publishers.

Exbrayat, J. M., and G. Collenot. 1983. Quelques aspects de l'evolution de l'ovarire de *Typhlonectes compressicaudus* (Duméril et Bibron, 1841), Batracien Apode vivipare. Etude quantitative et histochimique des corps jaunes. Reprod., Nutr., Dévt. 23:889–898.

Exbrayat, J. M., and M. Delsol. 1988. Oviparité et développement intrauterine chez les Gymnophiones. Bull. Soc. Zool. France 45:27–36.

Exbrayat, J. M., and J. Estabel. 2006. Anatomy with particular reference to the reproductive system. Pp. 79–155 in J. M. Exbrayat and B. G. M. Jamieson, eds., Reproductive Biology and Phylogeny of Gymnophiona (Caecilians). Enfield, NH: Science Publishers.

Fernandes, T. L., M. M. Antoniazzi, E. Sasso-Cerri, M. I. Egami, C. Lima, M. T. Rodrigues, and C. Jared. 2011. Carrying progeny on the back: reproduction in the Brazilian aquatic frog *Pipa carvalhoi*. South Amer. J. Herpet. 6:161–176.

Gilbert, S. F., and D. Epel. 2009. Ecological developmental biology: integrative epigenetics, medicine, and evolution. Sunderland, MA: Sinauer Associates.

Gould, S. J. 1980. Is a new theory of evolution emerging? Paleobiology 6:119–130.

Gregory, T. R. 2004. Macroevolution, hierarchy theory, and the C-value enigma. Paleobiology 30:179–202.

Grene, M. 1987. Hierarchies in biology. Amer. Sci. 75:504–510.

Greven, H. 1977. Comparative ultrastructural investigations of the uterine epithelium in the viviparous *Salamandra atra* (Laur.) and the ovoviviparous *Salamandra salamandra* (L.) (Amphibia, Urodela). Cell Tiss. Res. 181:215–237.

Greven, H. 1980. Licht- und elektronenmicrskopische Untersuchungen zur Struktur und Histochemie des Oviduktepithels (Pars recta und Pars convoluta I, II, III) von *Salamandra salamandra* (L.) (Amphibia, Urodela). Zeitschr. mikro-anat. Forsch., Leipzig 94:387–429.

Greven, H. 1984. Zona trophica and larval dentition in *Salamandra atra* Laur. (Amphibia, Urodela): Adaptation to intrauterine nutrition. Verh. Deut. Zool. Ges. 77:184.

Greven, H. 1998. Survey of the oviduct of salamandrids with special reference to the viviparous species. J. Exp. Zool. 282:507–525.

Greven, H. 2002. The urodele oviduct and its secretions in and after G. von Wahlert's doctoral thesis "Eileiter, Laich und Kloake der Lalamandriden." Bonner zool. Monogr. 50: 25–61.

Greven, H. 2003a. Larviparity and pueriparity. Pp. 447–475 in J. M. Exbrayat and B. G. M. Jamieson, eds., Reproductive Biology and Phylogeny of Gymnophiona (Caecilians). Enfield, NH: Science Publishers.

Greven, H. 2003b. Oviduct and egg-jelly. Pp. 151–181 in J. M. Exbrayat and B. G. M. Jamieson, eds., Reproductive Biology and Phylogeny of Gymnophiona (Caecilians). Enfield, NH: Science Publishers.

Greven, H., and G. D. Guex. 1994. Structural and physiological aspects of viviparity in *Salamandra salamandra*. Mertensiella 4:139–160.

Greven, H., and S. Richter. 2009. Morphology of skin incubation in *Pipa carvalhoi* (Anura: Pipidae). J. Morphol. 270:1311–1319.

Greven, H., and H. J. Rüterbories. 1984. Scanning electron microscopy of the oviduct of *Salamandra salamandra* (L.) (Amphibia, Urodela). Zeits. micr-anat. Forsch. 98:49–62.

Guex, G. D., and P. S. Chen. 1986. Epitheliophagy: intrauterine cell nourishment in the viviparous alpine salamander, *Salamandra atra*. Experientia 42:1205–1214.

Guex, G. D., and H. Greven. 1994. Structural and physiological aspects of viviparity in *Salamandra atra*. Mertensiella 4:161–208.

Hall, B. K. 1999. Evolutionary developmental biology. Dordrecht, The Netherlands: Kluwer.

Hall, B. K. 2002. Palaeontology and evolutionary developmental biology: a sciences of the nineteenth and twenty-first centuries. Palaeontology 45:647–699.

Hall, B. K. 2003. Descent with modification: the unity underlying homology and homoplasy as seen through an analysis of development and evolution. Biol. Rev. Camb. Phil. Soc. 78:400–433.

Hall, B. K. 2007. Homology or homoplasy: dichotomy or continuum? J. Hum. Evol. 52:473–479.

Hraoui-Bloquet, S., G. Escudie, and J.-M. Exbrayat. 1994. Aspects ultrastructuraux de l'evolution de la muqueuse uterine au cours de la gestation chez *Typhlonectes compressicaudus*, Amphibien Gymnophione. Bull. Soc. Zool. France 119:237–242.

Joly, J., F. Chesnel, and D. Boujard. 1994. Biological adaptations and reproductive strategies in the genus *Salamandra*. Mertensiella 4:255–269.

Joly, J., and B. Picheral. 1972. Ultrastructure, histochimie et physiologie du follicule préovulatoire et du corps jaune de l'urodèle ovo-vivipare *Salamandra salamandra* (L.). Gen. Comp. Endo. 18:235–259.

Jones, R. E., A. M. Gerrard, and J. J. Roth. 1973. Estrogen and brood pouch formation in the Marsupial Frog, *Gastrotheca riobambae*. J. Exp. Zool. 184:177–184.

Juberthie, C., J. Durand, and M. Dupuy. 1996. La reproduction des Protées (*Proteus anguinus*): bilan de 35 ans d'élevage dans les grottes-laboratoires de Moulis et d'Aulignac. Mém. Biospéol. 23:53–56.

Lankester, E. R. 1870. On the use of the term homology in modern zoology, and the distinction between homogenetic and homoplastic agreements. Ann. Mag. Nat. Hist. 4:34–43.

Lanza, B., and P. Leo. 2001. Prima osservazione sicura del riproduzione vivipara nel genere *Speleomantes* (Amphibia: Caudata: Plethodontidae). Pianura 13:317–319.

Leander, B. S. 2008. Different modes of convergent evolution reflect phylogenetic distances: a reply to Arendt and Reznick. Trends Ecol. Evol. 23:481–482.

Lostanlen, D., C. Boisseau, and J. Joly. 1976. Données ultrastructurales et physiologiques sur l'utérus d'un amphibien ovovivipare, *Salamandra salamandra* (L.) Ann. Sci. nat. Zool. Paris 18:113–144.

MacMahon, J. A., D. L. Phillips, J. V. Robinson, and D. J. Schimpf. 1978. Levels of biological organization: an organism-centered approach. BioScience 28:700–704.

Nixon, K., and J. Carpenter. 2012. On homology. Cladistics 28:160–169.

Olgun, K., C. Miaud, and P. Gautier. 2001. Age, growth and survivorship in the viviparous salamander *Mertensiella luschani* from southwestern Turkey. Can. J. Zool. 79:1559–1567.

Ozeti, N. 1979. Reproductive biology of the salamander *Mertensiella luschani anatolyana*. Herpetologica 35:193–197.

Packard, G. C., R. P. Elinson, J. Gavaud, L. G. Guillette, J. Lombardi, J. Schindler, R. Shine, H. Tyndale-Biscoe, M. H. Wake, F. J. Xavier and Z. Yaron. 1989. How are reproductive systems integrated and how has viviparity evolved? Pp. 281–293 in D. B. Wake and G. Roth, eds., Complex Organismal Functions: Integration and Evolution in Vertebrates. Chichester: John Wiley and Sons.

Polymeni, R. M., and H. Greven. 1992. Histology and fine structure of the oviduct in *Mertensiella luschani* (Steindachner, 1891) (Urodela, Salamandridae). Pp. 361–365 in Z. Korsoz and I. Kiss, eds., Proc. Sixth Ord. Gen. Meet. S. E. H.

Powell, R. 2007. Is convergence more than an analogy? homoplasy and its implications for macroevolutionary predictability. Biol. Philos. 22:565–578.

Rabb, G. B., and M. S. Rabb. 1960. On the mating and egg-laying behaviour of the Surinam toad *Pipa pipa*. Copeia 1960:271–176.

Rabb, G. B., and R. Snedigar. 1960. Observations on breeding and development of the Surinam toad, *Pipa pipa*. Copeia 1960:40–44.

Sammouri, R., S. Renous, J. M. Exbrayat, and J. Lescure. 1990. Développement embryonnaire de *Typhlonectes compressicaudus* (Amphibia, Gymnophiona). Ann. Sci. Nat. Zool., Paris, 13ème série, 11:135–163.

Sanderson, M. J., and M. J. Donoghue. 1989. Patterns of variation in levels of homoplasy. Evolution 43:1781–1795.

Sanderson, M. J., and L. Hufford. 1996. Homoplasy: The Recurrence of Similarity in Evolution. New York: Academic Press.

Sarasin, P., and F. Sarasin. 1887–1890. Ergebnisse naturwissenschaftlicher Forschungen auf Ceylon in den Jahren 1884–1886 Band II: Zur Entwicklungsgeschichte und Anatomie der Ceylonesischen Blindwühle *Ichthyophis glutinosus*. Wiesbaden: C. W. Kreidels Press.

Scotland, R. W. 2011. What is parallelism? Evol. Dev't. 13:214–227.

Sever, D. M., E. C. Moriarty, L. C. Rania, L. V. Diller, and W. C. Hamlett. 2001. Reproductive biology of the internal fertilizing frog *Ascaphus truei* (Anura: Leiopelmatidae). J. Morphol. 248: 1–21.

Shubin, N., C. Tabin, and S. Carroll. 2009. Deep homology and the origins of evolutionary novelty. Nature 457:818–823.

Townsend, D. S., M. M. Stewart, F. H. Pough, and P. F. Brussard. 1981. Internal fertilization in an oviparous frog. Science 212:469–471.

Trueb, L., and D. Massemin. 2000. The osteology and relationships of *Pipa aspera* (Amphibia: Anura: Pipidae), with notes on its natural history in French Guiana. Amphib.-Rept. 22:33–54.

Vilter, V., and A. Vilter. 1960. Sur la gestation de la salamander noire des Alpes, *Salamandra atra* Laur. Compt. Rend. Séanc. Soc. Biol. Paris 154:290–291.

Vilter, V., and A. Vilter. 1964. Sur l'évolution des corps jaunes ovariens chez *Salamandra atra* Laur. Des Alpes vaudoises. Compt. Rend. Séanc. Soc. Biol. Paris 158:457–461.

Vrba, E. S., and N. Eldredge. 1984. Individuals, hierarchies and processes: towards a more complete evolutionary theory. Paleobiology 10:146–171.

Wainwright, P. C., and S. M. Reilly. 1994. Ecological Morphology: Integrative Organismal Biology. Chicago: University of Chicago Press.

Wake, D. B. 1991. Homoplasy: the result of natural selection or evidence of design limitations? Amer. Nat. 138:543–567.

Wake, D. B. 1999. Homoplasy, homology, and the problem of "sameness" in biology. Pp. 25–46 in G. R. Boc and G. Cardew, eds., Novartis Foundation Symposium 222: Homology. New York: John Wiley.

Wake, D. B. 2003. Homology and homoplasy. Pp. 191–201 in B. K. Hall and W. M. Olson, eds., Key Words and Concepts in Evolutionary Developmental Biology. Cambridge, MA: Harvard University Press.

Wake, D. B., and A. Larson. 1987. A multidimensional analysis of an evolving lineage. Science 238:42–48.

Wake, D. B., M. H. Wake, and C. D. Specht. 2011. Homoplasy: from detecting pattern to determining process and mechanisms of evolution. Science 331:1031–1035.

Wake, M. H. 1968. Evolutionary morphology of the caecilian urogenital system. Part I. The gonads and fat bodies. J. Morphol. 126:291–332.

Wake, M. H. 1970. Evolutionary morphology of the caecilian urogenital system. Part II. The kidneys and urogenital ducts. Acta Anat. 75:321–358.

Wake, M. H. 1972. Evolutionary morphology of the caecilian urogenital system. Part IV. The cloaca. J. Morphol. 136:353:366.

Wake, M. H. 1977a. Fetal maintenance and its evolutionary significance in the Amphibia: Gymnophiona. J. Herpetol. 11:379–386.

Wake, M. H. 1977b. The reproductive biology of caecilians: an evolutionary perspective. Pp. 73–101 in D. H. Taylor and S. I. Guttman, eds., The Reproductive Biology of Amphibians. New York: Plenum Press.

Wake, M. H. 1978. An ovoviviparous *Eleutherodactylus*, with comments on the evolution of live-bearing systems. J. Herpetol. 12:121–133.

Wake, M. H. 1980a. Fetal tooth development and adult replacement in *Dermophis mexicanus* (Amphibia: Gymnophiona): fields versus clones. J. Morphol. 166:203–216.

Wake, M. H. 1980b. Reproduction, growth and population structure of *Dermophis mexicanus* (Amphibia: Gymnophiona). Herpetologica 36:244–256.

Wake, M. H. 1980c. The reproductive biology of *Nectophrynoides malcolmi* (Amphibia: Bufonidae), with comments on the evolution of reproductive modes in the genus *Nectophrynoides*. Copeia 1980:193–209.

Wake, M. H. 1985. Oviduct structure and function in non-mammalian vertebrates. Fortschr. Zool. 30:427–435.

Wake, M. H. 1989. Phylogenesis of direct development and viviparity. Pp. 235–250 in D. B. Wake and G. Roth, eds., Complex Organismal Functions: Integration and Evolution in Vertebrates. Chichester: John Wiley & Sons.

Wake, M. H. 1990. The evolution of integration of biological systems: an evolutionary perspective through studies of cells, tissues, and organs. Amer. Zool. 30:897–906.

Wake, M. H. 1993. Evolution of oviductal gestation in amphibians. J. Exp. Zool. 266:394–413.

Wake, M. H. 2003. What is "integrative biology"? Integ. Comp. Biol. 43:239–241.

Wake, M. H. 2008a. Integrative biology: science for the 21st century. BioScience 58:349–353.

Wake, M. H. 2008b. Organisms and organization. Biol. Theor. 3:213–223.

Wake, M. H., and R. Dickie. 1998. Oviduct structure and function, and reproductive modes in amphibians. J. Expt. Zool. 282:477–506.

Wake, M. H., and J. Hanken. 1982. The development of the skull of *Dermophis mexicanus* (Amphibia: Gymnophiona), with comments on skull kinesis and amphibian relationships. J. Morphol. 171:203–223.

Welsch, U., M. Muller, and C. Schubert. 1977. Electron-microscopical and histochemical observations on the reproductive biology of viviparous caecilians (*Chthonerpeton indistinctum*). Zool. Jahrb. Anat. 97:532–549.

Weygoldt, P. 1976. Beobachtungen zur Biologie und Ethologie von *Pipa* (*Hemipipa*) *carvalhoi* Mir. Rib. 1937 (Anura, Pipidae). Z. Tierpsychol. 40:70–99.

Wiens, J. J., C. A. Kuczynski, W. E. Duellman, and T. W. Reeder. 2007. Loss and re-evolution of complex life cycles in marsupial frogs: does ancestral trait reconstruction mislead? Evolution 61:1886–1899.

Wilkinson, M., D. San Mauro, E. Sherratt, and D. Gower. 2011. A nine-family classification of caecilians (Amphibia: Gymnophiona). Zootaxa 2874:41–64.

Xavier, F. 1977. An exceptional reproductive strategy in Anura: *Nectophrynoides occidentalis* Angel (Bufonidae), an example of adaptation in terrestrial life by viviparity. Pp. 545–572 in M. K. Hecht, P. C. Goody, and B. M. Hecht, eds., Major Patterns of Vertebrate Evolution. New York: Plenum.

Xavier, F. 1986. La réproduction des *Nectophrynoides*. Pp. 497–513 in P. P. Grasse and M. Delsol, eds., Traité de Zoologie Amphibiens. Vol. 14. Paris: Masson.

22

Rampant Homoplasy in Complex Characters: *Repetitive Convergent Evolution of Amphibian Feeding Structures*

David B. Wake,* David C. Blackburn,† and R. Eric Lombard‡

Introduction

Were the Great Transitions in the history of life unique events in which a lineage of organisms evolved from a primitive to a more derived condition? Or during these transitions did separate but related lineages make similar transitions from a more primitive configuration independently? If so, how often might these independent similar transitions have taken place? Homoplasy, the independent evolution of similarity in a given phenotype in different taxa, has long been a central issue in evolutionary biology (see M. Wake, this volume). Arising from convergence, parallelism, or reversal to ancestral states, homoplasy has the potential to frustrate attempts to generate robust phylogenetic hypotheses and can lead to incorrect scenarios for the evolution of functional and biomechanical systems. In many cases, researchers have proceeded on the assumption that homoplastic features are overwhelmed by true phylogenetic signal, for example, in cladistic analyses. However, in some taxa the balance between signal and what might be called "noise" (i.e., homoplastic traits) is so close that cladistic analyses yield problematic results, especially in cases in which an entire clade is characterized by homoplastic traits. In clades in which many lineages, sharing only a remote common ancestor, have

* Department of Integrative Biology and Museum of Vertebrate Zoology, University of California, Berkeley
† Department of Vertebrate Zoology and Anthropology, California Academy of Sciences
‡ Department of Organismal Biology and Anatomy, University of Chicago

independently addressed the same fundamental problems using the same pathways, phylogenetic analyses of phenotypes may be positively misled by homoplasy. For example, resolution of the salamander tree of life using morphological data remained impossible for decades because of the many instances of species that remain aquatic as adults by essentially prolonging larval life (larvamorph taxa, including among others mud puppies and axolotls; Wiens et al. 2005). With the advent of molecular approaches for the generation of data used to test phylogenetic hypotheses, a new era of interest in homoplasy has arrived. Scenarios of phenotypic evolution can now be evaluated in a rigorous framework in which the primary data (i.e., DNA sequences) for phylogenetic inference are at least largely independent of the phenotypes of interest. Such analyses can fundamentally reshape our understanding of morphological and life history evolution, especially in cases in which highly nested clades have "reversed" to phenotypes characterizing ancestral taxa. With robust phylogenies in hand, the study of homoplasy can lead to new interpretations of the underlying mechanisms, and their relative frequencies, associated with major transitions in form and function.

Major evolutionary transformations such as the origin of feathers or the fin-to-limb transition are traditionally thought of as the result of a "one-off" series of events. This is a reasonable working hypothesis as we look back disadvantaged by the passage of time, extinction of lineages, and a paucity of organismal remains in the fossil record. Were we closer in time to those events and graced by a multitude of associated taxa, would we still find a singular "one-off" evolutionary series? Living plethodontid salamanders provide a laboratory to examine this question. Within this clade, any reasonable hypotheses of relationship indicate that similar feeding modes have evolved multiple times, representing independent experiments with evolutionary transitions. The comparative framework for examining such homoplasy enables one to ask broad questions about the importance of these transformations. For example, how do we measure the success or importance of a particular transformation? Answers to this question could include the number of species, breadth of phenotypic diversity, or uniqueness of habitat or distribution that have evolved in a given clade subsequent to that transformation. Both neotropical plethodontid salamanders

(e.g., *Bolitoglossa* and related taxa) and the temperate web-footed *Hydromantes* of Europe and western North America have evolved similar complex tongue projection mechanisms. Yet, these two lineages differ dramatically in species number with only 11 species of *Hydromantes* and more than 270 tropical salamanders (i.e., Bolitoglossini of Wake 2012). Any approach to reconstructing the evolutionary history of transitions requires a robust phylogeny. Plethodontids appear to have arisen in the Late Jurassic (Vieites et al. 2007), but the weak fossil record offers little of value regarding relationships within the family. Our approach here is to use the framework of recent phylogenetic hypotheses based on DNA sequence data to examine cases of previously proposed hypotheses of levels of homoplasy and evolutionary transitions in morphology.

Evolution of Tongue Morphology in Plethodontid Salamanders

As early as the 19th century (Wiedersheim 1875) researchers were aware that different salamander taxa had tongues that were shot out of the mouth for a considerable distance to catch prey (fig. 22.1). Most frogs also project their tongue to catch prey, but it was apparent early on that different mechanisms were involved; in salamanders, the hyobranchial skeleton is a major element in the projected tongue and in the system controlling projection, whereas no skeletal element is projected in frogs. While salamanders in several families evolved effective mechanisms for tongue projection, members of the largest family, Plethodontidae, came under special scrutiny because of the high degree of specialization attained by many of its members (fig. 22.1).

The family Plethodontidae is by far the largest of the 10 families of Caudata (444 species out of a total of 671, AmphibiaWeb 2014), and accordingly there are abundant opportunities for homoplasy. The analysis of Lombard and Wake (1977) was unusual in that instead of atomizing morphology into characters, the tongue projection system was treated as an integrated functionally and morphologically complex character. The general biomechanical model of Lombard and Wake (1976) served as a point of departure, and historical treatments of feeding mechanisms and the then-prevailing phylogenetic hypothesis (Wake 1966) were

Hydromantes Tongue Organization and Function

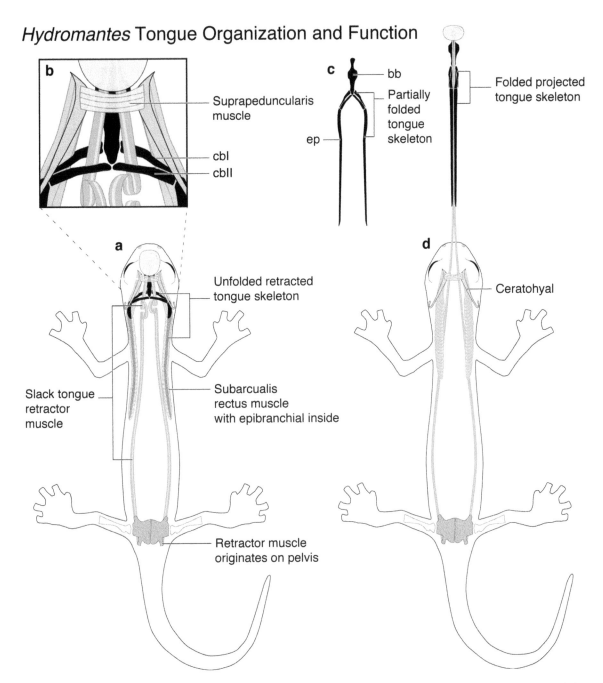

FIG. 22.1 Tongue organization and function in *Hydromantes*. a, Major skeletal and muscular elements with the tongue at rest. b, Detail of the major skeletal elements. c, Partly folded and projected tongue skeleton. d, Fully projected tongue with completely folded hyobranchial skeleton. Note the subarcualis rectus projector muscle originates on the ceratohyal, then sweeps back to enfold the epibranchial, the passage of the retractor muscle between the ceratobranchials, and the suprapeduncularis muscle, which is a major component of a muscular sheath that constrains and folds the hyobranchial skeleton during projection. bb, basibranchial; cbI, ceratobranchial I; cbII, ceratobranchial II; ep, epibranchial;

taken into account. Wake (1966) recognized four clades; a basal split separated Desmognathinae (a single, relatively small clade) from Plethodontinae (three large clades, Plethodontini, Hemidactyliini, and Bolitoglossini). Homoplasy was identified as a possibility by earlier

work, and Wake (1966) explicitly postulated that freely projectile tongues had evolved at least twice.

Lombard and Wake (1977) recognized eight functional feeding modes in plethodontids. Modes I (found in desmognathines, **DES** in the notation of Lombard and

Wake 1986: bold three-letter codes in figs. 22.2–22.4) and II (*Plethodon* and *Aneides,* **PLE**, both plethodontines) feature a relatively generalized morphology (seemingly less functionally specialized in I than in II), and represent what these authors termed an attached protrusible tongue. Mode III (*Ensatina*, **ENS**, a plethodontine) is biomechanically more specialized, with a more projectile tongue that is less firmly attached anteriorly (by elongated genioglossal muscles) to the lower jaw; its tongue was termed attached projectile. Mode IV (present-day tribe Spelerpini, then in the tribe Hemidactyliini and designated EUR now **SPE**) includes two subgroups, one highly specialized with freely projectile tongues (Mode IVa, **SPEa**) that utilizes a biomechanical system (folding option 1 of Lombard and Wake 1977)[1] necessitated by larval development and thought to constitute a constraint on functional specialization. These taxa all lack genioglossal muscles and the epibranchials are longer than in any taxa so far discussed. In two instances (*Stereochilus* and in one species currently deeply nested within *Eurycea, E. speleus*), a non-muscular attachment of the tongue to the lower jaw is retained (Mode IVb, **SPEb**); this attachment was thought to be ancestral in **SPE**, implying that freely projectile tongues (**SPEa**) have evolved more than once within spelerpines. Mode V (*Hemidactylium,* **HEM**) has a generalized morphology but with more components than other modes, and it also has a somewhat loose attachment (because of elongated genioglossal muscles) to the lower jaw. It was designated an attached projectile tongue. Mode VI (all of the tropical salamanders, i.e., the current bolitoglossines, encompassing more than 60% of plethodontid diversity, **BOL**) is a functionally highly specialized feeding system that utilizes folding option 2 and a tongue that is freely projectile (no genioglossal muscles). The epibranchials are very long. Mode VII (*Hydromantes*, **HYD**, then a bolitoglossine) is a highly specialized feeding mode featuring an extremely long (epibranchials greatly elongated), freely projectile tongue. While Lombard and Wake (1977) identified this mode as utilizing folding option 2, the two pairs of ceratobranchials are very short relative to the

basibranchials and the epibranchials and, as a result, folding may be a functionally inconsequential detail in this instance. The final mode (VIII, **BAT**, also a bolitoglossine) is found only in *Batrachoseps*, which has very long epibranchials and utilizes folding option 2, but at the same time has a tongue that is technically attached by means of extraordinarily elongated genioglossal muscles. Based on the phylogenetic hypothesis of Wake (1966), Lombard and Wake inferred (following Wake 1966) that tongue attachment of any kind was ancestral for plethodontid salamanders. Thus, attached projectile tongues had evolved at least three times and freely projectile tongues had evolved at least once within their hemidactyliines and twice within their bolitoglossines (fig. 22.2).

The hypotheses of Lombard and Wake (1977) were tested against a new phylogenetic analysis by Lombard and Wake (1986). The eight major tongue modes recognized were treated as operational taxonomic units; 30 characters, including 18 related directly to the adult feeding system, were evaluated. In an effort to conduct a "parsimonious analysis of a nonparsimonious system," they were unable to reject the earlier hypotheses and concluded that freely projectile tongues had evolved homoplastically in plethodontid salamanders. Their "working hypothesis" phylogeny (their figure 4, modified and elaborated here as fig. 22.2) envisioned a basal split that separated **DES** from all of the others and a second split that separated **SPE + HEM** from (**PLE + ENS**) + (**BOL + HYD + BAT**). According to this hypothesis, attached projectile tongues and free projectile tongues each evolved three times. They argued that some degree of anterior freedom from the lower jaw was a likely precursor to being fully unattached (i.e., loss of the genioglossal muscles in particular), but they thought that extreme elongation of the muscle, as in *Batrachoseps*, was an unlikely precursor to loss of the muscle and assumed that each had evolved from an attached, protrusible tongue. Thus, because they postulated a sister taxon relationship of **BOL** and **BAT**, Lombard and Wake (1986) argued that free tongues had evolved independently in *Hydromantes* (**HYD**) and the tropical salamanders (*Bolitoglossa* and relatives, **BOL**). Either the genioglossus muscle had been elongated in the common ancestors of **BOL, HYD** and **BAT**, and subsequently had superelongated in **BAT**, and been lost in **BOL** and **HYD**, or it had been lost in the common

---

1. In folding option 1, ceratobranchial I is the more robust, lies in the plane of the epibranchial during projection, and carries the projectile force to the basibranchial. In folding option 2, ceratobranchial II is the more robust, lies in the plane of the epibranchial during projection, and carries the projectile force to the basibranchial.

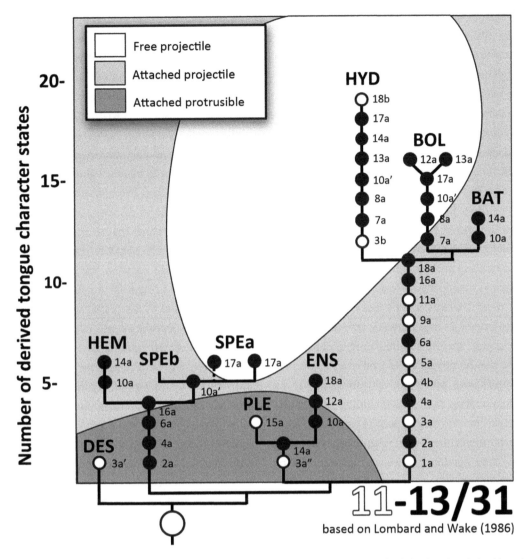

FIG. 22.2 Ground-plan diagram illustrating tongue and lineage evolution in salamanders of the family Plethodontidae, derived from Lombard and Wake (1986, figs. 4 and 7). Three general adaptive zones are recognized, based on use of the tongue in feeding: attached protrusible tongues, apparently close in structure and function to the ancestral state for the family (zone of heavy shading); attached projectile tongues (zone of light shading); and free projectile tongues (zone with no shading). Numbers at lower right indicate the number of uniquely derived character states (open circles on phylogeny), number of individual homoplastic characters, and number of homoplastic character states (closed circles on phylogeny). The characters are indicated in the phylogeny by integers and the character states by lowercase letters, from character lists in Lombard and Wake (1986); as in that work, the characters were ordered a priori and only presumptive derived states are shown. Uniquely derived synapomorphies indicated by a larger open circle on the stem signify four derived character states of the tongue unique to plethodontids.

ancestor and regained in **BAT** (a more parsimonious but seemingly less likely possibility because it required reacquisition of a muscle inferred to have been lost previously).

Lombard and Wake (1986) presented a ground-plan diagram (their figure 7, reproduced in revised form here in fig. 22.2) to illustrate tongue and lineage evolution. **SPEa, BOL,** and **HYD** all having free projectile tongues (open triangles in the unshaded zone of diagram). The

diagram was arranged so that both the phylogenetic relationships as well as the degree of specialization were displayed (as measured by numbers of apomorphic character states directly related to adult tongue features). The degree of specialization was displayed as character states thought to be convergently derived plus those thought to be uniquely derived, based on transformation arguments determined a priori. Their phylogeny (their figure 4) recognized four clades: desmognathines, hemidactyliines, plethodonines, and bolitoglossines, as proposed by Wake (1966). However, the interrelationships of these four were revised relative to Wake (1966). The desmognathine clade, **DES**, was not further subdivided. The hemidactyliine clade was divided into two subclades, *Eurycea* and relatives, our **SPE**, with alternative pathways for the basal morphological state into two derived states **SPEa** and **SPEb**, and *Hemidactylium* alone, **HEM**. The plethodonine clade was divided into two subclades, *Plethodon* and relatives (e.g., *Aneides*), **PLE** plus *Ensatina*, **ENS**; and the bolitoglossine clade was divided into three subclades, the tropical genera, **BOL**, *Hydromantes*, **HYD**, and *Batrachoseps*, **BAT**. The extent of homoplasy in character states (the characters were ordered, in that ancestral and derived states were identified; many characters were multistate) in relation to phylogeny of the eight feeding modes is mapped on the phylogeny. For the 18 adult tongue characters, there are 11 non-homoplastic derived states: four are autapomorphies and seven are synapomorphies, six of which characterize the clade including **HYD**, **BOL**, and **BAT**. *Hydromantes* (**HYD**) had an additional eight derived states, six of them homoplastic and two autapomorphic. Their hemidactyliines are characterized by four homoplastic character states and **SPEa** and **HEM** each by two additional derived states, all homoplastic. Overall, 13 of the 18 tongue character states were homoplastic, and there were 31 occurrences of homoplastic states. In many ways, the analysis of tongue evolution of Lombard and Wake (1986) only served to bolster previous arguments by Wake (1966) for a close relationship among the tropical plethodontids, *Batrachoseps*, and *Hydromantes*.

Reanalysis of the Evolution of Tongue Morphology in Plethodontids

The revolution in phylogenetic inference enabled by DNA sequence data has led to significantly revised perspectives on relationships within the Plethodontidae (e.g., Chippindale et al. 2004; Mueller et al. 2004; Min et al. 2005; Vieites et al. 2007; Vieites et al. 2011; Wake 2012). For this discussion, we use the higher-level taxonomy recently formalized by Wake (2012) based on analyses of Vieites et al. (2011), whose inferred phylogenies use molecular data alone. Two subfamilies, Plethodontinae (including the Desmognathinae of Wake 1966)

and Hemidactyliinae (including Bolitoglossini of Wake 1966) are recognized. Otherwise the most significant, and surprising, change is the discovery that *Hydromantes* (**HYD**) is not a close relative of **BOL** and **BAT**; it is instead nested within the Plethodontinae, in a subclade with *Karsenia*, and with **ENS**, **PLE**, and **DES**. The Asian plethodontid *Karsenia*, only recently discovered (Min et al. 2005), was not included in previous analyses. While *Karsenia* has yet to be studied in detail with respect to tongue morphology, the anatomy of the hyobranchial skeleton (Buckley et al. 2010) and preliminary dissections (by DBW) show that it can be included as a part of **PLE**, an already possibly paraphyletic assemblage that retains what may be the ancestral morphology for plethodontids. The 18 tongue characters identified by Lombard and Wake (1986) were plotted on recent phylogenetic hypotheses of Vieites et al. (2011). One of these is presented here as figure 22.3, the "reference phylogeny" (fig. 1A of Vieites et al. 2011), which shows *Batrachoseps* and *Hemidactylium* as sister taxa. Homoplasy in individual characters is surprisingly much more extensive than what was reported in the original analysis (Lombard and Wake 1986). For the "reference phylogeny" there are now only four unique states, each autapomorphic (i.e., no unique synapomorphies at the taxonomic level analyzed). The number of individual homoplastic states has increased from 13 to 20 (65%), and the total number of homoplastic states across all taxonomic units has risen from 31 to 52 (nearly 60%). An alternative phylogeny (fig. 2C of Vieites et al. 2011), which differs mainly in having a sister taxon relationship between *Batrachoseps* and Bolitoglossini, shows somewhat less homoplasy (20 homoplastic states but 47 instances of homoplasy) (fig. 22.4). A primary reason for the great increase in homoplasy is the assignment of *Hydromantes* to the Plethodontinae. **HYD** now displays 17 homoplasious states related to feeding compared with six in the 1986 phylogeny. In the "reference phylogeny" (fig. 22.3) **BOL** has 13 homoplasies, and in the alternative phylogeny (fig. 22.4) six homoplasies, compared with six in the 1986 phylogeny.

These analyses use the ordered characters of Lombard and Wake (1986). We also conducted a parsimony analysis of unordered characters to determine if homoplasy might be reduced. The answer is that homoplasy is reduced, but not by much, in relation to figure 22.3. The number of unique states is increased to five, one of

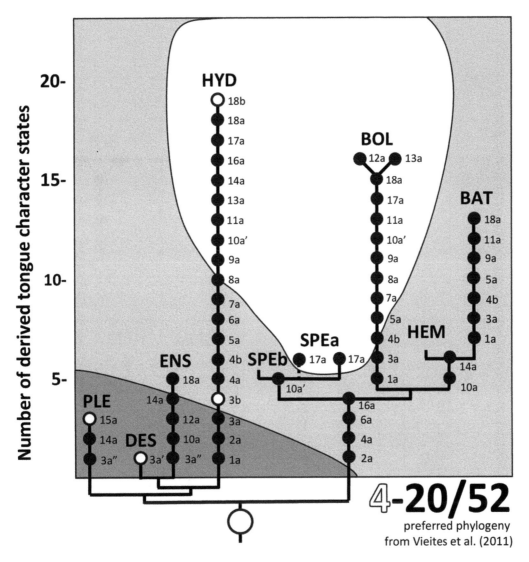

FIG. 22.3 Ground-plan diagram derived from figure 22.2, but using the "reference phylogeny" of Vieites et al. (2011, fig. 1A). Relative to figure 22.2 the number of uniquely derived states is greatly reduced and the number of individual homoplastic states and total number of homoplastic states across the phylogeny is greatly increased.

which is a synapomorphy for the Plethodontinae. There is a net reduction in the total number of homoplastic character states, from 52 to 48, and a reduction from 20 to 19 individual homoplasious characters; we consider these numbers to be inconsequential, especially in light of the fact that alternative phylogenetic hypotheses reduce the number of steps to 47 (while maintaining the number of homoplasious characters at 20). The additional uniquely derived fifth character is number 3 of Lombard and Wake (1986), the condition of the radii at the anterior ends of the basibranchial, a complex, multistate character with seven derived states. There are two equally parsimonious arrangements of this rather "messy" character, and we have selected 3a″, a detached, distally expanded condition, to represent the synapomorphy in this instance. The point of this exercise is only to show that either ordered or unordered characters require essentially the same levels of homoplasy, all much higher than previously inferred.

The greatest similarity between taxa with independently evolved freely projectile tongues is found in *Hydromantes* and Bolitoglossini. Characters involved in this example of homoplasy are provided in table 22.1 and noted as whether lost or elaborated (changed proportionally).

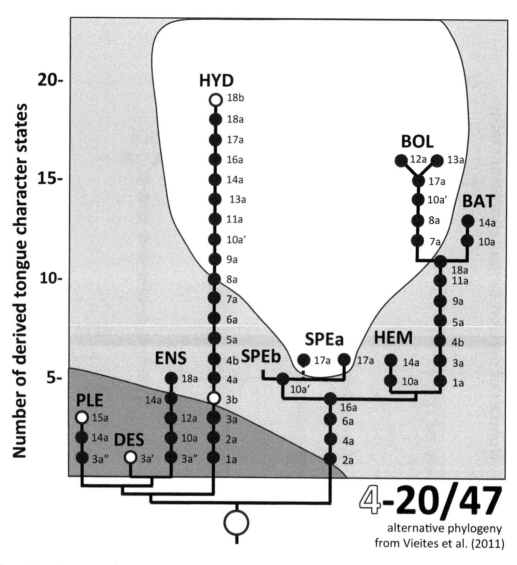

FIG. 22.4 Ground-plan diagram as in figure 22.2, except using an alternative phylogenetic hypothesis (Vieites et al. 2011, fig. 2C).

Discussion

Homoplasy is extraordinarily extensive in the evolution of the tongue of plethodontid salamanders. But the phenomenon is present more broadly if one extends the comparisons to other families of salamanders (Wake and Deban 2000). Highly projectile attached tongues are found in the Hynobiidae (*Onychodactylus*) and Salamandridae (*Salamandrina, Chioglossa*) (Özeti and Wake 1969). While all three genera have an aquatic larval stage, they are all also lungless, which appears to be a necessary, but not sufficient, precondition for the evolution of truly specialized tongue projection. The mechanisms of projection differ greatly from those found in plethodontids (e.g., the salamandrids flip a greatly elongated tongue pad using rotating radii), and one must count three additional instances of convergent evolution of projectile tongues.

The phylogenetic hypothesis of Lombard and Wake (1986) had a number of implications for interpreting the ecological setting of the extreme specializations of tongues of plethodontids, the evolution of freely projectile tongues, which enable "stealth" feeding by firing the tongue without movement of the body. Two constraints were envisioned (Roth and Wake 1985; Deban and Marks 2002): (1) the hyobranchial apparatus of nonplethodontids fulfills two functions—lung ventilation and tongue protraction—that are in conflict; and (2) the larval hyobranchial apparatus plays a critical role in larval biology and must be remodeled at metamorphosis.

TABLE 22.1 Characters shared between *Hydromantes* (Plethodontinae, Hydromantini) and *Bolitoglossa* (Hemidactylinae, Bolitoglossini) apparently due to convergence in feeding mode.

Form of convergence	Character number	Character description
Loss	1a	Urohyal
	7a	Lateral slip of rectus cervicis superficialis muscle
	8a	Omohyoid muscle
	10a′	Genioglossus muscle
	11a	Circumglossus muscle
	13a	Basiradialis muscle (lost in some Bolitoglossini)
	17a	Cutaneous attachments of tongue pad to floor of mouth
Proportional change	2a	Basibranchial expanded
	4b	Epibranchial increased in length, ceratobranchials decreased in length
	5a	Folding pattern of hyobranchial skeleton
	6a	Rectus cervicis profundus muscle lengthened and folded anteriorly
	9a	Muscular complex aiming cylinder
	12a	Intraglossus muscle attaching to glossal ligament
	16a	Suprapenduncularis muscle discrete

Source: Character numbers follow Lombard and Wake (1986).

With respect to the first, the lungless plethodontids are freed from lung ventilation constraint, permitting extreme specialization. With respect to the second, loss of the larval stage permits early and extensive remodeling and a high degree of specialization (e.g., abandonment of the less efficient folding option necessitated by the need for effective larval function). Because it was long assumed that ancestral plethodontids had larvae and that lungs evolved in some connection to aquatic habitat, skeletal folding option 1 (of Lombard and Wake 1977) involved less evolutionary change than did option 2.

Our analysis indicates that close phylogenetic relationships are not revealed by highly derived tongue morphology, patterns of force transmission (via ceratobranchial I or II), or the form of projectile tongue (attached vs. free). However, this statement needs qualification with respect to the alternative phylogeny (fig. 22.4) in which seven synapomorphic character states support the topology derived from molecular data. This finding lends support to that topology (in which **BAT** and **BOL** are sister taxa), which was only slightly less favored relative to the "reference phylogeny" by Vieites et al. (2011).

The increased homoplasy observed through reanalysis of the tongue morphologies of Lombard and Wake (1986) in the new phylogenetic contexts reveals that the independent evolution of freely projectile tongues by **HYD**, **SPE**, and **BOL** involves a greater number of changes than previously thought, especially since **HYD** and **BOL** are resolved as members of different major clades. Unlike the analysis of Lombard and Wake (1986), we can now better determine which characters might be shared due to shared evolutionary history and which are due to convergence. Those characters shared by **HYD** and **BOL** that are classified as convergent fall into two classes: loss, and proportional changes of existing structures (table 22.1). We consider it significant that none of these changes involves gaining novel structures. Rather, once the stage is "set," further specialization is readily attained.

The new phylogenetic inferences for Plethodontidae provide the surprising insight that the many synapomorphies proposed by Lombard and Wake (1986), especially those thought previously to unite **HYD**, **BAT**, and **BOL**, evolved convergently. The extent of homoplasy is further underlined because no new synapomorphies related to tongue morphology are revealed by the new phylogenies. The insight gained by the realization that homoplasy is more extensive than previously understood is that many morphological components must be integrated to produce the extreme freely projectile tongues observed in **HYD** and **BOL**. Evolutionary specializations required to produce a freely projectile tongue follow such predictable pathways of morphological change over evolutionary time that homoplasy seems inevitable.

Freely projectile tongues are one of several specializations that together comprise a key innovation of tropical salamanders—lunglessness, direct development, and feeding specialization. This complex of traits characterizes the entire clade **BOL**, containing more than 280 species, 90% of which are crowded into the narrow geographical boundaries of Middle America. These species represent more than 60% of plethodontids and more than 40% of all salamanders. They all have the extreme set of feeding specializations characteristic of **BOL**, but one subclade, *Thorius*, has evolved even further. In *Thorius*, comprised of more than 25 species, all miniatures, the hyobranchial skeleton disarticulates during projection, thus attaining freedom from the constraint of folding and achieving the highest level of specialization. We believe that this added specialization is facilitated by the reduction of anatomical structures associated with miniaturization. While **HYD** and **SPE** contain numerous species, neither approaches the "success" in number of species of **BOL**, although **HYD** is the most widely distributed of the plethodontid clades, occurring in both western North America and southern Europe. **SPE** has only 35 species, but because it has retained the more generalized folding option 1 and larval life history, some of its species have been able to exploit habitats such as ponds and streams in caves by remaining permanently larvamorph. The contrast with tongue projecting species in other families is striking: *Chioglossa* is monotypic, *Salamandrina* contains two species, and *Onychodactylus* contains seven.

Studies of great evolutionary transitions, the primary focus of this edited volume, emphasize morphological transformations, usually based on fossil evidence. While paleontologists are all well aware of the limitations of the fossil record, the lesson from our study of tongue evolution is that morphological transitions that give the appearance of being unique events may lead to an oversimplified interpretation. Do the findings of our study represent a special case, an exception to the general rule, or, instead, are great transformations likely to often involve homoplasy, leading to independently evolving, but related, lineages transitioning to the apparently "same" new functional states? We suggest that homoplasy is so common that one ignores it at one's peril.

Homoplasy is a deep problem for those developing phylogenetic hypotheses from morphological data. Once a robust phylogeny is developed, homoplasy might well be interpreted as error in coding of states of characters. However, we view homoplasy as an opportunity to gain insight into the mechanisms underlying morphological change, of any kind, but especially those relating to increased biomechanical, physiological, and behavioral performance. Whereas convergence in the evolution of freely projectile tongues in three independent clades of plethodontids can be interpreted as an indication of response to natural selection, one instead might profitably ask what underlying genetic and developmental mechanisms establish a biological framework in which convergence is expected, and almost inevitable (e.g., Wake 1991; Wake et al. 2011). Evolution of functional systems follows avenues of least resistance. In our example, the evolution of freely projectile tongues appears facilitated by convergent losses of multiple anatomical structures coupled with proportional changes in otherwise unmodified musculoskeletal elements. We have no understanding of the genetic mechanisms associated with the homoplastic events we have documented, but recent studies of genetic "hotspots," where loci of repeated evolution are located, may provide a path forward for understanding this phenomenon (Martin and Orgogozo 2013). These authors treat such hotspots as "loci of evolution" and state as a truism that specialized genes drive the evolution of specialized traits. Perhaps evolution has a limited directionality, following avenues of least resistance facilitated by specialized genes associated with particular phenotypes and special patterns of selection. Existing variational potential is reworked, rather than evolving novel structures. Repeated independent evolutionary events such as these can potentially point to underlying genetic signals and developmental networks, and lead to new avenues of research. As evolutionary biologists, we celebrate homoplasy as an opening to a new world of opportunity.

∗ ∗ ∗

Acknowledgments

We are honored to participate in this celebration of our friend and colleague, Farish Jenkins. One of us (DCB) owes much to Farish, from whom he learned what it means to truly teach and mentor. Together, the three authors of this manuscript represent a special confluence: DBW was REL's PhD advisor, who in turn was DCB's undergraduate instructor, who in turn was a PhD student of DBW's former student Jim Hanken and worked closely with Farish. We celebrate the frequent interconnectedness of our and our colleagues' careers and also recognize that single individuals such as Farish have powerful impacts on all of our careers across decades-long timescales. This manuscript benefited from comments by David Buckley and Stephen Deban. We also thank Marvalee Wake and Neil Shubin for comments on the manuscript. We thank Stephen Deban for permitting us to modify his artwork into our figure 22.1. The National Science Foundation provided support to one of us (DBW) for the molecular phylogenetic work on which our reanalyses are based.

References

AmphibiaWeb: Information on amphibian biology and conservation. [web application]. 2014. Berkeley, CA: AmphibiaWeb. Available at http://amphibiaweb.org/. Accessed November 11, 2014.

Buckley, D., M. H. Wake, and D. B. Wake. 2010. Comparative skull osteology of *Karsenia koreana* (Amphibia, Caudata, Plethodontidae). J. Morphol. 271:533–558.

Chippindale, P. T., R. M. Bonett, A. S. Baldwin, and J. J. Wiens. 2004. Phylogenetic evidence for a major reversal of life-history evolution in plethodontid salamanders. Evolution 5:2809–2822.

Deban, S. M., and S. B. Marks. 2002. Metamorphosis and evolution of feeding behaviour in salamanders of the family Plethodontidae. Zool. J. Linnean Soc. 134:375–400.

Lombard, R. E., and D. B. Wake. 1976. Tongue evolution in the lungless salamanders, family Plethodontidae. I. Introduction, theory and a general model of dynamics. J. Morphol. 148:265–286.

Lombard, R. E., and D. B. Wake. 1977. Tongue evolution in the lungless salamanders, family Plethodontidae. II. Function and evolutionary diversity. J. Morphol. 153:39–80.

Lombard, R. E., and D. B. Wake. 1986. Tongue evolution in the lungless salamanders, family Plethodontidae. IV. Phylogeny of plethodontid salamander and the evolution of feeding dynamics. Syst. Zool. 35:532–551.

Martin, A., and V. Orgogozo. 2013. The loci of repeated evolution: a catalog of genetic hotspots of phenotypic variation. Evolution 67:1235–1250.

Min, M. S., S. Y. Yang, R. M. Bonett, D. R. Vieites, R. A Brandon, and D. B. Wake. 2005. Discovery of the first Asian plethodontid salamander. Nature 435:87–90.

Mueller, R. L., J. R. Macey, M. Jaekel, D. B. Wake, and J. L. Boore. 2004. Morphological homoplasy, life history evolution, and historical biogeography of plethodontid salamanders inferred from complete mitochondrial genomes. Proc. Natl. Acad. Sci. USA 101:13820–13825.

Özetî, N., and D. B. Wake. 1969. The morphology and evolution of the tongue and associated structures in salamanders and newts (family Salamandridae). Copeia 1969:91–123.

Roth, G., and D. B. Wake. 1985. Trends in the functional morphology and sensorimotor control of feeding behavior in salamanders: an example of the role of internal dynamics in evolution. Acta Biotheor. 34:175–192.

Vieites, D. R., M. Mi-Sook, and D. B. Wake. 2007. Rapid diversification and dispersal during periods of global warming by plethodontid salamanders. Proc. Nat. Acad. Sci. 104:19903–19907.

Vieites, D. R., S. Nieto Román, M. H. Wake, and D. B. Wake. 2011. A multigenic perspective on phylogenetic relationships in the largest family of salamanders, the Plethodontidae. Mol. Phyl. Evol. 59:623–635.

Wake, D. B. 1966. Comparative osteology and evolution of the lungless salamanders, family Plethodontidae. Mem. Southern California Acad. Sci. 4:1–111.

Wake, D. B. 1991. Homoplasy: the result of natural selection or evidence of design limitations? Amer. Natural. 138:543–567.

Wake, D. B. 2012. Taxonomy of salamanders of the family Plethodontidae (Amphibia: Caudata). Zootaxa 3484:75–82.

Wake, D. B., and S. M. Deban. 2000. Terrestrial feeding in salamanders. Pp. 95–116 in K. Schwenk, ed., Feeding: Form, Function, Phylogeny. San Diego: Academic Press.

Wake, D. B., M. H. Wake, and C. D Specht. 2011. Homoplasy: from detecting pattern to determining process and mechanism of evolution. Science 331:1032–1035.

Wiedersheim, R. 1875. *Salamandrina perspicillata* und *Geotriton fuscus*; Versuch einer vergleichenden Anatomie des Salamandrinen mit besonderer Berücksichtigung der Skelet-Verhaeltnisse. Würzburg: Stahel'schen Buch-und Kunsthandlung. 205 pp., 17 plates.

Wiens, J. J., R. M. Bonett, and P. T. Chippindale. 2005. Ontogeny discombobulates phylogeny: paedomorphosis and higher-level salamander relationships. Syst. Biol. 54:91–110.

Contributors

Arhat Abzhanov
Department of Organismic and Evolutionary
Biology
Harvard University
Cambridge, MA 02138
aabzhanov@oeb.harvard.edu>

David B. Baier
Department of Biology
Providence College
Providence, RI 02918
dbaier@providence.edu

Andrew A. Biewener
Concord Field Station and Museum of
Comparative Zoology
Department of Organismic and Evolutionary
Biology
Harvard University
Cambridge, MA 02138
abiewener@oeb.harvard.edu

David C. Blackburn
Department of Vertebrate Zoology and
Anthropology
California Academy of Sciences
San Francisco, CA 94118
dblackburn@calacademy.org

Elizabeth L. Brainerd
Department of Ecology and Evolutionary
Biology
Brown University
Providence, RI 02912
elizabeth_brainerd@brown.edu

Ann Campbell Burke
Department of Biology
Wesleyan University
Middletown, CT 06459
acburke@wesleyan.edu

Leon Claessens
Department of Biology
College of the Holy Cross
Worcester, MA 01610
lclaesse@holycross.edu

A. W. Crompton
Museum of Comparative Zoology
Department of Organismic and Evolutionary
Biology
Harvard University
Cambridge, MA 02138
acrompton@oeb.harvard.edu

Edward B. Daeschler
Department of Vertebrate Biology
Academy of Natural Sciences of Drexel University
Philadelphia, PA 19103
daeschler@ansp.org

Kenneth P. Dial
Flight Laboratory
Division of Biological Sciences
University of Montana
Missoula, MT 59812
kdial@mso.umt.edu

Terry R. Dial
Department of Ecology and Evolutionary Biology
Brown University
Providence, RI 02912
terry_dial@brown.edu

John G. Fleagle
Department of Anatomical Sciences
Stony Brook University
Stony Brook, NY 11794-8081
john.fleagle@stonybrook.edu

Stephen M. Gatesy
Department of Ecology and Evolutionary Biology
Brown University
Providence, RI 02912
stephen_gatesy@brown.edu

Philip D. Gingerich
Museum of Paleontology
Department of Earth and Environmental Sciences
University of Michigan
Ann Arbor, MI 48109
gingeric@umich.edu

Ashley M. Heers
Division of Paleontology
American Museum of Natural History
Central Park W & 79th St
New York, NY 10024
ashmheers@gmail.com

James A. Hopson
Department of Organismal Biology and Anatomy
University of Chicago
Chicago, IL 60637
jhopson@uchicago.edu

Farish A. Jenkins Jr.
Museum of Comparative Zoology
Department of Organismic and Evolutionary Biology
Harvard University
Cambridge, MA 02138
(deceased)

Zerina Johanson
Earth Sciences.
Natural History Museum
Cromwell Road
London, SW75BD
UK
z.johanson@nhm.ac.uk

George V. Lauder
Museum of Comparative Zoology
Department of Organismic and Evolutionary Biology
Harvard University
Cambridge, MA 02138
glauder@oeb.harvard.edu

Daniel E. Lieberman
Department of Human Evolutionary Biology
Harvard University
Cambridge, MA 02138
danlieb@fas.harvard.edu

R. Eric Lombard
Department of Organismal Biology and Anatomy
University of Chicago
Chicago, IL 60637
elombard@uchicago.edu

Zhe-Xi Luo
Department of Organismal Biology and Anatomy
University of Chicago
Chicago, IL 60637
zxluo@uchicago.edu

Kevin M. Middleton
Department of Pathology and Anatomical Sciences
School of Medicine
University of Missouri
Columbia, MO 65211
middletonk@missouri.edu

Catherine Musinsky
Department of Organismic and Evolutionary Biology
Harvard University
Cambridge, MA, 02138
musinsky@oeb.harvard.edu

Tomasz Owerkowicz
Department of Biology
California State University, San Bernardino
San Bernardino, CA 92407
towerkow@csusb.edu

Kevin Padian
Museum of Paleontology
Department of Integrative Biology
University of California, Berkeley
Berkeley, CA 94720
kpadian@berkeley.edu

Michael D. Shapiro
Department of Biology
University of Utah
Salt Lake City, UT 84112
shapiro@biology.utah.edu

Neil Shubin
Department of Organismal Biology and Anatomy
University of Chicago
Chicago, IL 60637
nshubin@uchicago.edu

Kathleen K. Smith
Department of Biology
Duke University
Durham, NC 27708
kksmith@duke.edu

Moya Meredith Smith
Department of Craniofacial Development and Stem Cell Biology
Dental Institute
King's College
London SE1 9RT
UK
moya.smith@kcl.ac.uk

Sydney A. Stringham
Department of Biology
University of Utah
Salt Lake City, UT 84112
sydney.stringham@gmail.com

Hans-Dieter Sues
National Museum of Natural History
Department of Paleobiology
Smithsonian Institution
Washington, DC 20560
suesh@si.edu

Corwin Sullivan
Key Laboratory of Vertebrate Evolution and Human Origins
Institute of Vertebrate Paleontology and Paleoanthropology
Chinese Academy of Sciences
142 Xizhimenwai Dajie
100044 Beijing
China
csullivan@ivpp.ac.cn

David B. Wake
Museum of Vertebrate Zoology
Department of Integrative Biology
University of California, Berkeley
Berkeley, CA 94720
wakelab@berkeley.edu

Marvalee H. Wake
Museum of Vertebrate Zoology
Department of Integrative Biology
University of California, Berkeley
Berkeley, CA 94720
mhwake@berkeley.edu

Index

Page numbers ending in an *f* or a *t* indicate a figure or table, respectively.

Ernst von Baer, K., 318
erythrosuchians, 369
erythrosuchids, 363, 365
Erythrosuchus, 94, 111–12, 368
Euconodonta, 23
Eudimorphodon, 367
Eumaniraptora, 321, 323–24
Eunotosaurus, 83
Euoticus, 259
Euparkeria, 94, 96, 97, 113, 114, 116
Euparkeria capensis, 322
Eurycea, 398, 400. See also salamanders
Eusthenopteron, 74
Eusthenopteron foordi, 67, 71, 72
Eutheria (eutherians): debate over origins of reproductive
    differences, 206; nose evolution (*see* mammalian nose);
    placental evolution in, 215–18, 216f, 217t, 219–21; shoulder
    development (*see* mammalian shoulder)
"Evolutionary Aspects of Primate Locomotion" (Napier), 257
evolutionary developmental genetics: advantages of teleosts as
    model system, 334; central questions about morphological
    transformations, 334; cichlids, 342f, 342–43; coding versus
    regulatory mutations, 344; convergent evolution, 344–45;
    future directions, 345; genetic architecture of derived
    traits, 343–44; glossary of terminology, 345–46; limitations
    to studies, 333–34; Mexican cavefish, 339–42, 341f; micro-
    evolutionary transformations in teleost fishes, 334;
    sticklebacks, 334–39, 335f, 337f
Ewer, R. F., 113
Exaeretodon, 367, 369
Exbrayat, J. M., 386f, 388
Exocoetidae (flying fish), 289
extended hips and extended knees (EHEK), 269

falcon (*Falco peregrinus*), 329
Farmer, C. G., 99
faunal turnover, Triassic period: dominate faunas, 352–53, 354f,
    355f; indigenous fauna, 355–56; late Paleozoic holdover fauna,
    353, 355; living fauna, 356
feeding structures. See amphibian feeding structures; dentition;
    jaws
Ferae, 241
Fernandes, T. L., 380
Ferungulata, 242
finches (*Geospiza*), 325, 326f, 327, 328f
fins and fin rays. See ray-finned fishes
Fleagle, J. G., 257, 273
Flower, W., 241, 242
flying fish (Exocoetidae), 289
folded trabecular, 209t
Fordyce, E., 245
Fordyce, R. E., 249
Fraas, E., 241, 244
Fram Formation, Canada, 67
Fraser, G. J., 16, 18, 21
Fraser, N. C., 352, 355, 361
Freyer, C., 214
frogs, 318; buccal pumping and, 53; ovarian and uterine cycles
    in, 375f, 384–85; predator avoidance and, 289; in the Triassic
    period, 365
Fruitafossor, 177, 179–80, 181

fruit fly (*Drosophila melanogaster*), 318
Fucik, E., 80

Gaffney, E. S., 80, 82
gait and body size: correlation between body size and postures,
    228; EMA calculation, 228–29, 229f; evolutionary trend
    in locomotor limb posture, 227; evolution of gaits, 228;
    implications of shifts in locomotor limb posture, 228; joint
    torque calculation, 228–29, 229f; limb and muscle EMA
    during acceleration and grade changes, 233–34; muscle
    gearing changes with gait speed, 229; reduction in limb
    EMA with gait speed in humans, 233, 233f; scaling of muscle
    EMA within mammals, 229–31, 230f; scaling of muscle EMA
    within related taxonomic groups, 231–33, 232f; stance and
    gait in reptiles, 358–59; stance and gait in synapsids, 357–58;
    summary and future studies, 234–36, 235f. See also locomotor
    behavior of synapsids
Galago, 259, 272f
galagos, 261, 262
Galápagos and Cocos Islands, 325
Galeaspida, 12, 13, 22, 25
Galen, 54, 55t
Gallus gallus (chicken), 318, 320–21, 324
Gasterosteus aculeatus (threespine stickleback), 332. See also
    sticklebacks
Gasterosteus doryssus, 338–39
gastralia in the archosaur trunk, 100–101
Gastrotheca. See back-brooding in *Gastrotheca*
Gatesy, J. E., 244, 247, 249
Gatesy, S. M., 108, 303
Gaviacetus, 245
Gdf5/6/7, 12
Geisler, J. H., 249
Geist, N. R., 152
geladas, 263
Gemeroy, D., 242, 243, 247
genetic analyses: genomic imprinting, 218–19; *Hox* genes and,
    11–12, 82, 181–82, 220; jaw evolution, 11–13; mammalian
    shoulder, 181–82, 183; odontogenic genes and timing of
    interactions, 19–21; transformations in teleost fish (*see*
    evolutionary developmental genetics)
Georgiacetus vogtlensis, 246
Geospiza (finches), 325, 326f, 327, 328f
gharials (crocodylian), 93
gibbons (Hylobatidae), 264, 265
Gilbert, S. F., 85
Gill, T., 241
Gingerich, P. D., 239, 245, 247, 249
glenoacetabular length, 127
glenohumeral joint, 129, 173–75, 177, 179–80, 182–84, 308, 312, 314
glenoid: in extant therians, 173–74; in *Fruitafossor*, 179–80; in
    monotremes, 171f, 174–75; of multituberculates, 178; of spala-
    cotheroids, 178; *Tiktaalik roseae*, 67; of triconodonts, 177
gliders versus flyers, 367–68
gliding frogs (*Rhacophorus*), 289
gliding lemurs (Dermoptera), 289
gliding lizards (*Draco*), 289, 367
gliding snakes (*Chrysopelea*), 289
gliding squirrels (*Glaucomys*), 289
Gli genes, 182
Glires, 242

mechanisms examined, 378; pattern and process of evolution and, 377; *Pipa* compared to *Gastrotheca*, 381f, 382–83; potential continuum of reproductive process, 382; question of how similarity evolves, 375–76; research approach for analysis of homoplasy, 377–78; researchers' focus, 376–77

Reptilia (reptiles): aerial, in the Triassic, 367–68; aquatic carnivores, 364f, 364–65; archosaurian traits in, 318; crocodiles (*see* Crocodylia); erect stance and parasagittal gait in, 358–59; marine, 364t, 365–66; neck and shoulders (*see* neck and shoulders of tetrapods); olfaction in, 191f, 191–92; phylogenetic relationships, 352; skull phylogeny, 80; snakes, 3, 55, 100, 289, 318, 320f, 352, 356, 368; Squamata (*see* squamates); turtles (*see* turtle body plan); warm- versus cold-blooded and related terms, 359–60

respiration. *See* breathing mechanisms

respiratory turbinates: about, 143; conclusions, 158–59; criteria for structure to indicate endothermy, 145; endothermy and thermal independence, 144; future studies, 159–60; heat conservation and dissipation and, 157–58; as indicators of endothermy, 145, 147; in mammals versus birds, 149, 150f; mechanism of heat and water conservation, 147–48, 148f; morphology and histology, 146f, 148–49, 150f; pelage and plumage as indicators of endothermy, 144–45; posture as indicator of metabolic rate, 144; proposed origins of endothermy, 144; relationship with ectothermy, 155f, 156–57; scaling of RTSA, 149, 151f, 151–52; trachea and TCCE location and, 153–54, 155f, 156; tracheal surface area and, 143–44; trachea participation in TCCE, 152–53, 153f, 154f

respiratory turbinate surface area (RTSA), 149, 151f, 151–52

rete mirabile, 49, 50f, 50–51

rete mirabile ophthalmicum, 158

retroviral syncytin genes, 211

Review of the Archaeoceti, A (Kellogg), 241–42

Revueltosaurus callenderi, 366

rhynchosaurs, 120, 367, 369

Ribic, C. A., 251

Richardson, J., 24

Richter, S., 380, 381

Rieppel, O., 85

Riojasuchus, 115

river horses (*Hippopotamus*), 240

Roberts, R. B., 343

Roberts, T. J., 229, 234

Robinson, P. L., 370

Rodhocetus, 245, 246

Romer, A. S., 83, 125, 127, 130

rotoscoping, 110, 308, 309f, 310f

Rowe, T. B., 198

RTSA (respiratory turbinate surface area), 149, 151f, 151–52

Ruben, J. A., 145, 147, 152, 190, 191

Ruckes, H., 83, 85

Rücklin, M., 22, 23, 24

Rudapithecus, 265

rudimentary structures in juveniles: branching strategies among altricial species, 292–95, 293f; conceptual parallels about flight development, 295–96; conclusions and summary, 296–98; definition of rudimentary features, 285, 285f; differential maturity of hind limbs and forelimbs, 294f, 292–95; ecological transition zones importance, 288; fossil evidence of theropod-avian transition, 286; future directions, 295; locomotor strategies among precocial ground birds, 290f, 290–92, 291f; ontogenetic insights into evolutionary history, 295–96; ontogeny of locomotion with respect to bird ecology, 290; potential locomotor capacities of extinct theropods, 296, 297f; predation avoidance, 289; securing refuge and, 288–89; technology use, 296, 297f; transformational insights provided by juvenile birds, 284; transitional feathers of extinct theropods, 286, 287f; transitional locomotor behaviors of extinct theropods, 288; transitional skeletons of extinct theropods, 286–88; "trees-down" versus "ground-up" origins of avian flight, 288

Russell, A. P., 192

Russell, D., 245

Russell, E. M., 215

Saghacetus, 248

Sahelanthropus tchadensis, 267–69

Saimiri, 272f

salamanders: clades, 397, 399–400; control of air distribution, 51; dental development, 18, 22; development of aspiration breathing, 52, 52f, 54; functional feeding modes, 397–98, 399f, 401f, 402f; gas distribution in, 51; internal fertilization in, 384; juvenile transformations, 284; lungless (*see* plethodontid salamanders); obligate viviparity in, 385–86; presence in the Triassic, 368; tongue evolution (*see* amphibian feeding structures); viviparity in (*see* viviparity in amphibians)

Salamandra algira, 385

Salamandra atra, 382, 385–86, 386f, 388

Sánchez-Villagra, M. R., 182, 221

Sanders, A. E., 249

Sanderson, M. J., 376

Sarcophilus (Tasmanian devil), 213

Sarcopterygii (lobe-finned fish), 23, 25, 48, 49, 64–65, 65f, 66–70, 69f, 70f

Sarich, V. M., 244, 247

Saurischia, 101, 321, 327, 366, 368

Sauropoda, 94

Sauropodomorpha, 366, 369

Sauropterygia, 365

Saurosphargidae, 365

Savannah monitor (*Varanus exanthematicus*), 135

scapular blade, 177, 179, 182

scapulas, 169, 175, 178

scapulocoracoid: in mammaliaforms, 176; in monotremes, 174; in therians, 169; *Tiktaalik roseae*, 67, 71. *See also* coracoid process

Schachner, E. R., 120, 121

Schmidt-Nielsen, K., 148

scientific rotoscoping, 308, 309f, 310f

Scilla, A., 240

Scleromochlus, 362

Scotland, R. W., 377

sea squirt, 284

septospine, 169

Shapiro, M. D., 333, 338

sharks (chondrichthyans): dental evolution in, 19, 20, 22; fin ray design, 33, 35; jaw skeletal structures, 10f, 10–11; phylogenetic relationships, 49f; structural design of fin rays, 33; tooth development in, 12f, 13–16, 14f, 15f, 19

Sharovipteryx, 364f, 367

shell anatomy, turtle, 78–80

Shone, V., 24

therians: eutherians (*see* Eutheria); nose evolution (*see* mammalian nose); placental evolution in (*see* placental evolution); shoulder origin (*see* mammalian shoulder)

theriimorphs, 177-78. *See also* mammalian shoulder

therizinosaurs, 322

therocephalians, 192, 355

Thewissen, J. G. M., 245

Thomason, J. J., 192

Thorius, 404

threespine stickleback (*Gasterosteus aculeatus*), 336. *See also* sticklebacks

Thrinaxodon, 192, 196-97, 362

Tiktaalik roseae: anocleithrum, 68f, 68-69; clavicle, 71f, 71-72; cleithrum, 69-70f, 69-71; fossil records related to tetrapods, 64; functional trends during evolution of, 74; glenoid, 67; history, 67; interclavicle, 71f, 72; major discoveries from fossil records, 67; pectoral girdle overview, 66-67; scapulocoracoid, 69-70f, 71; supracleithrum, 67-68, 68f; transformation of pectoral girdle, 72-74, 73f, 74f; transitional elements of morphology, 74-75

tinamou, 324

Tobalske, B. W., 308

tongues in amphibians. *See* amphibian feeding structures

tooth evolution theories: both inside and outside, 15-16; hypotheses about, 12f, 13, 16f; "inside to outside," 13, 15, 17f, 18, 22-23; "outside to inside," 13, 15, 17f, 17-18, 23-24. *See also* dentition

trabecular, 210

tracheal surface area (TrSA), 143-44, 152

trachea participation in TCCE: constraints on TCCE location and, 153-54, 155f, 156; histology, scaling, and topography, 150f, 152, 153f; plasticity of the tracheal and, 153, 154f

Triadobatrachus (stem-frogs), 365, 368

Triassic faunal turnover: dominate faunas, 352-53, 354f, 355f; indigenous fauna, 355-56; late Paleozoic holdover fauna, 353, 355; living fauna, 356

Triassic period tetrapods: aerial reptiles, 367-68; aquatic carnivores, 364f, 364-65; conclusions, 371; ecosystem changes during the Triassic, 352; evolution of erect stance and parasagittal gait in reptiles, 358-59; evolution of erect stance and parasagittal gait in synapsids, 357-58; faunal changes causes, 370-71; faunal turnover, 352-56, 354f, 355f; feeding types, 361, 362-63f, 364t; first taxonomic shift, 352; functional ecology and structure of communities, 368; generalizations about ecological diversification, 361; generalized smaller carnivores, 361-62, 364t; herbivores, 366-67; hind limb posture evolution in archosauriforms (*see* hind limb posture); marine reptiles, 364t, 365-66; metabolic rate as determinate of growth rate, 359-60; phylogenetic relationships of, 352, 353f; principal changes during the Triassic, 351; size increase and ecological diversification, 368-69; smaller tetrapods and their ecological roles, 368; stability of vertebrate communities, 369; terrestrial macrocarnivores, 362-64, 364t; vertebrate bone as determinate of growth rate, 360f, 360-61; warm- versus cold-blooded and related terms, 359

Triassurus, 368

triconodonts, 171f, 177-78, 180, 182

trilophosaurs, 369

Trilophosaurus, 367

tritheledontid, 368

trophectoderm, 208t

trophoblast, 206, 208t, 215

Tropidosuchus, 112

trout, 35-36

TrSA (tracheal surface area), 143-44, 152

True, F., 241

true horses (*Equus*), 240

trunk evolution: bird trunk overview, 93f, 93-94; constraints and limitations for study of, 103; crocodylian trunk overview, 93, 93f; degrees of freedom of movement of the avian thorax, 97; diaphragmatic breathing pump, 99-100; diversity of the Archosauria, 91-92, 92f; in extinct species, 94; future studies, 103-4; gastralia, 100-101; lung ventilation with fused vertebral ribs in, 98; origins of tripartite construction of the crocodylian ribcage, 95; overview, 92; parapophyseal locations in alligators and birds, 96-97; pelvic girdle aspiration breathing, 93f, 99; postcranial skeletal pneumaticity and, 101, 102f; pulmonary apparatus structure, 101, 102f; shifts in locomotor-respiratory function, 96f, 97-98; structure of trunk skeleton, 94-95; transformations of ribcage, 94-97, 96f; uncinate processes, 96f, 98-99; utility of crocodylian model for basal archosaurian condition, 101, 103

turtle (*Chelyda serpentina*), 83, 84

turtle body plan: conclusions and future research, 84f, 85-86; developmental perspective, 79f, 83-85; fossil records of, 79f, 82-83, 84f; molecular data-derived phylogenetic hypotheses, 80t, 81f, 81-82; morphological data-derived phylogenetic hypotheses, 80t, 80-81, 81f; placement of turtles among amniotes, 82; shell anatomy, 78f, 78-80, 79f; skull anatomy, 80

Tyndale-Biscoe, H., 206

Typhlonectes compressicaudus, 386f, 388

Uhen, M. D., 249

Ukhaatherium, 179

Unguiculata, 241

Ungulata, 241

Utatsusaurus, 365

Vancleavea, 114, 364

Van Nievelt, A. F. H., 198

Van Valen, L. M., 243, 244, 247, 249

Varanus exanthematicus (Savannah monitor), 110, 111, 135

Varanus komodoensis (Komodo monitor), 127

Varecia, 272f

VCL. *See* vertical clinging and leaping (VCL) hypothesis

Velociraptor, 95, 295

Ventastega curonica, 74

vertebrate paleontology: approach to study, 2; challenges surrounding discovery, 1-2; emergent themes in transformation, 3-4; fundamental question surrounding evolution, 2; knowledge gained by linking the fossil and living worlds, 2-3; persistent fallacies regarding transformations, 4; value in interdisciplinary approaches, 4

vertical clinging and leaping (VCL) hypothesis: anatomical features distinguishing VCLs from other primates, 261, 261f; conclusions about, 261-62; ischium characteristics in leapers, 259-61, 260f; skeletal indicators of VCL in fossils, 259; VCL hypothesis, 259

Vesalius, A., 54

Vickaryous, M. K., 100

Victoriapithecus, 263

Vieites, D. R., 400

villous, 209t, 210, 217

Vilter, A., 386f

Vilter, V., 386f

Vincelestes, 178, 179

viviparity in amphibians, 206, 207, 208t; endocrine cycle's role in intraoviductal pregnancy, 388; independent evolution of in caecilians, 386–88, 387f; internal fertilization in salamanders, 384; obligate viviparity in salamanders, 385–86; ovarian and uterine cycles in frogs, 384–85, 385f; viviparity in *S. atra*, 386, 386f

vomer, 198

Vytshegdosuchus, 119

WAIR (wing-assisted incline running), 285f, 290, 291f, 292

Wake, D. B., 376, 395, 396, 397, 399, 400, 401, 403

Wake, M. H., 375, 376, 377, 386f

Walker, A. C., 259, 260, 262

Ward, P. D., 160

Washburn, S. J., 257, 267

Weber, M., 241

Weejasperaspis, 23

Wells, N., 245

Werneburg, I., 80

whales: blood serum protein analyses, 242, 244; classification of, 240–44, 243f; climatic events in transition from land to sea, 250f, 251; competing ideas on phylogeny and classification of, 240; contrasts between land-living Artiodactyla and marine Cetacea, 240f; current areas of study, 250–51; evidence of relationship to artiodactyls, 246–47, 249–50; final transitional stage from land to sea, 248, 249f; fossil records of, 241, 245–49; future studies, 252–53; historical observations about, 239–40, 240f; hypothesis on transition from land to sea, 240f, 251–52, 252f; knowledge gained from the *Pakicetus* find, 245–46, 246f; link to ungulates, 242–44; origin and early evolution, 250f; placement in its own family, 240f, 242, 243f, 244–45; placement of Cetacea in Mutica, 242; placement

with insectivore-carnivore ancestor, 241–42; reconstruction of *Maiacetus inuus* skeleton, 247f, 247–48, 248f; schematic representation of phylogenetic relationships, 243f; transition from Archaeoceti to Mysticeti, 248–49

White, T. D., 269

Whiteside, J. H., 369

white-tailed deer (*Odocoileus*), 251

Whitmore, F. C., 249

Wiens, J. J., 382

wildebeast (*Connochaetes*), 251

Williams, S. B., 234

Wilson, M. V. H., 23

Winchell, C. J., 82

wind tunnels, 296, 307f, 308, 310

wing-assisted incline running (WAIR), 285f, 290, 291f, 292

wombats, 214

Wood Jones, F., 257

Xavier, F., 384

Xenopus laevis (clawed frog), 318

Xilousuchus, 119

XROMM (X-ray Reconstruction of Moving Morphology), 57–59, 58f, 103, 296, 309, 311

Yablokov, A. V., 242

Yixianonis grabaui, 324

Yntema, C., 84

yolk sac, 207

yolk sac membrane, 209t

Zalmout, I., 246, 247

zebrafish, 17, 18, 20, 318, 334, 343

Zeuglodon, 241, 248

Zhangheotherium, 168f, 170f, 171f, 178

zonary placenta, 209t

Zygorhiza, 248